MECHANICS

MECHANICS

W. CHESTER
University of Bristol

London
GEORGE ALLEN & UNWIN
Boston Sydney

First published in 1979

GEORGE ALLEN & UNWIN LTD
40 Museum Street, London WC1A 1LU

British Library Cataloguing in Publication Data

Chester, W.
Mechanics.
1. Mechanics
I. Title
531 QC125.2 78-41320

ISBN 0-04-510058-6
ISBN 0-04-510059-4 Pbk

Typeset in 10 on 12 point Press Roman by Preface Ltd
Reproduced, printed and bound in Great Britain by
Fakenham Press Limited, Fakenham, Norfolk

CONTENTS

CHAPTER 5. IMPULSE, MOMENTUM, WORK AND ENERGY

CHAPTER 6. OSCILLATIONS

CHAPTER 7. CENTRAL FORCES

CHAPTER 8. SYSTEMS OF PARTICLES

CHAPTER 9. ANGULAR VECTORS

CHAPTER 10. MOMENTS

TABLES

Errata

p.xi, line 2: *for* 'May' *read* 'My'

p.40, line 6:

for $\,_{\mathrm{c}}\cos\theta\,\sin^2\theta\,\mathrm{d}(\sin\theta)$ *read* $\int_{\mathrm{c}}\cos\theta\,\sin^2\theta\,\mathrm{d}(\sin\theta)$

p.100, line 32: *for* 'coefficient of kinematic viscosity equal to 10^{-6} m^2 s^1 at a temperature of' *read*: 'coefficient. It does vary, but within the limits 0·1 to 1, again provided the'

p.107, line 2: *for* $a_1K_1 + a_2K_2^2$ *read* $a_1K_1^2 + a_2K_1^2$

p.112, Exercise 4.8: *for* $8'$ *read* g

p.127, 7 lines from end: *for* 'elastic' *read* 'inextensible'

p.138, line 8: *for* 'cross-section of the wire.' *read* 'cross-section (A) of the wire.'

p.140, Figure 6.12: *for* $2\mu mgl\ \lambda$ *read* $\dfrac{2\mu mgl}{\lambda}$

p.163, Exercise 6.12:

$$for\ n = \frac{k}{2}\left\{1 + \left(\frac{\pi}{\log\{(x_1 - x_2)/(x_3 - x_2)\}}\right)^2\right.^{1/2}$$

$$read\ n = \frac{k}{2}\left\{1 + \left(\frac{\pi}{\log\{(x_1 - x_2)/(x_3 - x_2)\}}\right)^2\right\}^{1/2}$$

p.169, Exercise 6.26: *for* $\sin\left(\left|\dfrac{\omega x}{u}\right|\right)$ *read* $\sin\left(\dfrac{\omega x}{u}\right)$

p.174, 10 lines from end: *for* $\mathbf{r}\times m\mathbf{r}$ *read* $\mathbf{r}\times m\dot{\mathbf{r}}$

p.203, Equation (7.7.3): *for* $\dot{a}$ *read* a

p.298, last line should read: 'angular momentum will arise only through variations in $\boldsymbol{\omega}$.'

p.321, line 1: *for* n *read* n^2

p.401, Equation (13.2.19) should read:
$x = At + B,\ v = A$

p.410, Equation (13.3.26): *for* $\dfrac{n_0}{n}^{1/2}$ *read* $\left(\dfrac{n_0}{n}\right)^{1/2}$

p.432, line 23: *for* 'line 2' (sub-entry under vectors) *read* 'line 2283'

PREFACE

When I began to write this book, I originally had in mind the needs of university students in their first year. May aim was to keep the mathematics simple. No advanced techniques are used and there are no complicated applications. The emphasis is on an understanding of the basic ideas and problems which require expertise but do not contribute to this understanding are not discussed. However, the presentation is more sophisticated than might be considered appropriate for someone with no previous knowledge of the subject so that, although it is developed from the beginning, some previous acquaintance with the elements of the subject would be an advantage. In addition, some familiarity with elementary calculus is assumed but not with the elementary theory of differential equations, although knowledge of the latter would again be an advantage.

It is my opinion that mechanics is best introduced through the motion of a particle, with rigid body problems left until the subject is more fully developed. However, some experienced mathematicians consider that no introduction is complete without a discussion of rigid body mechanics. Conventional accounts of the subject invariably include such a discussion, but with the problems restricted to two-dimensional ones in the books which claim to be elementary. The mechanics of rigid bodies is therefore included but there is no separate discussion of the theory in two dimensions. The argument for a preliminary treatment of two-dimensional mechanics is that it avoids the introduction of angular vectors. It is true that the student finds this notion difficult to grasp. But if he is to understand rigid body mechanics in any real sense, an understanding of angular vectors is a prerequisite and it is as well to accept this rather than delay the inevitable. In any case most of the interesting applications are three dimensional.

As already stated, the original aim was to produce an elementary text, but the decision to include three-dimensional rigid body mechanics promotes it to the intermediate category. Tensor calculus has, however, been omitted. This is not because I underestimate its importance, but because its inclusion would not be consistent with the aim to keep the mathematics simple. Furthermore, it is not usually taught at this level primarily with an eye on its application to mechanics. It is far more important for the student of mechanics to gain experience in the application of the theory to specific problems, and to learn from that experience how to choose by inspection the most appropriate set of axes for a given problem. On the other hand the book would now be incomplete without some introduction to Lagrange's equations. Again it seemed inappropriate to give a full discussion of the theory of analytical mechanics, which requires a knowledge of variational calculus and which, since it is now taught primarily as a prerequisite for quantum mechanics and field theory, requires a different development and point of view.

Chapter 1 deals with the elementary theory of vectors. The rather full treatment is in keeping with the present-day status in most universities, where it is taught in its own right and for application over a wider field than classical mechanics. The mechanics of a particle is introduced in Chapters 2–8 inclusive, which are genuinely elementary in the sense discussed above and which in my view form a satisfactory introduction to the subject.

Chapter 9 is exclusively devoted to a full discussion of angular vectors, including the theory of rotating axes, and Chapter 10 deals with the second prerequisite for an understanding of rigid body mechanics, namely the theory of moments. Although these preliminaries take some time, the subsequent derivation of the equations of motion is readily achieved and is discussed in Chapter 11. Whereas the derivation of the equations is relatively straightforward, their application to the many and varied problems can be confusing, and it takes further time and effort before the student becomes competent at solving such problems. For this reason the bulk of Chapter 11 is devoted to worked examples, beginning with simple two-dimensional problems and ending with the more sophisticated three-dimensional applications.

Virtual work and Lagrange's equations are introduced in Chapter 12 and finally Chapter 13, which is unconventional for a mechanics text, describes some important techniques used in non-linear mechanics and elsewhere. My own view is that these techniques are more useful at the elementary level than the conventional two-dimensional rigid body mechanics.

Illustrative examples are provided throughout each chapter, and a substantial number of exercises is given at the end of each chapter. I know of no subject where the working of examples is more important as an aid to understanding, and it should occupy the major part of a student's effort.

All practising mathematicians would agree that understanding is helped by a good notation and hampered by a bad one. I have tried to bear this in mind and the student is well advised to do likewise. It is a mistake to have a hard and fast rule, but generally speaking alliterative notation is to be preferred. Thus m for mass, m_{P} for the mass of a particle P, m_0 and m_1 for the masses of particles P_0 and P_1 are superior aids to identification than, say, m, μ, M, M'. This also allows the symbols μ and M to be reserved for the coefficient of friction and the moment of a force respectively. Since it is also desirable to use a notation which is widely accepted and therefore easily recognised, I have used the recommendations of the British Standards Institution (BS 1991, Part 1, 1967). The notation is also compatible with that recommended in the report of the symbols committee of the Royal Society of London (Quantities, Units and Symbols, second edition, 1975). For similar reasons of wide acceptance, the MKS system of units is used exclusively.

I am indebted to those authors, too numerous to mention, who have written on this subject before me and thereby helped to crystallise my own views. It is also a pleasure to acknowledge the helpful suggestions and advice from W. H. H. Banks, C. Illingworth, D. W. Moore and M. B. Zaturska. I owe a special debt of gratitude to Nancy Thorp for her indispensable help in the typing of the manuscript and for her tolerant acceptance of my seemingly endless revisions.

Any suggestions for improvement will be welcomed.

W. CHESTER

NOTATION

The following list shows the most common usage of the various symbols.

$\mathbf{a}$	acceleration
a	$\lvert \mathbf{a} \rvert$, radius, amplitude of oscillation
b	breadth
c	constant, speed of sound
d	distance, diameter
E	total energy
e	eccentricity, coefficient of restitution
$\mathbf{F}$	force
F	$\lvert \mathbf{F} \rvert$, frictional component of force
$\bar{\mathbf{F}}$	impulse
G	gravitational constant
$\mathbf{g}$	gravitational acceleration
H	Hamiltonian
h	height
I	moment of inertia
$\mathbf{i}, \mathbf{j}, \mathbf{k}$	unit vectors
k	radius of gyration, coefficient of resistance
$\mathbf{L}$	angular momentum
L	$\lvert \mathbf{L} \rvert$, Lagrangian
l	length
$\mathbf{M}$	moment
m	mass
N	normal component of force
n	frequency
P	power
$\mathbf{p}$	linear momentum
p	$\lvert \mathbf{p} \rvert$, generalised momentum
Q	generalised force
q	generalised coordinate
$\mathbf{R}$	resultant force
$\mathbf{r}$	variable vector, especially position vector
r, θ, ϕ	spherical polar coordinates
s	arc length
T	kinetic energy, tension, periodic time
t	time
$\mathbf{u}, \mathbf{v}, \mathbf{w}$	velocities
V	potential energy
W	work
x, y, z	rectangular cartesian coordinates
λ	modulus of elasticity
μ	coefficient of friction
ν	kinematic viscosity
ρ	density
$\boldsymbol{\omega}$	angular velocity

UNITS

length	metre	m
time	second	s
mass	kilogram	kg
force	newton	N
work	joule	J
power	watt	W
angle	radian	rad

1 Vectors

1.1 INTRODUCTION

Physical quantities in elementary mechanics are usefully classified as vectors or scalars according as they do or do not involve the notion of direction. Scalar quantities, such as mass, length, time, energy, density, pressure are completely specified by a number, measured according to some appropriate scale. When associated with the same scale of measurement, these numbers are manipulated according to the rules of elementary algebra. Indeed the motivation for these rules arises originally from the way in which scalar quantities combine. In particular we have the relations

$$a + b = b + a, \qquad ab = ba, \tag{1.1.1}$$

$$(a + b) + c = a + (b + c), \qquad (ab)c = a(bc), \tag{1.1.2}$$

$$a(b + c) = ab + ac. \tag{1.1.3}$$

The first pair of relations are the commutative rules of addition and multiplication. The second pair are the associative rules and the third is the distributive rule.*

A vector requires a direction, as well as a magnitude, for its complete specification. We also require certain rules of procedure which have a formal resemblance to (1.1.1)–(1.1.3). This is because it is convenient to adapt the familiar notation of scalar algebra in order to explain and manipulate the rules appropriate to vector algebra. In doing so, it is important to avoid the assumption that the relations which hold in scalar algebra are automatically valid in vector algebra simply because of the similarity of notation. While it is true that the basic rules are a matter of definition when regarded as the basis of a branch of mathematics, we are particularly concerned in mechanics with the behaviour and manipulation of physical quantities. Typical examples of vectors in mechanics are velocity, acceleration and force. It is the behaviour of physical quantities such as these which provides the motivation for vector algebra, and the rules are formulated with this in mind.

* It is worth noting that the theorems of elementary algebra need not be confined to the system of real numbers. They can be interpreted to apply to any set of entities which combine in a way that can be described by the rules of elementary algebra. Conversely algebras in which the basic rules are different can be, and are, studied.

1.2 DEFINITIONS

Free Vector

Let A, B be two points in three-dimensional space. Let C, D be two other points such that

(i) the length CD is equal to the length AB,
(ii) CD is parallel to AB and in the same sense.

The class of all such pairs of points (C, D) is defined as the free vector $\overline{AB}$. A particular pair of points such as (A, B) or (C, D) is called a representation of the vector. Any particular representation is sufficient to determine the vector. However the vector itself is the class of all its representations and defines a magnitude and direction, but no particular position in space. In Figure 1.2.1, (A, B), (C, D) and (E, F) are all representations of the same vector $\overline{AB}$ or, more briefly, **a**.

We say that two vectors **a**, **b** are equal if they refer to the same class and we then write $\mathbf{a} = \mathbf{b}$. Thus in Figure 1.2.1 we have*

$$\overline{AB} = \overline{CD} = \overline{EF} = \mathbf{a}.$$

Note that if two vectors are unequal they will have no representation in common. Hence if a representation of a vector **a** has the same magnitude and direction as a representation of a vector **b**, one can infer that $\mathbf{a} = \mathbf{b}$. This should be borne in mind in the subsequent development of the theory, for in the actual manipulation of vectors we are constantly dealing with their representations as

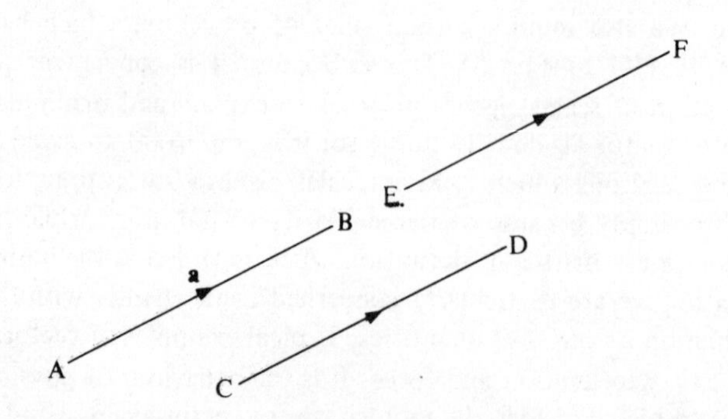

Figure 1.2.1 Representations of the free vector **a**.

* There are occasions when it is convenient to restrict a vector to a given line; it is then called a line vector. In Figure 1.2.1, for example, (A, B) and (E, F) are representations of the same line vector $\overline{AB}$ or $\overline{EF}$, which is to be distinguished from the line vector $\overline{CD}$. This distinction will not, however, be used here.

directed line segments. Where no confusion is likely to arise, the distinction between a vector and its representation will not always be stressed.

The magnitude of the vector $\overline{AB}$ is denoted by $|\overline{AB}|$ or AB; the magnitude of the vector **a** is denoted by $|\mathbf{a}|$ or a. It is a non-negative number obeying the laws of scalar algebra.

Note that $|\overline{AB}| = |\overline{BA}|$, but that $\overline{AB}$ and $\overline{BA}$ are different vectors since their directions are opposite in sense. We write

$$\overline{BA} = -\overline{AB},$$

so that the negative sign reverses the direction of a vector without changing its magnitude.

If λ is a scalar, then $\lambda\mathbf{a}$ defines a vector of magnitude $|\lambda||\mathbf{a}|$. Its direction is the direction of **a** if $\lambda > 0$ and the direction of $-\mathbf{a}$ if $\lambda < 0$. If either $\lambda = 0$ or $|\mathbf{a}| = 0$, it is a null vector.

Null Vector

The class of all pairs of coincident points (A, A) is called a null vector or zero vector. It has zero magnitude and no specific direction associated with it. It is comparable to the zero of scalar algebra although they are not identical concepts. If **a** is a null vector we write $\mathbf{a} = \mathbf{0}$.

Position Vector

If the point O is fixed, $\overline{OA}$ is called the position vector of the point A relative to O. Once the point O is chosen, the representation of any position vector relative to O is fixed, and other representations cease to be appropriate. If two points have the same position vector, then the points coincide.

The position vector will be used in the subsequent analysis as a useful way of specifying a point relative to a chosen origin, that is as an alternative to its specification by means of its coordinates relative to a chosen set of axes.

No special notation is used to distinguish a position vector and a corresponding free vector. It should be clear from the context which is intended, and where no restriction is imposed it is the free vector which is being discussed. In particular, the actual algebra of vectors, with which we are concerned in this chapter, refers to free vectors.

The Angle between two Vectors

Let (O, A), (O, B) be representations of the vectors $\overline{OA}$, $\overline{OB}$ respectively. The angle between $\overline{OA}$ and $\overline{OB}$ is defined as $A\hat{O}B$, with $0 \leqslant A\hat{O}B \leqslant \pi$.

This definition requires a choice of representations having a common point. This is always possible for free vectors, even if initially they are determined by

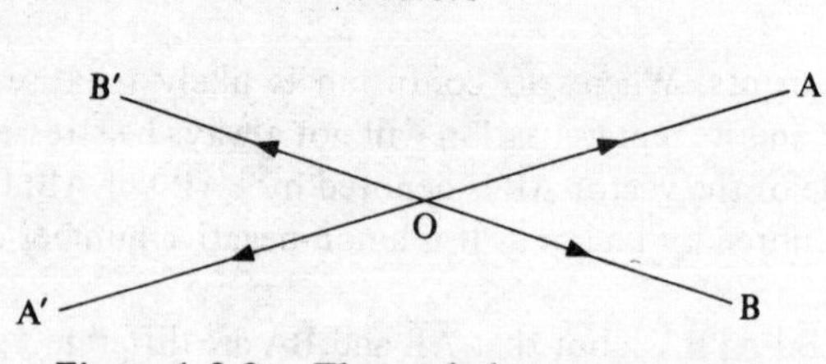

Figure 1.2.2 The angle between vectors.

representations which do not satisfy this condition. The angle is clearly independent of the particular representations chosen to define it.

We note the following properties:

(i) The angle between two vectors depends on their directions and not on their magnitudes.

(ii) The angle between $\overline{AO}$ and $\overline{BO}$ is the same as the angle between $\overline{OA}'$ and $\overline{OB}'$, where $\overline{OA'} = \overline{AO}$, $\overline{OB'} = \overline{BO}$. Thus it is $A'\hat{O}B' = A\hat{O}B$ and so is the same as the angle between $\overline{OA}$ and $\overline{OB}$ (see Fig. 1.2.2).

(iii) The angle between $\overline{OA}$ and $\overline{BO}$ is the same as the angle between $\overline{OA}$ and $\overline{OB'}$ and so is $A\hat{O}B' = \pi - A\hat{O}B$.

Coplanar Vectors

If the directions of several vectors are all parallel to a plane, the vectors are said to be coplanar. The plane of the vectors is any plane parallel to their directions.

Vector Addition

The sum, or resultant, of two vectors $\overline{AB}$ and $\overline{BC}$ is defined to be the vector $\overline{AC}$, as in Figure 1.2.3. We write

$$\overline{AB} + \overline{BC} = \overline{AC}. \quad (1.2.1)$$

In particular

$$\overline{AB} + \overline{BA} = \overline{AA} = \mathbf{0}. \quad (1.2.2.)$$

Again the representations of the two vectors must be appropriately chosen in order to apply this rule. The resultant, however, is independent of the particular representations chosen.

Note that the two vectors and their resultant are coplanar.

We also write

$$\mathbf{a} + (-\mathbf{b}) = \mathbf{a} - \mathbf{b},$$

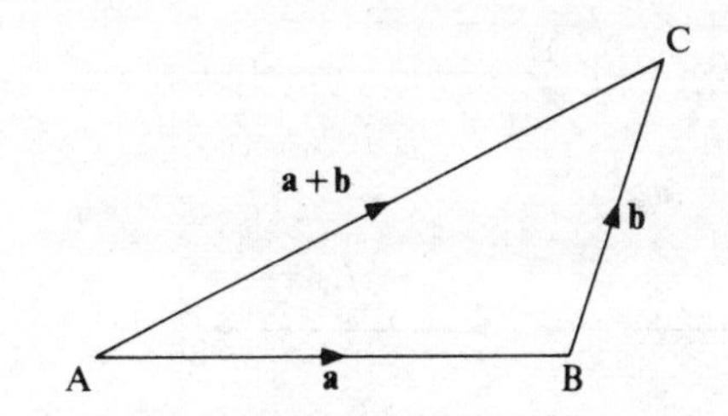

Figure 1.2.3 The sum of two vectors **a** and **b**.

and this serves to define the difference of two vectors. Thus, from Figure 1.2.3, we have

$$\overline{AC} - \overline{BC} = \overline{AC} + \overline{CB} = \overline{AB}. \tag{1.2.3}$$

Comparison of (1.2.1) and (1.2.3) shows that, as in scalar algebra, vectors can be transferred from one side of an equation to the other with a change of sign.

An immediate consequence of the rule of addition is the inequality

$$|\mathbf{a}+\mathbf{b}| \leqslant |\mathbf{a}| + |\mathbf{b}|, \tag{1.2.4}$$

since the shortest distance between two points is a straight line.

We conclude this section by pointing out that the notions of length, parallelism and direction, as used in the preceding definitions, are basic concepts as used in Euclidean geometry. It is assumed that we can always tell when two given lengths are equal, say by the use of a rigid measuring rod, and when two lines are parallel. The notion of direction can be introduced in various ways, but they all depend on the assumption of an underlying frame of reference relative to which the direction can be defined. The justification for these assumptions is based on their usefulness in practical applications. It is worth noting, however, that assumptions are being made and that they are not always applicable, for example in relativity and quantum mechanics, but for the purposes of Newtonian mechanics they are amply justified by the correspondence between theory and observation.

1.3 THE RULES OF VECTOR ALGEBRA

We are now in a position to deduce the following rules, which are a consequence of the above definitions.

(i)
$$\mathbf{a}+\mathbf{b} = \mathbf{b}+\mathbf{a}. \tag{1.3.1}$$

This is made clear in Figure 1.3.1. Let (A, B) and (D, C) be two representations of **a**. Let (A, D) and (B, C) be two representations of **b**. Then ABCD is a

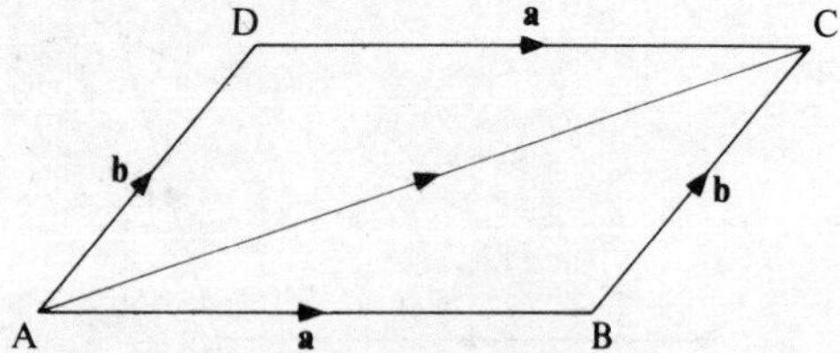

Figure 1.3.1 Illustration of the relation $\mathbf{a} + \mathbf{b} = \mathbf{b} + \mathbf{a}$.

parallelogram and

$$\mathbf{a} + \mathbf{b} = \overline{AB} + \overline{BC} = \overline{AC} = \overline{AD} + \overline{DC} = \mathbf{b} + \mathbf{a}.$$

(ii)
$$(\mathbf{a} + \mathbf{b}) + \mathbf{c} = \mathbf{a} + (\mathbf{b} + \mathbf{c}). \tag{1.3.2}$$

In Figure 1.3.2 let (A, B), (B, C), (C, D) be representations of **a**, **b** and **c** respectively. By successive application of the rule for addition we have

$$(\mathbf{a} + \mathbf{b}) + \mathbf{c} = (\overline{AB} + \overline{BC}) + \overline{CD} = \overline{AC} + \overline{CD} = \overline{AD},$$
$$\mathbf{a} + (\mathbf{b} + \mathbf{c}) = \overline{AB} + (\overline{BC} + \overline{CD}) = \overline{AB} + \overline{BD} = \overline{AD}.$$

Hence there is no confusion if we write $\mathbf{a} + \mathbf{b} + \mathbf{c}$ for this vector sum.

(iii)
$$\lambda(\mathbf{a} + \mathbf{b}) = \lambda\mathbf{a} + \lambda\mathbf{b}. \tag{1.3.3}$$

This rule follows from the fact that if $\mathbf{a}, \mathbf{b}, (\mathbf{a} + \mathbf{b})$ are represented by the sides of a triangle, then $\lambda\mathbf{a}, \lambda\mathbf{b}, \lambda(\mathbf{a} + \mathbf{b})$ can be represented by the sides of a similar triangle, which can then be used to define the sum of $\lambda\mathbf{a}$ and $\lambda\mathbf{b}$.

(iv)
$$(\lambda + \mu)\,\mathbf{a} = \lambda\mathbf{a} + \mu\mathbf{a}. \tag{1.3.4}$$

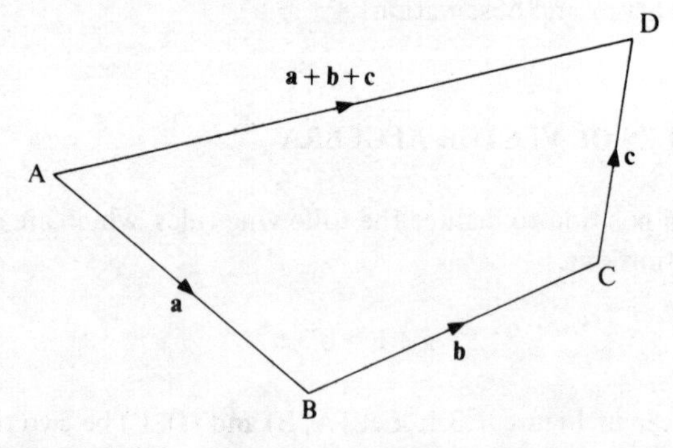

Figure 1.3.2 The sum of three vectors **a**, **b** and **c**.

Both sides of this relation represent a vector of magnitude $|\lambda + \mu| a$ in the direction of **a** if $(\lambda + \mu) > 0$ and of $-\mathbf{a}$ if $(\lambda + \mu) < 0$.

(v) $$\lambda(\mu\mathbf{a}) = (\lambda\mu)\mathbf{a}. \tag{1.3.5}$$

Both sides of this relation represent a vector of magnitude $|\lambda||\mu|\mathbf{a}$ in the direction of **a** if $\lambda\mu > 0$ and of $-\mathbf{a}$ if $\lambda\mu < 0$.

(vi) $$\mathbf{a} + \mathbf{0} = \mathbf{a}. \tag{1.3.6}$$

(vii) $$1\mathbf{a} = \mathbf{a}. \tag{1.3.7}$$

(viii) $$\mathbf{a} + (-\mathbf{a}) = \mathbf{0}. \tag{1.3.8}$$

The last three relations are merely statements of the properties of the null vector, the unit scalar, and the vector $-\mathbf{a}$. They are included in order to display the relations which are used as the basis of the theory of linear vector spaces. In this more general theory the quantities **a**, **b**, **c** . . . are not further defined beyond the requirement that they satisfy relations (1.3.1)–(1.3.8) above. The consequences of such a theory are quite versatile in that they can be applied to any set of quantities which obey the above rules. An elementary example is the set of real numbers; a less obvious example is the set of all polynomials with real coefficients. It is therefore worth bearing in mind that although we are concerned here with the development of a theory for geometric vectors as defined in §1.2, many results are capable of a more general interpretation in so far as they depend simply on a manipulation of relations (1.3.1)–(1.3.8). It should not, however, be assumed that vector algebra is appropriate in all applications where a physical concept is described by a magnitude and direction. One exception is the finite rotation of a rigid body; this has a magnitude, the angular displacement, and a direction defined by the axis of rotation. It does not, however, satisfy the rules of addition for vector algebra.*

Whenever a branch of mathematics is applied to a physical problem, there is an assumption that the physical quantities involved obey the rules of the mathematics used in their manipulation. The only justification for this assumption depends on the extent to which it leads to profitable conclusions, which can be satisfactorily confirmed by observation.

Example 1.3.1

Let **r** be the position vector of a point P relative to an origin O. Let **a** and **b** be constant vectors, and let λ be a variable scalar. We consider the locus of P when **r**

* For example let us define two perpendicular axes OX and OY and consider the displacement of a rod lying originally along OX. If it is given first a rotation through an angle $\pi/2$ about OY, so that it is perpendicular to both OX and OY, and then a rotation $\pi/2$ about OX, its final position will be along OY. If, however, the rotations are carried out in the reverse order, that is first a rotation $\pi/2$ about OX leaving the rod along OX, and then a rotation $\pi/2$ about OY, the final position of the rod will be perpendicular to OX and OY. Thus the commutative law of addition is violated.

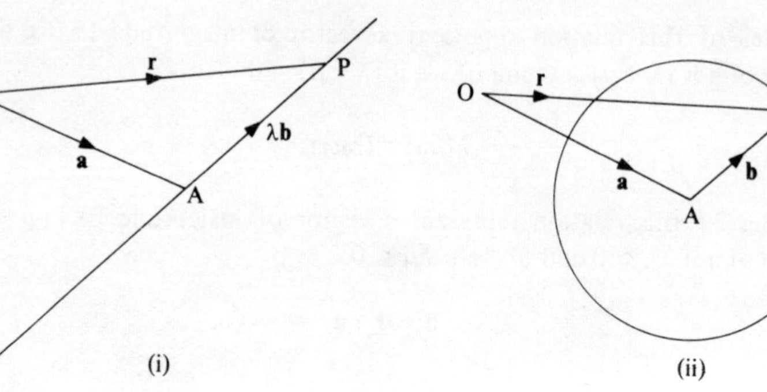

Figure 1.3.3 Illustration for Example 1.3.1: (i) $\mathbf{r} = \mathbf{a} + \lambda\mathbf{b}$; (ii) $|\mathbf{r} - \mathbf{a}| = b$.

varies in such a way that

$$\text{(i)}\quad \mathbf{r} = \mathbf{a} + \lambda\mathbf{b}, \quad \text{(ii)}\quad |\mathbf{r} - \mathbf{a}| = b,$$

(see Fig. 1.3.3).

(i) Here $\mathbf{r} - \mathbf{a} = \lambda\mathbf{b}$, which is a variable vector parallel to **b**. If **a** is the position vector of the point A, it follows that P lies on a line parallel to **b** and passing through A.

(ii) If **a** is the position vector of A, then $\mathbf{r} - \mathbf{a} = \overline{\text{AP}}$ and the given condition is equivalent to AP = b. Since b is constant this allows P to describe a sphere, centre A and radius b.

Example 1.3.2

Let the points A, G, B in Figure 1.3.4 lie on a straight line, in that order, and such that

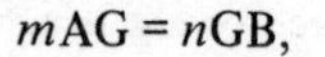

$$m\text{AG} = n\text{GB},$$

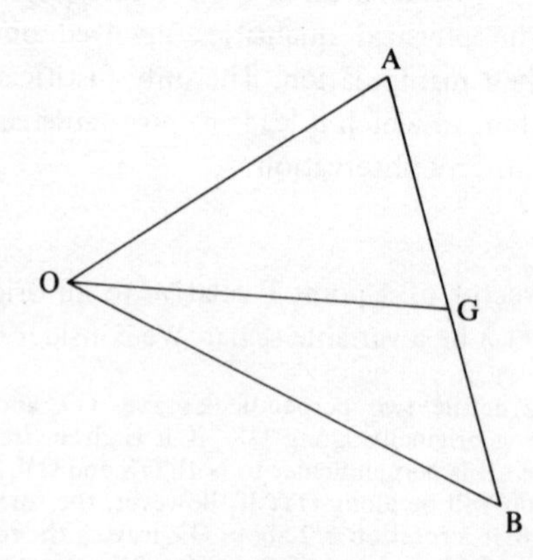

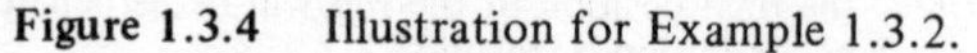

Figure 1.3.4 Illustration for Example 1.3.2.

where $m/n > 0$. Given that O is any other point, show that

$$m\overline{\mathrm{OA}} + n\overline{\mathrm{OB}} = (m+n)\overline{\mathrm{OG}}. \tag{1.3.9}$$

The information given is equivalent to the relation

$$m\overline{\mathrm{AG}} = n\overline{\mathrm{GB}},$$

since $m\overline{\mathrm{AG}}$ and $n\overline{\mathrm{GB}}$ have the same magnitude and the same direction. Hence

$$m(\overline{\mathrm{AO}} + \overline{\mathrm{OG}}) = n(\overline{\mathrm{GO}} + \overline{\mathrm{OB}}),$$

or, by rearrangement,

$$m\overline{\mathrm{OA}} + n\overline{\mathrm{OB}} = (m+n)\overline{\mathrm{OG}}.$$

This result is not restricted to points G which lie between A and B. The essential condition is that $m\overline{\mathrm{AG}} = n\overline{\mathrm{GB}}$ should be satisfied. This can include points G on the extensions of AB provided negative values of m/n are allowed.

Example 1.3.3
Vector algebra is particularly useful in the derivation of certain results in Euclidean geometry. The following example proves, among other things, that the medians of a triangle (that is the lines joining the vertices to the mid-points of the opposite sides) intersect at a single point, which is called the centroid of the triangle.

In Figure 1.3.5. let G be a point on the median AD of a triangle ABC such that

$$3\overline{\mathrm{AG}} = 2\overline{\mathrm{AD}}. \tag{1.3.10}$$

Show that

(i) $$\overline{\mathrm{AG}} + \overline{\mathrm{BG}} + \overline{\mathrm{CG}} = \mathbf{0}, \tag{1.3.11}$$

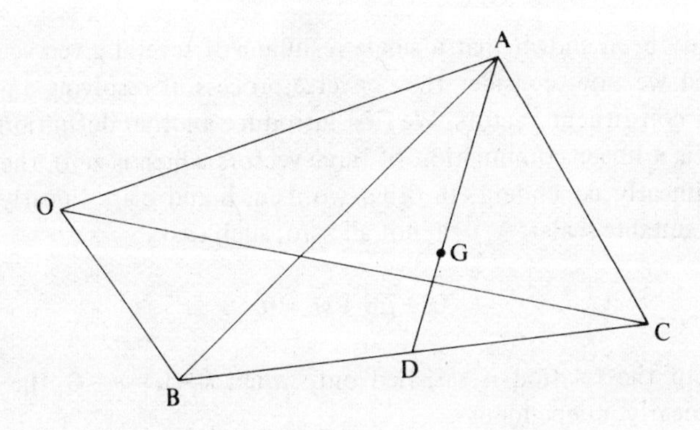

Figure 1.3.5 Illustration for Example 1.3.3.

(ii) if O is any other point (not necessarily in the plane of ABC)

$$3\overline{OG} = (\overline{OA} + \overline{OB} + \overline{OC}), \qquad (1.3.12)$$

(iii) G lies on the other medians.

(i) To prove the first result, we have

$$\begin{aligned} \overline{AG} = \overline{AB} + \overline{BG} &= \overline{AC} + \overline{CG}, && \text{by (1.2.1)},\\ 2\overline{AG} &= \overline{AB} + \overline{AC} + \overline{BG} + \overline{CG}, && \text{by addition},\\ &= 2\overline{AD} + \overline{BG} + \overline{CG}, && \text{by (1.3.9)},\\ &= 3\overline{AG} + \overline{BG} + \overline{CG}, && \text{by (1.3.10)}. \end{aligned}$$

Hence

$$\mathbf{0} = \overline{AG} + \overline{BG} + \overline{CG}.$$

(ii) Next we write

$$\begin{aligned} \overline{OA} + \overline{OB} + \overline{OC} &= (\overline{OG} + \overline{GA}) + (\overline{OG} + \overline{GB}) + (\overline{OG} + \overline{GC}), && \text{by (1.2.1)},\\ &= 3\overline{OG}, && \text{by (1.3.11)}. \end{aligned}$$

(iii) By a cyclic interchange of letters A, B, C in (1.3.12) we can infer that

$$\overline{OG'} = \tfrac{1}{3}(\overline{OB} + \overline{OC} + \overline{OA}) = \overline{OG},$$

where G′ is the point of trisection of the median BE that corresponds to G on AB. Hence G and G′ have the same position vector with respect to O and therefore coincide. A similar argument applies to the third median.

1.4 THE RESOLUTION OF A VECTOR

It has already been shown that a single resultant of several given vectors can be defined, and we now consider the converse process of resolving a given vector into several constituent vectors. We first introduce another definition.

If there is a linear combination of three vectors which is zero, the vectors are said to be linearly dependent. In other words **a**, **b** and **c** are linearly dependent if there are suitable scalars λ, μ, ν, not all zero, such that

$$\lambda\mathbf{a} + \mu\mathbf{b} + \nu\mathbf{c} = \mathbf{0}.$$

Conversely, if the relation is satisfied only when $\lambda = \mu = \nu = 0$, the vectors are said to be linearly independent.

Physically speaking the distinction is one between coplanar and non-coplanar vectors, apart from certain degenerate cases which are not exceptional if we accept the convention that any two vectors and a null vector are coplanar. The relation $\lambda\mathbf{a} + \mu\mathbf{b} + \nu\mathbf{c} = \mathbf{0}$ implies that $\lambda\mathbf{a}, \mu\mathbf{b}, \nu\mathbf{c}$ are coplanar since $-\lambda\mathbf{a}$ is the resultant of $\mu\mathbf{b}$ and $\nu\mathbf{c}$. If **a**, **b** and **c** are not coplanar there is a contradiction unless $\lambda = \mu = \nu = 0$. For if, say, $\lambda \neq 0$ we have $-\mathbf{a} = \lambda^{-1}\mu\mathbf{b} + \lambda^{-1}\nu\mathbf{c}$ which implies that **a** is coplanar with **b** and **c**.

Theorem

Any vector in three dimensions is equal to a linear combination of any three given linearly independent vectors.

Let (O, P) be a representation of **r**, and let Ox, Oy, Oz define the directions of the three given vectors (see Fig. 1.4.1).* It is possible to construct a parallelepiped whose adjacent edges OX, OY, OZ are in the given directions and whose diagonal is OP. First construct QP parallel to Oz so that Q lies in the plane of Ox, Oy. Then construct XQ parallel to Oy so that X lies on Ox. The vectors $\overline{\text{OX}}, \overline{\text{XQ}}, \overline{\text{QP}}$ define the parallelepiped and

$$\mathbf{r} = \overline{\text{OX}} + \overline{\text{XQ}} + \overline{\text{QP}} = \overline{\text{OX}} + \overline{\text{OY}} + \overline{\text{OZ}}. \tag{1.4.1}$$

This construction is always possible provided that the directions Ox, Oy, Oz are not coplanar.

We say that $\overline{\text{OX}}, \overline{\text{OY}}, \overline{\text{OZ}}$ are the *resolutes* of **r** in the three given directions.

If **a**, **b**, **c** are the three given vectors whose directions are defined by Ox, Oy, Oz respectively, it is possible to choose scalars λ, μ, ν such that

$$\lambda\mathbf{a} = \overline{\text{OX}}, \quad \mu\mathbf{b} = \overline{\text{OY}}, \quad \nu\mathbf{c} = \overline{\text{OZ}}. \tag{1.4.2}$$

Hence
$$\mathbf{r} = \lambda\mathbf{a} + \mu\mathbf{b} + \nu\mathbf{c}. \tag{1.4.3}$$

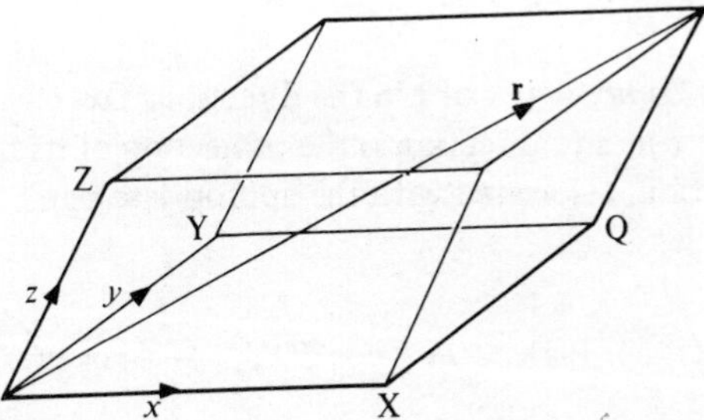

Figure 1.4.1 The resolutes of **r** parallel to Ox, Oy, Oz.

* In spite of the notation, the results of this section are not confirmed to the position vector. They apply to the free vector **r**, of which (O, P) is a representation based on O.

Since **a**, **b**, **c** are given to be linearly independent, it also follows that $\lambda = \mu = \nu = 0$ if $\mathbf{r} = \mathbf{0}$. This implies uniqueness of the resolutes. For if

$$\mathbf{r} = \lambda_1\mathbf{a} + \mu_1\mathbf{b} + \nu_1\mathbf{c} = \lambda_2\mathbf{a} + \mu_2\mathbf{b} + \nu_2\mathbf{c}, \tag{1.4.4}$$

we have

$$\mathbf{0} = (\lambda_1 - \lambda_2)\mathbf{a} + (\mu_1 - \mu_2)\mathbf{b} + (\nu_1 - \nu_2)\mathbf{c}, \tag{1.4.5}$$

and so

$$\lambda_1 = \lambda_2, \quad \mu_1 = \mu_2, \quad \nu_1 = \nu_2. \tag{1.4.6}$$

It follows that if two vectors are equal their resolutes are equal. This result can be useful in demonstrating vector identities.

Consider now the vector

$$\mathbf{r} = \mathbf{r}_1 + \mathbf{r}_2 + \ldots \mathbf{r}_s, \tag{1.4.7}$$

where

$$\mathbf{r}_i = \lambda_i\mathbf{a} + \mu_i\mathbf{b} + \nu_i\mathbf{c}, \quad (i = 1, 2, \ldots, s). \tag{1.4.8}$$

It is clear that, by addition,

$$\mathbf{r} = \sum_{i=1}^{s} \mathbf{r}_i = \left(\sum_{i=1}^{s} \lambda_i\right)\mathbf{a} + \left(\sum_{i=1}^{s} \mu_i\right)\mathbf{b} + \left(\sum_{i=1}^{s} \nu_i\right)\mathbf{c}, \tag{1.4.9}$$

and this defines, uniquely, the resolutes of the resultant **r** of several vectors $\mathbf{r}_i$. It shows that the sum of the resolutes in a given direction is the resolute of the resultant.

A most important application of the above results is to the situation in which Ox, Oy, Oz form a mutually orthogonal right-handed set of coordinate axes.* Let $\overline{\text{OP}} = \mathbf{r}$ and let P have coordinates (x, y, z) relative to such axes. Let **i**, **j**, **k** be vectors of unit magnitude in the directions Ox, Oy, Oz respectively. Then

$$\mathbf{r} = x\mathbf{i} + y\mathbf{j} + z\mathbf{k}, \qquad r = (x^2 + y^2 + z^2)^{1/2}. \tag{1.4.10}$$

We call x, y, z the *components* of **r** in the directions Ox, Oy, Oz. More generally the component of **r** in any direction is the projection of its representation on to a line in that direction, associated with the appropriate sign.

In Figure 1.4.2 let

$$l = \cos \text{P}\hat{\text{O}}\text{X}, \quad m = \cos \text{P}\hat{\text{O}}\text{Y}, \quad n = \cos \text{P}\hat{\text{O}}\text{Z}. \tag{1.4.11}$$

* A right-handed orthogonal set of axes Oxyz is such that the rotation of a right-handed screw through an angle $\pi/2$ from Oy to Oz, or Oz to Ox, or Ox to Oy propels it along the positive direction of the third axis. A left-handed set is obtained if the direction of one of the axes is reversed. It is conventional to choose a right-handed set. Such axes are also described as cartesian or rectangular.

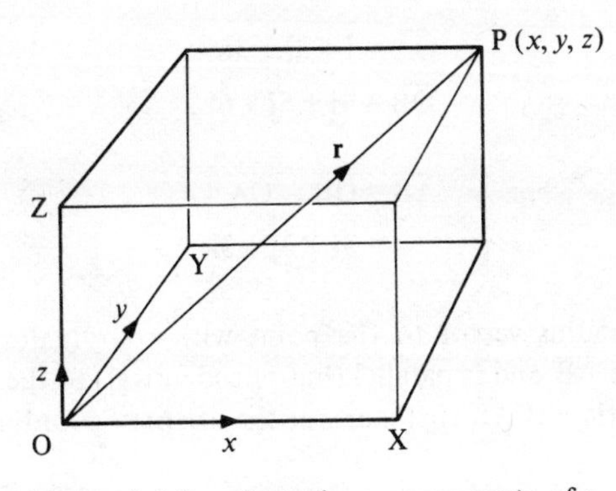

Figure 1.4.2 Cartesian components of **r**.

We then have, by projection

$$x = rl, \qquad y = rm, \qquad z = rn \tag{1.4.12}$$

and so

$$r = rl\mathbf{i} + rm\mathbf{j} + rn\mathbf{k} = r(l\mathbf{i} + m\mathbf{j} + n\mathbf{k}). \tag{1.4.13}$$

Since r is the magnitude of **r**, it follows that $l\mathbf{i} + m\mathbf{j} + n\mathbf{k}$ represents a vector of unit magnitude in the direction of **r**. In other words (l, m, n) are the coordinates of a point on OP at unit distance from O, and

$$l^2 + m^2 + n^2 = 1. \tag{1.4.14}$$

We call (l, m, n) the direction cosines of the line OP, relative to the given axes.

As a special case of (1.4.9) we note that the sum of the components, in a given direction, of a set of vectors is equal to the component of their resultant. That is

$$\left(\sum_{i=1}^{s} x_i\right)\mathbf{i} + \left(\sum_{i=1}^{s} y_i\right)\mathbf{j} + \left(\sum_{i=1}^{s} z_i\right)\mathbf{k} = \sum_{i=1}^{s} \mathbf{r}_i = \mathbf{r} = x\mathbf{i} + y\mathbf{j} + z\mathbf{k} \tag{1.4.15}$$

and this implies that

$$\sum_{i=1}^{s} x_i = x, \qquad \sum_{i=1}^{s} y_i = y, \qquad \sum_{i=1}^{s} z_i = z.$$

Henceforth the symbols **i**, **j**, **k** will denote vectors of unit magnitude parallel to the three coordinate axes of a right-handed orthogonal set. They will be referred to as unit vectors.

Example 1.4.1

Find the vector $\overline{\text{AB}}$, where A has coordinates $(1, 2, 3)$ and B has coordinates $(4, 5, 6)$.

Here
$$\overline{OA} = \mathbf{i} + 2\mathbf{j} + 3\mathbf{k},$$
$$\overline{OB} = 4\mathbf{i} + 5\mathbf{j} + 6\mathbf{k},$$

and
$$\overline{AB} = \overline{OB} - \overline{OA} = 3\mathbf{i} + 3\mathbf{j} + 3\mathbf{k}.$$

Note that the radius vector to the point with coordinates (3, 3, 3) has the same magnitude as $\overline{AB}$ and is parallel (but not identical) to the line segment AB. It is the representation of $\overline{AB}$ that starts at the origin of coordinates.

1.5 THE PRODUCTS OF VECTORS

There is no definition of the reciprocal of a vector, so that in no circumstances should one attempt to divide by a vector. The concept of multiplication is, however, part of vector algebra and, as in the case of addition, it needs to be defined. Apart from the multiplication of a vector by a scalar, already introduced in §1.2, there are two products which are used and which will now be discussed.

Scalar Product

The scalar product of two vectors is defined as the product of the magnitudes of the vectors and the cosine of the angle $\theta(0 \leqslant \theta \leqslant \pi)$ between the vectors (see Fig. 1.5.1). It is written

$$\mathbf{a} \cdot \mathbf{b} = ab \cos\theta \tag{1.5.1}$$

We note the following properties:

(i)
$$\mathbf{a} \cdot \mathbf{b} = ab \cos\theta = ba \cos\theta = \mathbf{b} \cdot \mathbf{a}. \tag{1.5.2}$$

(ii)
$$(\lambda\mathbf{a}) \cdot \mathbf{b} = \mathbf{a} \cdot (\lambda\mathbf{b}) = \lambda\mathbf{a} \cdot \mathbf{b}. \tag{1.5.3}$$

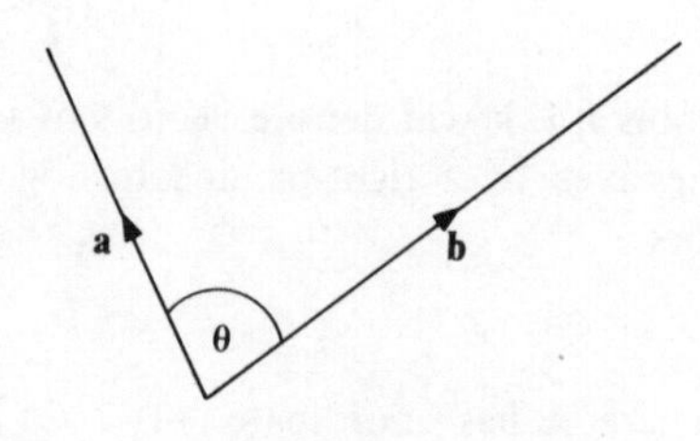

Figure 1.5.1 The angle θ between **a** and **b**.

This follows at once when $\lambda \geqslant 0$. It is also true when $\lambda < 0$ for then

$$(\lambda \mathbf{a}) \cdot \mathbf{b} = |\lambda| ab \cos(\pi - \theta) = -|\lambda| ab \cos\theta = \lambda \mathbf{a} \cdot \mathbf{b}.$$

(iii) If $\theta = 0$, then $\mathbf{a} \cdot \mathbf{b} = ab$. In particular

$$\mathbf{a} \cdot \mathbf{a} = a^2 \quad \text{and} \quad \mathbf{i} \cdot \mathbf{i} = \mathbf{j} \cdot \mathbf{j} = \mathbf{k} \cdot \mathbf{k} = 1. \tag{1.5.4}$$

(iv) If $\mathbf{a} \cdot \mathbf{b} = 0$ then one (or more) of a, b and $\cos\theta$ is zero. In particular

$$\mathbf{j} \cdot \mathbf{k} = \mathbf{k} \cdot \mathbf{i} = \mathbf{i} \cdot \mathbf{j} = 0. \tag{1.5.5}$$

(v) Since $\mathbf{a} \cdot \mathbf{b}$ is the magnitude of $\mathbf{b}$ multiplied by the projection of $\mathbf{a}$ in the direction of $\mathbf{b}$, it follows that $\mathbf{a} \cdot \mathbf{i}$ is the component of $\mathbf{a}$ in the direction of $\mathbf{i}$ and so

$$\mathbf{a} = (\mathbf{a} \cdot \mathbf{i})\mathbf{i} + (\mathbf{a} \cdot \mathbf{j})\mathbf{j} + (\mathbf{a} \cdot \mathbf{k})\mathbf{k}. \tag{1.5.6}$$

(vi)

$$\mathbf{a} \cdot (\mathbf{b} + \mathbf{c}) = \mathbf{a} \cdot \mathbf{b} + \mathbf{a} \cdot \mathbf{c}. \tag{1.5.7}$$

For

$$\mathbf{a} \cdot (\mathbf{b} + \mathbf{c}) = a \text{ (component of } (\mathbf{b} + \mathbf{c}) \text{ parallel to } \mathbf{a}).$$

By (1.4.9) this is equal to

$$a \text{ (sum of components of } \mathbf{b} \text{ and } \mathbf{c} \text{ parallel to } \mathbf{a}).$$

But scalar quantities satisfy the distributive law (1.1.3). Hence

$$\begin{aligned} \mathbf{a} \cdot (\mathbf{b} + \mathbf{c}) &= a \text{ (component of } \mathbf{b} \text{ parallel to } \mathbf{a}) \\ &\quad + a \text{ (component of } \mathbf{c} \text{ parallel to } \mathbf{a}) \\ &= \mathbf{a} \cdot \mathbf{b} + \mathbf{a} \cdot \mathbf{c}. \end{aligned}$$

(vii) The repeated application of (1.5.7) allows the expansion of the scalar product when the vectors are expressed in terms of their components. Thus if

$$\mathbf{a} = a_x\mathbf{i} + a_y\mathbf{j} + a_z\mathbf{k}, \qquad \mathbf{b} = b_x\mathbf{i} + b_y\mathbf{j} + b_z\mathbf{k},$$

it follows, with the help of (1.5.4), (1.5.5) and (1.5.7), that

$$\begin{aligned} \mathbf{a} \cdot \mathbf{b} &= (a_x\mathbf{i} + a_y\mathbf{j} + a_z\mathbf{k}) \cdot (b_x\mathbf{i} + b_y\mathbf{j} + b_z\mathbf{k}) \\ &= a_x b_x + a_y b_y + a_z b_z. \end{aligned} \tag{1.5.8}$$

Note, however, that in general $(\mathbf{a}\,.\,\mathbf{b})\mathbf{c} \neq \mathbf{a}(\mathbf{b}\,.\,\mathbf{c})$ and that $\mathbf{a}\,.\,\mathbf{b}\,.\,\mathbf{c}$ is not defined.

Vector Product

The second kind of product is a vector whose magnitude is defined as the product of the magnitudes of the vectors and the sine of the angle $\theta(0 \leqslant \theta \leqslant \pi)$ between them. The direction is perpendicular to the plane of the two vectors in the sense of the translation of a right-handed screw rotated from the first to the second vector. Note that the order of the vectors is now relevant. The notation is

$$\mathbf{a} \times \mathbf{b} = ab \sin \theta \, \mathbf{n}$$

where, in Figure 1.5.1, $\mathbf{n}$ is a unit vector perpendicular to the plane of the figure and into (rather than out of) this plane. We note the following properties:

(i) If $\mathbf{a} \times \mathbf{b} = \mathbf{0}$ then one (or more) of a, b and $\sin \theta$ is zero and vice versa. Thus $\mathbf{a} \times \mathbf{a} = \mathbf{0}$ and in particular

$$\mathbf{i} \times \mathbf{i} = \mathbf{j} \times \mathbf{j} = \mathbf{k} \times \mathbf{k} = \mathbf{0}. \tag{1.5.9}$$

(ii) If $\mathbf{a}$ and $\mathbf{b}$ are perpendicular then $|\,\mathbf{a} \times \mathbf{b}\,| = ab$. More specifically, for the unit vectors $\mathbf{i}, \mathbf{j}, \mathbf{k}$ we have

$$\mathbf{i} = \mathbf{j} \times \mathbf{k}, \qquad \mathbf{j} = \mathbf{k} \times \mathbf{i}, \qquad \mathbf{k} = \mathbf{i} \times \mathbf{j}. \tag{1.5.10}$$

(iii) Some care is necessary in the handling of this product since the commutative and associative laws are not satisfied. In fact, one can see from the definition that

$$\mathbf{a} \times \mathbf{b} = -\,\mathbf{b} \times \mathbf{a}.$$

Also by counter example it can be seen that, in general,

$$(\mathbf{a} \times \mathbf{b}) \times \mathbf{c} \neq \mathbf{a} \times (\mathbf{b} \times \mathbf{c}).$$

For example $(\mathbf{i} \times \mathbf{i}) \times \mathbf{j} = \mathbf{0}$ but $\mathbf{i} \times (\mathbf{i} \times \mathbf{j}) = \mathbf{i} \times \mathbf{k} = -\,\mathbf{j}$.

(iv) The distributive laws, namely,

$$(\lambda\mathbf{a}) \times \mathbf{b} = \mathbf{a} \times (\lambda\mathbf{b}) = \lambda\mathbf{a} \times \mathbf{b}, \tag{1.5.11}$$

$$\mathbf{a} \times (\mathbf{b} + \mathbf{c}) = \mathbf{a} \times \mathbf{b} + \mathbf{a} \times \mathbf{c}, \tag{1.5.12}$$

are, however, satisfied. The first relation (1.5.11) follows from the definition, with a little care necessary if λ is negative.

As a preliminary to the proof of (1.5.12) we note that if $a = 0$ the relation is clearly satisfied. When $a \neq 0$ it is sufficient to prove the result for $a = 1$, since for other values each side of the equation can be reduced by the factor a^{-1}.

In Figure 1.5.2 let the sides of the triangle PQR be representations of **b**, **c** and **b** + **c** as shown. Let the triangle P′Q′R′ be the projection of PQR on to a plane perpendicular to **a**. From the definition of the vector product it is clear that $|\mathbf{a} \times \mathbf{b}|$ is equal to the product of the magnitude of **a** and the component of **b** perpendicular to **a**. Since $a = 1$ here, it follows that $|\mathbf{a} \times \mathbf{b}| = \mathrm{P'Q'}$. The direction of $\mathbf{a} \times \mathbf{b}$ is perpendicular to **a** and **b**, it can therefore be obtained from $\overline{\mathrm{P'Q'}}$ by rotation through an angle $\pi/2$ about **a** in a direction appropriate to the right-handed screw rule. Similarly $\mathbf{a} \times \mathbf{c}$ and $\mathbf{a} \times (\mathbf{b} + \mathbf{c})$ can be constructed from $\overline{\mathrm{Q'R'}}$ and $\overline{\mathrm{P'R'}}$ by rotation through an angle $\pi/2$. Thus $\mathbf{a} \times \mathbf{b}$, $\mathbf{a} \times \mathbf{c}$ and $\mathbf{a} \times (\mathbf{b} + \mathbf{c})$ can be represented by $\overline{\mathrm{P'Q''}}$, $\overline{\mathrm{Q''R''}}$ and $\overline{\mathrm{P'R''}}$ respectively, where the triangle P′Q″R″ is obtained by the rotation of triangle P′Q′R′ about **a** through an angle $\pi/2$. It follows immediately, from the law of addition, that

$$\overline{\mathrm{P'Q''}} + \overline{\mathrm{Q''R''}} = \overline{\mathrm{P'R''}},$$

or

$$\mathbf{a} \times \mathbf{b} + \mathbf{a} \times \mathbf{c} = \mathbf{a} \times (\mathbf{b} + \mathbf{c}).$$

(v) By repeated application of (1.5.12), we can expand the vector product when the constituent vectors are defined in terms of their components. Thus, with the help of (1.5.9) and (1.5.10), we have

$$\begin{aligned} \mathbf{a} \times \mathbf{b} &= (a_x\mathbf{i} + a_y\mathbf{j} + a_z\mathbf{k}) \times (b_x\mathbf{i} + b_y\mathbf{j} + b_z\mathbf{k}) \\ &= (a_yb_z - a_zb_y)\mathbf{i} + (a_zb_x - a_xb_z)\mathbf{j} + (a_xb_y - a_yb_x)\mathbf{k}. \end{aligned} \tag{1.5.13}$$

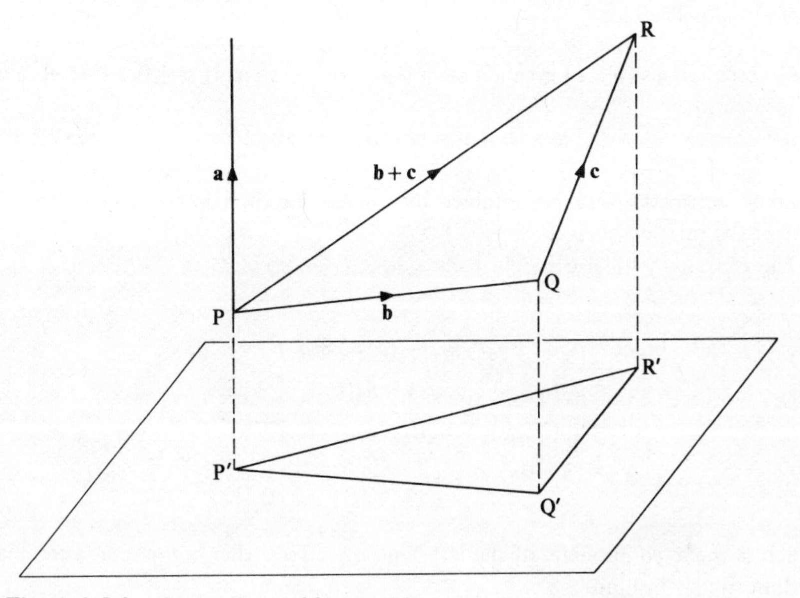

Figure 1.5.2 Projections of **b**, **c** and **b** + **c** on to a plane perpendicular to **a**.

Triple Scalar Product

The expression $\mathbf{a} \cdot (\mathbf{b} \times \mathbf{c})$ is called a triple scalar product. The equivalence of the expressions below follows at once from the properties of the scalar and vector products,

$$\mathbf{a} \cdot (\mathbf{b} \times \mathbf{c}) = (\mathbf{b} \times \mathbf{c}) \cdot \mathbf{a} = -\mathbf{a} \cdot (\mathbf{c} \times \mathbf{b}).$$

The following relations are also valid:

$$\mathbf{a} \cdot (\mathbf{b} \times \mathbf{c}) = \mathbf{b} \cdot (\mathbf{c} \times \mathbf{a}) = \mathbf{c} \cdot (\mathbf{a} \times \mathbf{b}). \tag{1.5.14}$$

These relations are easily demonstrated by expressing the vectors in their component form. Thus

$$\mathbf{a} \cdot (\mathbf{b} \times \mathbf{c}) = a_x(b_y c_z - b_z c_y) + a_y(b_z c_x - b_x c_z) + a_z(b_x c_y - b_y c_x). \tag{1.5.15}$$

It merely requires a judicious rearrangement of the various terms to demonstrate that (1.5.15) is invariant under a cyclic change of a, b, c, that is a change from a, b, c to b, c, a or c, a, b in the order indicated.

If $\mathbf{a}$, $\mathbf{b}$, $\mathbf{c}$ are linearly dependent, that is coplanar, then $\mathbf{a} \cdot (\mathbf{b} \times \mathbf{c}) = 0$ since $\mathbf{a}$ and $\mathbf{b} \times \mathbf{c}$ are perpendicular. Conversely if $\mathbf{a} \cdot (\mathbf{b} \times \mathbf{c}) \neq 0$ then $\mathbf{a}, \mathbf{b}, \mathbf{c}$ are linearly independent. This is a useful test for linear independence.

Triple Vector Product

The expression $\mathbf{a} \times (\mathbf{b} \times \mathbf{c})$ is called a triple vector product. It satisfies the relation

$$\mathbf{a} \times (\mathbf{b} \times \mathbf{c}) = (\mathbf{a} \cdot \mathbf{c})\mathbf{b} - (\mathbf{a} \cdot \mathbf{b})\mathbf{c}. \tag{1.5.16}$$

A direct verification can be obtained by considering the components of the two sides of the relation, as follows.

The x component of the right-hand side is

$$\begin{aligned}
&(a_x c_x + a_y c_y + a_z c_z)b_x - (a_x b_x + a_y b_y + a_z b_z)c_x \\
&= a_y(b_x c_y - b_y c_x) - a_z(b_z c_x - b_x c_z) \\
&= a_y(\mathbf{b} \times \mathbf{c})_z - a_z(\mathbf{b} \times \mathbf{c})_y \\
&= \{\mathbf{a} \times (\mathbf{b} \times \mathbf{c})\}_x,
\end{aligned}$$

which is the x component of the left-hand side. The other components are dealt with in similar fashion.

Example 1.5.1

We use the scalar product to establish the cosine rule for a triangle, that is, if $BC = a$, $CA = b$, $AB = c$, then

$$a^2 = b^2 + c^2 - 2bc \cos B\hat{A}C. \tag{1.5.17}$$

Since

$$\overline{BC} = \overline{BA} + \overline{AC},$$

we have

$$\begin{aligned}\overline{BC} \,.\, \overline{BC} &= (\overline{BA} + \overline{AC}) \,.\, (\overline{BA} + \overline{AC}),\\ &= \overline{AC} \,.\, \overline{AC} + \overline{BA} \,.\, \overline{BA} + 2\overline{AC} \,.\, \overline{BA}.\end{aligned}$$

Hence

$$\begin{aligned}a^2 &= b^2 + c^2 + 2bc \cos (\pi - B\hat{A}C),\\ &= b^2 + c^2 - 2bc \cos B\hat{A}C.\end{aligned}$$

Note that $B\hat{A}C$ is the angle between $\overline{AB}$ and $\overline{AC}$; the angle between $\overline{BA}$ and $\overline{AC}$ is $\pi - B\hat{A}C$ (see Fig. 1.5.3).

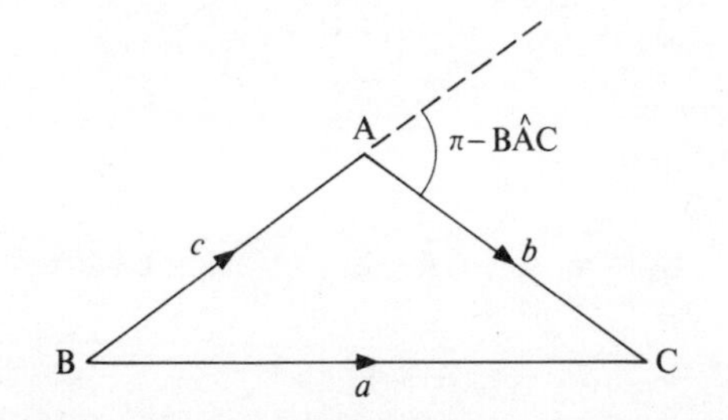

Figure 1.5.3 Illustration for Example 1.5.1.

Example 1.5.2

We prove that the altitudes of a triangle are concurrent.

Let the altitudes through B and C, of the triangle ABC, meet in H (see Fig. 1.5.4). Then

$$\overline{BH} \,.\, \overline{CA} = 0 \quad \text{or} \quad (\overline{BA} + \overline{AH}) \,.\, \overline{CA} = 0,$$

$$\overline{CH} \,.\, \overline{AB} = 0 \quad \text{or} \quad (\overline{CA} + \overline{AH}) \,.\, \overline{AB} = 0.$$

The addition of these two relations gives

$$\overline{AH} \,.\, \overline{CA} + \overline{AH} \,.\, \overline{AB} = 0 \quad \text{or} \quad \overline{AH} \,.\, \overline{CB} = 0$$

since

$$\overline{BA} \,.\, \overline{CA} + \overline{CA} \,.\, \overline{AB} = 0.$$

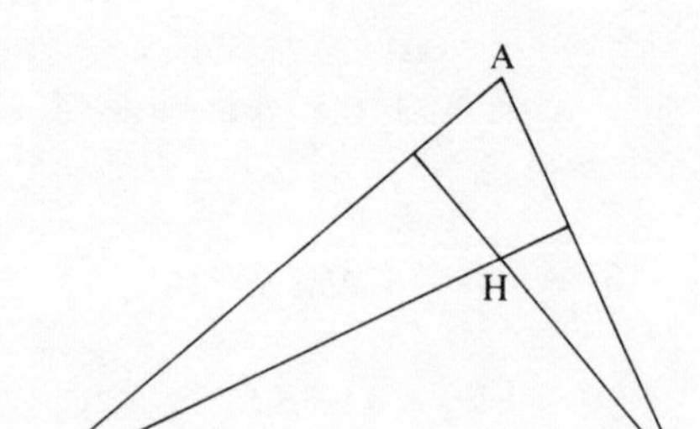

Figure 1.5.4 Illustration for Example 1.5.2.

The relation $\overline{\mathrm{AH}} \cdot \overline{\mathrm{CB}} = 0$ implies that $\overline{\mathrm{AH}}$ is perpendicular to $\overline{\mathrm{CB}}$, that is the altitude through A passes through H.

Example 1.5.3

We use the vector product to prove the sine rule for a triangle, namely that in any triangle ABC with sides a, b, c

$$\frac{\sin A}{a} = \frac{\sin B}{b} = \frac{\sin C}{c}. \tag{1.5.18}$$

In the triangle ABC, let

$$\overline{\mathrm{BC}} = \mathbf{a}, \qquad \overline{\mathrm{CA}} = \mathbf{b}, \qquad \overline{\mathrm{AB}} = \mathbf{c}.$$

Then

$$-\mathbf{c} = \mathbf{a} + \mathbf{b},$$

and hence

$$(\mathbf{a} + \mathbf{b}) \times \mathbf{c} = \mathbf{0}.$$

This is equivalent to

$$\mathbf{b} \times \mathbf{c} = \mathbf{c} \times \mathbf{a},$$

and similarly we can show that

$$\mathbf{b} \times \mathbf{c} = \mathbf{c} \times \mathbf{a} = \mathbf{a} \times \mathbf{b}.$$

Thus we have, since $\sin(\pi - A) = \sin A$ etc.,

$$bc \sin A = ca \sin B = ab \sin C,$$

or

$$\frac{\sin A}{a} = \frac{\sin B}{b} = \frac{\sin C}{c}.$$

Example 1.5.4

Let $\theta_1, \theta_2, \theta_3$ be the angles between the vectors $\mathbf{r}_2$ and $\mathbf{r}_3$, $\mathbf{r}_3$ and $\mathbf{r}_1$, $\mathbf{r}_1$ and $\mathbf{r}_2$ respectively. We consider the problem of finding the angle between $\mathbf{r}_3$ and $\mathbf{r}_1 \times \mathbf{r}_2$. Geometrically this is the angle between $\mathbf{r}_3$ and the perpendicular to the plane of $\mathbf{r}_1$ and $\mathbf{r}_2$. If this angle is ϕ, we have that

$$|\mathbf{r}_3 \times (\mathbf{r}_1 \times \mathbf{r}_2)| = |\mathbf{r}_3|\,|\mathbf{r}_1 \times \mathbf{r}_2| \sin\phi = r_1 r_2 r_3 \sin\theta_3 \sin\phi.$$

An alternative expression may be derived from the relation

$$\mathbf{r}_3 \times (\mathbf{r}_1 \times \mathbf{r}_2) = (\mathbf{r}_3 \,.\, \mathbf{r}_2)\mathbf{r}_1 - (\mathbf{r}_3 \,.\, \mathbf{r}_1)\mathbf{r}_2,$$

from which we can obtain the relation

$$\begin{aligned} |\mathbf{r}_3 \times (\mathbf{r}_1 \times \mathbf{r}_2)|^2 &= (\mathbf{r}_3 \,.\, \mathbf{r}_2)^2 r_1^2 + (\mathbf{r}_3 \,.\, \mathbf{r}_1)^2 r_2^2 - 2(\mathbf{r}_3 \,.\, \mathbf{r}_2)(\mathbf{r}_3 \,.\, \mathbf{r}_1)(\mathbf{r}_1 \,.\, \mathbf{r}_2) \\ &= r_1^2 r_2^2 r_3^2(\cos^2\theta_1 + \cos^2\theta_2 - 2\cos\theta_1\cos\theta_2\cos\theta_3). \end{aligned}$$

It follows that

$$\sin\theta_3 \sin\phi = (\cos^2\theta_1 + \cos^2\theta_2 - 2\cos\theta_1\cos\theta_2\cos\theta_3)^{1/2}.$$

Example 1.5.5

We have already seen, in §1.4, that

$$\mathbf{r} = \lambda\mathbf{a} + \mu\mathbf{b} + \nu\mathbf{c}$$

for linearly independent vectors **a**, **b**, **c** and suitably chosen scalars λ, μ, ν. It is clear that λ, μ, ν must be some functions of the four vectors **a**, **b**, **c** and **r**. We are now in a position to derive explicit expressions for λ, μ, ν.

Since $(\mathbf{b} \times \mathbf{c})$ is perpendicular to **b** and **c**, we have

$$\mathbf{r} \,.\, (\mathbf{b} \times \mathbf{c}) = \lambda\mathbf{a} \,.\, (\mathbf{b} \times \mathbf{c}).$$

Similarly

$$\begin{aligned} \mathbf{r} \,.\, (\mathbf{c} \times \mathbf{a}) &= \mu\mathbf{b} \,.\, (\mathbf{c} \times \mathbf{a}) = \mu\mathbf{a} \,.\, (\mathbf{b} \times \mathbf{c}), \\ \mathbf{r} \,.\, (\mathbf{a} \times \mathbf{b}) &= \nu\mathbf{c} \,.\, (\mathbf{a} \times \mathbf{b}) = \nu\mathbf{a} \,.\, (\mathbf{b} \times \mathbf{c}). \end{aligned}$$

It follows that

$$\{\mathbf{a} \,.\, (\mathbf{b} \times \mathbf{c})\}\mathbf{r} = \{\mathbf{r} \,.\, (\mathbf{b} \times \mathbf{c})\}\mathbf{a} + \{\mathbf{r} \,.\, (\mathbf{c} \times \mathbf{a})\}\mathbf{b} + \{\mathbf{r} \,.\, (\mathbf{a} \times \mathbf{b})\}\mathbf{c}. \quad (1.5.19)$$

This gives **r** as a linear combination of **a**, **b**, **c** if they are linearly independent. But if **a**, **b**, **c** are linearly dependent, the factor $\mathbf{a} \,.\, (\mathbf{b} \times \mathbf{c})$ on the left-hand side is zero. In this case, for any vector $\mathbf{r}$, we have

$$\mathbf{0} = \{\mathbf{r} \cdot (\mathbf{b} \times \mathbf{c})\}\mathbf{a} + \{\mathbf{r} \cdot (\mathbf{c} \times \mathbf{a})\}\mathbf{b} + \{\mathbf{r} \cdot (\mathbf{a} \times \mathbf{b})\}\mathbf{c}.$$

When $\mathbf{a}$, $\mathbf{b}$ and $\mathbf{c}$ are linearly independent, it is convenient to introduce the notation

$$\mathbf{A} = \frac{\mathbf{b} \times \mathbf{c}}{\mathbf{a} \cdot (\mathbf{b} \times \mathbf{c})}, \qquad \mathbf{B} = \frac{\mathbf{c} \times \mathbf{a}}{\mathbf{a} \cdot (\mathbf{b} \times \mathbf{c})}, \qquad \mathbf{C} = \frac{\mathbf{a} \times \mathbf{b}}{\mathbf{a} \cdot (\mathbf{b} \times \mathbf{c})}, \tag{1.5.20}$$

so that

$$\mathbf{r} = (\mathbf{r} \cdot \mathbf{A})\mathbf{a} + (\mathbf{r} \cdot \mathbf{B})\mathbf{b} + (\mathbf{r} \cdot \mathbf{C})\mathbf{c}. \tag{1.5.21}$$

There are certain reciprocal relations satisfied between the two sets of vectors $\mathbf{a}, \mathbf{b}, \mathbf{c}$ and $\mathbf{A}, \mathbf{B}, \mathbf{C}$. It follows from their definition that

$$\begin{aligned} &\mathbf{a} \cdot \mathbf{A} = \mathbf{b} \cdot \mathbf{B} = \mathbf{c} \cdot \mathbf{C} = 1, \\ &\mathbf{a} \cdot \mathbf{B} = \mathbf{a} \cdot \mathbf{C} = \mathbf{b} \cdot \mathbf{C} = \mathbf{b} \cdot \mathbf{A} = \mathbf{c} \cdot \mathbf{A} = \mathbf{c} \cdot \mathbf{B} = 0, \end{aligned} \tag{1.5.22}$$

$$\mathbf{B} \times \mathbf{C} = \frac{(\mathbf{c} \times \mathbf{a}) \times (\mathbf{a} \times \mathbf{b})}{\{\mathbf{a} \cdot (\mathbf{b} \times \mathbf{c})\}^2} = \frac{\{(\mathbf{c} \times \mathbf{a}) \cdot \mathbf{b}\}\mathbf{a}}{\{\mathbf{a} \cdot (\mathbf{b} \times \mathbf{c})\}^2} = \frac{\mathbf{a}}{\mathbf{a} \cdot (\mathbf{b} \times \mathbf{c})}, \tag{1.5.23}$$

$$\mathbf{A} \cdot (\mathbf{B} \times \mathbf{C}) = \frac{1}{\mathbf{a} \cdot (\mathbf{b} \times \mathbf{c})}. \tag{1.5.24}$$

Hence

$$\mathbf{a} = \frac{\mathbf{B} \times \mathbf{C}}{\mathbf{A} \cdot (\mathbf{B} \times \mathbf{C})}, \qquad \mathbf{b} = \frac{\mathbf{C} \times \mathbf{A}}{\mathbf{A} \cdot (\mathbf{B} \times \mathbf{C})}, \qquad \mathbf{c} = \frac{\mathbf{A} \times \mathbf{B}}{\mathbf{A} \cdot (\mathbf{B} \times \mathbf{C})}. \tag{1.5.25}$$

From (1.5.24), it follows that $\mathbf{A} \cdot (\mathbf{B} \times \mathbf{C}) \neq 0$ if $\mathbf{a} \cdot (\mathbf{b} \times \mathbf{c}) \neq 0$. Hence $\mathbf{A}$, $\mathbf{B}$ and $\mathbf{C}$ are linearly independent whenever $\mathbf{a}$, $\mathbf{b}$ and $\mathbf{c}$ are linearly independent, and an alternative expression for $\mathbf{r}$ can be given as a linear combination of $\mathbf{A}$, $\mathbf{B}$ and $\mathbf{C}$. Reference to (1.5.19) shows that the appropriate representation is

$$\{\mathbf{A} \cdot (\mathbf{B} \times \mathbf{C})\}\mathbf{r} = \{\mathbf{r} \cdot (\mathbf{B} \times \mathbf{C})\}\mathbf{A} + \{\mathbf{r} \cdot (\mathbf{C} \times \mathbf{A})\}\mathbf{B} + \{\mathbf{r} \cdot (\mathbf{A} \times \mathbf{B})\}\mathbf{C}$$

which, with the help of (1.5.25), can be written in the form

$$\mathbf{r} = (\mathbf{r} \cdot \mathbf{a})\mathbf{A} + (\mathbf{r} \cdot \mathbf{b})\mathbf{B} + (\mathbf{r} \cdot \mathbf{c})\mathbf{C}. \tag{1.5.26}$$

This last relation shows that any vector $\mathbf{r}$ is specified by its scalar products with three linearly independent vectors. A corollary of this result is that if the scalar products of $\mathbf{r}$ with three linearly independent vectors are zero, then $\mathbf{r}$ is a null vector.

A useful expression for the magnitude of $\mathbf{r}$ can be deduced from the use of relations (1.5.22), since

$$r^2 = \{(\mathbf{r}\,.\,\mathbf{A})\mathbf{a} + (\mathbf{r}\,.\,\mathbf{B})\mathbf{b} + (\mathbf{r}\,.\,\mathbf{C})\mathbf{c}\}\,.\,\{(\mathbf{r}\,.\,\mathbf{a})\mathbf{A} + (\mathbf{r}\,.\,\mathbf{b})\mathbf{B} + (\mathbf{r}\,.\,\mathbf{c})\mathbf{C}\}$$
$$= (\mathbf{r}\,.\,\mathbf{A})(\mathbf{r}\,.\,\mathbf{a}) + (\mathbf{r}\,.\,\mathbf{B})(\mathbf{r}\,.\,\mathbf{b}) + (\mathbf{r}\,.\,\mathbf{C})(\mathbf{r}\,.\,\mathbf{c}).$$

Example 1.5.6

Sometimes a vector is defined only through a vector equation. We take as an example the equation

$$\alpha\mathbf{x} + \mathbf{x} \times \mathbf{a} = \mathbf{b}, \tag{1.5.27}$$

from which an explicit expression for $\mathbf{x}$ is required.

We consider first the case $\alpha \neq 0$. We take the vector product of (1.5.27) with $\mathbf{a}$ to get

$$\alpha(\mathbf{a} \times \mathbf{x}) + \mathbf{a} \times (\mathbf{x} \times \mathbf{a}) = \mathbf{a} \times \mathbf{b}.$$

With the help of (1.5.16) and (1.5.27) this becomes

$$\alpha^2\mathbf{x} - \alpha\mathbf{b} + a^2\mathbf{x} - (\mathbf{a}\,.\,\mathbf{x})\mathbf{a} = \mathbf{a} \times \mathbf{b}. \tag{1.5.28}$$

This would be sufficient to determine $\mathbf{x}$ if we knew the value of $\mathbf{a}\,.\,\mathbf{x}$. To find this we take the scalar product of (1.5.27) with $\mathbf{a}$ to get

$$\alpha\mathbf{a}\,.\,\mathbf{x} = \mathbf{a}\,.\,\mathbf{b}.$$

Substitution in (1.5.28) then gives

$$(\alpha^2 + a^2)\mathbf{x} = \alpha^{-1}(\mathbf{a}\,.\,\mathbf{b})\mathbf{a} + \alpha\mathbf{b} + \mathbf{a} \times \mathbf{b},$$

which gives $\mathbf{x}$ as a linear combination of $\mathbf{a}$, $\mathbf{b}$ and $\mathbf{a} \times \mathbf{b}$ provided that $\alpha \neq 0$.

Now consider the case $\alpha = 0$, so that

$$\mathbf{x} \times \mathbf{a} = \mathbf{b}.$$

There is clearly no solution unless $\mathbf{a}$ and $\mathbf{b}$ are perpendicular. If this condition is satisfied and $\mathbf{x}_0$ is a particular solution, then so is $\mathbf{x}_0 + \lambda\mathbf{a}$ for all λ. The solution is therefore not unique. Suppose $\mathbf{x}_0$ is chosen to satisfy $\mathbf{x}_0\,.\,\mathbf{a} = 0$. Then

$$\mathbf{a} \times (\mathbf{x}_0 \times \mathbf{a}) = a^2\mathbf{x}_0 - (\mathbf{a}\,.\,\mathbf{x}_0)\mathbf{a} = a^2\mathbf{x}_0 = \mathbf{a} \times \mathbf{b}.$$

The general solution is therefore

$$\mathbf{x} = a^{-2}(\mathbf{a} \times \mathbf{b}) + \lambda\mathbf{a}.$$

1.6 THE DERIVATIVE OF A VECTOR

We now introduce the notion of a vector which varies as a function of some scalar variable t, and we write $\mathbf{r} = \mathbf{r}(t)$. Many of its properties can be discussed by analogy with the notion of a scalar function, of which it is a generalisation. However, it is important to realise that the variation of a vector is not absolute but relative to an implied frame of reference, and a change of frame will be accompanied by a change in the relative variation. For example, if $|\mathbf{r}|$ is constant then $\mathbf{r}$ itself is a constant vector relative to a frame of reference in which one of the axes is always parallel to $\mathbf{r}$. In particular the unit vectors $\mathbf{i}, \mathbf{j}, \mathbf{k}$ are constant vectors relative to the frame which defines them, although they may vary relative to some other frame. In any application of vector analysis, a suitable frame of reference must always be defined. In the development of the theory it is assumed that this has already been done.

We define the derivative of a vector function with a definition based on a generalisation of the one used in elementary calculus, namely

$$\frac{d\mathbf{r}}{dt} = \lim_{\delta t \to 0} \frac{\mathbf{r}(t + \delta t) - \mathbf{r}(t)}{\delta t} \tag{1.6.1}$$

provided such a limit exists. Let us choose a representation of $\mathbf{r}(t)$ based on some origin O, so that $\mathbf{r}(t)$ is the position vector of some point P relative to O. This is not essential to the above definition, but it does allow us to give it a geometrical interpretation. In Figure 1.6.1 we have

$$\mathbf{r}(t + \delta t) - \mathbf{r}(t) = \overline{\mathrm{PP}}', \tag{1.6.2}$$

and so we can write

$$\frac{d\mathbf{r}}{dt} = \lim_{\delta t \to 0} \frac{\overline{\mathrm{PP}}' \, . \, \mathrm{PP}'}{\mathrm{PP}' \, . \, \delta t}, \tag{1.6.3}$$

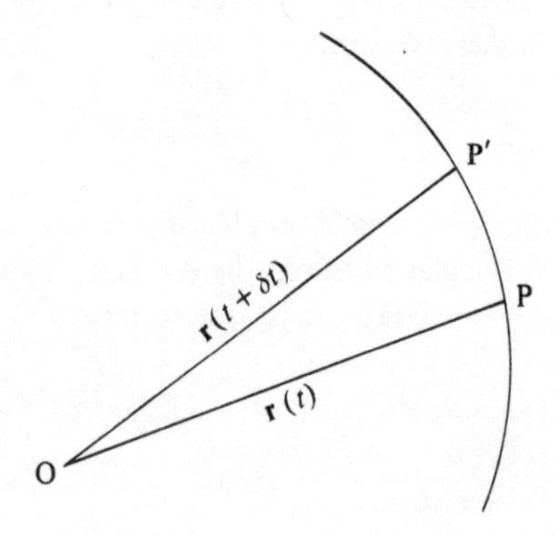

Figure 1.6.1 The locus of a point P whose position vector $\mathbf{r}(t)$ varies continuously with t.

where PP′ refers to the length of the chord PP′. We note that $\overline{\mathrm{PP'}}/\mathrm{PP'}$ represents a unit vector and that, as $\delta t \to 0$, the direction of this unit vector tends to coincide with that of the tangent at P. Also

$$\lim_{\delta t \to 0} \frac{\mathrm{PP'}}{\delta t} = \frac{\mathrm{d}s}{\mathrm{d}t}, \tag{1.6.4}$$

where s is the arc length. This is so because

$$\mathrm{PP'} = \delta s(1 + \epsilon), \tag{1.6.5}$$

where δs is the arc length from P to P′, and $\epsilon \to 0$ as $\delta s \to 0$. In fact one can introduce the notion of arc length along a curve as the function whose derivative is defined by (1.6.4), the arc length itself then being obtained by integration of the resulting function. For future reference we derive the following result from this definition. Let

$$\mathbf{r}(t) = x(t)\mathbf{i} + y(t)\mathbf{j} + z(t)\mathbf{k}$$

so that $x(t)$, $y(t)$, $z(t)$ are the coordinates of P, and

$$(\mathrm{PP'})^2 = \{x(t+\delta t) - x(t)\}^2 + \{y(t+\delta t) - y(t)\}^2 + \{z(t+\delta t) - z(t)\}^2. \tag{1.6.6}$$

When substituted in (1.6.4) this gives

$$\begin{aligned} \frac{\mathrm{d}s}{\mathrm{d}t} &= \lim_{\delta t \to 0} \left\{ \left(\frac{x(t+\delta t) - x(t)}{\delta t} \right)^2 + \left(\frac{y(t+\delta t) - y(t)}{\delta t} \right)^2 + \left(\frac{z(t+\delta t) - z(t)}{\delta t} \right)^2 \right\}^{1/2} \\ &= \left\{ \left(\frac{\mathrm{d}x}{\mathrm{d}t} \right)^2 + \left(\frac{\mathrm{d}y}{\mathrm{d}t} \right)^2 + \left(\frac{\mathrm{d}z}{\mathrm{d}t} \right)^2 \right\}^{1/2}. \end{aligned} \tag{1.6.7}$$

We have seen that the geometrical interpretation of the derivative of the position vector $\mathbf{r}(t)$ of a point P is a vector of magnitude $\mathrm{d}s/\mathrm{d}t$ parallel to the tangent to the locus of P. Two immediate consequences are that $\mathrm{d}\mathbf{r}/\mathrm{d}s$ is a vector of unit magnitude, and that if $\mathbf{r}$ is a constant vector (that is constant in magnitude and direction) then $\mathrm{d}\mathbf{r}/\mathrm{d}t = \mathbf{0}$. Furthermore the rules for differentiation are generalisations of the rules used in elementary calculus. In particular, we have

$$\frac{\mathrm{d}}{\mathrm{d}t}(\mathbf{r}_1 + \mathbf{r}_2) = \frac{\mathrm{d}\mathbf{r}_1}{\mathrm{d}t} + \frac{\mathrm{d}\mathbf{r}_2}{\mathrm{d}t}, \tag{1.6.8}$$

$$\frac{\mathrm{d}}{\mathrm{d}t}(\lambda\mathbf{r}) = \lambda\frac{\mathrm{d}\mathbf{r}}{\mathrm{d}t} + \frac{\mathrm{d}\lambda}{\mathrm{d}t}\mathbf{r}, \tag{1.6.9}$$

$$\frac{\mathrm{d}}{\mathrm{d}t}(\mathbf{r}_1 \cdot \mathbf{r}_2) = \mathbf{r}_1 \cdot \frac{\mathrm{d}\mathbf{r}_2}{\mathrm{d}t} + \frac{\mathrm{d}\mathbf{r}_1}{\mathrm{d}t} \cdot \mathbf{r}_2, \tag{1.6.10}$$

$$\frac{\mathrm{d}}{\mathrm{d}t}(\mathbf{r}_1 \times \mathbf{r}_2) = \mathbf{r}_1 \times \frac{\mathrm{d}\mathbf{r}_2}{\mathrm{d}t} + \frac{\mathrm{d}\mathbf{r}_1}{\mathrm{d}t} \times \mathbf{r}_2. \tag{1.6.11}$$

The proofs of these results are straightforward generalisations of the proofs used for differentiation involving ordinary functions. For example,

$$\begin{aligned}\frac{\mathrm{d}}{\mathrm{d}t}(\mathbf{r}_1 \times \mathbf{r}_2) &= \lim_{\delta t \to 0} \frac{1}{\delta t}\{\mathbf{r}_1(t+\delta t) \times \mathbf{r}_2(t+\delta t) - \mathbf{r}_1(t) \times \mathbf{r}_2(t)\} \\ &= \lim_{\delta t \to 0} \frac{1}{\delta t}\mathbf{r}_1(t+\delta t) \times \{\mathbf{r}_2(t+\delta t) - \mathbf{r}_2(t)\} \\ &\quad + \lim_{\delta t \to 0} \frac{1}{\delta t}\{\mathbf{r}_1(t+\delta t) - \mathbf{r}_1(t)\} \times \mathbf{r}_2(t).\end{aligned}$$

Now

$$\mathbf{r}_1(t+\delta t) \to \mathbf{r}_1(t) \qquad \text{as } \delta t \to 0,$$

$$\frac{1}{\delta t}\{\mathbf{r}_2(t+\delta t) - \mathbf{r}_2(t)\} \to \frac{\mathrm{d}\mathbf{r}_2}{\mathrm{d}t} \qquad \text{as } \delta t \to 0,$$

and so

$$\frac{\mathrm{d}}{\mathrm{d}t}(\mathbf{r}_1 \times \mathbf{r}_2) = \mathbf{r}_1 \times \frac{\mathrm{d}\mathbf{r}_2}{\mathrm{d}t} + \frac{\mathrm{d}\mathbf{r}_1}{\mathrm{d}t} \times \mathbf{r}_2.$$

In the final relation (1.6.11) it should be noted that the terms in the vector products are not commutative.

One important application of the rules allows us to express the derivative of a vector in terms of its components relative to three fixed directions. Thus if

$$\mathbf{r} = \lambda(t)\mathbf{a} + \mu(t)\mathbf{b} + \nu(t)\mathbf{c}, \tag{1.6.12}$$

where **a**, **b** and **c** are constant vectors, we have

$$\frac{\mathrm{d}\mathbf{r}}{\mathrm{d}t} = \frac{\mathrm{d}\lambda}{\mathrm{d}t}\mathbf{a} + \frac{\mathrm{d}\mu}{\mathrm{d}t}\mathbf{b} + \frac{\mathrm{d}\nu}{\mathrm{d}t}\mathbf{c}. \tag{1.6.13}$$

In particular, if

$$\mathbf{r} = x(t)\mathbf{i} + y(t)\mathbf{j} + z(t)\mathbf{k}, \tag{1.6.14}$$

where **i**, **j**, **k** are fixed, mutually orthogonal unit vectors, we have

$$\frac{\mathrm{d}\mathbf{r}}{\mathrm{d}t} = \frac{\mathrm{d}x}{\mathrm{d}t}\mathbf{i} + \frac{\mathrm{d}y}{\mathrm{d}t}\mathbf{j} + \frac{\mathrm{d}z}{\mathrm{d}t}\mathbf{k}. \tag{1.6.15}$$

It is important to note that the validity of (1.6.13) depends on the vectors **a**, **b** and **c** being fixed both in magnitude and direction. When only the magnitude of the vector is constant, the derivative will not in general be zero.

Let P describe a circle of unit radius, so that its position vector is a unit vector **a**, say (see Fig. 1.6.2). According to the interpretation given earlier, the rate of change of **a** is a vector of magnitude $\mathrm{d}s/\mathrm{d}t$ parallel to the tangent to the

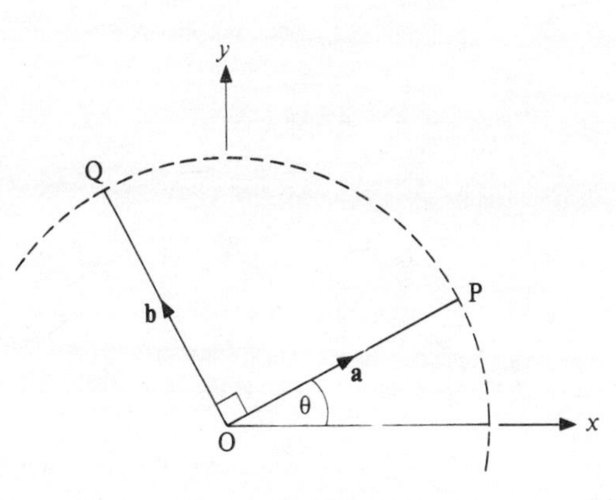

Figure 1.6.2 Perpendicular unit vectors **a** and **b**.

unit circle at P. Since the arc length here is just the angular distance θ, measured in radians from some initial line, we have the result that

$$\frac{d\mathbf{a}}{dt} = \frac{d\theta}{dt}\mathbf{b}, \tag{1.6.16}$$

where **b** is a unit vector perpendicular to **a** in the sense of increasing θ. Since **b** is also the position vector of some point Q on the unit circle at an angular distance $(\theta + \pi/2)$ from the initial line, a similar argument applied to **b** gives the result

$$\frac{d\mathbf{b}}{dt} = \frac{d}{dt}(\theta + \pi/2)(-\mathbf{a}) = -\frac{d\theta}{dt}\mathbf{a}. \tag{1.6.17}$$

Alternatively one can choose fixed unit vectors **i** and **j** along and perpendicular to the initial line, so that

$$\mathbf{a} = \cos\theta\mathbf{i} + \sin\theta\mathbf{j}, \tag{1.6.18}$$

$$\mathbf{b} = -\sin\theta\mathbf{i} + \cos\theta\mathbf{j}. \tag{1.6.19}$$

The rules for differentiation then give immediately

$$\frac{d\mathbf{a}}{dt} = -\sin\theta\frac{d\theta}{dt}\mathbf{i} + \cos\theta\frac{d\theta}{dt}\mathbf{j} = \frac{d\theta}{dt}\mathbf{b}, \tag{1.6.20}$$

$$\frac{d\mathbf{b}}{dt} = -\cos\theta\frac{d\theta}{dt}\mathbf{i} - \sin\theta\frac{d\theta}{dt}\mathbf{j} = -\frac{d\theta}{dt}\mathbf{a}, \tag{1.6.21}$$

as before.

The above results can be used to calculate the derivative $d\mathbf{r}/dt$ of a vector $\mathbf{r}(t)$ which always remains parallel to a fixed plane. Let **a**, **b** be unit vectors along and

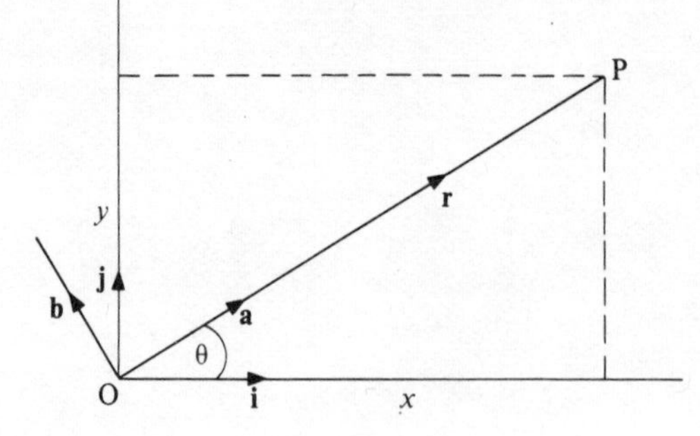

Figure 1.6.3 Cartesian coordinates (x, y) and polar coordinates (r, θ).

perpendicular to **r** and let the direction of **r** make an angle θ with a fixed axis, as shown in Figure 1.6.3. We have

$$\mathbf{r} = r\mathbf{a}, \tag{1.6.22}$$

$$\frac{\mathrm{d}\mathbf{r}}{\mathrm{d}t} = \frac{\mathrm{d}r}{\mathrm{d}t}\mathbf{a} + r\frac{\mathrm{d}\mathbf{a}}{\mathrm{d}t} = \frac{\mathrm{d}r}{\mathrm{d}t}\mathbf{a} + r\frac{\mathrm{d}\theta}{\mathrm{d}t}\mathbf{b}. \tag{1.6.23}$$

Also

$$\begin{aligned}\frac{\mathrm{d}^2\mathbf{r}}{\mathrm{d}t^2} &= \frac{\mathrm{d}^2 r}{\mathrm{d}t^2}\mathbf{a} + \frac{\mathrm{d}r}{\mathrm{d}t}\frac{\mathrm{d}\mathbf{a}}{\mathrm{d}t} + \frac{\mathrm{d}r}{\mathrm{d}t}\frac{\mathrm{d}\theta}{\mathrm{d}t}\mathbf{b} + r\frac{\mathrm{d}^2\theta}{\mathrm{d}t^2}\mathbf{b} + r\frac{\mathrm{d}\theta}{\mathrm{d}t}\frac{\mathrm{d}\mathbf{b}}{\mathrm{d}t} \\ &= \frac{\mathrm{d}^2 r}{\mathrm{d}t^2}\mathbf{a} + \frac{\mathrm{d}r}{\mathrm{d}t}\frac{\mathrm{d}\theta}{\mathrm{d}t}\mathbf{b} + \frac{\mathrm{d}r}{\mathrm{d}t}\frac{\mathrm{d}\theta}{\mathrm{d}t}\mathbf{b} + r\frac{\mathrm{d}^2\theta}{\mathrm{d}t^2}\mathbf{b} - r\left(\frac{\mathrm{d}\theta}{\mathrm{d}t}\right)^2\mathbf{a} \\ &= \left\{\frac{\mathrm{d}^2 r}{\mathrm{d}t^2} - r\left(\frac{\mathrm{d}\theta}{\mathrm{d}t}\right)^2\right\}\mathbf{a} + \left\{r\frac{\mathrm{d}^2\theta}{\mathrm{d}t^2} + 2\frac{\mathrm{d}r}{\mathrm{d}t}\frac{\mathrm{d}\theta}{\mathrm{d}t}\right\}\mathbf{b}.\end{aligned} \tag{1.6.24}$$

Equations (1.6.23) and (1.6.24) give the components of $\mathrm{d}\mathbf{r}/\mathrm{d}t$ and $\mathrm{d}^2\mathbf{r}/\mathrm{d}t^2$ along and perpendicular to **r**. In some applications these components can be more useful than the expressions obtained using cartesian components, namely,

$$\frac{\mathrm{d}\mathbf{r}}{\mathrm{d}t} = \frac{\mathrm{d}x}{\mathrm{d}t}\mathbf{i} + \frac{\mathrm{d}y}{\mathrm{d}t}\mathbf{j}, \qquad \frac{\mathrm{d}^2\mathbf{r}}{\mathrm{d}t^2} = \frac{\mathrm{d}^2x}{\mathrm{d}t^2}\mathbf{i} + \frac{\mathrm{d}^2y}{\mathrm{d}t^2}\mathbf{j}.$$

When **r** is used as the position vector of a point P in a plane, as in Figure 1.6.3, the variables (r, θ) are called the polar coordinates of P. In some applications, the geometry is such that polar coordinates are a preferable alternative to cartesian coordinates for specifying the position of P. If θ is measured relative to the x-axis of a system of cartesian coordinates, the relations between the two systems are

$$x = r\cos\theta, \qquad y = r\sin\theta,$$
$$r = (x^2 + y^2)^{1/2}, \qquad \theta = \tan^{-1}(y/x).$$

Example 1.6.1

If $\mathbf{r}$ and $\mathbf{r}'$ are two parallel vectors, the following relation holds:

$$\mathbf{r}' \cdot \frac{d\mathbf{r}}{dt} = r' \frac{dr}{dt}.$$

For if $\mathbf{r}$ and $\mathbf{r}'$ are parallel vectors we can write $\mathbf{r}' = \lambda(t)\mathbf{r}$ for some scalar variable $\lambda(t)$. It is then sufficient to prove that

$$\mathbf{r} \cdot \frac{d\mathbf{r}}{dt} = r \frac{dr}{dt}.$$

This result follows at once by differentiating the relation

$$\mathbf{r} \cdot \mathbf{r} = r^2$$

to get

$$\mathbf{r} \cdot \frac{d\mathbf{r}}{dt} + \frac{d\mathbf{r}}{dt} \cdot \mathbf{r} = 2\mathbf{r} \cdot \frac{d\mathbf{r}}{dt} = 2r \frac{dr}{dt}.$$

Example 1.6.2

If $\mathbf{a}, \mathbf{b}, \mathbf{c}$ are three mutually perpendicular unit vectors, their derivatives $d\mathbf{a}/dt$, $d\mathbf{b}/dt$, $d\mathbf{c}/dt$ are linearly dependent. To demonstrate this, it is sufficient to show that

$$\frac{d\mathbf{a}}{dt} \cdot \left(\frac{d\mathbf{b}}{dt} \times \frac{d\mathbf{c}}{dt}\right) = 0.$$

If $\mathbf{a}$, $\mathbf{b}$ and $\mathbf{c}$ form a right-handed triad, we have $\mathbf{a} = \mathbf{b} \times \mathbf{c}$. Hence

$$\begin{aligned}
\frac{d\mathbf{a}}{dt} \cdot \left(\frac{d\mathbf{b}}{dt} \times \frac{d\mathbf{c}}{dt}\right) &= \left(\mathbf{b} \times \frac{d\mathbf{c}}{dt} + \frac{d\mathbf{b}}{dt} \times \mathbf{c}\right) \cdot \left(\frac{d\mathbf{b}}{dt} \times \frac{d\mathbf{c}}{dt}\right) \\
&= \frac{d\mathbf{c}}{dt} \cdot \left\{\left(\frac{d\mathbf{b}}{dt} \times \frac{d\mathbf{c}}{dt}\right) \times \mathbf{b}\right\} + \frac{d\mathbf{b}}{dt} \cdot \left\{\mathbf{c} \times \left(\frac{d\mathbf{b}}{dt} \times \frac{d\mathbf{c}}{dt}\right)\right\} \\
&= -\left(\frac{d\mathbf{b}}{dt} \cdot \frac{d\mathbf{c}}{dt}\right)\left(\mathbf{b} \cdot \frac{d\mathbf{c}}{dt}\right) - \left(\frac{d\mathbf{b}}{dt} \cdot \frac{d\mathbf{c}}{dt}\right)\left(\mathbf{c} \cdot \frac{d\mathbf{b}}{dt}\right) \\
&= -\left(\frac{d\mathbf{b}}{dt} \cdot \frac{d\mathbf{c}}{dt}\right) \frac{d}{dt}(\mathbf{b} \cdot \mathbf{c}) = 0
\end{aligned}$$

since $\mathbf{b} \cdot \mathbf{c} = 0$. Use has also been made of the relations

$$\mathbf{b} \cdot \frac{d\mathbf{b}}{dt} = \mathbf{c} \cdot \frac{d\mathbf{c}}{dt} = 0.$$

1.7 THE INTEGRAL OF A VECTOR

A vector function of a scalar variable t can also be integrated. We interpret this as the reverse of differentiation. Let

$$\mathbf{v}(t) = \frac{d\mathbf{r}}{dt} = \lim_{\delta t \to 0} \frac{\delta \mathbf{r}}{\delta t}, \tag{1.7.1}$$

and write

$$\delta \mathbf{r} = (\mathbf{v}(t^*) + \boldsymbol{\epsilon})\delta t, \tag{1.7.2}$$

where t^* is any value of t satisfying $t \leqslant t^* \leqslant t + \delta t$ so that $t^* \to t$ as $\delta t \to 0$. If (1.7.2) is divided by δt and compared with (1.7.1), we see that $\epsilon \to 0$ as $\delta t \to 0$. Now let†

$$\mathbf{r}_i - \mathbf{r}_{i-1} = \delta \mathbf{r}_i, \tag{1.7.3}$$

where $\mathbf{r}_i = \mathbf{r}(t_i)$, and let

$$t_i - t_{i-1} = \delta t_i. \tag{1.7.4}$$

It follows, by reference to Figure 1.7.1 and equation (1.7.2), that

$$\mathbf{r}_n - \mathbf{r}_0 = \sum_{i=1}^{n} \delta \mathbf{r}_i = \sum_{i=1}^{n} (\mathbf{v}(t_i^*) + \boldsymbol{\epsilon}_i)\delta t_i, \tag{1.7.5}$$

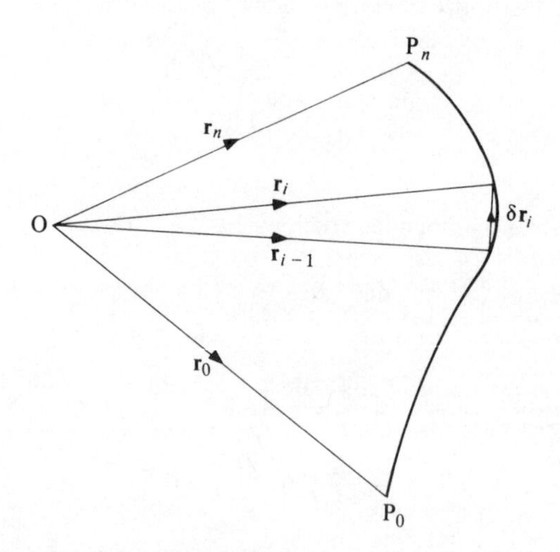

Figure 1.7.1 A typical increment $\delta \mathbf{r}_i = \mathbf{r}_i - \mathbf{r}_{i-1}$ of a continuously varying vector $\mathbf{r}$.

† There is a possible confusion of notation here in that δr is used conventionally to denote the increment $r(t + \delta t) - r(t)$ in the scalar function $r(t)$, that is OP′ – OP in Figure 1.6.1. On the other hand $\delta \mathbf{r}$ is used to denote $\mathbf{r}(t + \delta t) - \mathbf{r}(t)$, or the vector $\overline{\text{PP}'}$ of Figure 1.6.1. Accordingly δr is not the same as $|\delta \mathbf{r}|$ and the former should not be used as an alternative for the latter.

where $\epsilon_i \to 0$ as $\delta t_i \to 0$ for all i such that $1 \leqslant i \leqslant n$. Now with the help of a generalisation of (1.2.4), namely

$$|\,\mathbf{a}_1 + \mathbf{a}_2 + \ldots + \mathbf{a}_n\,| \leqslant |\,\mathbf{a}_1| + |\,\mathbf{a}_2\,| + \ldots + |\,\mathbf{a}_n\,|,$$

we have
$$\left|\sum_{i=1}^{n} \epsilon_i \delta t_i\right| \leqslant \sum_{i=1}^{n} \epsilon_i \delta t_i \leqslant \epsilon_{\mathrm{m}} \sum_{i=1}^{n} \delta t_i = \epsilon_{\mathrm{m}}(t_n - t_0), \tag{1.7.6}$$

where ϵ_{m} is the maximum value of ϵ_i for $1 \leqslant i \leqslant n$. Hence, as all the $\delta t_i \to 0$, so does $\epsilon_{\mathrm{m}}(t_n - t_0)$ and therefore

$$\mathbf{r}_n - \mathbf{r}_0 = \lim_{\delta t_i \to 0} \sum_{i=1}^{n} \mathbf{v}_i(t_i^*)\delta t_i. \tag{1.7.7}$$

When this limit exists, and is independent of the way of proceeding to the limit, it is called the Riemann integral of **v** and we write

$$\mathbf{r}_n - \mathbf{r}_0 = \lim_{\delta t_i \to 0} \sum_{i=1}^{n} \mathbf{v}_i(t_i^*)\delta t_i = \int_{t_0}^{t_n} \mathbf{v}(t)\,\mathrm{d}t. \tag{1.7.8}$$

The indefinite integral is written

$$\int \mathbf{v}(t)\,\mathrm{d}t = \mathbf{r} + \mathbf{c} \tag{1.7.9}$$

where **c** is a constant vector of integration. In summary, if a vector **r** can be found such that $\mathrm{d}\mathbf{r}/\mathrm{d}t = \mathbf{v}$, then the integral of **v** is defined by (1.7.9).

Example 1.7.1
If **d** is a constant vector we have

$$\int \mathbf{d}\frac{\mathrm{d}}{\mathrm{d}t} f(t)\,\mathrm{d}t = \int \frac{\mathrm{d}}{\mathrm{d}t}(\mathbf{d}f)\,\mathrm{d}t = \mathbf{d}f(t) + \mathbf{c},$$

where **c** is a constant vector of integration. In particular

$$\int\left(\mathbf{i}\frac{\mathrm{d}r_x}{\mathrm{d}t} + \mathbf{j}\frac{\mathrm{d}r_y}{\mathrm{d}t} + \mathbf{k}\frac{\mathrm{d}r_z}{\mathrm{d}t}\right)\mathrm{d}t = \int \frac{\mathrm{d}\mathbf{r}}{\mathrm{d}t}\,\mathrm{d}t = \mathbf{r} + \mathbf{c}.$$

Thus if $\quad \mathbf{a} = \mathbf{i}\cos\omega t + \mathbf{j}\sin\omega t, \qquad \mathbf{b} = -\,\mathbf{i}\sin\omega t + \mathbf{j}\cos\omega t,$

where ω is a constant, we have that

$$\int \mathbf{a}\,\mathrm{d}t = \frac{1}{\omega}(\mathbf{i}\sin\omega t - \mathbf{j}\cos\omega t) + \mathbf{c} = -\frac{1}{\omega}\mathbf{b} + \mathbf{c},$$

$$\int \mathbf{b}\,\mathrm{d}t = \frac{1}{\omega}(\mathbf{i}\cos\omega t + \mathbf{j}\sin\omega t) + \mathbf{c}' = \frac{1}{\omega}\mathbf{a} + \mathbf{c}'.$$

Example 1.7.2

$$\int\left(\mathbf{r}_1 \cdot \frac{d\mathbf{r}_2}{dt} + \mathbf{r}_2 \cdot \frac{d\mathbf{r}_1}{dt}\right) dt = \int \frac{d}{dt}(\mathbf{r}_1 \cdot \mathbf{r}_2)\, dt = \mathbf{r}_1 \cdot \mathbf{r}_2 + \mathbf{c}.$$

Example 1.7.3

$$\int \mathbf{r} \times \frac{d^2\mathbf{r}}{dt^2}\, dt = \int \frac{d}{dt}\left(\mathbf{r} \times \frac{d\mathbf{r}}{dt}\right) dt = \mathbf{r} \times \frac{d\mathbf{r}}{dt} + \mathbf{c}.$$

1.8 SCALAR AND VECTOR FIELDS

We may also consider a vector $\mathbf{r}$ as an independent variable and introduce the idea of a scalar funtion, $\phi(\mathbf{r})$, or a vector function, $\mathbf{F}(\mathbf{r})$, of $\mathbf{r}$. These functions represent mappings of $\mathbf{r}$ into associated scalar or vector quantities according to a prescribed rule. If an origin O is chosen, $\mathbf{r}$ can be considered as the position vector of some point P in the space of $\mathbf{r}$, and then $\phi(\mathbf{r})$ and $\mathbf{F}(\mathbf{r})$ can be associated with the appropriate point P. When these functions are known at all points of the space of $\mathbf{r}$, they are called scalar and vector fields respectively. In some applications the space of $\mathbf{r}$ may be restricted, such as a bounded region or a two-dimensional space. It may also be convenient in applications to define a set of cartesian axes based on O, and then ϕ and $\mathbf{F}$ can be regarded as functions of the components (x, y, z) of $\mathbf{r}$ or, equivalently, the coordinates of P. We then write $\phi(x, y, z)$ or $\mathbf{F}(x, y, z)$. Note, however, that the values of ϕ or $\mathbf{F}$ are not dependent on the *particular* choice of axes, in the same way that $\mathbf{r}$ is not so dependent.

A useful way to display a scalar field, and one that is often used, is by means of the family of surfaces ϕ = constant. In a two-dimensional space these surfaces become curves, a well-known and common example being the contours shown on the two-dimensional space represented by a map. These contours represent the curves of equal height above sea level, and with a little experience the topography of the area can be deduced from such contours. The curves of constant pressure, called isobars, found on meteorological maps are another example. It is clear that the height (or pressure) at a particular location is not dependent on any coordinates that might be used to define that location, although it is often useful in practice to define coordinates, such as those based on the grid reference system of ordnance survey maps.

As a simple mathematical example, consider the scalar field $\phi(|\mathbf{r}|)$, which is constant when the magnitude of $\mathbf{r}$ is constant. This is displayed as a family of concentric circles in two dimensions, or concentric spheres in three dimensions. A second example is $\phi = \mathbf{a} \cdot \mathbf{r}$, where $\mathbf{a}$ is a constant vector. Here ϕ = constant is equivalent to the condition that the component of $\mathbf{r}$ in the direction of $\mathbf{a}$ is constant, and the corresponding surfaces are the family of planes perpendicular to $\mathbf{a}$. Note that if axes are chosen for which $\mathbf{r}$ and $\mathbf{a}$ have the components (x, y, z)

and (a_x, a_y, a_z) respectively, then the relation ϕ = constant can be written

$$\phi = a_x x + a_y y + a_z z = \text{constant},$$

and this is the usual equation for a plane in coordinate geometry. In particular, if the axes are chosen so that $a_y = a_z = 0$, then **a** is parallel to the x-axis and the surfaces ϕ = constant are equivalent to the surfaces x = constant. These are clearly planes perpendicular to the x-axis.

With a vector function $\mathbf{F}(\mathbf{r})$ we are interested in the direction of $\mathbf{F}$ as well as the magnitude. In many applications the direction is of particular interest and is displayed by a family of lines for which the tangent at each point is in the direction of $\mathbf{F}$. Such lines are called field lines. Note that the vector function $\phi(\mathbf{r})\mathbf{F}(\mathbf{r})$ has the same field lines as $\mathbf{F}(\mathbf{r})$ for any scalar function $\phi(\mathbf{r})$ which is not zero, since ϕ affects only the magnitude of the vector. For example, the vector function $\phi(\mathbf{r})\mathbf{r}$ is always in the direction of the radius vector, and any straight line through the origin is a field line.

If $\mathbf{F} = F_x(x, y)\mathbf{i} + F_y(x, y)\mathbf{j}$, then the inclination of $\mathbf{F}$ at any point is F_y/F_x and so the field lines in two dimensions are represented by the integral of the equation

$$\frac{\mathrm{d}y}{\mathrm{d}x} = \frac{F_y}{F_x}. \qquad (1.8.1)$$

For the vector field $\mathbf{F} = y\mathbf{i} + x^2\mathbf{j}$, for example, this gives

$$\frac{\mathrm{d}y}{\mathrm{d}x} = \frac{x^2}{y}$$

which integrates to

$$\tfrac{1}{2}y^2 - \tfrac{1}{3}x^3 = \text{constant}.$$

The family of field lines is obtained by plotting these curves for different values of the constant (see Fig. 1.8.1).

The extension to three dimensions is straightforward and is most conveniently described parametrically. We suppose that the position vector $\mathbf{r}$ is itself defined by a parameter t so that, as t varies, $\mathbf{r}(t)$ defines a point P which moves along a prescribed path. At all points on this path $\mathrm{d}\mathbf{r}/\mathrm{d}t$ is parallel to the tangent (see §1.6). Now the tangent to the field lines of $\mathbf{F}(\mathbf{r})$ is everywhere parallel to $\mathbf{F}(\mathbf{r})$, and this is satisfied by the locus of P if

$$\frac{\mathrm{d}\mathbf{r}}{\mathrm{d}t} = \mathbf{F}(\mathbf{r}). \qquad (1.8.2)$$

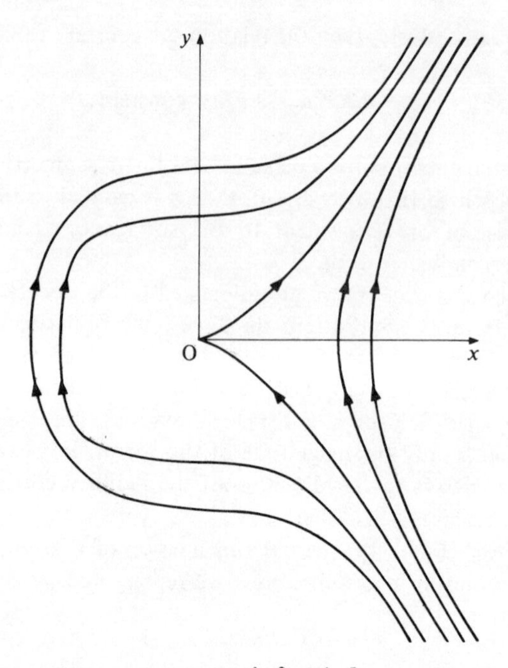

Figure 1.8.1 The family of curves $\phi = \frac{1}{2}y^2 - \frac{1}{3}x^3$ = constant for the scalar field ϕ. The curves also represent the field lines of the vector field $\mathbf{F} = y\mathbf{i} + x^2\mathbf{j}$. The arrows indicate the direction of **F**.

When **F**(**r**) is known as a function of **r**, equation (1.8.2) defines the slope of the field lines at each point. In principle, this is sufficient information to enable the field lines to be constructed. In practice it entails the integration of the set of equations

$$\frac{\mathrm{d}x}{\mathrm{d}t} = F_x(x, y, z), \qquad \frac{\mathrm{d}y}{\mathrm{d}t} = F_y(x, y, z), \qquad \frac{\mathrm{d}z}{\mathrm{d}t} = F_z(x, y, z) \qquad (1.8.3)$$

in order to obtain x, y, z as functions of the parameter t. Even in two dimensions this is not always possible in terms of elementary functions and numerical integration is sometimes required to obtain a detailed picture.

Since $\phi(\mathbf{r})\mathbf{F}(\mathbf{r})$ has the same direction as **F**(**r**), it can sometimes be convenient to replace equation (1.8.2) by

$$\frac{\mathrm{d}\mathbf{r}}{\mathrm{d}t} = \phi(\mathbf{r})\mathbf{F}(\mathbf{r}),$$

with an appropriate choice for $\phi(\mathbf{r})$, for the purpose of calculating the field lines.

Example 1.8.1
For the vector function $\mathbf{a} \times \mathbf{r}$, where $\mathbf{a}$ is a constant vector, the field lines are everywhere perpendicular to $\mathbf{a}$ and $\mathbf{r}$. Hence they are circles in planes perpendicular to $\mathbf{a}$ and with $\mathbf{a}$ as the common axis.

To derive the result formally, the analysis is simplified if the x axis is chosen to coincide with the direction of $\mathbf{a}$, so that $\mathbf{a} = a\mathbf{i}$. Then $\mathbf{a} \times \mathbf{r}$ has the components $a(0, -z, y)$ and the field lines are given by

$$\frac{\mathrm{d}x}{\mathrm{d}t} = 0, \quad \frac{\mathrm{d}y}{\mathrm{d}t} = -az, \quad \frac{\mathrm{d}z}{\mathrm{d}t} = ay.$$

Since the last two equations imply that

$$y\frac{\mathrm{d}y}{\mathrm{d}z} + z = 0,$$

or

$$y^2 + z^2 = \text{constant},$$

the field lines are the circles

$$x = \text{constant}, \quad y^2 + z^2 = \text{constant}.$$

Example 1.8.2
For the vector function

$$\mathbf{F} = yz\mathbf{i} + zx\mathbf{j} - xy\mathbf{k},$$

the field lines are obtained from the equations

$$\frac{\mathrm{d}x}{\mathrm{d}t} = yz, \quad \frac{\mathrm{d}y}{\mathrm{d}t} = zx, \quad \frac{\mathrm{d}z}{\mathrm{d}t} = -xy.$$

These equations are not immediately integrable, but sometimes it is possible to find simple integrals by suitable rearrangement. Here, for example, we can infer that

$$y\frac{\mathrm{d}y}{\mathrm{d}t} + z\frac{\mathrm{d}z}{\mathrm{d}t} = 0, \qquad z\frac{\mathrm{d}z}{\mathrm{d}t} + x\frac{\mathrm{d}x}{\mathrm{d}t} = 0.$$

These equations integrate immediately to give

$$y^2 + z^2 = c_1^2, \quad z^2 + x^2 = c_2^2,$$

where c_1^2 and c_2^2 are constants of integration. Since $y^2 + z^2 = c_1^2$ generates a circular cylinder of radius c_1 whose axis coincides with the x axis, it follows that the field lines are given by the lines of intersection of two families of circular cylinders, whose axes coincide with the x-axis and the y-axis respectively.

1.9 LINE INTEGRALS

When the scalar field $\phi(\mathbf{r})$ is defined as a function of the vector $\mathbf{r}$, and $\mathbf{r}$ is itself defined as a function of the scalar variable t, ϕ is given implicitly as a function of t and we can operate on it as such. If $\mathbf{r}(t)$ is interpreted as the position vector of a point P which describes a curve as t varies, we are in effect defining $\phi(\mathbf{r})$ at points of this curve. For this reason the integral

$$\int \phi(\mathbf{r}(t))\,\mathrm{d}t \tag{1.9.1}$$

is called a line integral. Indeed the parameter t may, in some applications, be just the arc length along this curve from a given origin. In practice it is evaluated in the usual way by expressing ϕ explicitly as a function of t. When $\mathbf{r}$ is defined in terms of its components, we can also express (1.9.1) in the equivalent form

$$\int \phi(x(t), y(t), z(t))\,\mathrm{d}t. \tag{1.9.2}$$

If, as a simple example,

$$\phi = xyz \qquad \text{and} \qquad x = t^3, \quad y = t^2, \quad z = t, \tag{1.9.3}$$

then

$$\int_0^1 \phi\,\mathrm{d}t = \int_0^1 xyz\,\mathrm{d}t = \int_0^1 t^6\,\mathrm{d}t = \tfrac{1}{7}. \tag{1.9.4}$$

A particular form of (1.9.2), which is often met in applications, is the integral

$$\int \phi\,\mathrm{d}x, \tag{1.9.5}$$

where the parameter is one of the variables defining the coordinates. This integral can be evaluated either by expressing y and z as functions of x to get

$$\int \phi(x, y(x), z(x))\,\mathrm{d}x, \tag{1.9.6}$$

or by using the usual formula for a change of variable from x to a more convenient parameter t, so that

$$\int \phi\,\mathrm{d}x = \int \phi(x(t), y(t), z(t))\,\frac{\mathrm{d}x}{\mathrm{d}t}\,\mathrm{d}t. \tag{1.9.7}$$

In the example defined by (1.9.3) we have

$$y = x^{2/3}, \qquad z = x^{1/3}$$

and so

$$\int_0^1 \phi \, \mathrm{d}x = \int_0^1 xyz \, \mathrm{d}x = \int_0^1 x^2 \, \mathrm{d}x = \tfrac{1}{3}. \tag{1.9.8}$$

Alternatively, since $\mathrm{d}x/\mathrm{d}t = 3t^2$, it can be evaluated as

$$\int_0^1 \phi \, \mathrm{d}x = \int_0^1 xyz \frac{\mathrm{d}x}{\mathrm{d}t} \, \mathrm{d}t = \int_0^1 3t^8 \, \mathrm{d}t = \tfrac{1}{3}. \tag{1.9.9}$$

It is important to appreciate that the line integral

$$\int \phi\,(x, y, z) \, \mathrm{d}x \tag{1.9.10}$$

is not in general to be interpreted as the integral of ϕ as a function of x with y and z regarded as constant parameters. Such an interpretation corresponds to the special case when the line of integration is a straight line parallel to the x-axis, along which y and z are constant. In fact integrals of the type (1.9.10) cannot in general be integrated until the line of integration is specified. To emphasise this fact the notation

$$\int_C \phi(x, y, z) \, \mathrm{d}x \tag{1.9.11}$$

is often used, where C denotes the curve involved along which the integration is taken. Only when ϕ is given as a function of x alone, independent of y and z, can (1.9.11) be integrated without further information.

The generalisation to the line integral of a vector

$$\int \mathbf{F}(\mathbf{r}(t)) \, \mathrm{d}t \tag{1.9.12}$$

is dealt with in similar fashion. When **F** is expressed as a function of t, (1.9.12) becomes the integral of a vector with respect to a parameter, as discussed in §1.7. In applications where **F** is defined in terms of its components, say,

$$\mathbf{F} = X\mathbf{i} + Y\mathbf{j} + Z\mathbf{k}, \tag{1.9.13}$$

(1.9.12) can be expressed in the form

$$\int \mathbf{F} \, \mathrm{d}t = \int X \, \mathrm{d}t\mathbf{i} + \int Y \, \mathrm{d}t\mathbf{j} + \int Z \, \mathrm{d}t\mathbf{k}. \tag{1.9.14}$$

Each component of (1.9.14) can be evaluated in accordance with the discussion of scalar line integrals.

We shall be particularly concerned with the line integral

$$\int \mathbf{F}(\mathbf{r}) \cdot \frac{\mathrm{d}\mathbf{r}}{\mathrm{d}t} \, \mathrm{d}t, \tag{1.9.15}$$

which is often written in the more concise form

$$\int \mathbf{F} \cdot \mathrm{d}\mathbf{r}. \tag{1.9.16}$$

When $\mathbf{r}$ is given as a function of t there is a vector $\mathbf{F}(\mathbf{r}(t))$ associated with each point P on the curve $\mathbf{r} = \mathbf{r}(t)$ (see Fig. 1.9.1). The integrand of (1.9.15) is therefore a scalar function of t to be integrated in the usual way. If, for example, the parameter t is taken to be the arc length s, we have from §1.6 that $\mathrm{d}\mathbf{r}/\mathrm{d}s$ is a unit vector along the tangent to the locus of P, and so

$$\int \mathbf{F} \cdot \frac{\mathrm{d}\mathbf{r}}{\mathrm{d}s} \, \mathrm{d}s = \int F_s \, \mathrm{d}s,$$

where F_s is the tangential component of $\mathbf{F}$.

We call the line integral (1.9.15), or (1.9.16), the work of the vector $\mathbf{F}$ associated with the appropriate displacement along the curve. Here again, when appropriate, the integrand of (1.9.15) can be expressed in terms of the components of $\mathbf{F}$ and $\mathrm{d}\mathbf{r}/\mathrm{d}t$. Since

$$\frac{\mathrm{d}\mathbf{r}}{\mathrm{d}t} = \frac{\mathrm{d}x}{\mathrm{d}t}\mathbf{i} + \frac{\mathrm{d}y}{\mathrm{d}t}\mathbf{j} + \frac{\mathrm{d}z}{\mathrm{d}t}\mathbf{k} \tag{1.9.17}$$

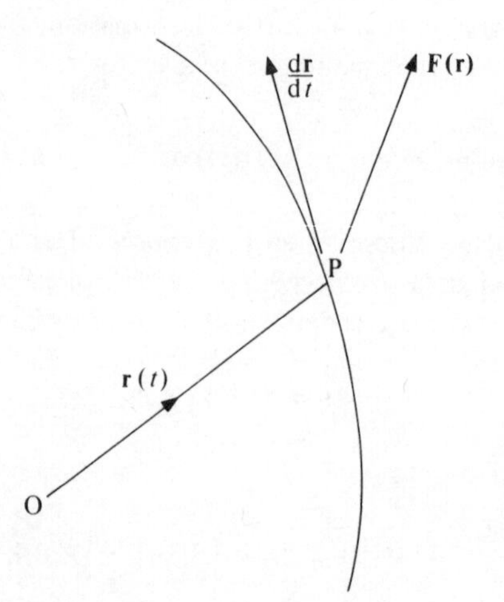

Figure 1.9.1 The vector $\mathbf{F}(\mathbf{r})$ associated with the point P whose position vector is $\mathbf{r}$.

we have, with **F** given by (1.9.13),

$$\int \mathbf{F} \cdot \frac{d\mathbf{r}}{dt}\, dt = \int \left(X(x, y, z)\frac{dx}{dt} + Y(x, y, z)\frac{dy}{dt} + Z(x, y, z)\frac{dz}{dt} \right) dt. \quad (1.9.18)$$

In (1.9.18), if the parameter t is chosen to be x, y and z respectively in each of the three terms, we get

$$\int \mathbf{F} \cdot d\mathbf{r} = \int \{X(x, y, z)\, dx + Y(x, y, z)\, dy + Z(x, y, z)\, dz\}, \quad (1.9.19)$$

which is suggested by the shorthand form of (1.9.16) used here on the left-hand side of (1.9.19). Whether or not the different contributions are actually evaluated with the above choices for t, (1.9.19) is the usual expression used for the work in terms of the components of **F**.

We repeat the warning here that, in general, expressions of the type (1.9.19) cannot be integrated until the curve of integration is specified. One obvious exception is the integral

$$\int \{X(x)\, dx + Y(y)\, dy + Z(z)\, dz\},$$

which can be evaluated as three straightforward integrals whose values depend only on the initial and final values of x, y and z respectively and not on any particular path of integration. As we shall see, there are less obvious expressions which do not require a knowledge of the path of integration for their evaluation. Although they are mathematically exceptional, they are of particular significance in many branches of applied mathematics, including mechanics.

Example 1.9.1
Evaluate the line integral

$$\int_C xy^2\, dy,$$

where C is the circle $x^2 + y^2 = 1$ traversed once anticlockwise.

Note that in this example the path of integration is closed, but it should not be assumed that the integral is therefore zero.

It is possible in such an example to use y as the variable of integration, but care must be taken to ensure that the correct sign is used for $x = \pm(1 - y^2)^{1/2}$. If the two halves of the circle for which $-1 \leqslant y \leqslant 1$ are considered, then x is positive over one half and negative over the other. Hence

$$\int_C xy^2\, dy = \int_{-1}^{1} (1 - y^2)^{1/2} y^2\, dy + \int_{1}^{-1} -(1 - y^2)^{1/2} y^2\, dy$$

$$= 2 \int_{-1}^{1} (1 - y^2)^{1/2} y^2\, dy.$$

The integral can now be evaluated with the substitution $y = \sin\theta$. But it is more direct to make the transformation

$$x = \cos\theta, \quad y = \sin\theta$$

in the original integral, and integrate directly with respect to the parameter θ. Then

$$\int_c xy^2\, dy = \int_c \cos\theta \sin^2\theta \, d(\sin\theta) = \int_0^{2\pi} \cos^2\theta \sin^2\theta \, d\theta$$

$$= \tfrac{1}{8}\int_0^{2\pi} (1 - \cos 4\theta)\, d\theta = \pi/4.$$

Example 1.9.2
Evaluate the line integral

$$\int_c x^n\, ds, \qquad (n \geqslant 0)$$

along the path defined by $x = t, y = 2^{-1/2}t^2, z = \frac{1}{3}t^3$ from $t = 0$ to $t = t_1$.
In this integral s refers to the arc length along the path, so that

$$\frac{ds}{dt} = \left\{\left(\frac{dx}{dt}\right)^2 + \left(\frac{dy}{dt}\right)^2 + \left(\frac{dz}{dt}\right)^2\right\}^{1/2} = (1 + 2t^2 + t^4)^{1/2}$$

$$= t^2 + 1.$$

The integral is therefore

$$\int_c x^n \frac{ds}{dt}\, dt = \int_0^{t_1} t^n(t^2 + 1)\, dt = \frac{t_1^{n+3}}{n+3} + \frac{t_1^{n+1}}{n+1}.$$

When $n = 0$ this represents the arc length along the path.

Example 1.9.3
We consider the work of $\mathbf{F} = y^2\mathbf{i} + x^2\mathbf{j}$ along the following (two-dimensional) curves between the points $(0, 0)$ and $(1, 2)$ in Figure 1.9.2, namely:

(i) The two straight lines between $(0, 0)$ and $(1, 0)$ and between $(1, 0)$ and $(1, 2)$.

We evaluate the integral in two parts. Along Ox we can take the parameter to be x itself and write

$$\int \mathbf{F} \cdot d\mathbf{r} = \int (y^2\, dx + x^2\, dy) = \int_0^1 \left(y^2 + x^2 \frac{dy}{dx}\right) dx = 0$$

since both y and dy/dx are zero along Ox.

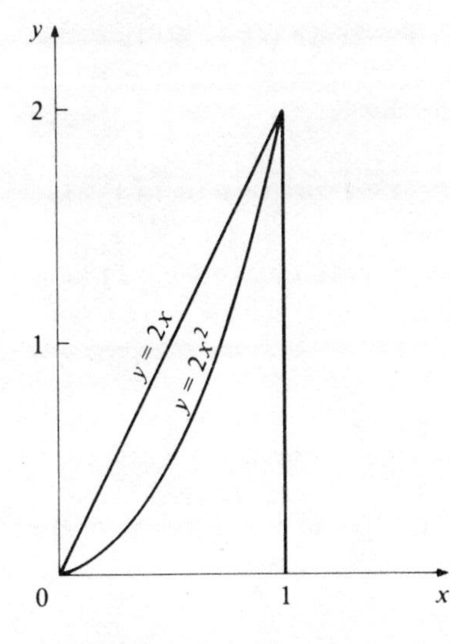

Figure 1.9.2 The various paths of integration in Example 1.9.3.

For the second contribution we take y as the parameter and write

$$\int \mathbf{F} \,.\, \mathrm{d}\mathbf{r} = \int_0^2 \left(y^2 \frac{\mathrm{d}x}{\mathrm{d}y} + x^2 \right) \mathrm{d}y = 2$$

since $\mathrm{d}x/\mathrm{d}y = 0$ and $x = 1$ along this path. Thus for the integral from $(0, 0)$ to $(1, 2)$ we have

$$\int \mathbf{F} \,.\, \mathrm{d}\mathbf{r} = 0 + 2 = 2.$$

(ii) The straight line joining the points $(0, 0)$ and $(1, 2)$. On this line $y = 2x$ and so, with x as the parameter, we get

$$\int \mathbf{F} \,.\, \mathrm{d}\mathbf{r} = \int (y^2 \,\mathrm{d}x + x^2 \,\mathrm{d}y) = \int_0^1 (4x^2 \,\mathrm{d}x + 2x^2 \,\mathrm{d}x) = 2.$$

(iii) The parabola $y = 2x^2$. On this curve

$$\int \mathbf{F} \,.\, \mathrm{d}\mathbf{r} = \int (y^2 \,\mathrm{d}x + x^2 \,\mathrm{d}y) = \int_0^1 (4x^4 \,\mathrm{d}x + x^2 \,.\, 4x \,\mathrm{d}x) = 9/5.$$

Example 1.9.4
Evaluate

$$\int_C \mathbf{F} \,.\, \frac{\mathrm{d}\mathbf{r}}{\mathrm{d}t} \,\mathrm{d}t$$

where
$$\mathbf{F} = 2xy\mathbf{i} + (x^2 + z)\mathbf{j} + (y + z^2)\mathbf{k}$$

and C is the curve defined by
$$\mathbf{r} = a\cos t\,\mathbf{i} + b\sin t\,\mathbf{j} + ct\,\mathbf{k}.$$

Since
$$\frac{\mathrm{d}\mathbf{r}}{\mathrm{d}t} = -a\sin t\,\mathbf{i} + b\cos t\,\mathbf{j} + c\,\mathbf{k}$$

we have
$$\mathbf{F}\,.\,\frac{\mathrm{d}\mathbf{r}}{\mathrm{d}t} = -2a^2b\cos t\sin^2 t + (a^2\cos^2 t + ct)b\cos t + (b\sin t + c^2t^2)c$$
$$= a^2b(1 - 3\sin^2 t)\cos t + bc(t\cos t + \sin t) + c^3t^2.$$

Hence
$$\int \mathbf{F}\,.\,\frac{\mathrm{d}\mathbf{r}}{\mathrm{d}t}\,\mathrm{d}t = \int\{a^2b(1 - 3\sin^2 t)\cos t + bc(t\cos t + \sin t) + c^3t^2\}\,\mathrm{d}t$$
$$= (a^2\cos^2 t + ct)b\sin t + \tfrac{1}{3}c^3t^3 + \text{constant}.$$

Note, however, that in this example
$$\frac{\mathrm{d}}{\mathrm{d}t}\{(x^2 + z)y + \tfrac{1}{3}z^3\} = 2xy\frac{\mathrm{d}x}{\mathrm{d}t} + (x^2 + z)\frac{\mathrm{d}y}{\mathrm{d}t} + (y + z^2)\frac{\mathrm{d}z}{\mathrm{d}t}$$
$$= \mathbf{F}\,.\,\frac{\mathrm{d}\mathbf{r}}{\mathrm{d}t}.$$

Hence
$$\int \mathbf{F}\,.\,\frac{\mathrm{d}\mathbf{r}}{\mathrm{d}t}\,\mathrm{d}t = \int\frac{\mathrm{d}}{\mathrm{d}t}\{(x^2 + z)y + \tfrac{1}{3}z^3\}\,\mathrm{d}t = (x^2 + z)y + \tfrac{1}{3}z^3 + \text{constant}.$$

This agrees with the previous calculation, but this alternative way of proceeding does not use any information concerning the particular path of integration, and the result is true whatever path of integration is chosen. We shall see eventually under what conditions this alternative procedure is possible.

1.10 THE GRADIENT OPERATOR

We have seen that the work of a vector may or may not depend on the path of integration. When it does not, the vector field is called conservative, and such fields play an important role in applied mathematics. It is useful to consider the conditions which a vector field must satisfy in order that it should be conservative. As a preliminary to this investigation we introduce the gradient of a scalar function.

If $\phi(x, y, z)$ is a function of the coordinate variables (x, y, z), and if C is a curve defined by specifying x, y and z as functions of a parameter t, then ϕ can be regarded as a function of t on C and we can form the derivative

$$\frac{d\phi}{dt} = \lim_{\delta t \to 0} \frac{1}{\delta t}\{\phi(x(t+\delta t), y(t+\delta t), z(t+\delta t)) - \phi(x(t), y(t), z(t))\}. \tag{1.10.1}$$

If we write

$$x(t+\delta t) = x(t) + \delta x,$$

and similarly for the other variables, the above relation for $d\phi/dt$ can be expressed in the form

$$\frac{d\phi}{dt} = \lim_{\delta t \to 0} \frac{\phi(x+\delta x, y+\delta y, z+\delta z) - \phi(x, y+\delta y, z+\delta z)}{\delta x}\frac{\delta x}{\delta t}$$

$$+ \lim_{\delta t \to 0} \frac{\phi(x, y+\delta y, z+\delta z) - \phi(x, y, z+\delta z)}{\delta y}\frac{\delta y}{\delta t}$$

$$+ \lim_{\delta t \to 0} \frac{\phi(x, y, z+\delta z) - \phi(x, y, z)}{\delta z}\frac{\delta z}{\delta t}.$$

Since $\delta x, \delta y, \delta z \to 0$ as $\delta t \to 0$, it will be seen that

$$\frac{d\phi}{dt} = \frac{\partial\phi}{\partial x}\frac{dx}{dt} + \frac{\partial\phi}{\partial y}\frac{dy}{dt} + \frac{\partial\phi}{\partial z}\frac{dz}{dt}, \tag{1.10.2}$$

where $\partial\phi/\partial x$, $\partial\phi/\partial y$ and $\partial\phi/\partial z$ are the partial derivatives of ϕ with respect to x, y and z respectively. Integration of equation (1.10.2) gives the indefinite integral expression for ϕ,

$$\phi(x, y, z) = \int\left(\frac{\partial\phi}{\partial x}\frac{dx}{dt} + \frac{\partial\phi}{\partial y}\frac{dy}{dt} + \frac{\partial\phi}{\partial z}\frac{dz}{dt}\right)dt, \tag{1.10.3}$$

where the integrand is a known function of t on the curve C. By analogy with previous integrals of this type we write equation (1.10.3) in the more compact form

$$\phi = \int\left(\frac{\partial\phi}{\partial x}dx + \frac{\partial\phi}{\partial y}dy + \frac{\partial\phi}{\partial z}dz\right) \tag{1.10.4}$$

on the understanding that, in practice, it may be conveniently evaluated in the form (1.10.3)

The importance of the above relations lies in the fact that although the value of the integrands in (1.10.3) or (1.10.4) depends on the choice of the curve of integration C, the value of the integral between two points of C depends only on the end points, and not on the curve joining them. This must clearly be so since the integral represents the change in ϕ between the two end points, and ϕ is a function only of position.*

A comparison of (1.10.4) with the expression (1.9.19) for the work of a vector **F** shows that if **F** is of the form

$$\mathbf{F} = \frac{\partial \phi}{\partial x}\mathbf{i} + \frac{\partial \phi}{\partial y}\mathbf{j} + \frac{\partial \phi}{\partial z}\mathbf{k} \tag{1.10.5}$$

for some scalar function ϕ, then the work of **F** depends only on the end points and not on the path. Hence **F** is a conservative vector field.

The converse statement is also true, that if

$$\int (X\,\mathrm{d}x + Y\,\mathrm{d}y + Z\,\mathrm{d}z) \tag{1.10.6}$$

depends only on the end points and not on the path of integration, then a function ϕ exists for which

$$X = \partial\phi/\partial x, \quad Y = \partial\phi/\partial y, \quad Z = \partial\phi/\partial z.$$

For if we choose an origin and consider all line integrals starting at the origin, the integral (1.10.6) depends only on the end point (x, y, z) and so defines a function

$$\phi(x, y, z) = \int (X\,\mathrm{d}x + Y\,\mathrm{d}y + Z\,\mathrm{d}z) = \int_0^t \left(X\frac{\mathrm{d}x}{\mathrm{d}t} + Y\frac{\mathrm{d}y}{\mathrm{d}t} + Z\frac{\mathrm{d}z}{\mathrm{d}t} \right)\mathrm{d}t \tag{1.10.7}$$

say, when the integral is taken along some path $\mathbf{r} = \mathbf{r}(t)$. From (1.10.2) and (1.10.7) we have

$$\frac{\mathrm{d}\phi}{\mathrm{d}t} = \frac{\partial \phi}{\partial x}\frac{\mathrm{d}x}{\mathrm{d}t} + \frac{\partial \phi}{\partial y}\frac{\mathrm{d}y}{\mathrm{d}t} + \frac{\partial \phi}{\partial z}\frac{\mathrm{d}z}{\mathrm{d}t} = X\frac{\mathrm{d}x}{\mathrm{d}t} + Y\frac{\mathrm{d}y}{\mathrm{d}t} + Z\frac{\mathrm{d}z}{\mathrm{d}t}. \tag{1.10.8}$$

But this must hold for all paths of integration and so the relation is true whatever the dependence of $x, y,$ and z on t. This is only possible if

$$X = \partial\phi/\partial x, \quad Y = \partial\phi/\partial y, \quad Z = \partial\phi/\partial z. \tag{1.10.9}$$

For example, the first relation follows from (1.10.8) with $\mathrm{d}y/\mathrm{d}t = \mathrm{d}z/\mathrm{d}t = 0$ and $\mathrm{d}x/\mathrm{d}t \neq 0$.

The scalar function ϕ is called the potential of the vector field **F**. The vector

* More correctly ϕ must be a single valued function of position. We shall consider only such functions here.

represented by (1.10.5) is called the gradient of ϕ, or briefly grad ϕ, and is written

$$\mathbf{F} = \nabla \phi = \frac{\partial \phi}{\partial x}\mathbf{i} + \frac{\partial \phi}{\partial y}\mathbf{j} + \frac{\partial \phi}{\partial z}\mathbf{k}. \tag{1.10.10}$$

The operator

$$\nabla = \mathbf{i}\frac{\partial}{\partial x} + \mathbf{j}\frac{\partial}{\partial y} + \mathbf{k}\frac{\partial}{\partial z} \tag{1.10.11}$$

has many of the properties of vectors. It can also occur in expressions of the type

$$\nabla . \mathbf{F} = \frac{\partial X}{\partial x} + \frac{\partial Y}{\partial y} + \frac{\partial Z}{\partial z}, \tag{1.10.12}$$

called the divergence of $\mathbf{F}$, or div $\mathbf{F}$. The vector product

$$\nabla \times \mathbf{F} = \left(\frac{\partial Z}{\partial y} - \frac{\partial Y}{\partial z}\right)\mathbf{i} + \left(\frac{\partial X}{\partial z} - \frac{\partial Z}{\partial x}\right)\mathbf{j} + \left(\frac{\partial Y}{\partial x} - \frac{\partial X}{\partial y}\right)\mathbf{k} \tag{1.10.13}$$

is called curl $\mathbf{F}$. The last expression provides a useful test for a conservative vector field. For if $\mathbf{F} = \nabla \phi$, or

$$X = \partial\phi/\partial x, \qquad Y = \partial\phi/\partial y, \qquad Z = \partial\phi/\partial z,$$

it follows by direct substitution that

$$\frac{\partial Y}{\partial x} - \frac{\partial X}{\partial y} = \frac{\partial^2 \phi}{\partial x \partial y} - \frac{\partial^2 \phi}{\partial y \partial x} = 0$$

and similarly for the other components of curl $\mathbf{F}$. Thus for a conservative vector field we must have

$$\nabla \times \mathbf{F} = \mathbf{0}. \tag{1.10.14}$$

The converse statement is also true, namely that $\nabla \times \mathbf{F} = \mathbf{0}$ implies that $\mathbf{F} = \nabla\phi$ for some scalar function $\phi(x, y, z)$. In order to prove this, we choose a suitable origin and cartesian axes, and define a function $\phi(x, y, z)$ by the relation

$$\phi(x, y, z) = \int_0^x X(\xi, 0, 0)\, d\xi + \int_0^y Y(x, \eta, 0)\, d\eta + \int_0^z Z(x, y, \zeta)\, d\zeta, \tag{1.10.15}$$

where X, Y, Z are the components of $\mathbf{F}$. Note that ϕ is defined as a line integral of $\mathbf{F}$ with the path of integration chosen as the three straight lines joining the points $(0, 0, 0)$, $(x, 0, 0)$, $(x, y, 0)$ and (x, y, z). Hence the path of the line integral is specially chosen, but nevertheless clearly defines a function of x, y, z.

We now wish to show that the function ϕ so defined is the potential of the vector field $\mathbf{F} = (X, Y, Z)$ whenever $\nabla \times \mathbf{F} = \mathbf{0}$. This will also show incidentally that, although a special path of integration has been chosen initially to define ϕ, the integral is in fact independent of the path of integration since $\mathbf{F}$ is a conservative field.

By differentiation of (1.10.15), we see that

$$\nabla \phi = \frac{\partial \phi}{\partial x}\mathbf{i} + \frac{\partial \phi}{\partial y}\mathbf{j} + \frac{\partial \phi}{\partial z}\mathbf{k}$$
$$= \left(X(x, 0, 0) + \int_0^y \frac{\partial}{\partial x} Y(x, \eta, 0)\, \mathrm{d}\eta + \int_0^z \frac{\partial}{\partial x} Z(x, y, \zeta)\, \mathrm{d}\zeta \right) \mathbf{i}$$
$$+ \left(Y(x, y, 0) + \int_0^z \frac{\partial}{\partial y} Z(x, y, \zeta)\, \mathrm{d}\zeta \right) \mathbf{j} + Z(x, y, z)\mathbf{k}. \qquad (1.10.16)$$

Now, by assumption, $\nabla \times \mathbf{F} = \mathbf{0}$ and so

$$\left.\begin{aligned} \frac{\partial}{\partial x} Y(x, \eta, 0) &= \frac{\partial}{\partial \eta} X(x, \eta, 0), \\ \frac{\partial}{\partial x} Z(x, y, \zeta) &= \frac{\partial}{\partial \zeta} X(x, y, \zeta), \\ \frac{\partial}{\partial y} Z(x, y, \zeta) &= \frac{\partial}{\partial \zeta} Y(x, y, \zeta). \end{aligned}\right\} \qquad (1.10.17)$$

Substitution of these relations into (1.10.16) gives

$$\nabla \phi = \left(X(x, 0, 0) + \int_0^y \frac{\partial}{\partial \eta} X(x, \eta, 0)\, \mathrm{d}\eta + \int_0^z \frac{\partial}{\partial \zeta} X(x, y, \zeta)\, \mathrm{d}\zeta \right) \mathbf{i}$$
$$+ \left(Y(x, y, 0) + \int_0^z \frac{\partial}{\partial \zeta} Y(x, y, \zeta)\, \mathrm{d}\zeta \right) \mathbf{j} + Z(x, y, z)\mathbf{k}. \qquad (1.10.18)$$

The integrals can now all be evaluated to give, finally,

$$\nabla \phi = X(x, y, z)\mathbf{i} + Y(x, y, z)\mathbf{j} + Z(x, y, z)\mathbf{k} \qquad (1.10.19)$$

and this is the required result.

It follows that $\nabla \times \mathbf{F} = \mathbf{0}$ is both a necessary and sufficient condition for $\mathbf{F}$ to be a conservative vector field.

Example 1.10.1

Show that the vector field $f(r)\mathbf{r}/r$ is conservative and find the potential.

We first check that the curl of this vector field is zero. Since

$$\frac{f(r)}{r}\mathbf{r} = \frac{f(r)}{r}(x\mathbf{i} + y\mathbf{j} + z\mathbf{k}) = X\mathbf{i} + Y\mathbf{j} + Z\mathbf{k},$$

say, we have

$$\frac{\partial Z}{\partial y} - \frac{\partial Y}{\partial z} = \frac{\partial}{\partial y}\left(\frac{zf(r)}{r}\right) - \frac{\partial}{\partial z}\left(y\frac{f(r)}{r}\right) = z\frac{\partial r}{\partial y}\frac{\mathrm{d}}{\mathrm{d}r}\left(\frac{f(r)}{r}\right) - y\frac{\partial r}{\partial z}\frac{\mathrm{d}}{\mathrm{d}r}\left(\frac{f(r)}{r}\right),$$

Now

$$r^2 = x^2 + y^2 + z^2$$

and so

$$r\frac{\partial r}{\partial y} = y, \quad r\frac{\partial r}{\partial z} = z.$$

It follows that

$$z\frac{\partial r}{\partial y} = y\frac{\partial r}{\partial z}$$

and hence that

$$\frac{\partial Z}{\partial y} - \frac{\partial Y}{\partial z} = 0,$$

and similarly for the two remaining components of the curl of this vector field. There is therefore a potential function ϕ such that

$$\frac{\partial \phi}{\partial x} = x\frac{f(r)}{r}, \quad \frac{\partial \phi}{\partial y} = y\frac{f(r)}{r}, \quad \frac{\partial \phi}{\partial z} = z\frac{f(r)}{r}.$$

Since for any function $F(r)$ of r, we have

$$\frac{\partial F}{\partial x} = \frac{x}{r}\frac{\mathrm{d}F}{\mathrm{d}r},$$

it follows, from

$$\frac{\partial \phi}{\partial x} = \frac{x}{r}f(r) = \frac{\partial}{\partial x}\int^r f(\xi)\,\mathrm{d}\xi,$$

that

$$\phi = \int^r f(\xi)\,\mathrm{d}\xi + g_1(y, z)$$

where g_1 is at most a function of y and z only.

Similarly, from

$$\frac{\partial \phi}{\partial y} = y\frac{f(r)}{r} \quad \text{and} \quad \frac{\partial \phi}{\partial z} = z\frac{\phi(r)}{r}$$

we have that

$$\phi = \int^r f(\xi)\,\mathrm{d}\xi + g_2(z, x) = \int^r f(\xi)\,\mathrm{d}\xi + g_3(x, y).$$

For consistency, therefore, we must have

$$g_1(y, z) \equiv g_2(z, x) \equiv g_3(x, y) = c,$$

say, where c is clearly at most a constant. The potential function is then

$$\phi = \int^r f(\xi)\,\mathrm{d}\xi + c$$

and, since the constant will have no effect on the gradient of ϕ, we have

$$\frac{f(r)\mathbf{r}}{r} = \nabla \int^r f(\xi)\,\mathrm{d}\xi.$$

1.11 EQUIPOTENTIALS AND FIELD LINES

For a conservative vector field $\mathbf{F} = \nabla\phi$, the field lines are everywhere orthogonal to the equipotential surfaces ϕ = constant.

To prove this result, consider the locus of a point P whose position vector $\mathbf{r}(t)$ is defined as a function of the parameter t. If this curve lies on the equipotential surface ϕ = constant, we have

$$\phi(\mathbf{r}(t)) = \phi(x(t), y(t), z(t)) = \text{constant}.$$

Hence
$$\frac{\mathrm{d}\phi}{\mathrm{d}t} = \frac{\partial\phi}{\partial x}\frac{\mathrm{d}x}{\mathrm{d}t} + \frac{\partial\phi}{\partial y}\frac{\mathrm{d}y}{\mathrm{d}t} + \frac{\partial\phi}{\partial z}\frac{\mathrm{d}z}{\mathrm{d}t} = \nabla\phi\,.\,\frac{\mathrm{d}\mathbf{r}}{\mathrm{d}t} = 0. \quad (1.11.1)$$

Now $\nabla\phi = \mathbf{F}$ is in the direction of the field lines, and $\mathrm{d}\mathbf{r}/\mathrm{d}t$ is a vector along the tangent to the locus of P and so lies in the tangent plane to ϕ = constant. Since equation (1.11.1) imples that $\nabla\phi$ and $\mathrm{d}\mathbf{r}/\mathrm{d}t$ are orthogonal, the result follows.

To take a simple example, the potential function $\phi(r)$, where $r = (x^2 + y^2 + z^2)^{1/2}$, has equipotentials given by r = constant. These are concentric spherical surfaces with their centres at the origin. Also, we have that

$$\nabla\phi(r) = \frac{\mathrm{d}\phi}{\mathrm{d}r}\nabla r = \frac{\mathrm{d}\phi}{\mathrm{d}r}\frac{\mathbf{r}}{r}.$$

Hence $\nabla\phi$ is everywhere in the direction of the radius vector, and the field lines are straight lines through the origin.

EXERCISES

1.1 Let $\mathbf{a}$ and $\mathbf{b}$ be elements of a vector field satisfying relations (1.3.1)–(1.3.8). Let $\mathbf{f}(\mathbf{a})$ be a vector function of $\mathbf{a}$ satisfying the following relations:

(i) for any scalar λ, $\lambda\mathbf{f}(\mathbf{a}) = \mathbf{f}(\lambda\mathbf{a})$, and, in particular, $-f(\mathbf{a}) = f(-\mathbf{a})$;
(ii) $\mathbf{f}(\mathbf{a}) + \mathbf{f}(\mathbf{b}) = \mathbf{f}(\mathbf{a} + \mathbf{b})$.

Show that $\mathbf{f}(\mathbf{a})$ is also an element of a vector field.

1.2 The vertices of a regular hexagon are A, B, C, D, E, F in that order. If

$$\overline{AB} = \mathbf{b} \quad \text{and} \quad \overline{AC} = \mathbf{c},$$

express $\overline{AD}$, $\overline{AE}$, $\overline{AF}$ in terms of **b** and **c**.
(Ans.: $2\mathbf{c} - 2\mathbf{b}$, $2\mathbf{c} - 3\mathbf{b}$, $\mathbf{c} - 2\mathbf{b}$.)

1.3 (i) The vertices of a parallelogram are A, B, C, D and G is the mid point of AC. Use a vector argument to show that $\overline{DG} = \overline{GB}$ and hence that the diagonals of a parallelogram bisect each other.
(ii) Show that if E, F are the mid points of BC, CD respectively,

$$3(\overline{AB} + \overline{AC} + \overline{AD}) = 4(\overline{AE} + \overline{AF}).$$

1.4 Let **a**, **b**, **c**, **d** be fixed vectors, and let λ, μ, ν be variable scalars. Let **r** be the position vector of a point P. Show that the loci of P represented by

$$\mathbf{r} = \mathbf{a} + \mu\mathbf{b}, \quad \mathbf{r} = \lambda\mathbf{a} + \mu\mathbf{b}$$

are respectively a line and a plane.
Show that if the lines

$$\mathbf{r} = \mathbf{a} + \mu\mathbf{b}, \quad \mathbf{r} = \mathbf{c} + \nu\mathbf{d}$$

intersect, then $\mathbf{a} - \mathbf{c}$ is parallel to the plane of **b** and **d**.

1.5 In the triangle ABC let BC = a, CA = b, AB = c. Let D be a point on BC such that AD bisects the angle A. Let I be a point on AD such that

$$\frac{\text{AI}}{\text{ID}} = \frac{b + c}{a}.$$

Show that

(i) $$\frac{\text{BD}}{\text{DC}} = \frac{c}{b},$$

(ii) $$(b + c)\overline{OD} = b\overline{OB} + c\overline{OC},$$

(iii) $$(a + b + c)\overline{OI} = a\overline{OA} + b\overline{OB} + c\overline{OC},$$

where O is some arbitrary origin.
Deduce that I also lies on the bisectors of the angles B and C (the point I is called the incentre of the triangle ABC and is the centre of the circle which touches all three sides).

1.6 Prove that the lines joining the vertices of a tetrahedron to the centroids of the opposite faces concur at a point of quadrisection of each of them.

1.7 The position vectors of the four points A, B, C, D, are **a**, **b**, $2\mathbf{a} + 3\mathbf{b}$ and $\mathbf{a} - 2\mathbf{b}$ respectively. Express $\overline{AC}$, $\overline{DB}$, $\overline{BC}$ and $\overline{DC}$ in terms of **a** and **b**.
(Ans.: $\mathbf{a} + 3\mathbf{b}$, $-\mathbf{a} + 3\mathbf{b}$, $2\mathbf{a} + 2\mathbf{b}$, $\mathbf{a} + 5\mathbf{b}$.)

1.8 The three vectors $\overline{OA}$, $\overline{OB}$, $\overline{OC}$ are linearly independent, and the three vectors $\overline{AD}$, $\overline{BD}$, $\overline{CD}$ are linearly dependent. Show that

$$\overline{OD} = \alpha\overline{OA} + \beta\overline{OB} + \gamma\overline{OC},$$

where $\alpha + \beta + \gamma = 1$.

1.9 Find the angle between the two non-zero vectors $\mathbf{a}$ and $\mathbf{b}$ in each of the following cases:

(i) $2\mathbf{a} + 3\mathbf{b} = \mathbf{0}$,
(ii) $|\mathbf{a}| = |\mathbf{b}| = |\mathbf{a} + \mathbf{b}|$,
(iii) $\mathbf{a} \cdot (\mathbf{a} + \mathbf{b}) = 0$ and $|\mathbf{b}| = 2|\mathbf{a}|$,
(iv) $(\mathbf{a} \times \mathbf{b}) \times \mathbf{a} = \mathbf{b} \times (\mathbf{b} \times \mathbf{a})$.

(Ans.: π, $2\pi/3$, $2\pi/3$, 0 or π.)

1.10 The three vectors $\mathbf{a}, \mathbf{b}, \mathbf{c}$ are such that $\mathbf{a}$ is perpendicular to $\mathbf{c} - \alpha\mathbf{a}$ and $\mathbf{b}$ is perpendicular to $\mathbf{c} - \beta\mathbf{b}$. Find the conditions for $\mathbf{a} + \gamma\mathbf{b}$ to be perpendicular to $\mathbf{c} - \alpha\mathbf{a} - \beta\mathbf{b}$.
(Ans.: Either $\mathbf{a} \cdot \mathbf{b} = 0$ or $\gamma = -\beta/\alpha$.)

1.11 Find two vectors $\mathbf{a}$ and $\mathbf{b}$ satisfying the following six conditions:

(i) $\mathbf{a} \cdot \mathbf{k} = 1$,
(ii) $\mathbf{b} \cdot \mathbf{k} = 1$,
(iii) $\mathbf{a}$ is perpendicular to $\mathbf{b}$,
(iv) $\mathbf{a}$ is perpendicular to $\mathbf{i} + \mathbf{j} + \mathbf{k}$,
(v) $\mathbf{b}$ is perpendicular to $\mathbf{i} + \mathbf{j} + \mathbf{k}$,
(vi) $\mathbf{b}$ is perpendicular to $\mathbf{i}$,

where $\mathbf{i}, \mathbf{j}, \mathbf{k}$ are unit vectors parallel to the coordinate axes.
(Ans.: $\mathbf{a} = -2\mathbf{i} + \mathbf{j} + \mathbf{k}$, $\mathbf{b} = -\mathbf{j} + \mathbf{k}$.)

1.12 Prove that:

(i) $\mathbf{a} \times (\mathbf{b} \times \mathbf{c}) + \mathbf{b} \times (\mathbf{c} \times \mathbf{a}) + \mathbf{c} \times (\mathbf{a} \times \mathbf{b}) = \mathbf{0}$,
(ii) $(\mathbf{a} \times \mathbf{b}) \cdot (\mathbf{c} \times \mathbf{d}) + (\mathbf{b} \times \mathbf{c}) \cdot (\mathbf{a} \times \mathbf{d}) + (\mathbf{c} \times \mathbf{a}) \cdot (\mathbf{b} \times \mathbf{d}) = \mathbf{0}$.

1.13 If $\mathbf{a} \times \mathbf{b} = \mathbf{a} \times \mathbf{c}$, show that $\mathbf{b} = \mathbf{c} + \lambda\mathbf{a}$, where λ is an arbitrary scalar.

1.14 Let $\mathbf{a}, \mathbf{b}, \mathbf{c}$ be a set of right-handed, mutually orthogonal, unit vectors. Let $\mathbf{a}$ be along $\mathbf{i} + \mathbf{j} + \mathbf{k}$ and let $\mathbf{b}$ be perpendicular to $\mathbf{k}$. Express $\mathbf{a}, \mathbf{b}$ and $\mathbf{c}$ in terms of $\mathbf{i}, \mathbf{j}$ and $\mathbf{k}$. Conversely, express $\mathbf{i}, \mathbf{j}, \mathbf{k}$ in terms of $\mathbf{a}, \mathbf{b}$ and $\mathbf{c}$.
(Ans.: $3^{-1/2}(\mathbf{i} + \mathbf{j} + \mathbf{k})$, $2^{-1/2}(\mathbf{i} - \mathbf{j})$, $6^{-1/2}(\mathbf{i} + \mathbf{j} - 2\mathbf{k})$;
$3^{-1/2}\mathbf{a} + 2^{-1/2}\mathbf{b} + 6^{-1/2}\mathbf{c}$, $3^{-1/2}\mathbf{a} - 2^{-1/2}\mathbf{b} + 6^{-1/2}\mathbf{c}$, $3^{-1/2}\mathbf{a} - 2.6^{-1/2}\mathbf{c}$.)

1.15 Show that any vector $\mathbf{r}$ can be written in the form

$$a^2\mathbf{r} = (\mathbf{a} \cdot \mathbf{r})\mathbf{a} + (\mathbf{a} \times \mathbf{r}) \times \mathbf{a},$$

where $\mathbf{a}$ is a given vector.

This result gives $\mathbf{r}$ in terms of its components parallel and perpendicular to $\mathbf{a}$.

1.16 Find a unit vector which is at right angles to both of the vectors $\mathbf{i} - 3\mathbf{j} + \mathbf{k}$ and $2\mathbf{i} + \mathbf{j} - \mathbf{k}$.
(Ans.: $62^{-1/2}(2\mathbf{i} + 3\mathbf{j} + 7\mathbf{k}$.)

1.17 The three vectors **a**, **b**, **c** are non-zero and satisfy the relation

$$\mathbf{a} \times (\mathbf{b} \times \mathbf{c}) = \mathbf{b} \times (\mathbf{a} \times \mathbf{c}).$$

Show that one possible relation between them is that **a** and **b** are parallel, and find a second possibility. Show that the magnitudes of the vector $\mathbf{a} \times (\mathbf{b} \times \mathbf{c})$ in the two cases are $|\mathbf{a}|\,|\mathbf{b} \times \mathbf{c}|$ and $|\mathbf{a} \cdot \mathbf{b}|\,|\mathbf{c}|$ respectively.

1.18 Prove that the four points with position vectors

$$4\mathbf{i} + 5\mathbf{j} + \mathbf{k}, \quad -3\mathbf{i} + \mathbf{k}, \quad 3\mathbf{i} + 9\mathbf{j} + 4\mathbf{k}, \quad -4\mathbf{i} + 4\mathbf{j} + 4\mathbf{k}$$

are coplanar.

1.19 The points A, B, C lie on the surface of a sphere of unit radius and centre O. The angles BOC, COA, AOB are denoted by α, β, γ respectively. Show that

$$\overline{OB} \cdot \overline{OC} = \cos \alpha, \qquad |\overline{OB} \times \overline{OC}| = \sin \alpha.$$

Show also that the angle between the planes OAB and OAC is

$$\cos^{-1}\left(\frac{\cos \alpha - \cos \beta \cos \gamma}{\sin \beta \sin \gamma}\right).$$

1.20 The two triplets **a**, **b**, **c** and **i**, **j**, **k** are mutually orthogonal right-handed sets of unit vectors such that

$$\mathbf{a} = l_{11}\mathbf{i} + l_{12}\mathbf{j} + l_{13}\mathbf{k},$$
$$\mathbf{b} = l_{21}\mathbf{i} + l_{22}\mathbf{j} + l_{23}\mathbf{k},$$
$$\mathbf{c} = l_{31}\mathbf{i} + l_{32}\mathbf{j} + l_{33}\mathbf{k}.$$

Show that $$\sum_{\gamma=1}^{3} l_{\alpha\gamma} l_{\beta\gamma} = \begin{cases} 1 & \text{if } \alpha = \beta, \\ 0 & \text{if } \alpha \neq \beta. \end{cases}$$

Show also that
$$\mathbf{i} = l_{11}\mathbf{a} + l_{21}\mathbf{b} + l_{31}\mathbf{c},$$
$$\mathbf{j} = l_{12}\mathbf{a} + l_{22}\mathbf{b} + l_{32}\mathbf{c},$$
$$\mathbf{k} = l_{13}\mathbf{a} + l_{23}\mathbf{b} + l_{33}\mathbf{c},$$

and hence that $$\sum_{\gamma=1}^{3} l_{\gamma\alpha} l_{\gamma\beta} = \begin{cases} 1 & \text{if } \alpha = \beta, \\ 0 & \text{if } \alpha \neq \beta. \end{cases}$$

1.21 The two triplets **a**, **b**, **c** and **i**, **j**, **k** are mutually orthogonal sets of unit vectors such that

$$l\mathbf{i} \times \mathbf{a} + m\mathbf{j} \times \mathbf{b} + n\mathbf{k} \times \mathbf{c} = \mathbf{0}.$$

If no two of l^2, m^2, n^2 are equal, prove that

$$\mathbf{a} = \pm\mathbf{i}, \quad \mathbf{b} = \pm\mathbf{j}, \quad \mathbf{c} = \pm\mathbf{k}.$$

1.22 Given that $\mathbf{x} + \mathbf{a} \times (\mathbf{x} \times \mathbf{a}) = \mathbf{b}$, show that

$$\mathbf{x} = \{(\mathbf{a} \cdot \mathbf{b})\mathbf{a} + \mathbf{b}\}/(1 + a^2).$$

1.23 Given that $\mathbf{x} + (\mathbf{x} \cdot \mathbf{b})\mathbf{a} = \mathbf{c}$, evaluate $\mathbf{x}$ when

(i) $1 + \mathbf{a} \cdot \mathbf{b} \neq 0$,
(ii) $1 + \mathbf{a} \cdot \mathbf{b} = 0$.

(Ans.: $\mathbf{x} = \mathbf{c} + \lambda\mathbf{a}$ where: (i) $(1 + \mathbf{a} \cdot \mathbf{b})\lambda = -\mathbf{b} \cdot \mathbf{c}$, (ii) λ is arbitrary provided $\mathbf{b} \cdot \mathbf{c} = 0$.)

1.24 Multiply the equation

$$\alpha\mathbf{x} + \mathbf{a} \times \mathbf{x} + (\mathbf{x} \cdot \mathbf{b})\mathbf{a} = \mathbf{a}$$

scalarly and vectorially by $\mathbf{a}$. Use the resulting relations to show that

$$\alpha\mathbf{x} + (\mathbf{x} \cdot \mathbf{b})\mathbf{a} = \mathbf{a},$$

and use this result to find $\mathbf{x}$ given that $\alpha \neq 0$ and $\alpha + \mathbf{a} \cdot \mathbf{b} \neq 0$.
(Ans.: $(\alpha + \mathbf{a} \cdot \mathbf{b})\mathbf{x} = \mathbf{a}$.)

1.25 The position vector of a moving point P with respect to a fixed origin is $\mathbf{r}$, and the unit vector in the direction of $\mathbf{r}$ is $\mathbf{a}$. Show that

$$(\mathbf{r} \times \dot{\mathbf{r}}) \times \mathbf{r} = r^3 \dot{\mathbf{a}}.$$

1.26 The position vectors of the two points A and B relative to an origin O are $\mathbf{a}(t)$ and $\mathbf{b}(t)$ respectively. Given that

$$\mathbf{a} \times \frac{d\mathbf{b}}{dt} = \mathbf{b} \times \frac{d\mathbf{a}}{dt},$$

show that

$$\frac{d}{dt}(\mathbf{a} \times \mathbf{b}) = \mathbf{0}.$$

Hence show that the two points A and B move in a plane and in such a way that the area of the triangle OAB remains constant.

1.27 A moving point P has position vector $\mathbf{r}$ and polar coordinates (r, θ). Calculate the radial and transverse components of the vector $d^3\mathbf{r}/dt^3$.

Given that:

(i) $d^3\mathbf{r}/dt^3$ is parallel to $\mathbf{r}$,
(ii) $d\theta/dt = \omega$, where ω is constant,
(iii) $\mathbf{r} = \mathbf{r}_0$, $\theta = 0$, $d\mathbf{r}/dt = \mathbf{u}$ when $t = 0$,

show that the locus of P is given by

$$r = r_0 \cosh(\theta/3^{1/2}) + 3^{1/2}\,\frac{\mathbf{r}_0 \,.\, \mathbf{u}}{r_0\omega}\sinh(\theta/3^{1/2}).$$

Show also that

$$\omega^2 = \frac{u^2}{r_0^2} - \frac{(\mathbf{u}\,.\,\mathbf{r}_0)^2}{r_0^4}.$$

Find the locus of P for large t when $r_0^2\omega \neq -3^{1/2}\mathbf{r}_0 \,.\, \mathbf{u}$, and show that $d\mathbf{r}/dt$ ultimately makes a fixed angle with the radius vector.

1.28 The position vector of a moving point P is $\mathbf{r}$.
Show that

$$\frac{d}{dt}(\mathbf{r} \times \dot{\mathbf{r}}) = \mathbf{r} \times \ddot{\mathbf{r}}.$$

Given that $$\ddot{\mathbf{r}} = E\mathbf{r} + H\dot{\mathbf{r}} \times \mathbf{r}/r^3,$$

where E is a scalar function of $\mathbf{r}$ and t, and H is a scalar constant, show that

$$\mathbf{r} \times \dot{\mathbf{r}} = H\mathbf{r}/r + \mathbf{a},$$

where $\mathbf{a}$ is a constant vector which has component $-H$ in the direction of $\mathbf{r}$. Hence show that $\mathbf{r}$ lies on the surface of a circular cone with vertex at the origin and axis $\mathbf{a}$.

1.29 Show that the solution of

$$\mathbf{a} \times \ddot{\mathbf{r}} = \mathbf{b},$$

where $\mathbf{a}$ and $\mathbf{b}$ are constant vectors satisfying $\mathbf{a} \,.\, \mathbf{b} = 0$, is

$$\mathbf{r} = \lambda\mathbf{a} - \mathbf{a} \times (\tfrac{1}{2}\mathbf{b}t^2 + \mathbf{c}t + \mathbf{d})/a^2,$$

where λ is an arbitrary scalar function of t, and $\mathbf{c}$, $\mathbf{d}$ are constant vectors.

1.30 Find the equations of the field lines for each of the following two-dimensional vector fields:

(i) $y\mathbf{i}$,
(ii) $xy\mathbf{i} + x^2\mathbf{j}$,
(iii) $x\mathbf{i} + (x + y)\mathbf{j}$.

(Ans.: (i) $y = c$, (ii) $x^2 - y^2 = c$, (iii) $y = x(\log x + c)$.)

1.31 Show that the slope of the plane curves $\phi(x, y) = \text{constant}$ is

$$-\frac{\partial\phi/\partial x}{\partial\phi/\partial y}.$$

Use this result to deduce that the field lines of the conservative vector field $\nabla\phi$ are orthogonal to the curves $\phi = \text{constant}$.

1.32 Sketch the curves ϕ = constant for the following two-dimensional scalar fields $\phi(x, y)$:

(i) $x + y$, (ii) $|x| + |y|$, (iii) $x^2 - y$, (iv) $x^2 + y$,
(v) $x^2 - y^2$, (vi) $x^2 + y^2$, (vii) $x^3 + y^3$, (viii) $x^4 + y^4$.

1.33 Evaluate the following integrals along the straight-line paths joining the end points:

(i) $\int_{(0,0)}^{(2,2)} y^2 \, dx$, (ii) $\int_{(2,1)}^{(1,2)} y \, dx$, (iii) $\int_{(1,1)}^{(2,1)} x \, dy$.

(Ans.: 8/3, −3/2, 0.)

1.34 Evaluate the following line integrals:

(i) $$\int_C (y^2 \, dx + x^2 \, dy),$$

where C is the semi-circle $x = (1 - y^2)^{1/2}$ between (0, −1) and (0, 1).

(ii) $$\int_C (y \, dx + x \, dy),$$

where C is the parabola $y = x^2$ between (0, 0) and (2, 4).

(iii) $$\int_C \frac{x \, dy - y \, dx}{x^2 + y^2},$$

where C is the arc of the curve $x = \cos^n t, y = \sin^n t \; (n > 0)$ for $0 \leqslant t \leqslant \pi/2$.

(Ans.: 4/3, 8, $\pi/2$.)

1.35 Evaluate the following line integrals:

(i) $$\int_C (y^2 \, dx + xy \, dy),$$

where C is the square with vertices (1, 1), (−1, 1), (−1, −1), (1, −1),

(ii) $$\int_C (y \, dx - x \, dy),$$

where C is the circle $x^2 + y^2 = 1$,

(iii) $$\int_C (x^2 y^2 \, dx - xy^3 \, dy),$$

where C is the triangle with vertices (0, 0), (1, 0), (1, 1),

(iv) $$\int_C \{(x^2 - z^2) \, dx - y^2 \, dy + xy \, dz)\},$$

where C is *any* closed curve in the plane z = constant.
(Ans.: 0, -2π, $-\frac{1}{4}$, 0.)

1.36 Evaluate the following line integrals:

(i) $$\int_C (x^2 - y^2)\,ds,$$

where C is the circle $x^2 + y^2 = 4$,

(ii) $$\int_C x\,ds,$$

where C is the line $y = x$ between (0, 0) and (1, 1).

(iii) $$\int_C ds,$$

where C is the parabola $y = x^2$ between (0, 0) and (1, 1).

(iv) $$\int_C ds,$$

where C is the curve defined by $x = \sin t$, $y = \cos t$, $z = t$ for $0 \leqslant t \leqslant 2\pi$.
(Ans.: 0, $2^{-1/2}$, $\frac{1}{4}\{\log(2 + 5^{1/2}) + 2.5^{1/2}\}$, $2^{3/2}\pi$.)

1.37 Which of the following vector fields are conservative? Give the corresponding potential functions for those that are.

(i) $x^{-1}\mathbf{i} + y^{-1}\mathbf{j}$,
(ii) $y^{-1}\mathbf{i} + x^{-1}\mathbf{j}$,
(iii) $yz\mathbf{i} + zx\mathbf{j} + xy\mathbf{k}$,
(iv) $(2xy + z^3)\mathbf{i} + (x^2 + 2y)\mathbf{j} + (3xz^2 - 2)\mathbf{k}$,
(v) $e^x z^{-2}\{yz(1 + x)\mathbf{i} + (zx + z^2 e^{-x})\mathbf{j} - xy\mathbf{k}\}$.

(Ans.: (i) $\phi = \log xy$, (iii) $\phi = xyz$, (iv) $\phi = x^2y + xz^3 + y^2 - 2z$, (v) $\phi = y(1 + z^{-1}xe^x)$.)

1.38 Prove that

(i) $$\operatorname{grad} f(r) = \frac{1}{r}\frac{df}{dr}\mathbf{r},$$

where $\mathbf{r}$ is a position vector and r is its magnitude;

(ii) $$\operatorname{grad}\{f(r_1) - f(r_2)\} = \frac{1}{r_1}\frac{df}{dr_1}\mathbf{r}_1 - \frac{1}{r_2}\frac{df}{dr_2}\mathbf{r}_2,$$

where $\mathbf{r}_1 = \mathbf{r} + \mathbf{a}$ and $\mathbf{r}_2 = \mathbf{r} - \mathbf{a}$. Show that, when $\mathbf{r}\,.\,\mathbf{a} = 0$, $\operatorname{grad}\{f(r_1) - f(r_2)\}$ is parallel to $\mathbf{a}$. Here $\mathbf{a}$ is a constant vector.

1.39 A vector field $\mathbf{F}$, which is well behaved near the origin of a system of cartesian coordinates, can be approximated near the origin in the form

$$F_x = a_1 + b_{11}x + b_{12}y + b_{13}z,$$
$$F_y = a_2 + b_{21}x + b_{22}y + b_{23}z,$$
$$F_z = a_3 + b_{31}x + b_{32}y + b_{33}z,$$

where the a_i's and the b_{ij}'s are constants. Show that such a field is locally conservative if $b_{ij} = b_{ji}$ for all i, j, and that the associated potential is then

$$\phi = (a_1x + a_2y + a_3z) + \tfrac{1}{2}(b_{11}x^2 + b_{22}y^2 + b_{33}z^2 + 2b_{23}yz + 2b_{31}zx + 2b_{12}xy).$$

1.40 Show that

(i) $\nabla\{\phi(x, y, z)\psi(x, y, z)\} = \phi\nabla\psi + \psi\nabla\phi$,
(ii) $\nabla(\mathbf{a}\,.\,\mathbf{r}) = \mathbf{a}$,
(iii) $\nabla\{(\mathbf{a}\,.\,\mathbf{r})(\mathbf{b}\,.\,\mathbf{r})\} = (\mathbf{a}\,.\,\mathbf{r})\mathbf{b} + (\mathbf{b}\,.\,\mathbf{r})\mathbf{a}$,
(iv) $\nabla\phi(\theta) = \dfrac{\mathbf{k}\times\mathbf{r}}{x^2+y^2}\dfrac{\mathrm{d}\phi}{\mathrm{d}\theta}$,

where $\mathbf{a}$, $\mathbf{b}$ are constant vectors, $\mathbf{r} = x\mathbf{i} + y\mathbf{j} + z\mathbf{k}$ and $\tan\theta = y/x$.

2 Kinematics

2.1 INTRODUCTION

A physical theory is based on a set of concepts, which are descriptions of certain relevant quantities associated with the real world, and on rules for their measurement. In addition, there is a set of hypotheses, which are idealisations of how we believe the real world behaves in terms of these concepts. When the conclusions of such a theory are seen to agree substantially with what is actually observed, the hypotheses become accepted as physical laws. It is important to realise, however, that these laws are always open to modification in the light of more sophisticated experiments, and that this is, in fact, the way in which science progresses.

The term kinematics refers to the study of the motion of bodies in space, without regard to the mechanisms which cause the motion to take place. Thus we can discuss the trajectory of a projectile as a part of kinematics. But an investigation into the mechanism which causes the projectile to move in a prescribed fashion involves a discussion of the physical laws which govern the attraction of one body for another, and this is regarded as a part of dynamics.

The concepts required in kinematics are those of length and time. These are fundamental concepts and, as such, do not have a formal definition. We rely on our everyday experience for an understanding of them. This understanding involves certain assumptions about their properties, for example the very existence of standards by which length and time can be measured and which mean the same thing to different observers, is a basic assumption without which theoretical physics could not exist. Indeed the transactions of everyday life would be difficult without agreed measures of length and time. Fortunately there is no reason to doubt these assumptions. The correspondence between the physical notion of measurement and the mathematical notion of addition is one of the reasons why mathematics is so useful in application.

We do have to state how length and time are to be measured before we can proceed further, but in the first instance the actual units used are arbitrary and there are several units of length in existence. For obvious reasons of convenience, however, it is desirable to have an internationally accepted unit, and in the '*Système International*' (SI) which we shall use, this is the metre (m). For practical purposes fractions and multiples of this unit are often employed, the most common of which are the millimetre ($1 \text{ mm} = 10^{-3} \text{m}$) and the kilometre ($1 \text{ km} = 10^3 \text{m}$).

The definition of the metre was originally referred to a material standard kept in Paris. This was, in turn, intended to be 10^{-7} times a quadrant of the earth's meridian. The agreement, though close, is not exact by modern measurements, the quadrant of the meridian being $1{\cdot}02 \times 10^7 \text{m}$.

The above mentioned standard is not conveniently reproducible with the accuracy required for present day measurements, and so the definition is now referred to the wavelength of the radiation from krypton (1 650 763·73 wavelengths of radiation = 1 m).

To define a unit of time requires a process which is generally accepted as regular, repeatable and a suitable reference against which other motions can be compared. Originally this was based on the rotation of the earth, more specifically on the mean solar day which is the average interval between two successive transits of the sun across a given meridian. Nowadays the unit is referred to the radiation from caesium, again for reasons of accuracy of reproduction (9 192 631 770 periods of radiation = 1 s). In the SI system the unit is the second (s), multiples of which are the minute (min), hour (h), day (d), year (a).

In Newtonian mechanics it is further assumed that the geometry of space is Euclidean, that we can always recognise when two lengths are equal and when two events are simultaneous and that, in principle, there is no limit to the accuracy of our measurements. In the theories of relativity and quantum mechanics these assumptions are all modified, but for a large class of everyday problems there is ample evidence that the assumptions of Newtonian mechanics form the basis of a useful theory.

In Newtonian mechanics we do not assert that a body can be in the same position at two different times. We refer to motion relative to a chosen frame of reference without implying that the frame is at rest in an absolute sense. There is indeed no kinematical criterion to distinguish one frame as preferable to another, and the choice is one of convenience. For example the motion of a projectile is conveniently described with reference to axes fixed relative to the observer, which will usually mean fixed on the earth's surface. However, we shall see in the next chapter that dynamical considerations are properly discussed with reference to a particular choice of axes.

In addition to the basic concepts we introduce further derived concepts, and these are defined with reference to the idealised notion of a particle. This is a body whose kinematical behaviour is completely described by the specification of its position vector. When the theory is applied to a practical situation, the results for a particle might be expected to be good approximations where the size of the body is small compared with the other linear dimensions of the problem. Typical examples are the trajectory of a missile or the motion of the earth round the sun, where the size of the body is small compared with the radius of curvature of the path. Where the size of the body is important, as in the motion of a gyroscope, the theory is a generalisation of that for a particle, so that particle dynamics is in any case fundamental to the subject as a whole.

2.2 VELOCITY AND ACCELERATION

Let a particle P have position vector $\mathbf{r}$ relative to a chosen frame of reference.

Then its velocity $\mathbf{v}$ and its acceleration $\mathbf{a}$ are defined by the relations

$$\mathbf{v} = \frac{d\mathbf{r}}{dt} = \dot{\mathbf{r}}, \tag{2.2.1}$$

$$\mathbf{a} = \frac{d\mathbf{v}}{dt} = \frac{d^2\mathbf{r}}{dt^2} = \ddot{\mathbf{r}}, \tag{2.2.2}$$

where t is the time.

In particular, when the motion of the particle is in a straight line, both the velocity and acceleration are in the direction of motion and their magnitudes are respectively equal to the first and second derivatives of distance travelled with respect to the time. When the motion is not one dimensional, the velocity and acceleration can be determined from their components. In two dimensions, for example, the velocity components are $(\dot{x}, \dot{y})$ if P has cartesian coordinates (x, y), or $(\dot{r}, r\dot{\theta})$ along and transverse to the radius vector if P has polar coordinates (r, θ); see equation (1.6.23). Similarly the corresponding components of acceleration are $(\ddot{x}, \ddot{y})$ or $(\ddot{r} - r\dot{\theta}^2, r\ddot{\theta} + 2\dot{r}\dot{\theta})$; see equation (1.6.24).

The magnitude of the velocity is called the speed. The rate of change of angle, $\dot{\theta}$, is called the angular speed. There is no special name for the magnitude of the acceleration.

Let a particle P′ have position vector $\mathbf{r}'$. The velocity of P′ relative to P is defined to be

$$\frac{d}{dt}\overline{PP'} = \frac{d\mathbf{r}'}{dt} - \frac{d\mathbf{r}}{dt}. \tag{2.2.3}$$

Similarly the acceleration of P′ relative to P is

$$\frac{d^2}{dt^2}\overline{PP'} = \frac{d^2\mathbf{r}'}{dt^2} - \frac{d^2\mathbf{r}}{dt^2}. \tag{2.2.4}$$

Note that the velocity of P′ is the vector sum of the velocity of P and the velocity of P′ relative to P. This can be a useful rule for the calculation of the velocity of a particle. In the same way the acceleration of P′ can be calculated from the vector sum of the acceleration of P and the acceleration of P′ relative to P.

Example 2.2.1

An aeroplane, flying in a north-east wind,* travels a distance d due north and returns to its original position. The time for the outward journey is t_o and for the return journey t_r. Find the speeds v and w of the aeroplane in still air and of the wind, respectively.

* A north-east wind blows from the north-east to the south-west.

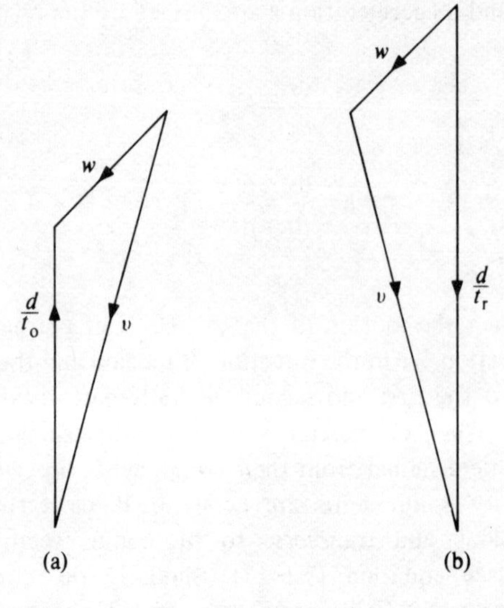

Figure 2.2.1 Illustration for Example 2.2.1: (a) outward journey; (b) return journey.

The velocity of the plane relative to the wind is of magnitude v in a direction such that the two velocities combined carry the plane due north on the outward journey with speed d/t_o and due south on the return journey with speed d/t_r (see Fig. 2.2.1).

We use the cosine rule (1.5.17) in the triangular law of addition for the velocities to get the two relations

$$v^2 = w^2 + \frac{d^2}{t_o^2} + \frac{2wd}{t_o} \cos \tfrac{1}{4}\pi,$$

$$v^2 = w^2 + \frac{d^2}{t_r^2} - \frac{2wd}{t_r} \cos \tfrac{1}{4}\pi.$$

By subtraction, we have

$$2^{1/2}\, dw \left(\frac{1}{t_o} + \frac{1}{t_r}\right) = d^2 \left(\frac{1}{t_r^2} - \frac{1}{t_o^2}\right),$$

or

$$w = 2^{-1/2}\, d \left(\frac{1}{t_r} - \frac{1}{t_o}\right).$$

Then a straightforward calculation for v shows that

$$v = 2^{-1/2}\, d \left(\frac{1}{t_r^2} + \frac{1}{t_o^2}\right)^{1/2}.$$

Example 2.2.2

A plane P takes off with constant velocity **u** and climbs at a fixed angle α to the horizontal. A helicopter H is in steady horizontal flight with velocity **v** at a height h, and the two flight paths are in the same vertical plane. If the horizontal distance between the helicopter and the plane is d when the latter is about to take off, find the distance of nearest approach.

If a velocity $-\mathbf{v}$ is superimposed on both H and P, then H is reduced to rest and the direction of the combined velocities $-\mathbf{v}$ and $\mathbf{u}$ will give the flight path of P relative to H. If this path is PN, and if HN is the perpendicular from H on to this path, then HN is the distance of nearest approach (see Fig. 2.2.2).

Let β be the inclination of PN to the horizontal and let γ be the inclination of PH to the horizontal. We then have that the distance HN, without regard to sign, is

$$\begin{aligned} \mathrm{HN} &= |\,\mathrm{PH}\sin(\beta-\gamma)\,| = |\,\mathrm{PH}\sin\beta\cos\gamma - \mathrm{PH}\cos\beta\sin\gamma\,| \\ &= |\,d\sin\beta - h\cos\beta\,| \end{aligned}$$

and a knowledge of HN depends on the calculation of β. To find β note that the velocities $-\mathbf{v}$ and $\mathbf{u}$ must combine so that there is no component perpendicular to PN. This implies that

$$u\sin(\alpha-\beta) = v\sin\beta,$$

or

$$\sin\alpha\cot\beta - \cos\alpha = v/u,$$

$$\cot\beta = \frac{\cos\alpha + v/u}{\sin\alpha}.$$

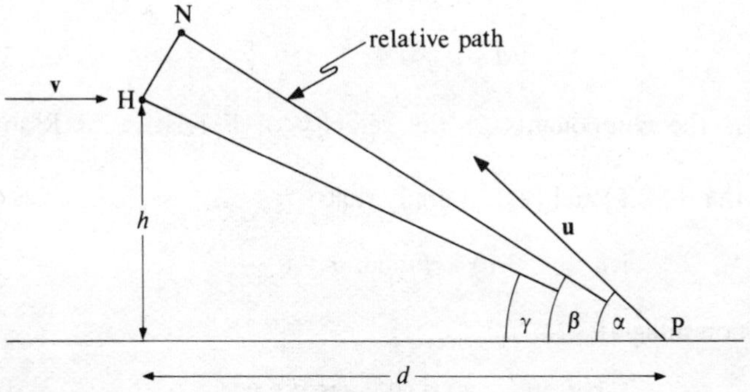

Figure 2.2.2 Illustration for Example 2.2.2.

Example 2.2.3

A rabbit runs in a straight line with constant speed u. A fox chases the rabbit with velocity **v** whose magnitude $v(>u)$ is constant and is always directed towards the rabbit. At time $t = 0$ the fox is at a distance d from the rabbit and

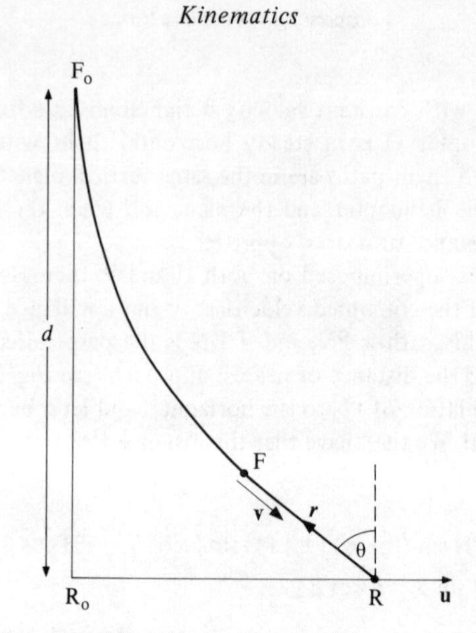

Figure 2.2.3 Illustration for Example 2.2.3.

the velocities of the two animals are perpendicular. Show that the fox catches the rabbit after a time $vd/(v^2 - u^2)$.

The curve followed by the fox is called the curve of pursuit and can be calculated by considering the velocity of the fox relative to the rabbit. Denote the rabbit by R and the fox by F, and let their initial positions be R_0 and F_0 (see Fig. 2.2.3). Let $\overline{RF} = \mathbf{r}$, and let the angle between $\mathbf{r}$ and $\mathbf{u}$ be $\theta + \pi/2$. Then

$$\dot{r} = -v + u \sin \theta, \tag{2.2.5}$$

$$r\dot{\theta} = u \cos \theta. \tag{2.2.6}$$

these are just the components of the velocity of F relative to R in polar coordinates.

First note that (2.2.5) and (2.2.6) imply that

$$\dot{r}(u \sin \theta + v) + r\dot{\theta}u \cos \theta = u^2 - v^2$$

and this equation integrates to

$$r(u \sin \theta + v) = (u^2 - v^2)t + vd, \tag{2.2.7}$$

since $r = d$, $\theta = 0$ when $t = 0$. This is all that is required to find the time taken to catch the rabbit, for then $r = 0$ and hence the time is $vd/(v^2 - u^2)$.

In order to find the path, it is necessary to integrate the relation

$$\frac{1}{r}\frac{\mathrm{d}r}{\mathrm{d}\theta} = \frac{-v + u \sin \theta}{u \cos \theta}, \tag{2.2.8}$$

which is implied by the ratio of (2.2.5) and (2.2.6). The integral of (2.2.8) is

$$\log \frac{r}{d} = -\frac{v}{u} \log(\sec\theta + \tan\theta) - \log\cos\theta,$$

or

$$\frac{r}{d} = \frac{1}{\cos\theta(\sec\theta + \tan\theta)^{v/u}}. \tag{2.2.9}$$

Equations (2.2.7) and (2.2.9) contain all the information required to find the path taken by the fox. They determine, at least implicitly, r and θ (and hence $\mathbf{r}$) as functions of t. Thus the position vector $\overline{R_0F}$ of F, relative to R_0, can be determined as a function of t since

$$\overline{R_0F} = \overline{R_0R} + \overline{RF} = \mathbf{u}t + \mathbf{r}.$$

2.3 CONSTANT ACCELERATION

There are several problems of practical interest where the acceleration is constant in both magnitude and direction. The relations governing such motion are easily derived by integration of the relation

$$\frac{\mathrm{d}^2\mathbf{r}}{\mathrm{d}t^2} = \mathbf{a}. \tag{2.3.1}$$

Since $\mathbf{a}$ is a constant vector we get

$$\mathbf{v} = \frac{\mathrm{d}\mathbf{r}}{\mathrm{d}t} = \mathbf{u} + \mathbf{a}t, \tag{2.3.2}$$

$$\mathbf{r} = \mathbf{r}_0 + \mathbf{u}t + \tfrac{1}{2}\mathbf{a}t^2, \tag{2.3.3}$$

where $\mathbf{r}_0$ and $\mathbf{u}$ are constant vectors of integration which will be determined by the initial conditions. Another useful formula, which is implied by (2.3.2) and (2.3.3), is obtained as follows. From (2.3.2) we have

$$\begin{aligned} v^2 &= u^2 + a^2t^2 + 2\mathbf{u}\,.\,\mathbf{a}t \\ &= u^2 + 2\mathbf{a}\,.\,(\mathbf{u}t + \tfrac{1}{2}\,\mathbf{a}t^2) \\ &= u^2 + 2\mathbf{a}\,.\,(\mathbf{r} - \mathbf{r}_0). \end{aligned} \tag{2.3.4}$$

The most general motion of this kind is two dimensional. For we can without loss of generality choose the origin so that $\mathbf{r}_0 = \mathbf{0}$. Then the plane of the motion is that plane of the constant vectors $\mathbf{u}$ and $\mathbf{a}$ which passes through the origin. When $\mathbf{u}$ and $\mathbf{a}$ are parallel the motion is one dimensional and if

$$\mathbf{r} = x\mathbf{i},\ \mathbf{r}_0 = \mathbf{0},\ \mathbf{u} = u\mathbf{i},\ \mathbf{v} = v\mathbf{i},\ \mathbf{a} = a\mathbf{i}, \tag{2.3.5}$$

the equations (2.3.2), (2.3.3) and (2.3.4) reduce to

$$v = u + at, \tag{2.3.6}$$

$$x = ut + \tfrac{1}{2}at^2, \tag{2.3.7}$$

$$v^2 = u^2 + 2ax. \tag{2.3.8}$$

2.4 PROJECTILES

The theory of the motion of projectiles is based on the following experimental observation. All bodies in the neighbourhood of the earth's surface move when unrestrained with a vertical acceleration **g** relative to the earth. Its magnitude varies with locality, but in a local region in the vicinity of the earth's surface it is constant to a good approximation in magnitude and direction. In England its magnitude is

$$g = 9{\cdot}81 \text{ m s}^{-2}.$$

This result is actually a consequence of a much more general law of gravitation, first postulated by Newton and now accepted as one of the most far-reaching results of theoretical physics. According to this law, which will be discussed in more detail in Chapter 7, a spherical body (to which the earth is a good approximation) induces an acceleration in a neighbouring particle which is towards the centre of the sphere and inversely proportional to the square of the distance from the centre. From this we can estimate the accuracy of the approximation that g is constant. If g is the acceleration at the earth's surface and r_e is the radius of the earth, then the acceleration at a distance r from the centre of the earth will be gr_e^2/r^2. Now r_e is 6378 km at the equator and 6357 km at the poles. If we round this off to 6400 km, the acceleration at a height of 10 km above the earth's surface will be

$$g\left(\frac{6400}{6410}\right)^2 = 0{\cdot}997\,g.$$

There would hence be a 0·3% error at this height if the variation in g were ignored.

The fact that the acceleration is towards the centre of the earth means that the assumption that it is constant in direction is also approximate. Two points 10 km apart on the earth's surface subtend an angle of 10/6400 radians or 0·1° of arc at the centre of the earth. This represents the departure from parallelism between two such points.

The flattening of the earth towards the poles means that the earth is not a true sphere and that consequently there is a slight variation in g along a meridian. This variation is adequately represented by the formula

$$g = 9{\cdot}806\,(1 - 0{\cdot}0027 \cos 2\phi)\ \text{km s}^{-2}$$

where ϕ is the latitude, so that there is a total variation in g of 0·5% between the poles and the equator.

Finally the accuracy of the result will depend on the extent to which additional mechanisms, which influence the motion of the projectiles, can be ignored. In practice the most important of these is the resistance of the air. The modifications to which this gives rise will be discussed in Chapter 4. For the moment we investigate the consequences of the assumption that the gravitational acceleration is constant in magnitude and direction and is the only operative mechanism. Then the motion is described by equations (2.3.2), (2.3.3) and (2.3.4) with $\mathbf{a}$ equal to the gravitational acceleration. In particular

$$\mathbf{r} = \mathbf{u}t + \tfrac{1}{2}\mathbf{g}t^2 \tag{2.4.1}$$

if the particle starts from the origin with velocity $\mathbf{u}$ at time $t = 0$.

We take axes Ox horizontally and Oy vertically in the plane of the motion. The gravitational acceleration then has components $(0, -g)$ and the initial velocity $\mathbf{u}$ makes an angle α with the horizontal as in Figure 2.4.1. The governing equations can be written

$$\dot{x} = u \cos\alpha, \qquad \dot{y} = u \sin\alpha - gt, \tag{2.4.2}$$

$$x = ut \cos\alpha, \qquad y = ut \sin\alpha - \tfrac{1}{2}gt^2, \tag{2.4.3}$$

$$v^2 = u^2 - 2gy. \tag{2.4.4}$$

In particular, if $\alpha = \pi/2$, the motion is one-dimensional with $x = 0$ and $y = ut - \frac{1}{2}gt^2$.

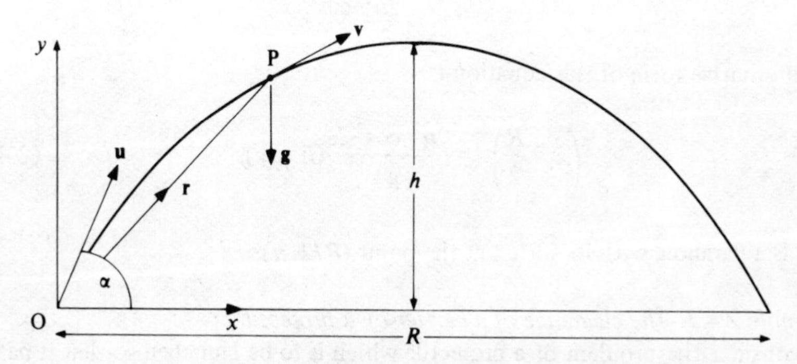

Figure 2.4.1 The path of a projectile moving under the influence of gravity.

The maximum height, h, is attained when $\dot{y} = 0$ or, from (2.4.2),

$$t = \frac{u \sin \alpha}{g} \quad (= \tfrac{1}{2} T, \text{ say}).$$

Then, from (2.4.3),

$$h = \frac{u^2 \sin^2 \alpha}{g} - \frac{1}{2}\frac{u^2 \sin^2 \alpha}{g} = \frac{u^2 \sin^2 \alpha}{2g}. \tag{2.4.5}$$

The y coordinate of the projectile is zero when

$$0 = ut \sin \alpha - \tfrac{1}{2}gt^2,$$

and so

$$t = 0 \quad \text{or} \quad \frac{2u \sin \alpha}{g} \quad (=T).$$

Thus T is the time of flight and the range, R, on the horizontal plane through the origin is therefore, from (2.4.3),

$$R = u \cos \alpha \left(\frac{2u \sin \alpha}{g}\right) = \frac{u^2 \sin 2\alpha}{g}. \tag{2.4.6}$$

It follows that the maximum range, for a fixed value of u and varying values of α, is $R_m = u^2/g$ achieved when $\alpha = \pi/4$. Any intermediate range can be achieved with two possible values for α since, if $\alpha = \theta$ gives the required range, so also does $\alpha = \frac{1}{2}\pi - \theta$. These two directions lie on either side of $\alpha = \pi/4$ and are equally inclined to it.

Equations (2.4.3) are the parametric equations of the path of the projectile, with t as the parameter. The cartesian equation of the path can be obtained by the elimination of t. It is

$$y = x \tan \alpha - \frac{gx^2}{2u^2 \cos^2 \alpha}. \tag{2.4.7}$$

An alternative form of this equation is

$$\left(x - \frac{R}{2}\right)^2 = \frac{2u^2 \cos^2 \alpha}{g}(h - y). \tag{2.4.8}$$

This is a parabola with its vertex at the point $(R/2, h)$.

Example 2.4.1. The clearance of a barrier by a projectile.
We consider the problem of a projectile which is to be launched so that it passes over the point whose coordinates are (x_0, y_0).

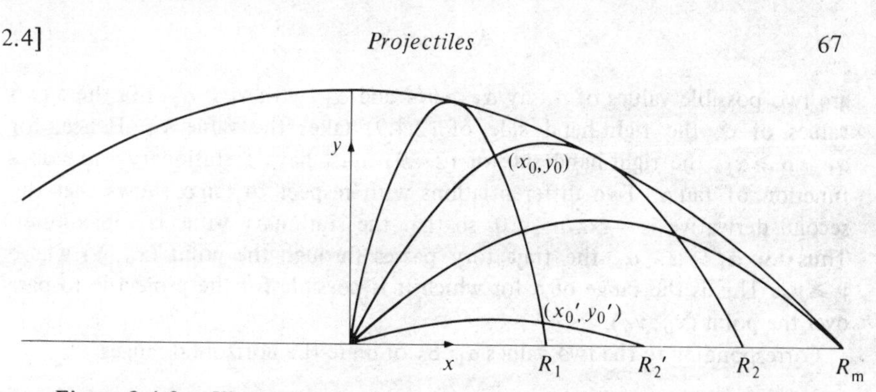

Figure 2.4.2 Illustration for Example 2.4.1: the parabola of safety.

The initial inclination α which takes the projectile through the point (x_0, y_0) must, from (2.4.7), satisfy

$$y_0 = x_0 \tan\alpha - \frac{gx_0^2}{2u^2\cos^2\alpha} = x_0 \tan\alpha - \frac{gx_0^2}{2u^2}(1 + \tan^2\alpha). \qquad (2.4.9)$$

This is a quadratic equation for $\tan\alpha$ which can be solved to give

$$\tan\alpha = \left[1 \pm \left\{1 - \frac{2g}{u^2}\left(y_0 + \frac{gx_0^2}{2u^2}\right)\right\}^{1/2}\right]\frac{u^2}{gx_0}. \qquad (2.4.10)$$

For real values of α, and hence possible trajectories, the part of the expression within the square root must not be negative, and so we have

$$1 \geqslant \frac{2g}{u^2}\left(y_0 + \frac{gx_0^2}{2u^2}\right),$$

or

$$y_0 \leqslant \frac{u^2}{2g}\left(1 - \frac{g^2x_0^2}{u^4}\right). \qquad (2.4.11)$$

The point (x_0, y_0) must therefore not lie above the curve

$$y = \frac{u^2}{2g}\left(1 - \frac{g^2x^2}{u^4}\right) \qquad (2.4.12)$$

if there is to be a possible trajectory. This curve is called the parabola of safety and is shown as the enveloping curve in Figure 2.4.2.

It is sufficient to discuss points for which $x_0 > 0$, $y_0 \geqslant 0$. Then, from (2.4.11) we must have $u^2/gx_0 \geqslant 1$. This latter inequality, together with (2.4.10), implies that at least one value of α is greater than or equal to $\pi/4$.

To reach a point on the parabola of safety, there is one value of $\tan\alpha = u^2/gx_0 \geqslant 1$ or $\alpha \geqslant \pi/4$. To reach a point within the parabola of safety there are two possible values of α, say $\alpha_1 > \pi/4$ and α_2, with $\alpha_1 > \alpha_2$. For these two values of α, the right-hand side of (2.4.9) takes the value y_0. Hence, for $\alpha_1 \geqslant \alpha \geqslant \alpha_2$, the right-hand side of (2.4.9) must have a stationary value as a function of $\tan \alpha$. Two differentiations with respect to $\tan \alpha$ shows that the second derivative is $-gx_0^2/u^2 < 0$, so that the stationary value is a maximum. Thus for $\alpha_1 \geqslant \alpha \geqslant \alpha_2$ the trajectory passes through the point (x_0, y) where $y \geqslant y_0$. This is the range of α for which it is possible for the projectile to pass over the point (x_0, y_0).

Corresponding to the two values α_1, α_2 of α are the horizontal ranges

$$R_1 = \frac{u^2 \sin 2\alpha_1}{g}, \quad R_2 = \frac{u^2 \sin 2\alpha_2}{g} \tag{2.4.13}$$

with $R_1 \leqslant R_2$. This inequality follows because the two trajectories can only meet in one other point in addition to the origin. Since this point is (x_0, y_0) with $x_0 \geqslant 0, y_0 \geqslant 0$, a contradiction is implied unless $R_1 \leqslant R_2$.

If $\alpha_2 > \pi/4$, so that (x_0, y_0) lies between the trajectory for maximum range and the parabola of safety, then $R_1 \leqslant R \leqslant R_2$ represents the interval of ranges on the horizontal plane accessible to a projectile which has to pass over the point (x_0, y_0). Only one trajectory is possible for a given range R, for a second trajectory would imply that $\alpha_2 < \pi/4$ which contradicts the assumption that $\alpha_2 > \pi/4$.

If $\alpha_2 < \pi/4$, the maximum range $R_m = u^2/g$ is attainable. Here $R_1 \leqslant R \leqslant R_m$ is the interval of accessible ranges. For $R_1 \leqslant R \leqslant R_2$ only one trajectory is possible, for which $\alpha > \pi/4$. For $R_2 < R < R_m$ there are two trajectories possible, one with $\alpha > \pi/4$, the other with $\alpha < \pi/4$.

Figure 2.4.2 shows two trajectories through the point (x_0, y_0) for which $\alpha_1 > \alpha_2 > \pi/4$. The accessible points on the horizontal plane for a projectile passing over (x_0, y_0) lie between R_1 and R_2. The two trajectories through the point (x_0', y_0') are such that $\alpha_1 > \pi/4 > \alpha_2$. In this case the accessible points lie between R_1 and R_m.

Example 2.4.2. The range of a projectile on an inclined plane.
The original equations (2.3.2), (2.3.3) and (2.3.4) are still valid, but it is now simpler to take axes along and perpendicular to the plane, as shown in Figure 2.4.3. Let α be the inclination of $\mathbf{u}$ to the plane and β the inclination of the plane to the horizontal. Then $\mathbf{u}$ has components $(u \cos \alpha, u \sin \alpha)$ and $\mathbf{g}$ has components $(-g \sin \beta, -g \cos \beta)$. Hence

$$\dot{x} = u \cos \alpha - gt \sin \beta, \quad \dot{y} = u \sin \alpha - gt \cos \beta, \tag{2.4.14}$$

$$x = ut \cos \alpha - \tfrac{1}{2}gt^2 \sin \beta, \quad y = ut \sin \alpha - \tfrac{1}{2}gt^2 \cos \beta. \tag{2.4.15}$$

The y coordinate of the particle is zero when

$$0 = ut \sin \alpha - \tfrac{1}{2}gt^2 \cos \beta$$

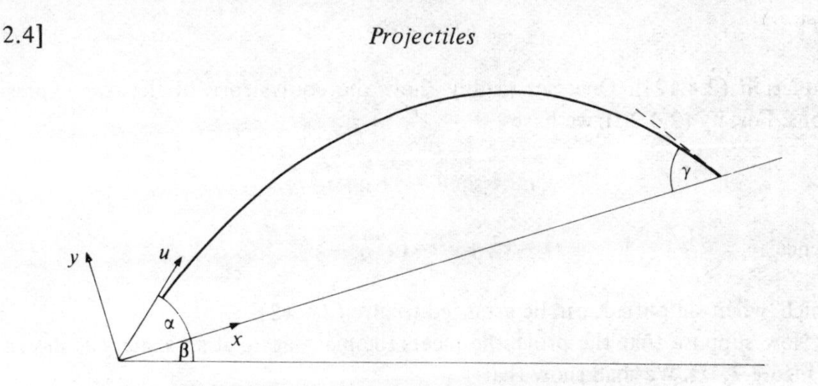

Figure 2.4.3 Illustration for Example 2.4.2.

and so the time of flight is

$$T = 2u \sin \alpha/(g \cos \beta). \tag{2.4.16}$$

The value of x for this value of t gives the range, and so

$$\begin{aligned} R &= u \cos \alpha\left(\frac{2u \sin \alpha}{g \cos \beta}\right) - \tfrac{1}{2}g \sin \beta\left(\frac{2u \sin \alpha}{g \cos \beta}\right)^2 \\ &= \frac{2u^2 \sin \alpha \cos (\alpha + \beta)}{g \cos^2 \beta} \qquad (2.4.17) \\ &= \frac{u^2\{\sin (2\alpha + \beta) - \sin \beta\}}{g \cos^2 \beta}. \end{aligned}$$

The maximum range, for varying values of α, will occur when $\sin (2\alpha + \beta) = 1$; the inclination for maximum range is therefore

$$\alpha_m = \tfrac{1}{2}(\tfrac{1}{2}\pi - \beta) \tag{2.4.18}$$

and the maximum range is

$$R_m = \frac{u^2(1 - \sin \beta)}{g \cos^2 \beta} = \frac{u^2}{g(1 + \sin \beta)}. \tag{2.4.19}$$

Note that since this gives the maximum distance that can be attained in a direction inclined at an angle β to the horizontal, only points within the curve whose polar equation is

$$r = \frac{u^2}{g(1 + \sin \theta)} \tag{2.4.20}$$

can be reached by a projectile starting from the origin. Equation (2.4.20) must therefore be an alternative expression for the equation of the parabola of safety (equation (2.4.12)). One can readily show the equivalence of the two expressions. For, by (2.4.20), we have

$$r + r \sin \theta = r + y = u^2/g.$$

Hence $$r^2 = x^2 + y^2 = (u^2/g - y)^2,$$

which, when simplified, can be arranged to give (2.4.12).

Now suppose that the projectile meets the plane again at an angle γ as shown in Figure 2.4.3. We shall show that

$$\cot \alpha - \cot \gamma = 2 \tan \beta. \tag{2.4.21}$$

Where the projectile strikes the plane we have, from (2.4.14) and (2.4.16),

$$\dot{x} = u \cos \alpha - g \sin \beta \left(\frac{2u \sin \alpha}{g \cos \beta}\right) = u \cos \alpha - 2u \sin \alpha \tan \beta,$$

$$\dot{y} = u \sin \alpha - g \cos \beta \left(\frac{2u \sin \alpha}{g \cos \beta}\right) = -u \sin \alpha.$$

But we are given that at this point $\dot{x}/\dot{y} = -\cot \gamma$. Hence

$$\cot \gamma = \frac{u \cos \alpha - 2u \sin \alpha \tan \beta}{u \sin \alpha} = \cot \alpha - 2 \tan \beta,$$

from which (2.4.21) follows.

Example 2.4.3

A particle is projected from a horizontal plane to pass over two objects at heights h and k, and a slant distance d apart. Show that the least possible speed of projection is $\{g(h + k + d)\}^{1/2}$.

Since, from equation (2.4.4),

$$u_p^2 = u_o^2 - 2gh,$$

where u_o, u_p are respectively the speeds at O and P (see Fig. 2.4.4), it follows that when the speed of projection u_o is a minimum so is u_p. Now u_p is a minimum when the maximum range along PQ is equal to PQ, and so

$$\text{PQ} = \frac{u_p^2}{g(1 + \sin \beta)} \quad \text{or} \quad u_p^2 = g(d + k - h).$$

It follows that $$u_o^2 = u_p^2 + 2gh = g(h + k + d).$$

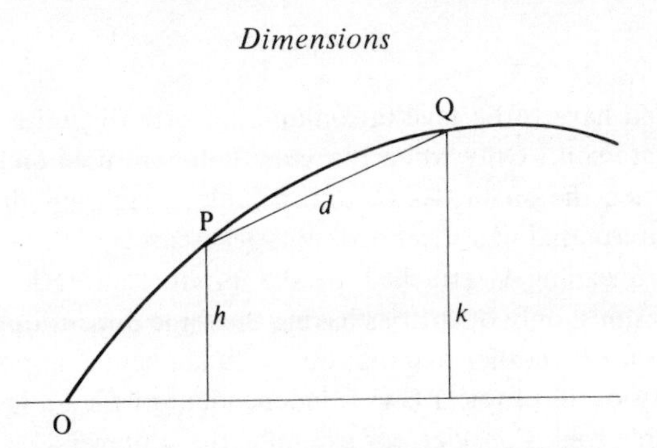

Figure 2.4.4 Illustration for Example 2.4.3.

In this problem the point of projection O is not fixed, but has to be chosen to be consistent with the solution.

2.5 DIMENSIONS

Derived quantities, such as velocity and acceleration, are measured in terms of the fundamental units, in this case the units of length and time. For example velocity is measured as a distance divided by a time and acceleration as a velocity divided by a time (if necessary involving a limiting process when they are not uniform). We say that velocity has the dimensions of (length)/(time), or briefly LT^{-1}, and that acceleration has the dimensions LT^{-2}. This is a useful shorthand description of the measure of any physical quantity in terms of the measures of the fundamental quantities from which it is derived. It enables us to see how the measure of a quantity will change if the fundamental units are changed. For example, since velocity depends linearly on the unit of length, its measure will also change linearly with the measure of length. Thus

$$x \text{ km} = 1000x \text{ m},$$

and so

$$v \text{ km s}^{-1} = 1000v \text{ m s}^{-1}. \tag{2.5.1}$$

On the other hand the measure of area, which has dimensions L^2, will change in proportion to the square of the measure of length, so that

$$A \text{ km}^2 = (1000)^2 A \text{ m}^2. \tag{2.5.2}$$

For acceleration we have

$$a \text{ km h}^{-2} = 1000a \text{ m h}^{-2} = \frac{1000a}{(3600)^2} \text{ m s}^{-2}. \tag{2.5.3}$$

Note the special use of the equality sign in relations such as (2.5.1), (2.5.2), (2.5.3). The actual numbers on the two sides of the relation

$$1 \text{ km s}^{-1} = 1000 \text{ m s}^{-1}$$

are different and have to be read in conjunction with the units in order to make sense of the statement. Only when the same units are used on both sides of the equation, or when the quantities are dimensionless, can the units be ignored and the equation interpreted in a strict mathematical sense.

Although a meaning is attached to the product or ratio of quantities of different dimensions, only quantities having the same dimensions are added. This is a consequence of another assumption of fundamental importance, which is that the validity of all physical laws is independent of the units used to measure the quantities involved. Consider, for example, the statement

$$x = ut + \tfrac{1}{2}at^2,$$

where x, t, u, a represent respectively distance, time, velocity and acceleration. It is clear that if such a relation holds for one set of units of length and time, then it will hold for all sets of units since the dimensions of the two sides of the equation agree. On the other hand a statement such as

$$x = u$$

can have no such general validity since it will cease to hold if the unit of time is altered.

All the relations subsequently derived will be consistent in the above sense, and this dimensional consistency is a very useful check of the algebra, or even of the basic assumptions, in any physical problem.

The law of dimensional consistency can even be used to find the form of a solution, and this can be put to significant practical use where problems are too complex to be treated in full mathematical detail. As a simple example, suppose we consider the distance travelled from rest by a particle given a constant acceleration a. Since the distance travelled can depend only on a and the time t, and since it must have the dimensions of a length, it must be proportional to at^2 and we have the result, apart from a numerical factor. If we introduce further parameters, for example if the particle is given an initial velocity u, the correct dimensional combination at^2 is still required, but the factor of proportionality can now be a function of any non-dimensional combination of the parameters. Since u/at is a dimensionless combination (and there is no other independent combination) the distance travelled must be expressible in the form $at^2 F(u/at)$, where F is some unknown function. Of course in this simple example the formula can be calculated explicitly (see equation (2.3.7)). It is

$$x = ut + \tfrac{1}{2}at^2 = at^2\left(\frac{u}{at} + \frac{1}{2}\right),$$

and is indeed of the right form. In more complex problems arguments of this kind can bring to light the non-dimensional parameters on which a problem depends. From the practical point of view it is clearly useful to know that a certain function $F(u, a, t)$ of three parameters appearing in the problem is in fact of the form $F(u/at)$.

EXERCISES

2.1 Show that the hour and minute hands of a 12 hour clock coincide every 12/11 hours.

The hour, minute and second hands of the clock have a common axis and they coincide at 12 o'clock (mid-day and mid-night). Do they coincide at any other time?

2.2 An aeroplane flies at a constant speed v and has a range of action (out and home) R in calm weather. Prove that in a north wind of constant speed w its range of action is

$$\frac{R(v^2 - w^2)}{v(v^2 - w^2 \sin^2 \phi)^{1/2}}$$

in a direction inclined at an angle ϕ to the north. In what direction is the range a maximum?
(Ans.: $\phi = \pi/2$.)

2.3 On board a ship travelling due east the wind appears to blow from an angle α west of north. Show that if the actual direction of the wind is θ west of north,

$$v(\mathbf{i} \sin \alpha - \mathbf{j} \cos \alpha) = w(\mathbf{i} \sin \theta - \mathbf{j} \cos \theta) - u\mathbf{i},$$

where $\mathbf{i}, \mathbf{j}$ are unit vectors pointing east and north respectively, u and w are the true speeds of the ship and the wind respectively, and $\mathbf{v}$ is the velocity of the wind relative to the ship.

When the ship turns and travels due north (at the same speed) the wind appears to come from a direction β west of north. Write down a similar equation and hence show that

$$\tan \theta = \frac{\tan \alpha - 1}{1 - \cot \beta}.$$

Find the ratio of the speeds of the ship and the wind in terms of α and β.
(Ans.: $u/w = (\tan \alpha \cot \beta - 1)\{(\tan \alpha - 1)^2 + (1 - \cot \beta)^2\}^{-1/2}$.)

2.4 A torpedo, which drives itself through the water at a constant speed u, is fired from O at an enemy ship P whose constant speed through the water is v. At the moment of firing the direction of motion of the enemy ship makes an angle ϕ with OP (when the ship is moving directly away $\phi = 0$). Prove that, in order to allow for the motion of the enemy ship, the axis of the torpedo must be aimed ahead of the ship by an angle δ, where

$\sin \delta = v \sin \phi/u$. Prove that this result is unaffected by a uniform tidal current.

Prove that relative to the sea bed the locus of positions of torpedos fired simultaneously in all horizontal directions from a fixed point is a circle whose radius is independent of the tidal current. Find the centre of the circle.

2.5 A boat, which can travel at speed u in still water, makes a trip from point A to point B and back again. Because of the tide the shortest time for the journey is achieved with a true speed v_o for the outward journey and v_r for the return journey. Show that if w is the speed of the tide,

$$u^2 - w^2 = v_o v_r.$$

2.6 A river, of uniform width d, flows along a straight channel and its speed at a distance y from the bank is $u(y)$. A boat travels from a point O ($x = y = 0$) on one bank of the river to the opposite bank. The speed of the boat relative to the current is v, and the boat is set at an angle θ to the upstream direction.

Show that the velocity $\mathbf{w}$ of the boat is given by

$$\mathbf{w} = \mathbf{i}(u - v \cos \theta) + \mathbf{j}v \sin \theta$$

relative to a coordinate system based on O with Ox in the downstream direction.

(i) Show that if the boat travels directly across the river, so that $\mathbf{w} = w\mathbf{j}$, the time taken to reach the opposite bank is given by

$$\int_0^d \frac{\mathrm{d}y}{(v^2 - u^2)^{1/2}}.$$

If, further,

$$v = \text{constant},$$

$$u = \begin{cases} vy/d & \text{for} \quad 0 \leqslant y \leqslant \tfrac{1}{2}d, \\ v(d-y)/d & \text{for} \quad \tfrac{1}{2}d \leqslant y \leqslant d, \end{cases}$$

show that the time is $\pi d/(3v)$.

(ii) Show that if $\theta = \text{constant}$ ($<\pi/2$), $v = \text{constant}$, $u = ay(d - y)$, where a is a positive constant, when the boat has reached the opposite bank it has travelled a distance downstream equal to

$$\frac{ad^3}{6v \sin \theta} - d \cot \theta.$$

On the return journey the boat is set at a constant angle ϕ with the upstream direction. Prove that it returns to the original point of departure provided that

$$6v > ad^2, \quad \tan\tfrac{1}{2}\phi = \frac{6v - ad^2}{6v + ad^2}\cot\tfrac{1}{2}\theta.$$

(iii) Show that if both u and v are constant, and if the boat is always directed towards a point O_1 on the opposite bank directly across from the starting point O, the differential equation for the path is

$$\frac{dx}{dy_1} = \frac{x - \alpha(x^2 + y_1^2)^{1/2}}{y_1}$$

where $\alpha = u/v$, and $y_1 = d - y$.

Solve the differential equation with the substitution $x = y_1 z$ and deduce the condition for the boat to reach the opposite bank.
(Ans.: $x + (x^2 + y_1^2)^{1/2} = d^\alpha y_1^{1-\alpha}, \alpha < 1$.)

2.7 A rod PQ of length l moves with its ends P and Q on two straight lines OA, OB inclined to each other at an angle α. The velocities of P and Q are of magnitude u and v and in the directions OA, OB respectively. Show that $u\cos\beta = -v\cos\gamma$, where β and γ are the angles OPQ and OQP.

Show also that if the motion of a particular point R of the rod is wholly along the rod, then

$$PR = l\sin\beta\cos\gamma/\sin\alpha.$$

2.8 Two particles P, Q describe, at constant rates, concentric circles with centre at O and radii a and $2a$ respectively. At time $t = 0$, P and Q lie on the x-axis in the order O, P, Q. Show that, if the times taken to describe a complete circle are $\pi/2\omega$ and $2\pi/\omega$ for P and Q respectively, the velocity of P relative to Q will be parallel to PQ after time t such that $\cos 3\omega t = 4/5$.

2.9 A ladder AB is in contact with a horizontal floor OA and a vertical wall OB. The foot A of the ladder is pulled away from the wall with constant speed u, so that $OA = ut$ at time t. Show that the mid point of the ladder describes an arc of a circle with speed $\tfrac{1}{2}u$ AB/OB.

2.10 Particles $P_1, P_2, \ldots, P_n$ are arranged in that order and symmetrically placed around the circumference of a circle of radius a. For time $t \geqslant 0$ each particle moves in such a way that the motion of P_j is always directed towards the adjacent particle P_{j+1} if $j < n$ or P_1 if $j = n$. The speed of each particle is constant and equal to u.

What is the path of P_j and how long does it take before the particles all collide at the centre of the circle?
(Ans.: $r = a\exp\{-\theta\tan(\pi/n)\}$, $a/\{u\sin(\pi/n)\}$.)

2.11 In the rectilinear motion of a point the time t and position x satisfy the equation $t = ax^2 + bx + c$, where a, b and c are constants. Prove that: (i) the velocity is given by $(2ax + b)^{-1}$; (ii) the acceleration is inversely proportional to the cube of the distance from the point $x = -b/2a$.

2.12 A particle moves in a straight line with an acceleration $a - bx$, where x is the distance travelled and a and b are positive constants. Show that if it starts from rest at the origin $x = 0$, it next comes to rest when $x = 2a/b$ after a time $\pi/b^{1/2}$

2.13 A particle moves with constant speed u along the spiral curve $r = ae^{-k\theta}$, where a and k are positive constants. Show that if the particle starts with $\theta > 0$ at time $t = 0$,

$$e^{-k\theta} = 1 - \frac{kut}{a(1+k^2)^{1/2}}.$$

Calculate the components of the acceleration vector as functions of t and show that this vector makes a constant angle with the radius vector.

2.14 A crank OQ revolves about a fixed point O with constant angular speed ω, and a connecting rod QP is hinged to it at Q, whilst P is constrained to move along a straight line through O. Prove that if OQ $= a$, QP $= b$, and θ is the angle QOP, the acceleration of P is

$$a\left(\cos\theta + \frac{a}{b}\cos 2\theta\right)\omega^2$$

approximately, if a/b is small.

2.15 A particle describes the curve $r = ae^{\theta}$, where a is a positive constant and (r, θ) are polar coordinates, in such a way that the radial component of acceleration is equal to a quarter of the transverse component. At time $t = 0$ the particle is at $\theta = 0$ and moving with speed u in such a way that θ is increasing. Show that

$$2^{1/2}a\dot{\theta} = ue^{2\theta/3}$$

Hence show that, as $r \to \infty$, $t \to 3a/2^{1/2}u$.

2.16 A particle, initially at rest at the origin, moves in a plane with an acceleration that is the resultant of a radial acceleration a outwards, and an acceleration in a direction perpendicular to the velocity equal to 2ω times the speed, where a and ω are positive constants.

Show that its angular speed about the origin is equal to ω, and that the polar equation of its path, taking the direction of the acceleration at the origin as the initial line, is $\omega^2 r = a(1 - \cos\theta)$.

2.17 A particle P_1 is projected from a point O with speed u and thereafter moves along a straight line with a constant acceleration a. One second later a particle P_2 is projected from O with speed $\frac{1}{2}u$ and it then follows P_2, along the same straight line, with acceleration $2a$. When P_2 overtakes P_1 their speeds are 31 and 22 m s^{-1} respectively. Prove that the distance traversed is 48 m.

2.18 A particle moves with constant acceleration in a straight line. Explain what is meant by the mean velocity with respect to: (i) distance, (ii) time, and show that the difference between the means is equal to

$$\frac{(v-u)^2}{6(v+u)},$$

where u and v are the initial and final speeds.

2.19 The maximum acceleration of a train is a_1, its maximum retardation is a_2, and its maximum speed is u. Show that it cannot run a distance d from rest to rest in a time less than

$$\left(\frac{2d(a_1+a_2)}{a_1a_2}\right)^{1/2} \quad \text{if } d \leqslant \frac{u^2(a_1+a_2)}{2a_1a_2}$$

or less than

$$\frac{u(a_1+a_2)}{2a_1a_2}+\frac{d}{u} \quad \text{if } d \geqslant \frac{u^2(a_1+a_2)}{2a_1a_2}.$$

2.20 The depth of a well is to be measured by dropping in a stone and observing the time T which then elapses before the impact at the bottom is heard. Prove that the depth of the well is given by the smaller root of the equation

$$gx^2 - 2x(c^2 + gcT) + gc^2T^2 = 0,$$

where c is the speed of sound in air. Why is the larger root not appropriate?

Show that the depth of the well is approximately $\frac{1}{2}gT^2(1 - gT/c)$, if $gT \ll c$.

Given that $T = 1{\cdot}5$ s and $c = 330$ m s^{-1} show that the small correction term alters the result by 4·4%.

2.21 A projectile is fired from ground level at A with the least speed needed to clear an obstacle of height h at a horizontal distance a from A. Show that if it reaches the ground at B, where AB $= a + b$, then

$$a^2h^2 = b^2(a^2 + h^2).$$

2.22 A fort and a ship are both armed with guns which give their projectiles a muzzle velocity $(2ga)^{1/2}$, and the guns in the fort are at a height h above the guns in the ship. The greatest (horizontal) ranges at which the fort and ship can engage are d_1 and d_2 respectively. Show that

$$\frac{d_1}{d_2} = \left(\frac{a+h}{a-h}\right)^{1/2}.$$

2.23 A particle is projected at time $t = 0$ in a vertical plane from a given point with given velocity $(2ga)^{1/2}$, of which the upward vertical component is v. Show that at time $t = 2a/v$ the particle is on a fixed parabola (independent of v), that its path touches the parabola, and that its direction of motion is then perpendicular to its direction of projection.

2.24 A particle is projected with velocity $(2ga)^{1/2}$ from a point at a height h above a plane of inclination β. Prove that the maximum ranges up and down the plane are increased by

$$2a \sec^2 \beta \left\{ \left(1 + \frac{h \cos^2 \beta}{a}\right)^{1/2} - 1 \right\}.$$

relative to the ranges when $h = 0$.

Find a first approximation to this result when $h \cos^2 \beta \ll a$.

2.25 Two projectiles P_1, P_2 moving under gravity have velocities $\mathbf{v}_1$, $\mathbf{v}_2$. At one stage in their motion $\mathbf{v}_1 = k\overline{P_1 Q}$ and $\mathbf{v}_2 = k\overline{P_2 Q}$, where k is a positive constant. Show that they will eventually collide.

2.26 A particle is projected under gravity from a point O with speed u at an inclination α to the horizontal. The polar coordinates of the particle, relative to O, are (r, θ), where θ is the inclination of the radius vector $\mathbf{r}$ to the horizontal. Show that the polar equation of the trajectory is

$$r = \frac{2u^2 \cos \alpha \sin(\alpha - \theta)}{g \cos^2 \theta}.$$

Show that r increases monotonically if $\cos \alpha > \frac{1}{3}$. If $\cos \alpha < \frac{1}{3}$ show that r decreases as θ decreases through the range

$$\tfrac{1}{2}\alpha + \tfrac{1}{2} \cos^{-1}(3 \cos \alpha) > \theta > \tfrac{1}{2}\alpha - \tfrac{1}{2} \cos^{-1}(3 \cos \alpha).$$

For the critical case

$$\cos \alpha = \tfrac{1}{3} \quad (\alpha = 1{\cdot}23 \text{ rad}), \qquad \theta = \alpha/2$$

show that r is less than the horizontal range.

Show that the first stationary value of r will become equal to the horizontal range when

$$\tan \alpha = 2^{-1/2}(5^{3/2} + 11)^{1/2}, \quad (\alpha = 1{\cdot}28 \text{ rad}).$$

2.27 A projectile is fired from O with initial speed u at a target in the same horizontal plane as O. Show that if a small error ϵ radians is made in the angle of elevation, and an error 2ϵ in azimuth, the shot will strike the plane at a distance from the mark $2u^2 \epsilon / g$.

Show also that if the angle of elevation α is such that $\tan 2\alpha < 2 (\alpha < 0{\cdot}55 \text{ rad})$ an error in elevation will cause the shot to miss the target by an amount greater than would be caused by an equal error in azimuth.

2.28 Mud is thrown off from all points of a tyre of radius a, which is rolling along the ground with constant velocity of magnitude u. Show that, if

$u^2 \geqslant ga$, none of the mud reaches a height greater than

$$a + \tfrac{1}{2}a^2 g/u^2 + \tfrac{1}{2}u^2/g.$$

Discuss also the case $u^2 < ag$.

2.29 A particle moves in a plane, its coordinates at time t referred to rectangular axes in the plane being (x, y). The component acceleration in the direction Oy has the constant value $-g$, and the component acceleration in the direction Ox is $k(y - h)$, where k and h are positive constants. The particle is projected from the origin in the direction Oy with speed $(3gh)^{1/2}$. Discuss the motion, and prove that the path of the particle is a parabola.

2.30 Given that the periodic time for the swing of a pendulum depends only on its length l, the magnitude of the gravitational acceleration g and the maximum angular displacement α, use a dimensional argument to deduce that the periodic time is given by

$$t_0\left(\frac{lg_0}{l_0 g}\right)^{1/2} \frac{F(\alpha)}{F(\alpha_0)}.$$

Here l_0, g_0, α_0 are a particular combination of l, g, α for which the periodic time is t_0, and F is some function of α.

3 Dynamics

3.1 NEWTON'S LAWS

We now introduce the laws of motion first postulated by Newton, which form the basis of the whole of classical mechanics. We shall first state the laws in the form in which they are usually quoted, and then proceed to discuss them.

1. Every body continues in its state of rest or of uniform motion in a straight line, unless it is compelled to change that state by external impressed forces.
2. The rate of change of momentum is proportional to the impressed force, and takes place in the direction of the straight line in which the force acts.
3. To every action there is an equal and opposite reaction.

It should be noted that the above three statements are formulated in terms of certain concepts, namely force, momentum, action, which it is necessary to define before the significance of the statements can be appreciated. It will appear that their definitions depend, in turn, on the additional concept of mass, this being the third and last fundamental concept of mechanics, after length and time.

The first law associates the notion of force with that of acceleration, and attaches significance to the state of zero acceleration. But before any statement can be made about the acceleration of a particle a reference system is required, since a particle may have zero acceleration relative to one set of axes, but not relative to a different set. There is therefore an implication in this first law that certain axes are preferential. This is connected with the belief that the appropriate way to describe the acceleration of a particle is in terms of the influence of other bodies, and that if a particle were isolated and free from any external influence it would move with zero acceleration. A preferential, or inertial, set of axes should be chosen to be consistent with such a motion. These axes, though not arbitrary, are also not unique, for axes moving with constant velocity relative to an inertial set also constitute an inertial set. The laws of mechanics are in fact identical in the two frames, as will subsequently appear. This is called the Galilean relativity principle.

As far as the idealised mathematical model is concerned, it is quite legitimate to postulate the existence of inertial frames of reference. The theory, however, does not explain how such a frame is to be chosen in practical applications, since we are not able to isolate a particle from the influence of other bodies. Here we must appeal to our experience to decide what is reasonable. Thus a system of axes fixed on the earth's surface is adequate for many problems, in the sense that a theory of Newtonian mechanics using these axes gives results which compare satisfactorily with observation. Strictly speaking, however, a point on the earth's surface does have an acceleration because the earth rotates about its polar axis. At a latitude ϕ the acceleration is $\omega_e^2 r_e \cos\phi$ towards the polar axis, where r_e, ω_e are respectively the radius of the earth and its angular speed of rotation. Since, approximately, $r_e = 6{\cdot}4 \times 10^6$ m and $\omega_e = 2\pi$ radians per day, we have

$$\omega_e^2 r_e \cos\phi = (2\pi)^2 6{\cdot}4 \times 10^6 \cos\phi \text{ m d}^{-2} = 3{\cdot}4 \times 10^{-2} \cos\phi \text{ m s}^{-2}.$$

This acceleration is small compared with the gravitational acceleration of $9{\cdot}81$ m s^{-2} at the earth's surface. It can, if necessary, be allowed for in the equations of motion. Otherwise a different set of axes must be chosen and certainly, foı a discussion of motion in the solar system, the simplest satisfactory modification is to choose a system of axes with origin at the centre of the sun and directions fixed relative to the 'fixed' stars. More sophisticated problems may require a more sophisticated set of axes. In all cases the ultimate justification is through a comparison of the resulting theory with observation.

To proceed with the theory, we envisage a region in which a single particle, when isolated, has zero acceleration. Into this region we introduce two particles P_0 and P_1 with position vectors $\mathbf{r}_0$ and $\mathbf{r}_1$, each particle being influenced by the other, but by no other mechanism. We then assert that the vector accelerations $\ddot{\mathbf{r}}_0$, $\ddot{\mathbf{r}}_1$ are in the line P_0P_1 and are opposite in sense, and that the ratio $|\ddot{\mathbf{r}}_1| / |\ddot{\mathbf{r}}_0|$ is invariant for different positions, different velocities, and different mechanisms of influence (gravitational, electrical, connection by a spring, and so on).

Choose a number m_0 and associate it with the particle P_0. Associate with P_1 a number m_1 determined by the relation

$$m_1 = m_0 \, |\ddot{\mathbf{r}}_0| / |\ddot{\mathbf{r}}_1| \tag{3.1.1}$$

so that

$$m_0\ddot{\mathbf{r}}_0 = -m_1\ddot{\mathbf{r}}_1. \tag{3.1.2}$$

In the same way, by the substitution of any other particle P_2 for P_1, we can determine a number m_2 associated with P_2, which is invariant for all reactions with P_0.

We further assert that the result is transitive, so that if P_1 and P_2 are chosen as the two interacting particles, then

$$-m_1\ddot{\mathbf{r}}_1 = m_2\ddot{\mathbf{r}}_2.$$

It follows similarly that there is associated with any particle P a number m which is characteristic of P in its interactions with other particles. It is called the mass of P. The above procedure explains how it can be measured once the mass m_0 of the reference particle P_0 has been fixed. Then for any two particles the vectors like $m\ddot{\mathbf{r}}$ associated with the particles are equal and opposite under all conditions in which the particles interact. We define the quantity $m\ddot{\mathbf{r}}$ associated with P to be the force **F** acting on P and write

$$\mathbf{F} = m\ddot{\mathbf{r}}. \qquad (3.1.3)$$

We also define the momentum of the particle to be $m\dot{\mathbf{r}}$, so that (3.1.3) is equivalent to Newton's second law. However, it has been presented here as a definition, and not as a relation which is to be justified experimentally.

So far, the important dynamical postulate is expressed by the relation (3.1.2), which is the equivalent of Newton's third law. A relation such as (3.1.3) does not add to the postulates, but simply recognises the significance of the quantity $m\ddot{\mathbf{r}}$ in the interactions between particles. However, before (3.1.3) can be used to make predictions it is necessary to formulate further empirical laws, again based on experience, which describe how **F** varies when a particle is subject to the various mechanisms appearing in any given problem. In other words an independent rule is required for the calculation of the force acting on the particle. One example is the law of gravitation, which says that the force on a particle P_1 arising from the gravitational attraction between P_1 and P_2 is

$$\frac{Gm_1m_2(\mathbf{r}_2 - \mathbf{r}_1)}{|\mathbf{r}_2 - \mathbf{r}_1|^3},$$

where G is the gravitational constant. By Newton's third law, there is an equal and opposite force on P_2. When this information is combined with (3.1.3) we get the pair of differential equations

$$m_1\ddot{\mathbf{r}}_1 = -m_2\ddot{\mathbf{r}}_2 = \frac{Gm_1m_2(\mathbf{r}_2 - \mathbf{r}_1)}{|\mathbf{r}_2 - \mathbf{r}_1|^3}.$$

These can now be integrated to give $\mathbf{r}_1$ and $\mathbf{r}_2$ as functions of the time, and hence to predict the motion of the particles if the appropriate initial conditions are given. There is a similar rule for the electrical force between charged particles. For particles connected to a spring or elastic string there is a different rule which says that the force is proportional to the extension of the spring (see Chapter 6). Whenever a new mechanism is introduced a corresponding rule will always be required before the calculations can proceed.

If the theory is to be generally applicable it is necessary to consider situations more complex than the interaction between two particles. A particle may, for example, be subjected to several forces (measured by the accelerations they would produce if acting separately). We then assert that the resultant force, obtained from the vector sum of the individual forces, is equivalent to the product of the mass and the acceleration of the particle. This is the postulate of the independence of forces. It enables us to use the empirical laws to calculate the forces on a given particle, irrespective of the fact that there may be other factors influencing its motion. We may also speak of an unaccelerated particle being subject to several forces whose vector sum is zero.

Finally, for those who find it difficult to reconcile the idealised assertions which have been made with their intuitive idea of what actually happens in practice, the following quotation is the answer given by H. Lamb (*Dynamics* (1923) Cambridge University Press).

> 'The laws which are to be imposed on these ideal representations are in the first instance largely at our choice, since we are dealing now with mental objects. Any scheme of abstract dynamics constructed in this way, provided it be self-consistent, is mathematically legitimate; but from the physical point of view we require that it should help us to picture the sequence of phenomena as they actually occur. The success or failure in this respect can only be judged *a posteriori*, by comparison of the results to which it leads with the facts. It is to be noticed, moreover, that available tests apply only to the scheme as a whole, since, owing to the complexity of real phenomena, we cannot subject any one of its postulates to verification apart from the rest.'

3.2 UNITS

The unit of mass, as for all fundamental units, is a matter of choice. In the SI system it is the kilogram (kg), which is 10^3 gram (g), defined in terms of a material standard kept in Paris. In dynamics there is no particular reason for the preference of the kilogram over the original unit of the gram, apart perhaps from the fact that it tends to be more useful in everyday life. For the theoretical reason for choosing the kilogram, reference must be made to the complete system of SI units, which includes those used in electromagnetism. In this discussion it is used simply because it is internationally acceptable.

Force is here a derived concept with the dimensions of mass times acceleration, or MLT^{-2}. The unit in the SI system is the newton (N), which is the force represented by 1 kg having an acceleration of 1 m s^{-2}.

Since it is observed that the attraction of the earth gives all bodies an acceleration $\mathbf{g}$, it follows that the associated force of attraction is $m\mathbf{g}$. This will vary from place to place because of the variation of $\mathbf{g}$, but at a given place $\mathbf{g}$ is constant and the mass of a body is proportional to the gravitational force, called the weight (wt) of the body. Since the weights of two bodies can be compared on a balance, this also gives a practical means of comparing masses. In fact the force of magnitude mg newtons is sometimes referred to as m kilogram weight (kg-wt), but it should always be kept in mind that it has the dimensions MLT^{-2} and refers to the weight of a mass m kg. In problems possible confusion is best avoided by working entirely in terms of the force mg newtons rather than m kg-wt.

3.3 APPLICATIONS

In the following examples certain assumptions and idealisations are made, the chief of which are discussed below.

Some bodies, which are manifestly of finite size, can be treated as if their equations of motion are identical with those of a particle subject to the same forces. Strictly speaking this requires a study of rigid body dynamics, but at this stage such treatment may be accepted, at least for one-dimensional motion or where the size is clearly not an important parameter.

When two bodies are in contact they exert, by Newton's third law, equal and opposite forces on each other, called contact forces. A contact force may vary according to the dynamical situation, but when bodies are smooth this force is always perpendicular to the plane of contact of two such bodies. This is an idealisation which neglects the frictional force (see Chapter 4).

When two particles are connected by an inelastic string, the string transmits a force, called the tension of the string. Consider an element δs of a straight string moving parallel to itself with acceleration a (see Fig. 3.3.1). The contact forces between δs and the rest of the string are $T(s + \delta s)$ and $T(s)$, say, so that the resultant force on δs is

$$T(s + \delta s) - T(s) = \rho \delta s \, a,$$

where ρ is the mass per unit length of the string. Division by δs gives, in the limit as $\delta s \to 0$,

$$\mathrm{d}T/\mathrm{d}s = \rho a. \tag{3.3.1}$$

If the string passes round a smooth peg or pulley this result is not affected since the contact force between the string and the pulley is perpendicular to the string and hence does not affect the equation of motion along the string.

There is an extra term if the string is subject also to a gravitational force. For example, for a string pulled vertically upwards with acceleration a, equation (3.3.1) is modified to

$$\frac{\mathrm{d}T}{\mathrm{d}s} - \rho g = \rho a. \tag{3.3.2}$$

A light string is an idealisation in which the mass of the string can be neglected. With $\rho = 0$ we have $\mathrm{d}T/\mathrm{d}s = 0$ and the tension is constant throughout the string in this case.

As a more general observation, we do not attempt to describe a dynamical problem in all its detail. Certain features are deemed to be important and hence

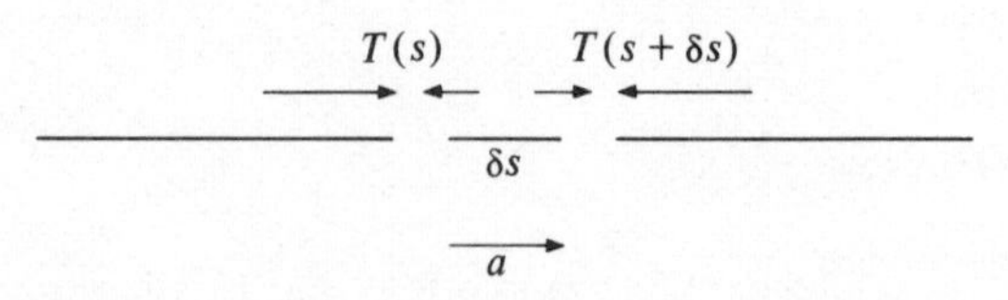

Figure 3.3.1 The forces on an element δs of a string.

to be emphasised, others are regarded as extraneous and ignored, or postponed to be discussed at a later stage. To achieve this some idealisation is required; indeed provided this is done with discretion, it is desirable and perhaps even necessary in order to describe the problem in reasonably simple terms.

Example 3.3.1

A particle, of mass m_1, is free to slide on the inclined face of a smooth wedge, of mass m_2 and angle α. The wedge is itself free to slide on a smooth horizontal plane. Find the acceleration of the particle and the wedge.

Let the acceleration of the particle relative to the wedge be a_1 down the face of the wedge, and let the acceleration of the wedge be a_2, as shown in Figure 3.3.2.

The forces consist of the normal reaction N_1 between the particle and the wedge, which is a mutual force perpendicular to the wedge face, the reaction N_2 between the wedge and the plane, and the gravitational forces on both the wedge and the particle.

The equations of motion, when resolved horizontally and vertically, are, for the wedge,

$$N_1 \sin \alpha = m_2 a_2,$$

$$m_2 g - N_2 + N_1 \cos \alpha = 0,$$

and, for the particle,

$$N_1 \sin \alpha = m_1(a_1 \cos \alpha - a_2),$$

$$m_1 g - N_1 \cos \alpha = m_1 a_1 \sin \alpha.$$

The four equations are to be solved for the four unknowns a_1, a_2, N_1 and N_2. In particular

$$a_1 = \frac{(m_1 + m_2)g \sin \alpha}{m_1 \sin^2 \alpha + m_2}, \tag{3.3.3}$$

$$a_2 = \frac{m_1 g \sin \alpha \cos \alpha}{m_1 \sin^2 \alpha + m_2}. \tag{3.3.4}$$

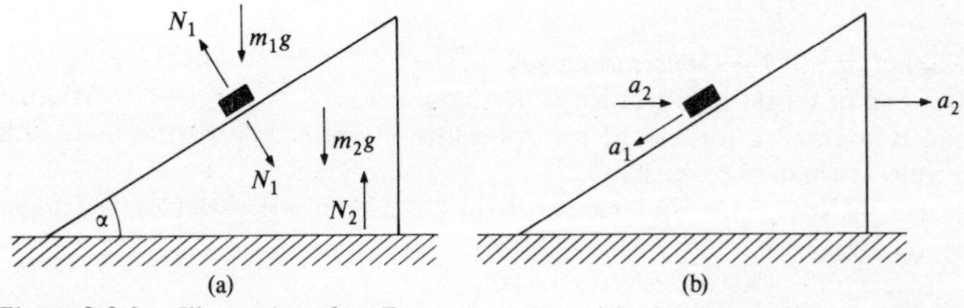

Figure 3.3.2 Illustration for Example 3.3.1: (a) the forces; (b) the accelerations.

Example 3.3.2
A light inextensible string passes over a smooth pulley and particles, of mass m_1 and m_2 respectively, are suspended from the ends of the string. The pulley has a constant acceleration a_0 vertically upwards. Find the accelerations of the masses.

The accelerations of the particles relative to the pulley are equal and opposite since this is true of their displacements. Let these be $\pm a$ so that m_1 has an acceleration $a_0 + a$ and m_2 has an acceleration $a_0 - a$ as shown in Figure 3.3.3.

The forces involved are the tension in the string, which is constant if the mass of the string is neglected, and the gravitational forces.

The equations of motion are

$$(T - m_1 g) = m_1(a_0 + a), \quad (T - m_2 g) = m_2(a_0 - a).$$

Elimination of T gives

$$a = \frac{(m_2 - m_1)(a_0 + g)}{m_1 + m_2}. \tag{3.3.5}$$

from which the accelerations $a_0 + a$ of m_1 and $a_0 - a$ of m_2 follow.

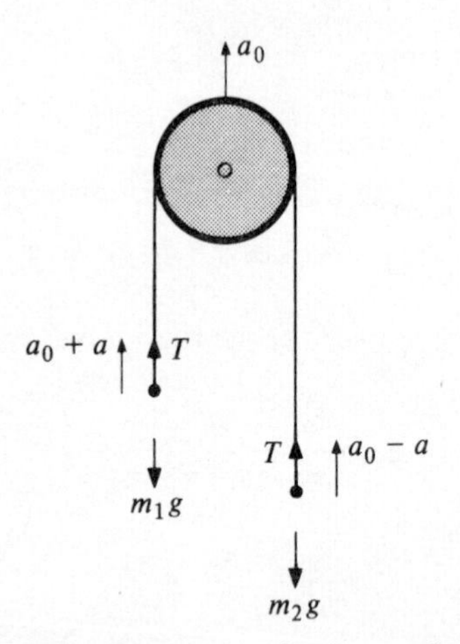

Figure 3.3.3 Illustration for Example 3.3.2.

Example 3.3.3. The Conical Pendulum
One end of a light string, of length l, is attached to a fixed point. To the other end is attached a particle, of mass m, which describes a horizontal circle with angular speed ω (see Fig. 3.3.4).

Let the string make an angle α with the downward vertical and let the tension in the string be T.

There is no vertical acceleration, and so

$$T \cos \alpha = mg.$$

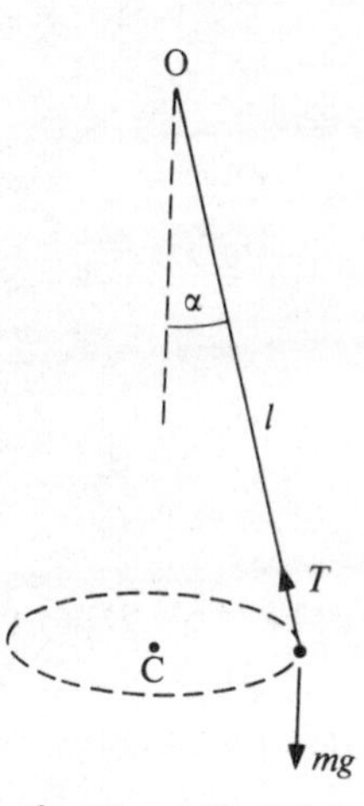

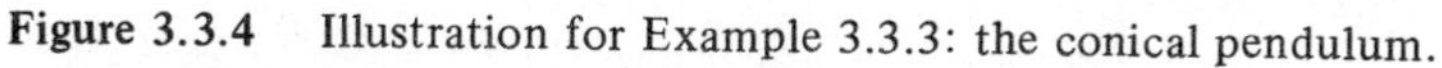

Figure 3.3.4 Illustration for Example 3.3.3: the conical pendulum.

In the horizontal plane there is no component of force tangential to the path of the particle, and since the radius of the path is constant this means that ω is constant.

Along the radius of the particle's path, we have

$$T \sin \alpha = m\omega^2 l \sin \alpha$$

and so

$$mg \sec \alpha = m\omega^2 l$$

or

$$\omega^2 = \frac{g \sec \alpha}{l} = \frac{g}{\mathrm{OC}}, \tag{3.3.6}$$

where OC is the depth of the particle below the fixed point. The time for one complete revolution is

$$\frac{2\pi}{\omega} = 2\pi\left(\frac{\mathrm{OC}}{g}\right)^{1/2}. \tag{3.3.7}$$

Note that ω cannot be less than $(g/l)^{1/2}$.

Example 3.3.4

A particle is slightly displaced from rest in its position of equilibrium at the highest point of a smooth fixed sphere of radius a. Determine the point where the particle leaves the sphere.

When the radius vector from the centre of the sphere to the particle makes an angle θ with the vertical as in Figure 3.3.5, the components of acceleration along and perpendicular to the radius vector are $(-a\dot{\theta}^2, a\ddot{\theta})$. If the normal reaction

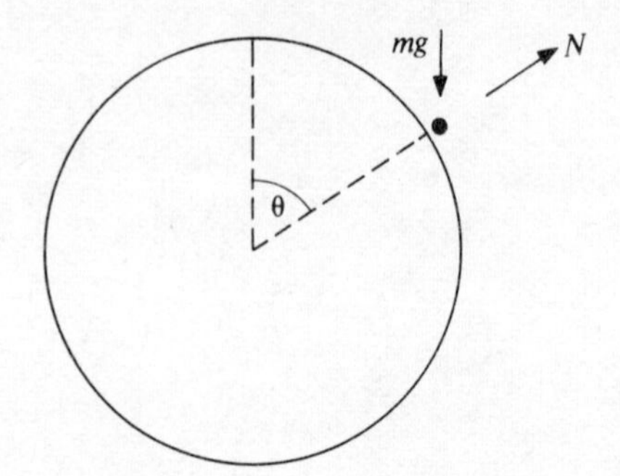

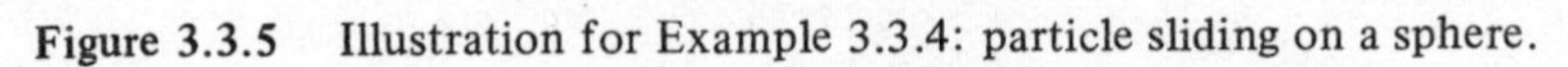

Figure 3.3.5 Illustration for Example 3.3.4: particle sliding on a sphere.

between the sphere and the particle is N, the equations of motion are

$$mg \cos \theta - N = ma\dot{\theta}^2, \quad mg \sin \theta = ma\ddot{\theta}.$$

If the second equation is multiplied by $\dot{\theta}$ it can be integrated to give

$$g(1 - \cos \theta) = \tfrac{1}{2}a\dot{\theta}^2 \tag{3.3.8}$$

since $\dot{\theta} = 0$ when $\theta = 0$. Then

$$\begin{aligned} N &= mg \cos \theta - 2mg(1 - \cos \theta) \\ &= mg(3 \cos \theta - 2). \end{aligned} \tag{3.3.9}$$

Since N cannot become negative, the particle will lose contact with the sphere when $N = 0$, that is when $\cos \theta = 2/3$. It will then begin to move freely under gravity.

Example 3.3.5

A bead is free to slide on a smooth, straight, horizontal wire which rotates with constant angular speed ω about a vertical axis through a point O of the wire (see Fig. 3.3.6).

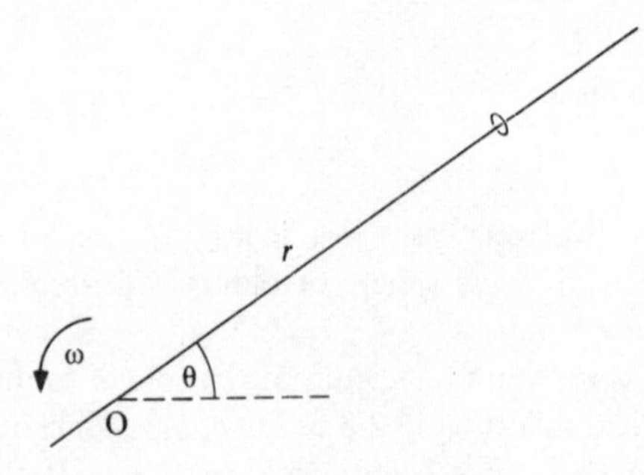

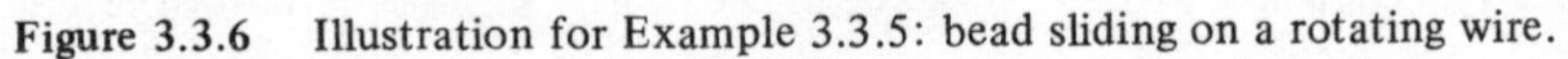

Figure 3.3.6 Illustration for Example 3.3.5: bead sliding on a rotating wire.

We discuss the displacement of the bead as a function of the time.

Since the wire is smooth and horizontal, there is no force on the bead along the length of the wire. Hence the component of acceleration in this direction is zero. Let (r, θ) be the polar coordinates of the particle relative to 0. Then $\dot{\theta} = \omega$ and the radial acceleration is

$$\ddot{r} - r\omega^2 = 0. \tag{3.3.10}$$

This is a second order linear differential equation with constant coefficients. The solutions of such equations are, in general, of exponential type, possibly with complex exponents, and so are of the form $r = Ae^{\lambda t}$, where A and λ are constants. Substitution in (3.3.10) shows that

$$\lambda^2 - \omega^2 = 0,$$

which gives the two solutions $A_1 e^{\omega t}, A_2 e^{-\omega t}$. For a linear differential equation, such solutions can be combined linearly to give the general solution

$$r = A_1 e^{\omega t} + A_2 e^{-\omega t}.$$

The arbitrary constants A_1 and A_2 will be determined by appropriate initial conditions. For example, if $r = 0, \dot{r} = u$ when $t = 0$, the displacement at time t is given by

$$r = \frac{u}{2\omega}(e^{\omega t} - e^{-\omega t}) = \frac{u}{\omega} \sinh \omega t.$$

EXERCISES

3.1 A particle of unit mass is subject to the force

$$\mathbf{F} = A\mathbf{i} + Bt\mathbf{j} + Ct^2\mathbf{k},$$

where t is the time and A, B and C are constants. Initially the particle is at rest at the origin. Find its position vector $\mathbf{r}$ at any subsequent time and show that

$$\int \mathbf{F} \,.\, d\mathbf{r} = \tfrac{1}{2}A^2 + \tfrac{1}{8}B^2 + \tfrac{1}{18}C^2$$

where the integral is taken along the path of the particle from the origin to its position after unit time.

3.2 A particle of mass m is moving with constant velocity $\mathbf{u}$ relative to an inertial frame. A second frame of reference has its axes always parallel to those of the inertial frame but its origin has the position vector $a\mathbf{i} + bt\mathbf{j} + ct^2\mathbf{k}$, where a, b and c are constants and t is the time. Find the displacement of the particle relative to the second frame and show that the particle appears to be subject to a force $-2mc\mathbf{k}$.

3.3 A particle rests in equilibrium under the action of three forces. Show that each force is proportional to the sine of the angle between the other two.

3.4 For time $t < 0$ a particle rests at the top of a smooth hemisphere of radius a. The hemisphere rests with its plane face on a horizontal plane. For $t \geqslant 0$ the hemisphere is constrained to move with constant velocity, of magnitude u, along the plane. Show that for $u^2 \leqslant ga$ the particle leaves the hemisphere when

$$3 \cos \theta - 2 - u^2/ga = 0,$$

where θ is the angle between the radius vector to the particle and the upward vertical. Deduce that if $u^2 \geqslant ga$ the particle will hit the horizontal plane at a point vertically below its initial position.

3.5 A man of mass m is holding on to a light rope which passes over a smooth fixed pulley and has a counterpoise of mass m attached to the other end. Initially the system is at rest. Show that if the man begins to climb his portion of the rope the man and the counterpoise move equal distances.

3.6 Over a smooth pulley A, of mass 2 kg, passes a light inelastic string having masses 6 and 9 kg at its ends. Over a smooth fixed pulley B passes a light inelastic string having a mass 15 kg at one end and being attached to the axis of the pulley A at the other end. The system moves so that the parts of the string not in contact with the pulleys move vertically. Show that the acceleration of the 15 kg mass is $0{\cdot}44 \text{ m s}^{-1}$.

3.7 Two particles, of masses m_1, m_2, are joined by a light inelastic string and are initially close together on a smooth table. A smooth ring, of mass m_3, is threaded by the string and hangs over the edge of the table and the portions of the string on the table are at right angles to the edge. Find the acceleration of the ring.
(Ans.: $m_3(m_1 + m_2)\, g/\{4m_1 m_2 + m_3(m_1 + m_2)\}$.)

3.8 A bead slides under gravity, on a smooth straight wire, from rest at a point O to a point X. Show that if X lies on a prescribed circle having O as its highest point, the time of descent is independent of the position of X on the circle. Hence find the straight line of quickest descent from:
(i) a point to a straight line,
(ii) a straight line to a point,
(iii) a point to a circle.

3.9 A smooth cone of angle 2α has its axis vertical. A particle moves in a horizontal circle on the inner surface of the cone at a height h above the vertex. What is the speed of the particle and what is the period of the motion?
(Ans.: $(gh)^{1/2}$, $2\pi(h/g)^{1/2} \tan \alpha$.)

3.10 A light string ABC passes through a small, smooth ring B, and has its ends fixed at two points A, C such that C is vertically below A. If the ring is whirled round so as to describe a horizontal circle with constant angular speed ω, show that

$$2(a + c)bg = \{(c + a)^2 - b^2\}(c - a)\omega^2,$$

where a, b, c denote respectively the lengths of BC, CA, AB.

3.11 A bead can slide freely on a smooth straight wire OA, of length l, which is rotated in a horizontal plane with constant angular speed ω about its end O. Initially the bead is projected along the wire with speed u from O. When the bead leaves the wire, show that the angle between the line of the wire and the direction of motion of the bead is $\tan^{-1}\{l\omega/(u^2+l^2\omega^2)^{1/2}\}$.

3.12 A smooth straight wire OA, of which O is fixed, is made to rotate in a fixed vertical plane with constant angular speed ω. A bead of mass m, which is free to move along the wire, is initially at rest at O when OA points vertically downwards. Show that if r is the distance of the bead from O at time t

$$\ddot{r} - \omega^2 r = g \cos \omega t$$

and hence that

$$r = g(\cosh \omega t - \cos \omega t)/2\omega^2.$$

Show further that the reaction of the wire on the bead is

$$mg(\sinh \omega t + 2 \sin \omega t).$$

3.13 A ring of mass m_1 is free to slide along a fixed horizontal smooth wire. An inelastic string of length l has one end fastened to a fixed point of the wire, is threaded through the ring and has a particle of mass m_2 fastened to its other end. The particle is allowed to fall from an initial position in which the string is taut and horizontal with the ring at its mid point.

Write down the accelerations of the ring and the particle when the portion of the string between the ring and the particle is of length $r(< l)$ and has turned through an angle $\theta(< \pi/2)$ from the horizontal.

Hence show that:

(i) $T(1 - \cos\theta) = m_1\ddot{r}$,
(ii) $g\cos\theta = r\ddot{\theta} + 2\dot{r}\dot{\theta} + \ddot{r}\sin\theta$,
(iii) $m_2 g \sin\theta - T = m_2\ddot{r}(1 - \cos\theta) - m_2 r\dot{\theta}^2$,

where T is the tension in the string.

To a first approximation when θ is small show:

(i) from the first equation that $r = l/2$,
(ii) from the second equation that $\theta = gt^2/l$,
(iii) from the third equation that $T = 3m_2 g\theta$.

Hence from the first equation show that, to a higher approximation,

$$r = \frac{l}{2}\left(1 + \frac{3}{56}\frac{m_2}{m_1}\theta^4\right).$$

3.14 A ring, of mass m_1, is free to slide along a fixed smooth wire inclined at an angle α to the horizontal. An inelastic string has one end fastened to a fixed point of the wire, is threaded through the ring and has a particle of mass m_2 fastened to its other end.

(i) Let the system be released from a position of rest in which the string is taut and that portion of it between the ring and the particle inclined at an angle $\alpha + \theta$ to the horizontal. Show that the initial accelerations of the particle along and perpendicular to the string are $\ddot{r}(1 - \cos\theta)$ and $g\cos(\alpha + \theta)$, where

$$\{m_1 + m_2(1 - \cos\theta)^2\}\ddot{r} = \{m_2 \sin(\alpha + \theta)(1 - \cos\theta) - m_1 \sin\alpha\}g.$$

(ii) Show that a motion is possible in which the ring moves with constant acceleration up the wire provided θ is constant and such that

$$m_1 \cos\alpha \cos\theta + m_2(1 - \cos\theta)\{\cos(\alpha + \theta) - \cos\alpha\} = 0.$$

Show that this acceleration has the same sign as

$$\{m_2 - (m_1 + m_2)\cos\theta\}.$$

3.15 A smooth circular cylinder of radius a is fixed, with its axis vertical, to a smooth, horizontal table so that its plane base is in contact with the table. A light inextensible string of length $2l$ is fastened to a point of the circumference of the base of the cylinder and a length l of the string is wrapped round the cylinder. A particle of mass m is attached to the other end of the string and moves on the table in such a way that the string always remains taut. Initially the particle is given a horizontal velocity u perpendicular to the unwound part of the string in such a direction that the string starts to unwind.

Show that the position vector of the particle relative to the axis of the cylinder at any instant can be expressed as

$$\mathbf{r} = a\mathbf{p} + (l + a\theta)\mathbf{q},$$

where $\mathbf{p}$ and $\mathbf{q}$ are unit vectors along a radius of the cylinder and along the string respectively. Hence prove that the string is completely unwound after a time $3l^2/2au$, and calculate the tension in the string as a function of the time t $(<3l^2/2au)$.
(Ans.: $mu^2(l^2 + 2aut)^{-1/2}$.)

4 Resisting Forces

4.1 SOLID FRICTION

We have seen in the previous chapter that there are mutual reactions between two bodies in contact and that these reactions are equal in magnitude and opposite in sense according to Newton's third law. When the bodies are ideally smooth the reactions are perpendicular to the plane of contact, which is the common tangential plane for bodies of continuous curvature. It is, however, a matter of everyday experience that when one body slides on another there is a force which opposes the relative motion, and that a body at rest in contact with a fixed body requires a certain minimum impressed force before it will move. This means that, in addition to a component N normal to the plane of contact, the reaction $\mathbf{R}$ has a component F in the plane of contact, called the frictional force. Account is taken of the frictional force according to the following idealised laws:

(i) Friction opposes relative motion between surfaces in contact.
(ii) When there is no relative motion, F can take any value up to a certain limiting value, which depends on N and is roughly proportional to N for specific materials. The coefficient of proportionality is called the coefficient of static friction μ_s, so that for equilibrium $F < \mu_s N$.
(iii) When there is relative motion, F/N is roughly equal to a constant, called the coefficient of kinetic friction μ_k. In general it is less than μ_s.

That the frictional resistance is proportional to the load and independent of the area of the sliding surfaces was known to Leonardo da Vinci in the fifteenth century. The explanation of this rule is, however, of more recent origin. Accurate experimental techniques show that even carefully prepared smooth surfaces have in fact corrugations on them which are large on a molecular scale, and true contact between any two surfaces is confined to the highest peaks of these irregularities. The true area of contact is consequently very small and tends to be independent of the apparent area of contact. The local pressure at these isolated points of contact is very high since they have to withstand the total load and this causes a local deformation of the material until the contact area is sufficient to support the total load. For this reason the true contact area tends to be proportional to the load. For example the true area of contact between two nominally flat steel surfaces is of the order of 10^{-8} square metres per kilogram of load.

With many materials, including metals, glass and plastics, the deformation and flow which occur under the intense pressure result in a local welding of the two surfaces, a process which is assisted by the heat generated when there is a relative motion. It is this welding and subsequent shearing that is the source of the frictional force. Since the true area of contact, where the shearing and welding take place, tends to be proportional to the load, so does the frictional force. There are exceptions to the rule. For very hard substances, such as diamond, the true area of contact is not proportional to the load and μ_s decreases as the load increases. A substance such as mica can be molecularly smooth in some circumstances and then the true and apparent areas of contact are comparable. The frictional behaviour is then quite different from that for substances like metals.

The coefficient μ_s varies from one material to another, and even for the same material under different conditions. The degree of roughness of the contact surfaces is clearly a factor, and μ_s tends to decrease to a constant value as the surfaces become smoother. The trend can, however, be dramatically reversed for highly polished surfaces. For example $\mu_s = 1$ for glass under normal conditions, and does not vary appreciably between ground glass and plate glass surfaces. But if the surfaces are optically flat, highly polished and free from all surface impurities, the adhesion can be so great that the glass will fracture before sliding. Similarly two pieces of lead with flat and highly polished surfaces weld together over an appreciable area and seizure occurs on contact, whereas under normal conditions μ_s is about 1·5. Other metals also react in this way, though they may require very careful treatment to achieve substantial adhesion.

The law of proportionality can be violated for some metal surfaces because under normal conditions most metals have a coating of oxide, and μ_s is smaller for the contact of such surfaces than it is for the uncoated metal itself. For example $\mu_s = 0{\cdot}4$ for copper surfaces if $N < 10^{-3}$ kg and 1·6 if $N > 10^{-2}$ kg. Between these values of N there is a continuous transition. The reason is that for light loads friction arises from the oxide coatings, but as the load becomes heavier this coating is gradually broken through by the local high spots on the surfaces so that actual metallic contact increases. If the thickness of the oxide film is deliberately increased the transition is delayed but the range $0{\cdot}4 \leqslant \mu_s \leqslant 1{\cdot}6$ is essentially unchanged. For aluminium, which is a softer metal, $\mu_s = 1{\cdot}2$ and does not vary appreciably over the range 10^{-5} to 1 kg for N since the surface film is easily broken. By contrast $\mu_s = 0{\cdot}4$ for chromium over the same range for N because both the oxide and underlying metal are hard enough to resist deformation and remain intact.

The coefficient of kinetic friction is generally somewhat lower than μ_s, but the difference is not substantial. It arises because the welding of local high spots has ample time to take place when the surfaces are not in relative motion. When they move the welds are continually being broken before they have time to consolidate. However, the source of the frictional force, namely the local welding of the two surfaces, is essentially the same.

A change in the mechanism of friction can, of course, substantially affect the sliding properties. Thus the introduction of a thin film of lubricant can reduce the coefficient of friction between metals by a factor of 5, and is an important consideration in the design of the moving parts of most machines. If the relative

Table 4.1.1 Values of μ_s for some common substances

Metals in contact with the same metal	
flat, highly polished and free of surface contamination	10^2
dry and in air	1
oxidised surfaces which remain intact	0·4
lubricated	0·2
Glass on glass, clean and dry	1
Ice on ice	
0 °C	0·1
−50 °C	0·5
Wood on wood	
clean and dry	0·4
wet	0·2
Brake material on cast iron	
clean and dry	0·4
wet	0·2
greasy	0·1

speed between the surfaces is sufficiently high, frictional heating can result in lubrication by a melting of the contact surfaces themselves. This can occur at quite low speeds for ice on ice, for which $\mu_s = 0{\cdot}1$ at 0 °C and 0·5 at −50 °C, whereas $\mu_k = 0{\cdot}1$ over this range of temperature at a speed of 4 m s^{-1}. Similar decreases occur for metals, but the speeds required for melting are much higher.

The conclusion is that the law of proportionality between F and N is a reasonable approximation if attention is paid to the conditions under which it is being applied. It is not to be used indiscriminately.

Table 4.1.1 gives the values of μ_s for some common substances. It should be interpreted as a rough guide since experiment indicates that the coefficient can vary according to the conditions. The coefficient μ_k is rather less than μ_s and its variation with conditions is more marked.

Example 4.1.1

A particle is at rest on a rough horizontal plane. The plane is slowly tilted until the particle starts to move, and is then kept fixed. If the static and kinetic coefficients of friction are μ_s and μ_k ($<\mu_s$) respectively, find the velocity of the particle after it has travelled a distance d.

Let α be the inclination of the plane to the horizontal (see Fig. 4.1.1). In equilibrium we have

$$mg\cos\alpha - N = 0, \qquad mg\sin\alpha - F = 0.$$

Hence

$$\tan\alpha = \frac{F}{N} = \mu_s$$

when equilibrium is broken.

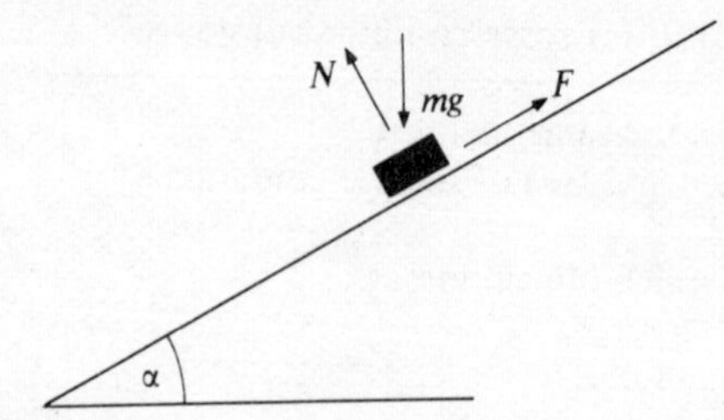

Figure 4.1.1 Illustration for Example 4.1.1.

During the subsequent motion we have

$$mg \cos \alpha - N = 0, \qquad mg \sin \alpha - F = ma.$$

Hence

$$\frac{mg \sin \alpha - ma}{mg \cos \alpha} = \mu_K,$$

and the acceleration is therefore given by

$$\begin{aligned} a &= (\tan \alpha - \mu_K) g \cos \alpha \\ &= \frac{\mu_S - \mu_K}{(1 + \mu_S^2)^{1/2}} g. \end{aligned}$$

Since the acceleration is constant and one-dimensional, the motion is described by equations (2.4.6), (2.4.7), (2.4.8) with the above value of a and zero initial velocity. In particular the speed will be given by

$$v^2 = \frac{2(\mu_S - \mu_K) g d}{(1 + \mu_S^2)^{1/2}}.$$

Example 4.1.2
A rough wedge, of mass m_2 and angle α, is free to move on a horizontal plane. A rough particle, of mass m_1, is free to move on the inclined face of the wedge.

The accelerations and forces are displayed in Figure 4.1.2. The horizontal and vertical components of the equation of motion for the wedge are

$$\left.\begin{aligned} N_1 \sin \alpha - F_1 \cos \alpha - F_2 &= m_2 a_2, \\ m_2 g - N_2 + N_1 \cos \alpha + F_1 \sin \alpha &= 0. \end{aligned}\right\} \qquad (4.1.1)$$

For the particle the equations are

$$\left.\begin{aligned} N_1 \sin \alpha - F_1 \cos \alpha &= m_1 (a_1 \cos \alpha - a_2), \\ m_1 g - N_1 \cos \alpha - F_1 \sin \alpha &= m_1 a_1 \sin \alpha. \end{aligned}\right\} \qquad (4.1.2)$$

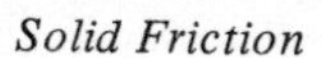

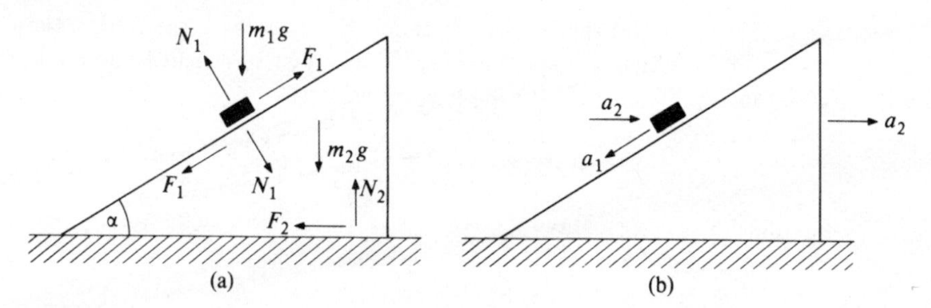

Figure 4.1.2 Illustration for Example 4.1.2: (a) the forces; (b) the accelerations.

There are only four independent relations to be obtained from the equations of motion, although different combinations may simplify the calculations. For example F_1 and N_1 can be eliminated by the substitution of (4.1.2) in (4.1.1) to get

$$\left.\begin{aligned} F_2 &= m_1 a_1 \cos\alpha - (m_1 + m_2)a_2, \\ N_2 &= (m_1 + m_2)g - m_1 a_1 \sin\alpha. \end{aligned}\right\} \tag{4.1.3}$$

In addition (4.1.2) can be solved to give F_1, N_1 in terms of a_1, a_2. The resulting equations are equivalent to resolving the equation of motion of the particle along and perpendicular to the wedge instead of horizontally and vertically. Thus we have

$$\left.\begin{aligned} m_1 g \sin\alpha - F_1 &= m_1(a_1 - a_2 \cos\alpha), \\ m_1 g \cos\alpha - N_1 &= m_1 a_2 \sin\alpha. \end{aligned}\right\} \tag{4.1.4}$$

In the four independent relations (4.1.1), (4.1.2) or (4.1.3), (4.1.4) there are six unknowns, namely F_1, N_1, F_2, N_2, a_1, a_2. Hence two supplementary conditions are required. There are four possibilities to consider:

(i) Both the wedge and the particle are in equilibrium. The two further conditions are $a_1 = a_2 = 0$. Equations (4.1.3) and (4.1.4) then give F_1, N_1, F_2, N_2 immediately.

(ii) The particle slides on the wedge and the wedge remains at rest. Here $a_2 = 0$ and $F_1 = \pm\mu_1 N_1$ are the two further conditions, where μ_1 is given and the sign depends on the direction in which the particle slides. If, for example, it starts from rest, then $F_1 = \mu_1 N_1$ and equations (4.1.4) show that

$$F_1 = \mu_1 N_1 = \mu_1 m_1 g \cos\alpha, \quad a_1 = g(\sin\alpha - \mu_1 \cos\alpha)$$

so that the particle slides with a uniform acceleration. With a_1 and a_2 known, equations (4.1.3) then serve to determine F_2 and N_2.

(iii) The wedge slides and the particle does not slide relative to the wedge. Here $a_1 = 0$ and $F_2 = \pm\mu_2 N_2$. Equations (4.1.3) are then sufficient to determine F_2, N_2 and a_2. If, for example, $F_2 = \mu_2 N_2$ we have

$$F_2 = \mu_2 N_2 = (m_1 + m_2)g, \quad a_2 = -\mu_2 g.$$

Equations (4.1.4) then give

$$F_1 = m_1 g(\sin\alpha - \mu_1 \cos\alpha), \quad N_1 = m_1 g(\cos\alpha + \mu_1 \sin\alpha).$$

(iv) Both the particle and the wedge slide, so that

$$F_1 = \pm\mu_1 N_1, \quad F_2 = \pm\mu_2 N_2$$

give the two further relations.

The coefficients μ_1 and μ_2 refer to the coefficients of kinetic friction at the appropriate points of contact.

Note that in cases (i)–(iii) there will be conditions to be satisfied by the coefficient of static friction arising from the fact that there is no slip. For example in case (i) $\mu_s > \tan\alpha$ for the particle.

Example 4.1.3

A light string is wrapped round a rough circular cylinder and is in a plane at right angles to the generators of the cylinder. Show that when the string is on the point of slipping the tensions at its two ends are in the ratio $e^{\mu_s\theta}$, where θ is the angle between the normals at the two ends of that part of the string in contact with the cylinder.

We consider an element of the string of length δs subtending an angle $\delta\theta$ at the axis of the cylinder. Let the tensions in this element at its two ends be T and $T + \delta T$. Let the normal component of the reaction be $N\delta s$, where N is the

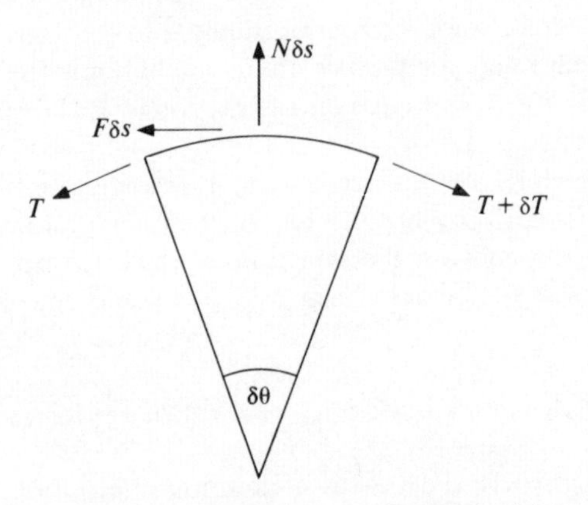

Figure 4.1.3 Illustration for Example 4.1.3.

reaction per unit length, and let the frictional force be $F\delta s$ (see Fig. 4.1.3). Since the string is in equilibrium, the components of the equation of motion along and perpendicular to the normal reaction give

$$(T+\delta T)\cos\tfrac{1}{2}\delta\theta - T\cos\tfrac{1}{2}\delta\theta - F\delta s = 0$$
$$(T+\delta T)\sin\tfrac{1}{2}\delta\theta + T\sin\tfrac{1}{2}\delta\theta - N\delta s = 0.$$

These equations are now divided by δs and considered in the limit as $\delta s \to 0$, and hence $\delta T \to 0$, $\delta\theta \to 0$.

Since $$\frac{\delta T}{\delta s} \to \frac{\mathrm{d}T}{\mathrm{d}s}, \quad \frac{\sin\frac{1}{2}\delta\theta}{\delta s} \to \frac{1}{2}\frac{\mathrm{d}\theta}{\mathrm{d}s}$$

we have that $$\frac{\mathrm{d}T}{\mathrm{d}s} = F, \quad T\frac{\mathrm{d}\theta}{\mathrm{d}s} = N,$$

or $$\frac{1}{T}\frac{\mathrm{d}T}{\mathrm{d}\theta} = \frac{F}{N}.$$

For limiting equilibrium, $F/N = \mu_s$, and the resulting equation integrates to

$$T = T_0 e^{\mu_s \theta},$$

where T_0 is the value of T for $\theta = 0$.

4.2 MOTION THROUGH A FLUID

When a body moves through a fluid, such as air or water, the fluid is set in motion and the body thereby experiences a force. For a particular body and a particular fluid the force depends on the velocity, but a detailed theoretical solution is one of the outstanding unsolved problems of fluid dynamics, although a great deal of experimental information is available. To simplify the analysis we resort to approximate expressions, but it should be kept in mind that there is no adequate universal formula.

We know from the law of dimensions that the force should be represented by a dimensionally correct combination of the parameters on which it depends. Let us consider what these parameters might be. The speed v and the size of the body, which we account for by a representative length l, are clearly involved. We might also expect the density ρ of the fluid to be a factor. It is easily verified that the correct dimensions of force are given by the combination $\rho l^2 v^2$. Moreover there is no non-dimensional combination, since ρ is the only parameter whose measure varies with the mass of the fluid, and v is the only parameter whose measure varies with time. Thus a practically useful representation of the force is $\rho l^2 v^2 \mathbf{k}$, where $\mathbf{k}$ is a non-dimensional vector varying only with possible non-dimensional factors which may influence the motion. As a matter of detail, in aeronautical engineering which is the source of most of the experimental information, it is written $\frac{1}{2}\rho A v^2 \mathbf{C}$, where A is some significant area defined by the body and the reason for the factor $\frac{1}{2}$ need not concern us. Then $\mathbf{C}$ is called the force coefficient. It can vary with the direction of flight, with the orientation of the body and with any non-dimensional shape parameter such as the ratio of two lengths, since one representative length is not sufficient to define a three-dimensional body. Usually $\mathbf{C}$ is resolved into a component in the downstream direction of motion called the drag coefficient C_D and a component perpendicular to the direction of motion called the lift coefficient C_L. These components are then independent of the direction of flight. They can of course vary substantially with the orientation of the body as is evident in the control of the flight of aircraft.

The expression $\frac{1}{2}\rho A v^2 \mathbf{C}$ implies that the force is proportional to v^2 unless $\mathbf{C}$ itself varies with v, that is unless there are further non-dimensional parameters which involve v and which can influence the value of $\mathbf{C}$. There are such parameters, and the most important arises from the viscous property of a fluid. It is a property more marked in oily fluids than in air or water and is associated with the resistance of the fluid to deformation, especially shear deformation. For a particular fluid, the property is characterised by a coefficient of kinematic viscosity ν which has the dimensions L^2T^{-1}. For example, water has a coefficient of kinematic viscosity equal to 10^{-6} m^2 s^{-1} at a temperature of 20 °C.

It is easily verified that ν/l has the dimensions LT^{-1} and hence that vl/ν is a non-dimensional combination, called the Reynolds number. The variation of $\mathbf{C}$ with this parameter is complicated, but limited for typical conditions. For example a circular disc, of area A and moving broadside on, has a drag coefficient of about one and there is little change with Reynolds number provided the latter is not too small (by symmetry there is no lift coefficient). A sphere with the same maximum cross-sectional area has a smaller drag coefficient of kinematic viscosity equal to 10^{-6} m^2 s^1 at a temperature of Reynolds number is not too small. It is even less for a streamlined body with the same maximum cross-sectional area. Such a body is specially designed to reduce the drag, with a rounded nose and a long tapering tail.

When the Reynolds number is small the situation is different. It is known, both theoretically and experimentally, that $|\mathbf{C}|$ is inversely proportional to the Reynolds number and hence inversely proportional to v. It follows that the force varies in proportion to v rather than v^2. This is known as viscous damping.

Since $\nu = 10^{-6}$ m^2 s^{-1} for water, the small-Reynolds-number regime is restricted to very small bodies or very small speeds, for example a sphere with radius $l = 1$ mm moving with a speed $v = 1$ mm s^{-1} has unit Reynolds number. For air ν is between 10 and 20 times larger than its value for water (it varies with the temperature for both fluids), but the condition for a small Reynolds number is still very restrictive. A grain of sand of radius 1 mm falling freely under gravity through air or water would achieve a Reynolds number of several hundred. If, however, the fluid were olive oil the linear law of resistance would be quite reasonable (see Exercise 4.10).

We mention one other non-dimensional factor which is velocity dependent and which may affect the value of **C**. This is the Mach number, which is the ratio of the speed of the body to the speed of sound. In the transonic regime, where the Mach number is of order unity, this can be a significant effect. But for speeds which are small compared with the speed of sound, which is about 330 m s^{-1} in air at ground level, this effect can be ignored. It is invariably ignored in liquids. The speed of sound in water, for example, is $1{\cdot}4 \times 10^3 \text{ m s}^{-1}$.

Apart then from certain exceptional regimes, it appears that the simplest rule which might be generally acceptable is one in which the force varies as the square of the speed provided, as will usually be the case, that the Reynolds number is not small. If the Reynolds number is small then a linear law of variation is appropriate. Unfortunately, apart from one-dimensional problems, it is only the linear law which is amenable to simple analysis. This is because it leads to linear differential equations which are easier to handle than non-linear equations.

In some of the following examples the linear law of resistance is used where the conditions do not seem to be appropriate. This is partly because a theory is best illustrated by examples which are not too complicated. But they should not be dismissed as completely artificial, because the effects of resisting forces of different kinds have certain features in common. It is often useful to have a simple solution where the gross features are qualitatively acceptable, even though quantitatively they are not (see also Exercise 4.11).

Example 4.2.1

We consider the one-dimensional motion of a particle subject to a resistance whose magnitude per unit mass is equal to k times the square of the speed, and a constant force f per unit mass.

(i) Suppose first that the constant force also opposes the motion, so that the total force is $-m(f + kv^2)$ as shown in Figure 4.2.1. The equation of motion is then

$$\frac{\mathrm{d}v}{\mathrm{d}t} = -f - kv^2. \tag{4.2.1}$$

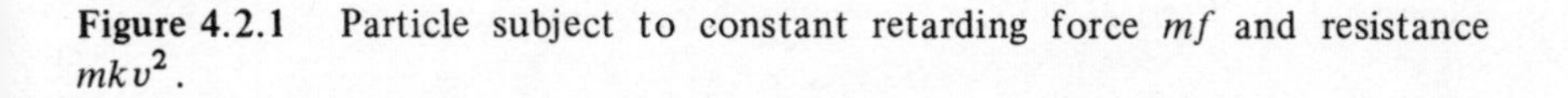

Figure 4.2.1 Particle subject to constant retarding force mf and resistance mkv^2.

Note that this case only arises when the particle is given an initial velocity u in the positive x direction. Eventually the particle will come to rest, and then begin to move in the opposite direction since the acceleration is negative. When this happens the resisting force will change direction, since it always opposes the motion. Equation (4.2.1) is valid only up to that point.

It is convenient to write $f = kV^2$, so that

$$\frac{dv}{dt} = -k(V^2 + v^2). \tag{4.2.2}$$

Equation (4.2.2) can be integrated to give

$$\tan^{-1}(v/V) = \tan^{-1}(u/V) - kVt, \tag{4.2.3}$$

where the constant of integration has been chosen so that $v = u$ when $t = 0$.

Alternatively we can write

$$\frac{dv}{dt} = \frac{dx}{dt}\frac{dv}{dx} = v\frac{dv}{dx} = -k(V^2 + v^2),$$

which can be integrated to give

$$\log\left(\frac{V^2 + u^2}{V^2 + v^2}\right) = 2kx. \tag{4.2.4}$$

Equations (4.2.3) and (4.2.4) show that the body will come to rest in a time $\{\tan^{-1}(u/V)\}/kV$ after travelling a distance $\{\log(1 + u^2/V^2)\}/2k$. For example, suppose a ship reverses its engines when going full speed ahead. Here $u = V$ (see below) and it will travel a distance $\frac{1}{2}k^{-1}\log 2$ in time $\pi/4kV$ before coming to rest.

(ii) When the particle moves in the direction of the propelling force, it is less confusing if x is measured in the same direction, as in Figure 4.2.2, so that the equation of motion is now

$$\frac{dv}{dt} = v\frac{dv}{dx} = f - kv^2 = k(V^2 - v^2). \tag{4.2.5}$$

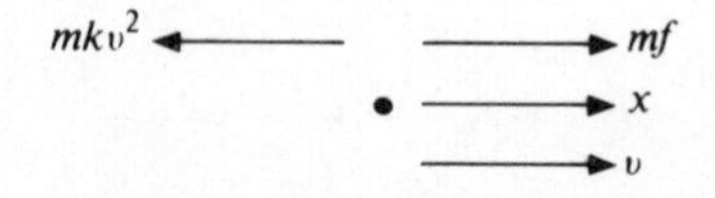

Figure 4.2.2 Particle subject to constant propelling force mf and resistance mkv^2.

Here V has a physical interpretation. As $dv/dt \to 0$, $v \to V$. If $u > V$ the acceleration is negative and $v \to V$ from above. If $u < V$ the acceleration is positive and $v \to V$ from below. This can be seen directly from the integrals of equation (4.2.5) which give the two relations

$$\log \left| \frac{(V+v)(V-u)}{(V-v)(V+u)} \right| = 2kVt, \tag{4.2.6}$$

$$\log \left| \frac{V^2 - u^2}{V^2 - v^2} \right| = 2kx \tag{4.2.7}$$

and show that $v \to V$ as $t \to \infty$ and as $x \to \infty$.

Consider the case of a particle projected vertically upwards with speed u. Here $f = g$ and equations (4.2.4), (4.2.3) show that it will reach a height

$$x = \frac{1}{2k} \log \left(1 + \frac{u^2}{V^2} \right) \tag{4.2.8}$$

in time

$$t_1 = \frac{1}{kV} \tan^{-1} \frac{u}{V}, \tag{4.2.9}$$

where $V = (g/k)^{1/2}$.

During the subsequent motion, with x now measured downward from the highest point, equations (4.2.6) and (4.2.7) apply and give, for zero initial speed,

$$v = V \tanh kVt, \tag{4.2.10}$$

$$x = \frac{1}{2k} \log \left(\frac{V^2}{V^2 - v^2} \right) = \frac{1}{k} \log \cosh kVt. \tag{4.2.11}$$

The time t_2 for the subsequent descent is obtained by equating (4.2.8) and (4.2.11). The result is

$$t_2 = \frac{1}{kV} \sinh^{-1} \frac{u}{V}. \tag{4.2.12}$$

Example 4.2.2

A particle, of mass m and position vector $\mathbf{r}$ relative to O, is projected from O with initial velocity $\mathbf{u}$ at an angle α to the horizontal. It is subject to a gravitational force mg and a resisting force $\mathbf{R} = -mk\mathbf{v}$, where $\mathbf{v}$ is its velocity (see Figure 4.2.3).

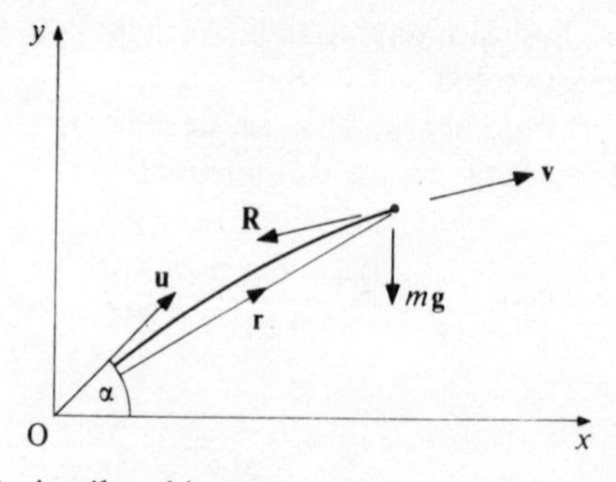

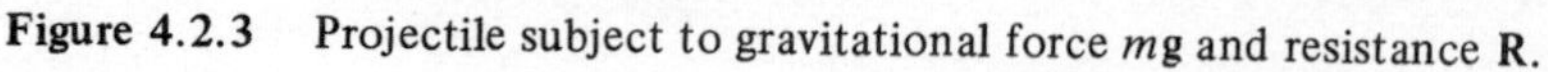
Figure 4.2.3 Projectile subject to gravitational force $m\mathbf{g}$ and resistance $\mathbf{R}$.

The equation of motion is

$$\ddot{\mathbf{r}} = \mathbf{g} - k\dot{\mathbf{r}}. \tag{4.2.13}$$

It can be written in the form

$$\frac{\mathrm{d}}{\mathrm{d}t}(\mathrm{e}^{kt}\dot{\mathbf{r}}) = \mathbf{g}\mathrm{e}^{kt}$$

and hence integrated to give

$$\dot{\mathbf{r}} = \frac{\mathbf{g}}{k} + \left(\mathbf{u} - \frac{\mathbf{g}}{k}\right)\mathrm{e}^{-kt}, \tag{4.2.14}$$

where the constant vector of integration is chosen so that $\dot{\mathbf{r}} = \mathbf{u}$ when $t = 0$. One further integration gives the equation

$$\mathbf{r} = \frac{\mathbf{g}t}{k} + \frac{1}{k}\left(\mathbf{u} - \frac{\mathbf{g}}{k}\right)(1 - \mathrm{e}^{-kt}), \tag{4.2.15}$$

where the constant vector of integration is chosen so that $\mathbf{r} = \mathbf{0}$ when $t = 0$. Relative to horizontal and vertical axes Ox, Oy, equation (4.2.15) has the components

$$x = \frac{u \cos \alpha}{k}(1 - \mathrm{e}^{-kt}), \tag{4.2.16}$$

$$y = -\frac{gt}{k} + \frac{1}{k^2}(ku \sin \alpha + g)(1 - \mathrm{e}^{-kt}). \tag{4.2.17}$$

An explicit equation for the trajectory follows by the elimination of t. It is

$$y = \frac{g}{k^2}\log\left(1 - \frac{kx}{u \cos \alpha}\right) + x\left(\tan \alpha + \frac{g}{ku \cos \alpha}\right). \tag{4.2.18}$$

Figure 4.2.4 shows the effect of resistance on the trajectory of the particle. The various trajectories shown in the figure were calculated using a non-dimensional form of equations (4.2.16) and (4.2.17). This preliminary reorganisation of the equations is an invariable procedure in all numerical work, for the reasons discussed in §2.5. It is always an advantage in practice to use non-dimensional variables whose magnitudes are of order unity, and this should be one of the guides in the choice of the variables. Now in the special simple case of no resistance we know from §2.4 that the values of t, x and y at the highest point of the trajectory are, respectively,

$$\frac{u \sin \alpha}{g}, \qquad \frac{u^2 \sin \alpha \cos \alpha}{g}, \qquad \frac{u^2 \sin^2 \alpha}{2g}.$$

Hence the transformations

$$t = \frac{u \sin \alpha}{g} \tau, \qquad x = \frac{u^2 \sin \alpha \cos \alpha}{g} \xi, \qquad y = \frac{u^2 \sin^2 \alpha}{g} \eta, \tag{4.2.19}$$

omitting the factor $\frac{1}{2}$ in y, will certainly yield non-dimensional variables τ, ξ, η which are of order unity for all values of the parameters u, g, α. Furthermore the equation of the trajectory, given by equation (2.4.7), transforms to

$$\eta = \xi - \tfrac{1}{2}\xi^2 \tag{4.2.20}$$

and hence to one basic shape, independent of u, g, α, in the plane of (ξ, η). This is another hopeful sign that (4.2.19) is a useful transformation, because another important consideration is to seek transformations which minimise the number of significant parameters on which the problem depends.

Substitution of relations (4.2.19) into (4.2.16), (4.2.17) and (4.2.18) gives the following equations:

$$\xi = K_1^{-1} (1 - e^{-K_1 \tau}), \tag{4.2.21}$$

$$\begin{aligned} \eta &= -K_1^{-1} \tau + K_1^{-2} (1 + K_1)(1 - e^{-K_1 \tau}), \\ &= K_1^{-2} \log(1 - K_1 \xi) + (1 + K_1^{-1})\xi, \end{aligned} \tag{4.2.22}$$

where

$$K_1 = \frac{ku \sin \alpha}{g} = \frac{u \sin \alpha}{V} \tag{4.2.23}$$

and $V = g/k$ is the terminal speed of the particle in free fall under gravity. In these variables, then, the reduced trajectory depends on the four parameters k, u, g, α only in the non-dimensional combination $K_1 = kug^{-1} \sin \alpha$, which we call the specific resistance (the suffix denotes that the resistance is linear). The advantage is that calculation can now be confined to the family of curves depending on the single parameter K_1. In any specific problem the trajectory will be given by that member of the family having the appropriate value of K_1, and the values of k, u, g, α merely determine the appropriate scales of length and time through the transformations (4.2.19).

When $K_1 \ll 1$, the expression for η in (4.2.22) can be expanded to give

$$\eta = -K_1^{-2} \sum_{n=1}^{\infty} \frac{1}{n} K_1^n \xi^n + (1 + K_1^{-1})\xi$$

$$= \xi - \tfrac{1}{2}\xi^2 - \sum_{n=3}^{\infty} \frac{1}{n} K_1^{n-2} \xi^n, \qquad (4.2.24)$$

from which the parabolic trajectory (4.2.20) is recovered in the limit $K_1 \to 0$. The series can be used to obtain an approximation to the range, which is the non-zero solution of the equation

$$0 = \xi - \tfrac{1}{2}\xi^2 - \sum_{n=3}^{\infty} \frac{1}{n} K_1^{n-2} \xi^n$$

or, equivalently,

$$\xi - 2 = -\sum_{n=3}^{\infty} \frac{2}{n} K_1^{n-2} \xi^{n-1}. \qquad (4.2.25)$$

Since ξ is approximately equal to 2 when K_1 is small, it follows that the right-hand side of (4.2.25) is equal to

$$-\tfrac{8}{3} K_1 + O(K_1^2)$$

and a better approximation for the range is

$$\xi = 2 - \tfrac{8}{3} K_1 + O(K_1^2). \qquad (4.2.26)$$

This process can be made systematic. If we write

$$\xi = 2 + \sum_{n=1}^{\infty} a_n K_1^n \qquad (4.2.27)$$

and substitute this expression into (4.2.25), we have

$$a_1K_1 + a_2K_2^2 + \ldots = -\tfrac{2}{3}K_1(2 + a_1K_1 + \ldots)^2 - \tfrac{1}{2}K_1^2(2 + \ldots)^3 - \ldots$$

$$= -\tfrac{8}{3}K_1 - (\tfrac{8}{3}a_1 + 4)K_1^2 - \ldots. \tag{4.2.28}$$

The coefficients a_n can now be found by comparison of the coefficients of like powers of K_1 on the two sides of the above equation. Thus we have

$$a_1 = -8/3, \quad a_2 = -4 - 8/3a_1 = 28/9 \tag{4.2.29}$$

and so on. The range is now given by

$$\xi = 2 - \tfrac{8}{3}K_1 + \tfrac{28}{9}K_1^2 + O(K_1^3). \tag{4.2.30}$$

While the trajectory for $K_1 \ll 1$ is a small perturbation of a parabola, that for $K_1 \gg 1$ is quite different. Provided $(1 - K_1\xi)$ is not small, the trajectory follows the straight line $\xi = \eta$ approximately. But it is clear from both (4.2.21) and (4.2.22) that K_1^{-1} is an upper bound for ξ, and as this bound is approached there are substantial changes in η for small changes in ξ. This means that ultimately the particle will be falling in a near vertical path with a total range of approximately K_1^{-1}. Since, by (4.2.22) with $\eta = 0$, the range satisfies the equation

$$K_1\xi = 1 - \exp\{-K_1(K_1 + 1)\xi\}, \tag{4.2.31}$$

it follows that, for $K_1 \gg 1$, it is given by

$$\xi = K_1^{-1}\,(1 + \text{exponentially small terms}). \tag{4.2.32}$$

The proportionate error in the approximation K_1^{-1} is therefore very small.

For arbitrary values of K_1 there is no simple expression for the range, which must be calculated from the implicit relation (4.2.31). In some problems, however, simple approximate formulae can be found. Here for example the range is given by

$$\xi = (K_1 + \tfrac{1}{2}e^{-2K_1/3})^{-1}(1 + \epsilon(K_1)), \tag{4.2.33}$$

where $\epsilon = 0{\cdot}0025$ for $K_1 = 2{\cdot}3$ and is smaller than this for all other values of K_1. There is no *a priori* reason why such a simple formula should exist and no general method of deducing such formulae, although to be successful they must have some gross features in common with the exact result. In this case, with $\epsilon = 0$, expression (4.2.33) agrees with (4.2.30) for $K_1 \ll 1$ and with (4.2.32) for $K_1 \gg 1$. Although the exponentially small terms are not modelled correctly, they are of no significance compared with K_1^{-1} when $K_1 \gg 1$.

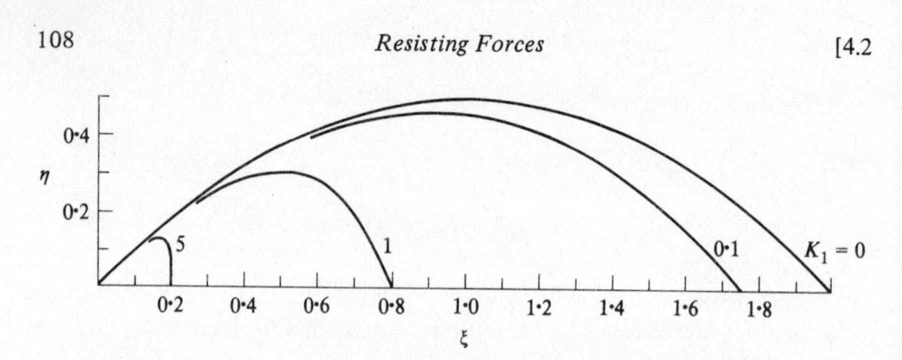

Figure 4.2.4 Linear law of resistance: trajectories for various values of K_1.

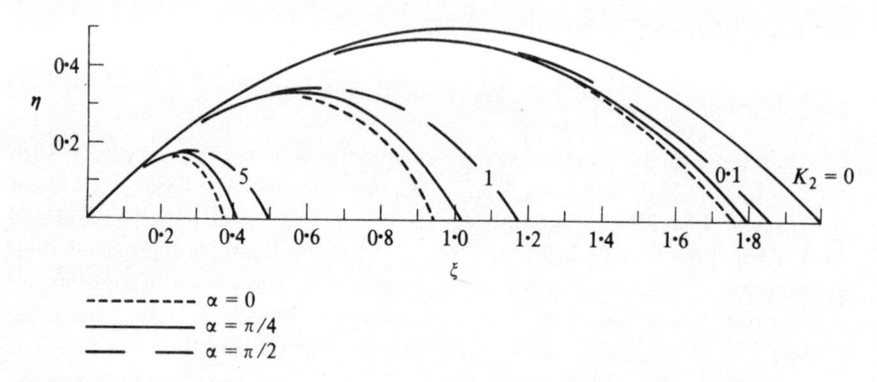

Figure 4.2.5 Quadratic law of resistance: trajectories for various values of K_2 and α.

Example 4.2.3

When the resistance $\mathbf{R}$ varies with the square of the speed the projectile problem is more complicated. Since the resistance is given, in magnitude and direction, by an expression of the form $-mkv\mathbf{v}$, the equation of motion is

$$\ddot{\mathbf{r}} = \mathbf{g} - kv\mathbf{v} \tag{4.2.34}$$

with components

$$\ddot{x} = -kv\dot{x}, \quad \ddot{y} = -g - kv\dot{y}, \tag{4.2.35}$$

where

$$v = |\mathbf{v}| = (\dot{x}^2 + \dot{y}^2)^{1/2}. \tag{4.2.36}$$

With the transformations defined by (4.2.19), equations (4.2.35) become

$$\frac{d^2\xi}{d\tau^2} = -K_2 \frac{d\xi}{d\tau}\left\{\cos^2\alpha\left(\frac{d\xi}{d\tau}\right)^2 + \sin^2\alpha\left(\frac{d\eta}{d\tau}\right)^2\right\}^{1/2}, \tag{4.2.37}$$

$$\frac{d^2\eta}{d\tau^2} = -1 - K_2 \frac{d\eta}{d\tau}\left\{\cos^2\alpha\left(\frac{d\xi}{d\tau}\right)^2 + \sin^2\alpha\left(\frac{d\eta}{d\tau}\right)^2\right\}^{1/2}, \tag{4.2.38}$$

where
$$K_2 = kg^{-1}u^2 \sin\alpha \tag{4.2.39}$$

and the initial conditions are that

$$\xi = \eta = 0, \quad \frac{\mathrm{d}\xi}{\mathrm{d}\tau} = \frac{\mathrm{d}\eta}{\mathrm{d}\tau} = 1 \quad \text{when } \tau = 0. \tag{4.2.40}$$

The problem therefore depends on the two non-dimensional parameters K_2 and α. There is no simple integral, and one must use numerical analysis for a full solution. If, however, the inclination of the trajectory to the horizontal is small, then the equations simplify. A bullet fired from a rifle, for example, has such a trajectory. In this approximation one should think of K_2 and α as independent parameters. The fact that K_2 contains a factor $\sin\alpha$ in its definition does not imply that K_2 is to be regarded as small whenever α is regarded as small.

The approximate equations are, when terms of order α^2 are neglected,

$$\frac{\mathrm{d}^2\xi}{\mathrm{d}\tau^2} = -K_2\left(\frac{\mathrm{d}\xi}{\mathrm{d}\tau}\right)^2, \quad \frac{\mathrm{d}^2\eta}{\mathrm{d}\tau^2} = -1 - K_2\frac{\mathrm{d}\xi}{\mathrm{d}\tau}\frac{\mathrm{d}\eta}{\mathrm{d}\tau}. \tag{4.2.41}$$

The first of these equations integrates to

$$\frac{\mathrm{d}\xi}{\mathrm{d}\tau} = (1 + K_2\tau)^{-1} \tag{4.2.42}$$

and then a second integration gives

$$K_2\xi = \log(1 + K_2\tau). \tag{4.2.43}$$

Next we use (4.2.42) to eliminate $\mathrm{d}\xi/\mathrm{d}\tau$ from the second of equations (4.2.41), which then becomes

$$\frac{\mathrm{d}^2\eta}{\mathrm{d}\tau^2} = -1 - \frac{K_2}{1 + K_2\tau}\frac{\mathrm{d}\eta}{\mathrm{d}\tau},$$

or

$$\frac{\mathrm{d}}{\mathrm{d}\tau}\left((1 + K_2\tau)\frac{\mathrm{d}\eta}{\mathrm{d}\tau}\right) = -(1 + K_2\tau). \tag{4.2.44}$$

Integration of (4.2.44), together with the boundary conditions (4.2.40), gives

$$\frac{\mathrm{d}\eta}{\mathrm{d}\tau} = \left(1 + \frac{1}{2K_2}\right)\frac{1}{1 + K_2\tau} - \frac{1}{2K_2}(1 + K_2\tau), \tag{4.2.45}$$

$$\eta = \frac{1}{K_2}\left(1 + \frac{1}{2K_2}\right)\log(1 + K_2\tau) - \frac{1}{4K_2^2}(1 + K_2\tau)^2 + \frac{1}{4K_2^2}$$

$$= \left(1 + \frac{1}{2K_2}\right)\xi - \frac{1}{4K_2^2}(e^{2K_2\xi} - 1). \qquad (4.2.46)$$

If the exponential term is first written as a power series, it is easily seen that (4.2.46) agrees with (4.2.20) in the limit $K_2 \to 0$. This means that, in addition to being correct for finite values of K_2 in the limit $\alpha \to 0$, (4.2.46) also yields the correct trajectory in the limit $K_2 \to 0$ without restriction on α. The range of validity is consequently greater than might otherwise have been expected. On the other hand it should be borne in mind that the validity of the approximation really depends on the inclination of the trajectory to the horizontal being small throughout the flight of the projectile. This is clear from equations (4.2.35), which are the dimensional form of (4.2.41) with the approximation $v = \dot{x}$. Equation (4.2.36) shows that this requires $\dot{y} \ll \dot{x}$ at all times. If the trajectory is bounded by the horizontal plane this is ensured by $\alpha \ll 1$. But if the projectile is allowed to continue through to negative values of y, the approximation will eventually fail.

For general values of K_2 and α it is not possible to integrate equations (4.2.37) and (4.2.38) in terms of elementary functions. However, it is a straightforward matter to integrate them numerically using a computer. The trajectories shown in Figure 4.2.5 were produced from such a numerical integration and should be compared with the equivalent Figure 4.2.4 for the linear law of resistance. Note that the trajectories labelled $\alpha = 0$ and $\alpha = \pi/2$ are limiting trajectories for the non-dimensional variables (ξ, η). Of course in the physical (x, y) plane these trajectories degenerate to a point and a vertical straight line respectively because the non-dimensionalisation by means of (4.2.19) also involves α. The trajectories in the (ξ, η) plane are preferred because they exhibit information in a more revealing fashion and, as in the previous example, the trajectory in the physical plane is readily obtained by the appropriate change of scale, which will involve other parameters as well as α. We see in particular that in the (ξ, η) plane the variation with α is less significant than the variation with K_2, and that the nearly vertical descent for $K_1 = 5$ is not duplicated in the trajectory for $K_2 = 5$.

EXERCISES

4.1 A rough plane is inclined at an angle α to the horizontal. A particle of mass m is attached to a fixed point of the plane by a light string, and rests in equilibrium with the string taut and making an acute angle θ with the line of greatest slope down the plane. Show that $\sin\theta \leqslant \mu_s \cot\alpha$, where μ_s is the coefficient of static friction.

When $\sin\theta = \mu_s \cot\alpha$, show that the tension in the string is

$$mg(\sin^2\alpha - \mu_s^2\cos^2\alpha)^{1/2}.$$

4.2 Two particles, each of mass m, lie at points A and B on a rough horizontal plane, and are connected by a taut light inelastic string AB. The particle at B is pulled by a horizontal force T applied to a second light inelastic string BC, where $A\hat{B}C = \pi - \alpha$ $(\alpha < \pi/4)$.

Show that the least force T which will cause both particles to be about to move is $T = 2\mu_s mg \cos \alpha$.

Explain the significance of the restriction $\alpha < \pi/4$.

4.3 Two particles of equal mass are connected by a light string. The first particle rests on the surface of a rough circular cylinder, whose axis is horizontal. Part of the string is in contact with the cylinder over an angular distance α, the remainder hangs vertically with the second particle suspended from its end. The two particles and the string are in equilibrium in a vertical plane. The coefficient of friction between the particle and the cylinder, and between the string and the cylinder, is μ_s.

Write down the equations for equilibrium. Show that when the friction is limiting,

$$(\mu_s \sin \alpha - \cos \alpha)e^{\mu_s \alpha} = 1.$$

Show that if μ_s is small, the approximate solution is $\alpha = \pi - (2\mu_s \pi)^{1/2}$.
Show that if μ_s is large, the approximate solution is $\alpha = a/\mu_s$, where

$$(a - 1)\,e^a = 1, \qquad \text{or} \qquad a = 1{\cdot}2785.$$

4.4 The mass of a train is 5×10^5 kg, the driving wheels support a weight of 4×10^4 kg-wt and the coefficient of friction between the driving wheels and the rails is 0·6. Show that at the end of 10 s after starting from rest the speed will be less than 17 km h^{-1}.

4.5 Two equally rough planes slope down at angles α and β to the horizontal from the horizontal ridge in which they meet. A particle is projected from the bottom of the first plane up the line of greatest slope with a speed v just sufficient to carry it over the ridge. After descending the other face it reaches the level from which it started with speed kv. Prove that, if $\mu_k < \tan \beta$,

$$\mu_k(\tan \alpha + k^2 \tan \beta) = (1 - k^2)\tan \alpha \tan \beta,$$

where μ_k is the coefficient of kinetic friction.

4.6 A bead slides on a fixed vertical rough circular hoop of radius a. It is projected from the lowest point of the hoop with speed $a\omega_0$. Let ω be the angular speed of the bead when the inclination of the radius vector to the downward vertical is θ. Show that, for a certain range of θ which should be specified,

$$a\left(\frac{d\omega^2}{d\theta} + 2\mu_k \omega^2\right) = -2g(\sin \theta + \mu_k \cos \theta).$$

Hence show that

$$a(4\mu_k^2 + 1)(\omega_0^2 e^{-2\mu_k \theta} - \omega^2) = 2g\{3\mu_k \sin \theta + (2\mu_k^2 - 1)(\cos \theta - e^{-2\mu_k \theta})\}.$$

4.7 A bead slides on a fixed horizontal rough circular hoop of radius a. If v is the speed of the bead after travelling an angular distance θ, show that

$$\frac{dv^2}{d\theta} = -2\mu_K(g^2a^2 + v^4)^{1/2}.$$

Show that the solution of this equation is given by

$$v^2 = ga \sinh\{2\mu_K(\alpha - \theta)\},$$

where α is the value of θ for which $v = 0$.

Show that the frictional force is of magnitude $\mu_K mg \cosh\{2\mu_K(\alpha - \theta)\}$. Show that the line integral of this force, between $\theta = 0$ and $\theta = \alpha$, is $\frac{1}{2}mv_0^2$, where v_0 is the value of v when $\theta = 0$.

4.8 Two particles, of masses m_1, m_2 respectively, are connected by a long light string which passes over a rough circular cylinder whose axis is horizontal, and lies in a plane at right angles to the axis of the cylinder.

Show that the two particles can hang in equilibrium if $m_2 < m_1 < m_2 e^{\mu_S \pi}$. Show that if $m_1 \geqslant m_2 e^{\mu_S \pi}$ the acceleration of the particles is

$$\frac{m_1 - m_2 e^{\mu_K \pi}}{m_1 + m_2 e^{\mu_K \pi}} g.$$

4.9 A particle slides on a rough plane inclined at an angle β to the horizontal. It starts from the origin of a set of cartesian coordinates on the inclined plane such that the x-axis is horizontal and the y-axis is down the line of greatest slope. The initial velocity is of magnitude u at an angle α ($<\pi/2$) to the y-axis.

Write down the equations of motion. In non-dimensional variables τ, ξ, η such that

$$t = \frac{u \cos \alpha}{g \sin \beta}\tau, \quad x = \frac{u^2 \sin \alpha \cos \alpha}{g \sin \beta}\xi, \quad y = \frac{u^2 \cos^2 \alpha}{g \sin \beta}\eta,$$

show that the equations of motion become

$$\ddot{\xi} = -\frac{\dot{\xi}\mu_K \cot \beta}{(\dot{\xi}^2 \tan^2 \alpha + \dot{\eta}^2)^{1/2}}, \quad \ddot{\eta} = 1 - \frac{\dot{\eta}\mu_K \cot \beta}{(\dot{\xi}^2 \tan^2 \alpha + \dot{\eta}^2)^{1/2}}$$

with the initial conditions

$$\xi = \eta = 0, \quad \dot{\xi} = \dot{\eta} = 1 \qquad \text{when } \tau = 0.$$

Integrate these equations, and discuss the subsequent motion of the particle, on the basis of the simplifying assumption that $\tan \alpha \ll 1$. In particular show that, if $\mu_K \cot \beta > \frac{1}{2}$, the ultimate value of ξ is $(2\mu_K \cot \beta - 1)^{-1}$ approximately.

Show also that if $\mu_k \cot \beta > 1$, the particle will finally come to rest but that if $\frac{1}{2} < \mu_k \cot \beta < 1$, the particle will ultimately slide in the y direction with constant acceleration.

4.10 (i) A spherical grain of sand, of radius $l = 10^{-3}$ m, falls freely under gravity in air. Show that its terminal speed V is 9 m s^{-1} and that the corresponding Reynolds number Vl/ν is 600.

If the medium is water instead of air, show that V is 0·23 m s^{-1} and that the Reynolds number is 230.

Assume that the drag coefficient $C_D = 0{\cdot}5$ and $g = 9{\cdot}8$ m s^{-2}.

(ii) When the medium is olive oil, it is found that $V = 2 \times 10^{-2}$ m s^{-1}. Show that the Reynolds number cannot be greater than 0·2 and deduce an appropriate law of variation of resistance with velocity.

Verify from the equation of motion that the terminal speed will never be achieved in finite time but that a speed of $99V/100$ will be achieved in about 0·02 s after starting from rest.

Use the following physical constants:

	Specific gravity	*Kinematic viscosity,* ν (m^2 s^{-1})
sand	2	
air	$1{\cdot}3 \times 10^{-3}$	$1{\cdot}5 \times 10^{-5}$
water	1	10^{-6}
olive oil	1	10^{-4}

4.11 A particle is in one-dimensional motion and experiences a resistance of magnitude $f(v/V)$ per unit mass, where v is the speed, V is a fixed parameter with the dimensions LT^{-1} and f is a differentiable function having the appropriate dimensions and such that $f(0) = 0$.

Let

$$v/V = W + w$$

where W is constant and w is small. Show that there is an approximation to the resistance per unit mass of the form $a + bv$, where a and b are constant and given by

$$a = f(W) - Wf'(W),$$
$$b = f'(W)/V.$$

In particular, deduce that for small values of v/V a resistance proportional to the speed is an appropriate approximation. What condition on the function f would give a resistance proportional to the square of the speed for small v/V?

4.12 A hollow cylindrical pipe of radius a is fixed with its axis vertical. A particle hangs by a string from a fixed point on the axis of the pipe. The particle is held against the rough internal wall of the pipe, with the string taut and at an angle α to the downward vertical, and projected horizontally with angular speed ω_0 about the axis of the pipe. The coefficient of kinetic friction between the particle and the pipe is μ and there is, in addition, a resistance to the motion equal to k times the square of the speed per unit mass. Given that the particle remains in contact with the pipe during the subsequent motion, write down the equations of motion. Show that the angular speed ω satisfies the equation

$$\omega \frac{d\omega}{d\theta} + (\mu + ka)(\omega^2 - \omega_1^2) = 0,$$

where θ is the angular displacement of the particle about the axis and ω_1 is a constant which should be determined. Hence determine ω as a function of θ.

Show that if $a\omega_0^2 > g \tan \alpha$, the particle does not lose contact with the pipe immediately, but that contact will cease when θ satisfies the equation

$$\exp\{2(\mu + ka)\theta\} = \frac{\mu}{ka}\left(\frac{\omega_0^2}{\omega_1^2} - 1\right).$$

$$\left(\text{Ans.: } \omega_1^2 = \frac{\mu}{\mu + ka}\,\frac{g \tan \alpha}{a}, \qquad \omega^2 = \omega_1^2 + (\omega_0^2 - \omega_1^2)\exp\{-2(\mu + ka)\theta\}.\right)$$

4.13 A vehicle of mass m experiences a constant frictional resistance ma and air resistance proportional to the square of its speed. It can exert a constant propelling force mb and can attain a maximum speed V. Show that, starting from rest, it can attain the speed $V/2$ in the time

$$\frac{V \log 3}{2(b - a)}$$

and that the friction and air resistance alone can then bring it to rest in a further time

$$\frac{V}{\{a(b - a)\}^{1/2}} \tan^{-1}\left(\frac{b - a}{4a}\right)^{1/2}.$$

4.14 A particle falls from rest under gravity in a medium offering a resistance $av + bv^2$ per unit mass, where a and b are positive constants and v is the speed of the particle. Find the velocity acquired and distance fallen in time t.

$$\left(\text{Ans.: } v = \frac{c}{b} \tanh T - \frac{a}{2b},\ y = \frac{1}{b} \log\left(\frac{\cosh T}{\cosh \alpha}\right) - \frac{at}{2b},\right.$$
$$\left.\text{where } c = (bg + \tfrac{1}{4}a^2)^{1/2},\ \tanh \alpha = a/2c,\ T = (ct + \alpha).\right)$$

4.15 A particle of soot is carried vertically upwards in a current of air, which has a constant vertical velocity of magnitude u. Initially the particle has the same velocity as the air. The force exerted on the particle by the air is gv^2/w^2 per unit mass, where g is the gravitational acceleration, v is the speed of the particle relative to the air, and w is a constant.

Show that the maximum height, above its initial level, attained by the particle is

$$\tfrac{1}{2}wg^{-1}\{(w+u)\log(1+u/w)+(w-u)\log(1-u/w)\}$$

provided $u < w$.

Show that the time taken to reach this height is $wg^{-1}\tanh^{-1}(u/w)$.

What happens if $u > w$?

4.16 A particle is projected vertically upwards with initial speed u in a medium offering a resistance whose magnitude is kv^2, per unit mass, when its speed is v. Prove that it returns to the starting point after a time $\tan^{-1}(ku^2/g)^{1/2} + \sinh^{-1}(ku^2/g)^{1/2}$. Show that if $ku^2 \ll g$ the time is approximately

$$\frac{2u}{g}\left(1-\frac{ku^2}{4g}\right).$$

4.17 A particle is projected vertically upwards with initial speed u and moves in a medium offering a resistance kv^4, per unit mass, when its speed is v. Prove that the greatest height reached above the point of projection is $\alpha/(4kg)^{1/2}$, where $g\tan^2\alpha = ku^4$. Show that if the particle has speed w on regaining the point of projection, $g\tanh^2\alpha = kw^4$.

4.18 A particle is projected from a fixed point O with speed u in a resisting medium in which the force per unit mass opposing the motion at any distance x from O is $2\alpha^2 x$ times the speed of the particle, α being constant. No other forces act on the particle. Given that the maximum distance of the particle from O is X, show by a dimensional argument that $X = ku^{1/2}/\alpha$, where k is a numerical constant.

Calculate x explicitly as a function of time, prove that X exists, and find k. Show that when x approaches X and is written in the form $x = (1-\epsilon)X$, where $\epsilon \ll 1$, the corresponding time t for the particle to reach x is given approximately by

$$t = \frac{\log(2/\epsilon)}{2\alpha u^{1/2}}.$$

(Ans.: $k = 1$, $X = u^{1/2}/\alpha$.)

4.19 A particle of mass m can move along a straight line. The only force acting on the particle opposes its motion and depends only on its speed. The time T taken for the particle to come to rest after it is set in motion with speed u is given by the formula $T = k\log(1+au)$ for all u, where k and a are positive constants. Find the force acting on the particle during the interval $0 \leqslant t \leqslant T$ as a function of its speed v.

Find also the distance the particle travels in time T.

(Ans.: $m(1+av)/ak$, $ku - T/a$).)

4.20 A particle of mass m is projected horizontally with initial speed u from the top of a tower which stands on a horizontal plane. It is subject to a constant gravitational force $m\mathbf{g}$ and a resisting force equal to $-mk\mathbf{v}$ when its velocity is $\mathbf{v}$. What is the height of the tower if the particle hits the plane at a distance x from the base of the tower?

Let x_0 be the distance from the base of the tower at which the particle would hit the plane if there were no resistance. Prove that, approximately, $x = x_0(1 - kx_0/3u)$ provided that $kx_0 \ll u$.

$$\left(\text{Ans.: } \frac{g}{k}\log\left(\frac{u}{u-kx}\right) - \frac{gx}{u}.\right)$$

4.21 A particle of mass m is fired with initial speed u at an inclination α $(0 < \alpha < \pi/2)$ to the horizontal, and into a horizontal wind which has constant speed v. The particle is subject to a constant gravitational force $m\mathbf{g}$ and a resisting force equal to $-mk$ times its velocity relative to the wind.

(i) Show that if the angle of inclination is such that $\tan\alpha = g/kv$, the particle eventually returns to its point of projection.
(ii) Show that the particle is directly above its point of projection at the highest point in its trajectory if the angle of inclination is such that

$$\left(1 + \frac{g}{ku\sin\alpha}\right)\log\left(1 + \frac{ku\sin\alpha}{g}\right) = 1 + \frac{u\cos\alpha}{v}.$$

Show that this condition is, approximately, $\tan\alpha = 2g/kv$ when $ku\sin\alpha \ll g$.

4.22 Consider the nearly vertical trajectory of a projectile with a quadratic law of resistance. Show that, in non-dimensional coordinates, the equations of motion are approximately

$$\frac{d^2\xi}{d\tau^2} = -K_2\frac{d\xi}{d\tau}\left|\frac{d\eta}{d\tau}\right|, \quad \frac{d^2\eta}{d\tau^2} = -1 - K_2\frac{d\eta}{d\tau}\left|\frac{d\eta}{d\tau}\right|,$$

where

$$K_2 = kg^{-1}u^2\sin\alpha,$$

$$\xi = \eta = 0, \quad d\xi/d\tau = d\eta/d\tau = 1 \qquad \text{when } \tau = 0,$$

and ξ, η are defined by equations (4.2.19). Hence show that

$$\frac{d\xi}{d\eta} = \begin{cases} \dfrac{K_2^{1/2}e^{-K_2\eta}}{\{(K_2+1)e^{-2K_2\eta} - 1\}^{1/2}} & \text{if } \dfrac{d\eta}{d\tau} \geqslant 0, \\ -\dfrac{K_2^{1/2}e^{K_2\eta}}{(K_2+1)\{1-(K_2+1)^{-1}e^{2K_2\eta}\}^{1/2}} & \text{if } \dfrac{d\eta}{d\tau} \leqslant 0. \end{cases}$$

Deduce that the value of ξ when the maximum height is attained ($d\eta/d\tau = 0$) is given by

$$K_2^{-1/2}(K_2+1)^{-1/2}\sinh^{-1}K_2^{1/2},$$

and that the value of ξ when $\eta = 0$ (the non-dimensional range) is given by

$$K_2^{-1/2}(K_2+1)^{-1/2}(\sinh^{-1}K_2^{1/2}+\tan^{-1}K_2^{1/2}).$$

4.23 The non-dimensional range ξ, for a quadratic law of resistance, satisfies the relations

$$0 = \left(1+\frac{1}{2K_2}\right)\xi - \frac{1}{4K_2^2}(e^{2K_2\xi}-1)$$

when $\alpha = 0$, and

$$\xi = K_2^{-1/2}(K_2+1)^{-1/2}(\sinh^{-1}K_2^{1/2}+\tan^{-1}K_2^{1/2})$$

when $\alpha = \pi/2$, where K_2 is the specific resistance.

(i) Show that the value of ξ when $\eta = 0$ is given by

$$\xi = 2 - \tfrac{8}{3}K_2 + \tfrac{40}{9}K_2^2 + O(K_2^3)$$

when $\alpha = 0$ and

$$\xi = 2 - \tfrac{3}{2}K_2 + \tfrac{51}{40}K_2^2 + O(K_2^3)$$

when $\alpha = \pi/2$.

(ii) Let ξ_0 be the value of ξ at the maximum height in the trajectory. Show that for both $\alpha = 0$ and $\alpha = \pi/2$, the value of ξ when $\eta = 0$ is of the form

$$\xi_0(1+\epsilon(K_2)),$$

where $\epsilon(K_2) \to 0$ as $K_2 \to \infty$.

4.24 A projectile experiences a resistance $R(v)$ per unit mass, where v is the speed.

When the trajectory is everywhere nearly horizontal, show that its equation can be written parametrically as

$$x = \int_v^u \frac{\xi d\xi}{R(\xi)}, \quad y = \int_v^u \left(\sin\alpha - \int_\eta^u \frac{g d\xi}{\xi R(\xi)}\right)\frac{\eta d\eta}{R(\eta)},$$

where α is the initial angle of projection and u is the initial speed.

4.25 A projectile is in two-dimensional motion under gravity and a resistance

$R(v)$ per unit mass, where v is the speed of the projectile. The velocity vector makes an angle ψ with the horizontal. Initially $v = u$ and $\psi = \alpha$.

(i) Show that

$$\frac{1}{v}\frac{dv}{d\psi} = \frac{g^{-1}R(v) + \sin\psi}{\cos\psi}.$$

(ii) For a resistance given by $R(v) = kv^2$, show that

$$\frac{d}{d\psi}(v\cos\psi)^{-2} = -2kg^{-1}\sec^3\psi.$$

Hence show that the speed at the highest point of the trajectory is $Au\cos\alpha$, where

$$A^{-2} = 1 + K_2\{1 + \cos\alpha\cot\alpha\log(\sec\alpha + \tan\alpha)\},$$
$$K_2 = kg^{-1}u^2\sin\alpha.$$

(iii) Verify that the result obtained in (ii) agrees with that obtained from the solution derived in Example 4.2.3 when $\alpha \to 0$, $K_2 \neq 0$.

(iv) Verify that the fractional variation in A, for $K_2 = 0{\cdot}2$ and $0 \leqslant \alpha \leqslant \pi/2$ is about 7%.

(v) Show that the corresponding value of A, when $R(v) = kv$, is $(1 + K_1)^{-1}$, where $K_1 = kg^{-1}u\sin\alpha$.

5 Impulse, Momentum, Work and Energy

5.1 IMPULSE AND MOMENTUM

In this chapter we discuss certain integrals of the equation of motion which can sometimes be used with advantage as alternatives to the basic equation.

Let a particle of mass m be subject to a force $\mathbf{F}$ and thereby experience an acceleration $\mathrm{d}\mathbf{v}/\mathrm{d}t$, where $\mathbf{v}$ is its velocity. The integral with respect to the time of the equation of motion

$$\mathbf{F} = m\,\mathrm{d}\mathbf{v}/\mathrm{d}t \tag{5.1.1}$$

gives immediately

$$\int_{t_1}^{t_2} \mathbf{F}\,\mathrm{d}t = m(\mathbf{v}_2 - \mathbf{v}_1). \tag{5.1.2}$$

The time integral of the force is called the impulse of the force. The product $m\mathbf{v}$ is called the momentum. Thus equation (5.1.2) states that the impulse of the force is equal to the change in the momentum of the particle.

Suppose that $\mathbf{F}$ is large in magnitude but acts for a short time interval only. We can idealise the situation by letting $F \to \infty$ and $(t_2 - t_1) \to 0$ in such a way that the expression

$$\bar{\mathbf{F}} = \lim_{t_2 \to t_1} \int_{t_1}^{t_2} \mathbf{F}\,\mathrm{d}t \tag{5.1.3}$$

remains finite. Then $\bar{\mathbf{F}}$ is called an impulse. It produces an instantaneous change in velocity without changing the position of the particle (for the velocity is finite and the time interval is infinitesimal). Physically such an impulse never occurs, but it is a useful idealisation to which many practical examples approximate. Typical examples are the blow of a hammer, the impact of one billiard ball on another and the impact of a missile on its target.

When dealing with problems involving impulses, note that any contribution to $\mathbf{F}$ which does not become correspondingly large as the time interval tends to zero will make no contribution to $\bar{\mathbf{F}}$.

Example 5.1.1

Consider the impact of two particles. By Newton's third law the impulses of the forces at contact are equal and opposite and so there is no net change in momentum. If the particles have masses m_1, m_2 and velocities $\mathbf{u}_1$, $\mathbf{u}_2$ before impact, their velocities $\mathbf{v}_1$, $\mathbf{v}_2$ after impact must be such that

$$m_1\mathbf{u}_1 + m_2\mathbf{u}_2 = m_1\mathbf{v}_1 + m_2\mathbf{v}_2.$$

One further condition is required to give $\mathbf{v}_1$ and $\mathbf{v}_2$ separately. If, for example, the two particles coalesce on impact, we have

$$\mathbf{v}_1 = \mathbf{v}_2 = \frac{m_1\mathbf{u}_1 + m_2\mathbf{u}_2}{m_1 + m_2}.$$

Other possibilities are discussed in Example 8.3.2.

Example 5.1.2

A light inelastic string OAB has one end attached to a fixed point O and particles of masses m_1, m_2 are attached to the string at A, B respectively. The system rests on a horizontal table with each of the portions OA, AB of the string straight and horizontal and inclined to each other at an angle $\pi - \alpha$ ($\alpha < \pi/2$), as

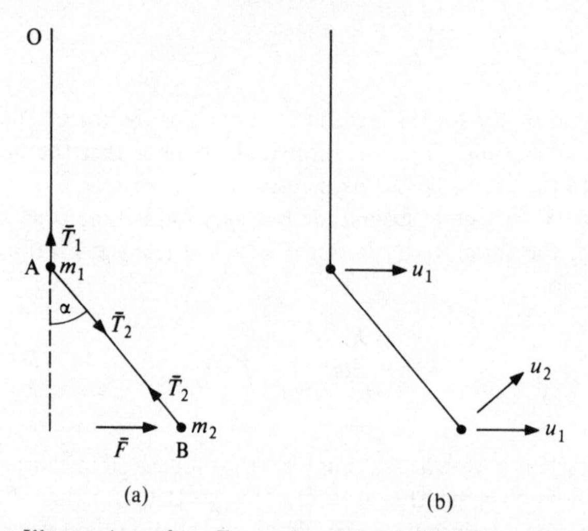

Figure 5.1.1 Illustration for Example 5.1.2: (a) the impulses; (b) the velocities.

in Figure 5.1.1. The particle B is given an impulsive blow $\bar{F}$ in the plane of OAB and in the direction perpendicular to and away from OA. Calculate the instantaneous velocities acquired by m_1 and m_2.

In addition to the applied impulse $\bar{F}$, there will be impulses $\bar{T}_1$ in OA and $\bar{T}_2$ in AB, say, since the string is inelastic.* The velocity of m_1 will be perpendicular

* In an elastic string no such impulse is produced because the tension, being proportional to the extension, cannot become large.

to OA, of magnitude u_1 say. The velocity of m_2 will consist of the velocity of m_1 and the velocity relative to m_1, say of magnitude u_2 perpendicular to AB.

The components of the impulsive equations of motion are

$$\bar{T}_2 \sin\alpha = m_1 u_1, \quad \bar{T}_1 - \bar{T}_2 \cos\alpha = 0,$$

$$\bar{F} - \bar{T}_2 \sin\alpha = m_2(u_1 + u_2 \cos\alpha), \quad \bar{T}_2 \cos\alpha = m_2 u_2 \sin\alpha.$$

These four equations are readily solved for the four unknowns u_1, u_2, $\bar{T}_1$, $\bar{T}_2$. The solutions for u_1 and u_2 are

$$u_1 = \frac{\bar{F}\sin^2\alpha}{m_1 + m_2 \sin^2\alpha}, \quad u_2 = \frac{m_1 \bar{F}\cos\alpha}{m_2(m_1 + m_2\sin^2\alpha)}.$$

5.2 WORK AND ENERGY

The next relation is obtained from the line integral of the basic equation of motion (5.1.1); this integral is evaluated along the path actually traversed by the particle. Let the position vector of the particle be $\mathbf{r}$ and let the initial and final values of $\mathbf{r}$ be $\mathbf{r}_1$ and $\mathbf{r}_2$. We have

$$\int_{\mathbf{r}_1}^{\mathbf{r}_2} \mathbf{F}\,.\,\mathrm{d}\mathbf{r} = m\int_{\mathbf{r}_1}^{\mathbf{r}_2} \frac{\mathrm{d}\mathbf{v}}{\mathrm{d}t}\,.\,\mathrm{d}\mathbf{r} = m\int_{t_1}^{t_2} \frac{\mathrm{d}\mathbf{v}}{\mathrm{d}t}\,.\,\mathbf{v}\,\mathrm{d}t = m\int_{t_1}^{t_2} \frac{\mathrm{d}v}{\mathrm{d}t} v\,\mathrm{d}t$$

$$= m\int_{t_1}^{t_2} \frac{\mathrm{d}}{\mathrm{d}t}(\tfrac{1}{2}v^2)\,\mathrm{d}t = \tfrac{1}{2}m(v_2^2 - v_1^2). \tag{5.2.1}$$

The line integral of the force is called the work done by the force. The product $\frac{1}{2}mv^2$ is called the kinetic energy of the particle. Equation (5.2.1) states that the work done by the force is equal to the change in kinetic energy of the particle.

A special simple case arises when $\mathbf{F}$ is a constant force, say $m\mathbf{a}$, so that the particle is moving with constant acceleration $\mathbf{a}$. Then

$$\int_{\mathbf{r}_1}^{\mathbf{r}_2} \mathbf{F}\,.\,\mathrm{d}\mathbf{r} = m\mathbf{a}\,.\,(\mathbf{r}_2 - \mathbf{r}_1) \tag{5.2.2}$$

and the energy equation (5.2.1) simplifies to

$$v_2^2 = v_1^2 + 2\mathbf{a}\,.\,(\mathbf{r}_2 - \mathbf{r}_1). \tag{5.2.3}$$

This is simply a derivation of equation (2.3.4) by different means.

The unit of work in the SI system is called the joule (J). It is equivalent to the work done by a constant force of 1 N moving a distance of 1 m in the direction of the force. Its dimensions are ML^2T^{-2}.

If for example a man of mass 80 kg climbs to a height of 1 m, the work done against the gravitational attraction is $80g$ J or about 785 J. To rise freely to the same height he would require an amount of initial kinetic energy equal to the work he does in climbing in order to overcome the gravitational attraction. This is equivalent to a speed of $(2g)^{1/2}$ or 4·4 m s^{-1}. In this example the work done by gravity is negative, because the man climbs upwards and gravity acts downwards. On the other hand, if the man were to fall, from rest, through a vertical distance of 1 m, the work done by gravity would be positive and the man would thereby acquire kinetic energy sufficient to give him a falling speed of 4·4 m s^{-1}. Thus the initial energy which produced the ascent is recovered when the man returns to his original position.

Example 5.2.1

A particle slides up a fixed rough slope inclined at an angle α to the horizontal. We consider the subsequent motion of the particle given that its initial speed is v_1 and that the coefficient of friction is μ.

The sum of the components of the forces perpendicular to the plane is zero, and the net force down the slope is $mg \sin\alpha + F$ (see Fig. 5.2.1). Since the particle is sliding, we have $F = \mu N = \mu mg \cos\alpha$, so that the force on the particle is $mg(\sin\alpha + \mu\cos\alpha)$ down the slope and this is constant. Hence the work done by this force when the particle travels a distance x up the slope is $-mg(\sin\alpha + \mu\cos\alpha)x$. If the particle then has speed v, it follows from (5.2.1) that

$$-mg(\sin\alpha + \mu\cos\alpha)x = \tfrac{1}{2}m(v^2 - v_1^2). \qquad (5.2.4)$$

Alternatively one can use the fact that the constant force $mg(\sin\alpha + \mu\cos\alpha)$ produces a constant acceleration $-g(\sin\alpha + \mu\cos\alpha)$ in the direction of x increasing. Equation (5.2.4) then follows from (2.3.8). The distance d travelled by the particle up the slope is obtained from (5.2.4) by putting $v = 0$, and so

$$d = \frac{v_1^2}{2g(\sin\alpha + \mu\cos\alpha)}. \qquad (5.2.5)$$

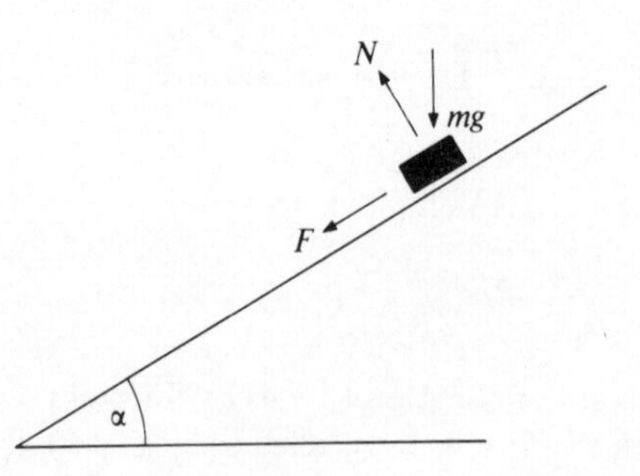

Figure 5.2.1 Illustration for Example 5.2.1: particle sliding up a rough inclined plane.

Let us suppose that the frictional force is sufficiently small to allow the particle to slide back down the plane. The magnitude of the frictional force will remain the same, but the sign will change. The net force on the particle is now $mg(\sin\alpha - \mu\cos\alpha)$ down the plane. When the particle reaches its initial position, the work done by this force is $mg(\sin\alpha - \mu\cos\alpha)d$. Accordingly the net work done during the whole motion is

$$mg(\sin\alpha - \mu\cos\alpha)d - mg(\sin\alpha + \mu\cos\alpha)d = -2mg\mu d\cos\alpha. \qquad (5.2.6)$$

The net loss in kinetic energy is $2mg\mu d\cos\alpha$, and if v_2 is the final speed, we have

$$\tfrac{1}{2}m(v_1^2 - v_2^2) = 2mg\,\mu\,d\cos\alpha. \qquad (5.2.7)$$

Note that the net work done on the particle by the gravitational force mg is zero. This is because the sign of this force does not change, therefore the negative work done by mg in the upward motion exactly balances the positive work done in the downward motion. All the contribution comes from the frictional force, which does change its sign so that it always opposes the motion, and hence does a negative amount of work leading to a loss in kinetic energy.

5.3 POTENTIAL ENERGY

Example 5.2.1 illustrates an essential difference between two types of forces, which it is now useful to classify. On the one hand there is the dissipative class of forces, typical examples of which are discussed in Chapter 4. The work done by such a force depends on the path of integration and is always negative. This is because a resistive force always opposes the motion of a particle on which it acts; it depends not only on the position of the particle, but also on the velocity.

On the other hand, in a force field the force is defined as a function only of position. The work of such a field along a path from A_1 to A_2 is then equal and opposite to the work evaluated from A_2 to A_1 along the *same* path. It follows that if the work is negative when evaluated in one direction, it will be positive when evaluated in the opposite direction along the same path. In general the work will depend on the path of integration between A_1 and A_2, but we have seen that there exist some vector fields with the property that the work depends only on the end points and not on the path (see §§1.9, 1.10). For such a field it also follows that when the work is a single valued function of the position of the end points, the net work evaluated around a closed path is zero. Vector fields of force which have this property are called conservative fields of force. A particle which traverses a closed trajectory under the influence of a conservative force returns to its original position with its kinetic energy unchanged, since the net work done is zero. Gravitational and electrostatic force fields are typical examples of conservative fields of force. The tension in an elastic string is also a good approximation to a conservative force.

The two classes of dissipative forces and conservative forces do not exhaust all the possibilities and there are others, in electromagnetism for example, which do not fit into either class. It is interesting to note the following implication for a force field in which the force is a function only of position but which does not have the property that the net work done along a closed path is zero. If the direction of traverse is chosen appropriately, the net work done will be positive. A particle under the influence of such a force could, by repeated traverse of such a path, acquire unlimited kinetic energy without permanently changing its position. If the gravitational field had this property we would have a readily available source of unlimited energy.

For the most part we shall be concerned with the dissipative and the conservative types of force. The latter plays a particularly important part in mechanics and its special properties can be exploited with advantage.

For a conservative force field the work done in moving from A_1 to A_2 by any path is equal and opposite to the work done in moving from A_2 to A_1 by any path. Equivalently the kinetic energy lost by a particle moving from A_1 to A_2 under the influence of such a force is equal to the kinetic energy gained in moving from A_2 to A_1. It is convenient to introduce the idea of a potential energy which is stored during the motion from A_1 to A_2 at the expense of the kinetic energy. This potential energy is released if the particle returns to A_1.

We define the potential energy at a point in a conservative field as the work done by a conservative force in bringing the particle from the given point with position vector $\mathbf{r}$, to some standard point with, say, position vector $\mathbf{r}_0$. Thus the potential energy V is given by

$$V = \int_{\mathbf{r}}^{\mathbf{r}_0} \mathbf{F} \, . \, d\mathbf{r} = -\int_{\mathbf{r}_0}^{\mathbf{r}} \mathbf{F} \, . \, d\mathbf{r}. \tag{5.3.1}$$

Conversely if the potential energy is known, the conservative force field is deducible, as in §1.10, from the relation

$$\mathbf{F} = -\nabla V.$$

Note the negative sign which arises from the way in which the potential energy is defined.

If the particle moves from $\mathbf{r}_1$ to $\mathbf{r}_2$, it follows that

$$\int_{\mathbf{r}_1}^{\mathbf{r}_2} \mathbf{F} \, . \, d\mathbf{r} = -\left(\int_{\mathbf{r}_2}^{\mathbf{r}_0} \mathbf{F} \, . \, d\mathbf{r} - \int_{\mathbf{r}_1}^{\mathbf{r}_0} \mathbf{F} \, . \, d\mathbf{r}\right) \tag{5.3.2}$$

irrespective of the path between $\mathbf{r}_1$ and $\mathbf{r}_2$. This is the loss in potential energy. By (5.2.1) it is equal to the gain in kinetic energy and so we have the conservation equation

$$\text{kinetic energy} + \text{potential energy} = \text{constant},$$

or
$$\tfrac{1}{2}mv^2 + \int_{\mathbf{r}}^{\mathbf{r}_0} \mathbf{F} \cdot d\mathbf{r} = \text{constant}. \qquad (5.3.3)$$

The simplest type of conservative field is a uniform force field such as gravity. Since $m\mathbf{g}$ is a constant vector, the potential energy is easily evaluated to be

$$\int_{\mathbf{r}}^{\mathbf{r}_0} m\mathbf{g} \cdot d\mathbf{r} = m\mathbf{g} \cdot (\mathbf{r}_0 - \mathbf{r}) = mg(y - y_0), \qquad (5.3.4)$$

where y is the vertical coordinate of the particle (remember that $\mathbf{g}$ has the component $-g$ vertically upwards). If y_0 is taken at the origin of coordinates, (5.3.4) reduces to mgy and so

$$\tfrac{1}{2}mv^2 + mgy = \text{constant},$$

or
$$v^2 = u^2 - 2gy$$

as in equation (2.4.4).

Example 1.10.1 shows that the force field $f(r)\mathbf{r}/r$, called a central force field, is conservative. The associated potential is

$$\int_{\mathbf{r}}^{\mathbf{r}_0} \frac{f(r)}{r} \mathbf{r} \cdot d\mathbf{r} = \int_{r}^{r_0} f(r)\, dr,$$

which clearly depends only on the distance from the origin. This type of force field will be studied in more detail in Chapter 7.

It is important to note that the concept of potential energy is appropriate only for conservative force fields, whereas equation (5.2.1), which equates the work done by the force along the prescribed path of the particle to the increase in kinetic energy, is quite general. Energy conservation is likewise restricted to problems involving conservative force fields. In Example 5.2.1, which involves dissipative frictional forces, the particle loses kinetic energy for which there is no compensation within the framework of Newtonian mechanics. In a wider context, however, energy conservation, which has been presented here as a deduction from the laws of motion, is regarded as a very basic law. Suitably modified, it is retained throughout the whole of modern physics. In Example 5.2.1 the loss of kinetic energy of the particle is equated to a dissipation into heat through the mechanism of the frictional forces. This is a form of energy not included in the mechanics of a single particle although Example 8.3.3 shows how it might be accounted for at a molecular level.

Example 5.3.1

A particle of mass m_2 hangs vertically from a light inextensible string, which passes over a smooth peg and is attached to a ring of mass m_1 ($<m_2$), free to slide on a smooth vertical wire. The horizontal distance between the peg and the

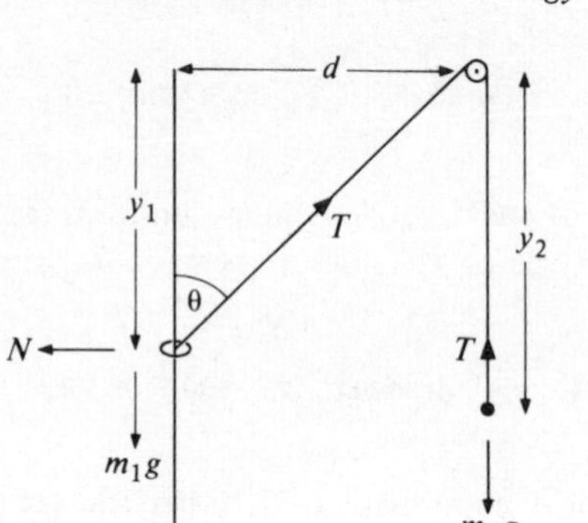

Figure 5.3.1 Illustration for Example 5.3.1.

wire is d. Initially that part of the string between m_1 and the peg is horizontal. If m_1 is released from rest in this position find the maximum depth to which it descends.

The configuration and the forces acting on m_1 and m_2 are shown in Figure 5.3.1. The vertical distances of m_1 and m_2 below the peg are denoted by y_1 and y_2 respectively. Because the wire is smooth, the reaction N between it and m_1 is horizontal. Hence there is no work done by N during the motion of m_1. There is, however, a tension T in the string which does work when the calculation is made separately for each mass. Moreover T is one of the unknown quantities in this problem.

Let us calculate the work equation for each mass separately. We have

$$\tfrac{1}{2}m_1\dot{y}_1^2 = \int_0^{y_1}(m_1 g - T\cos\theta)\,\mathrm{d}y_1, \tag{5.3.5}$$

for m_1, and

$$\tfrac{1}{2}m_2\dot{y}_2^2 = \int_{l-d}^{y_2}(m_2 g - T)\,\mathrm{d}y_2 \tag{5.3.6}$$

for m_2. Here θ is the inclination to the vertical of that part of the string between the ring and the peg, and l is the total length of the string, so that $y_2 = l - d$ initially. We also have the relations

$$\cos\theta = \frac{y_1}{(d^2 + y_1^2)^{1/2}}, \tag{5.3.7}$$

$$y_2 = l - (d^2 + y_1^2)^{1/2}, \tag{5.3.8}$$

and so

$$\mathrm{d}y_2 = -\frac{y_1\,\mathrm{d}y_1}{(d^2 + y_1^2)^{1/2}} = -\cos\theta\,\mathrm{d}y_1. \tag{5.3.9}$$

It follows that, whatever the magnitude of T,

$$\int_0^{y_1} T\cos\theta \, dy_1 = -\int_{l-d}^{y_2} T \, dy_2.$$

We can therefore eliminate T by the addition of (5.3.5) and (5.3.6) to get

$$\tfrac{1}{2}m_1\dot{y}_1^2 + \tfrac{1}{2}m_2\dot{y}_2^2 = \int_0^{y_1} m_1 g \, dy_1 + \int_{l-d}^{y_2} m_2 g \, dy_2 = m_1 g y_1 - m_2 g(l - d - y_2). \tag{5.3.10}$$

Equation (5.3.10) amounts to an energy equation for the system as a whole in that it equates the increase in kinetic energy of the system to the decrease of potential energy arising solely from the conservative gravitational field. This result is a consequence of the fact that the remaining forces N and T do no work during the motion of the complete system.

The maximum displacement is obtained when $\dot{y}_1$ and $\dot{y}_2$ are zero. Equations (5.3.8) and (5.3.10) then give

$$m_1 y_1 - m_2(l - d - y_2) = m_1 y_1 - m_2\{(d^2 + y_1^2)^{1/2} - d\} = 0$$

or

$$m_1 y_1 + m_2 d = m_2(d^2 + y_1^2)^{1/2}. \tag{5.3.11}$$

The solution is $y_1 = 0$, the initial position, or

$$y_1 = \frac{2m_1 m_2 d}{m_2^2 - m_1^2}. \tag{5.3.12}$$

One point to note here is that (5.3.12) is obtained from (5.3.11) by squaring each side of the relation. Hence the solution of (5.3.11) satisfies (5.3.12), but the converse is not necessarily true. In fact (5.3.12) satisfies (5.3.11) only if $m_2 > m_1$. If $m_2 < m_1$ there is no solution of (5.3.11) and in this case the ring, once released, never comes to rest while the string remains suspended over the peg.

The result that the tension in the string does no net work during the displacement of the system is capable of wider application in that it depends on the following conditions, which are satisfied for many other configurations involving light, elastic strings:

(i) the tensions at the two ends of the string are equal in magnitude, because the peg is smooth and the mass of the string is negligible;

(ii) when the two ends of the string suffer small displacements, the components of those displacements along the length of the string (and hence parallel to T) are equal in magnitude. This follows immediately from the assumption that the string is inextensible, and ensures the validity of (5.3.9).

5.4 POWER

Since the work done by a force $\mathbf{F}$ is

$$\int \mathbf{F} \cdot d\mathbf{r} = \int \mathbf{F} \cdot \mathbf{v}\, dt,$$

the time derivative of the work done, or the rate of working, is $\mathbf{F} \cdot \mathbf{v}$. This is referred to as power. The unit of power in the SI system is the watt (W), or $1\ \mathrm{J\,s^{-1}}$, and its dimensions are $\mathrm{ML^2\,T^{-3}}$. The kilowatt hour ($1\ \mathrm{kWh} = 3{\cdot}6 \times 10^6\ \mathrm{J}$) is often used as a practical measure of work or energy, particularly with reference to heat. The latter is also a form of energy although it plays no part in Newtonian mechanics.

Power is a useful practical measure of a machine's capacity for doing work. For example the powers of a small electric motor and a typical automobile engine are likely to be in the ranges 0·1 – 1 kW and 10–100 kW respectively.

Example 5.4.1
Water of density ρ is pumped from a depth d and delivered through a pipe of cross-section A. What is the power required to deliver a volume $\dot{V}$ per unit time?

There are two parts to this calculation. The first part represents the work done in raising a mass of water $\rho\dot{V}$ from a depth d in unit time, which is $\rho\dot{V}gd$. In addition there is the work required to generate the requisite kinetic energy. In unit time a volume $\dot{V}$ of water is delivered through a pipe of cross-section A. The speed is therefore $\dot{V}/A$ and since the mass is $\rho\dot{V}$ the kinetic energy is $\frac{1}{2}\rho\dot{V}(\dot{V}/A)^2$. The total work per unit time, which is the power required, is $\rho\dot{V}gd + \frac{1}{2}\rho\dot{V}^3/A^2$.

If we substitute the particular values $g = 9.81\ \mathrm{m\,s^{-2}}$, $\rho = 10^3\ \mathrm{kg\,m^{-3}}$, $d = 10\ \mathrm{m}$, $A = 10^{-2}\ \mathrm{m^2}$, $\dot{V} = 10^{-1}\ \mathrm{m^3\,s^{-1}}$ we get $(9{\cdot}81 + 5)10^3$ W or about 15 kW.

Example 5.4.2
A train of mass m runs on a level track at constant speed u. We assume that the power exerted by the engine is constant and that the resistance to the motion arises from two sources. The air resistance is modelled by a term Kv^2, where v is the speed and K is a constant, and frictional resistance is assumed to be equal to a constant fraction β of the weight of the train. We investigate the change in the speed of the train when it slips a coach of mass αm.

Initially the total resistance is $Ku^2 + \beta mg$ and, since the speed is constant, this must be balanced by the tractive force. The power is therefore $(Ku^2 + \beta mg)u$ and is constant. When the coach is slipped, let the subsequent speed be v. The tractive force is then $(Ku^3 + \beta mgu)/v$ and the resistance is $Kv^2 + \beta(1 - \alpha)mg$. It follows that the equation of motion is

$$(1 - \alpha)m\frac{dv}{dt} = \frac{Ku^3 + \beta mgu}{v} - Kv^2 - \beta(1 - \alpha)mg,$$

or
$$v\frac{\mathrm{d}v}{\mathrm{d}t} = -\frac{K}{m(1-\alpha)}(v^3 - u^3) + \frac{\beta g}{1-\alpha}(u - v + v\alpha).$$

In order to describe the details of the motion, it is necessary to integrate this equation. The integration is routine but the final result is not of a simple form and we shall proceed on the basis of a simplifying assumption. The equation has the special solution $v = u$ when $\alpha = 0$, as might be expected, so that presumably $|v - u| \ll u$ when $\alpha \ll 1$. To investigate this possibility we put $v = u(1 + \delta)$, so that

$$(1+\delta)\frac{\mathrm{d}\delta}{\mathrm{d}t} = -\frac{Ku}{m(1-\alpha)}\{(1+\delta)^3 - 1\} + \frac{\beta g}{(1-\alpha)u}\{\alpha - \delta(1-\alpha)\},$$

and approximate this equation by retaining only the first order terms in δ. The result is a linear equation of the form

$$\frac{\mathrm{d}\delta}{\mathrm{d}t} = -\tau^{-1}(\delta - \epsilon),$$

where
$$\tau^{-1} = \frac{\beta g}{u}\left(1 + \frac{3Ku^2}{m\beta g(1-\alpha)}\right), \quad \epsilon = \frac{\alpha}{1 - \alpha + 3Ku^2/m\beta g}.$$

The solution is
$$\delta = \epsilon(1 - \mathrm{e}^{-t/\tau})$$

and we can now see more clearly the conditions for which the approximation is valid. As $t \to \infty$, $\delta \to \epsilon$ and the terminal speed is $u(1 + \epsilon)$. In order that δ should be uniformly small, it is necessary that $\epsilon \ll 1$, or $\alpha \ll 1 + 3Ku^2/(m\beta g)$. The parameter τ, which has the dimensions of time, is a measure of the time interval over which the speed changes by a significant proportion of itself.

EXERCISES

5.1 A golf ball of mass m is struck by a club, the contact lasting for a time τ, and is thereby given a speed v. Assume that, during the contact, the force increases at a constant rate from zero to a maximum and then decreases at a constant rate from this maximum to zero. Find: (i) the impulse of the blow, (ii) the maximum force on the ball.
(Ans.: mv, $2mv/\tau$.)

5.2 A light rod rests on a smooth horizontal plane and has two particles, of masses m_1 and m_2 respectively, attached to it at A and B. A horizontal impulse $\bar{F}$ is given to A in a direction making an angle α with AB. Find the velocities of A and B, and show that the kinetic energy imparted to the system is equal to $\frac{1}{2}\bar{F}w$, where w is the component of the velocity of A in the direction of the impulse.

What would be the effect on the initial motion if the plane were rough? (Ans.: A component $\bar{F} \cos \alpha/(m_1 + m_2)$ along AB for both particles and a component $\bar{F} \sin \alpha/m_1$ perpendicular to AB for A.)

5.3 A man of mass m_1 is standing in a stationary lift of mass m_2, the counterpoise being of mass $m_1 + m_2$. Suddenly the man jumps with an impulse which would raise him to a height h if he were jumping from the ground.

Given that the inertial and frictional effects in the cable and pulley which connect the lift to the counterpoise may be neglected, calculate the velocity of the man, relative to the lift, immediately after the impulse. Find also the subsequent acceleration of the man relative to the lift.

Deduce that the height above the floor of the lift to which the man jumps is $2h(m_1 + m_2)/(m_1 + 2m_2)$.

5.4 If the earth is regarded as a sphere, of radius r_e, the gravitational force on a particle of mass m at a distance r ($>r_e$) from the centre of the earth is $mg(r_e/r)^2$.

Show that the work done in raising the particle from the earth's surface to a height r_e above the earth's surface is $\frac{1}{2}mgr_e$.

5.5 A mass of 200 kg is suspended at the lower end of a light vertical rope and is being hauled up vertically. Initially the mass is at rest and the pull on the rope is 300 kg-wt. The pull diminishes uniformly at the rate of 6 kg-wt for each metre through which the mass is raised.

Find the work done by the pull on the rope, and the speed of the mass after it has been raised 20 m.
(Ans.: $4800g$ J, $(8g)^{1/2}$ m s^{-1}.)

5.6 The tension in a stretched elastic string is given to be $\lambda x/l$ in magnitude, where l is the natural length of the string, x is the extension beyond its natural length, and λ is a constant for a particular string.

Show that the work done in increasing x from x_1 to x_2 is $\frac{1}{2}\lambda(x_2^2 - x_1^2)/l$.

A uniform elastic string is stretched in a straight line between two points A, B which are fixed. A point C of the string is then displaced so that the components of the displacement are x along AB and y perpendicular to AB.

Before the displacement let AC $= a$, CB $= b$ and let the tension in the string be T. Let the unstretched length be $k(a + b)$, where $0 < k < 1$.

Show that if the string remains taut during the displacement, the work done is

$$\frac{T(a+b)(x^2+y^2)}{2ab(1-k)} - \frac{Tk}{1-k}\left[\{(a+x)^2+y^2\}^{1/2} - a + \{(b-x)^2+y^2\}^{1/2} - b\right].$$

For $r^2 = x^2 + y^2 \ll a^2 + b^2$, show that this approximates to

$$\frac{T(a+b)}{2ab}\left(\frac{x^2}{1-k} + y^2\right).$$

When the tensions in AC and CB are the same after displacement, show that the work done can be written in the form

$$\frac{Tk(a+b)}{1-k}\left\{\frac{r^2}{2kab}-\left(1+\frac{r^2}{ab}\right)^{1/2}+1\right\}.$$

5.7 A particle moves along the x-axis under the action of a force $-\mu/x^2$ ($\mu > 0$) per unit mass in the x direction. Initially $\dot{x} > 0$. Establish the energy equation in the form

$$\tfrac{1}{2}\dot{x}^2 = E + \mu/x,$$

where E is a constant.

(i) Show that if $E = -\mu/2a$ ($a > 0$), the particle comes to rest and then falls into the origin. Use the substitution $x = a(1 + \cos\theta)$ to obtain a parametric relation between x and t.

(ii) Show that if $E = 0$, the particle moves off to infinity and

$$x^{3/2} = c^{3/2} + 3(\mu/2)^{1/2}t.$$

(iii) For $E = \mu/2a$ ($a > 0$), use the substitution $x = a(\cosh\phi - 1)$ to obtain a parametric relation between x and t.

5.8 A parcel, of mass m, rests on a horizontal conveyor belt. For time $t \geqslant 0$ the belt is constrained to move with a constant horizontal velocity of magnitude u. The coefficient of friction between the parcel and the belt is μ. Find:

(i) the time that elapses before the parcel is again at rest relative to the belt;
(ii) the distance the particle slides relative to the belt;
(iii) the energy dissipated during this sliding;
(iv) the impulse of the frictional force.

(Ans.: $u/\mu g$, $u^2/2\mu g$, $\frac{1}{2}mu^2$, mu.)

5.9 Show that the force field

$$\mathbf{F} = -F_0(x\mathbf{i} + y\mathbf{j} + c\mathbf{k})/l$$

where F_0, c and l are constants, is conservative. Deduce the potential energy.

A particle of mass m has the position vector

$$\mathbf{r} = r_0(\mathbf{i}\sin\omega t + \mathbf{j}\cos\omega t + \mathbf{k}n^2t^2)$$

at time t. Show that its motion is compatible with that arising when it is subject to the force $\mathbf{F}$, with appropriate values for F_0 and c. Show that the magnitude of the force remains constant during the motion of the particle.

What is the work done by the force in the interval $0 \leqslant t \leqslant T$? (Ans.: $2mr_0^2 n^4 T^2$.)

5.10 A particle of mass m moves under the influence of the force field

$$\mathbf{F} = F_0(y\mathbf{i} + x\mathbf{j} + z\mathbf{k})/l,$$

where F_0 and l are constants. Show that this is a conservative field. Find the potential energy and deduce the energy equation

$$v^2 = n^2(2xy + z^2) + u^2,$$

where $n^2 = F_0/ml$, v is the speed of the particle and u is the value of v at the origin.

Let $X = x + y$, $Y = x - y$. Show that the components of the equation of motion for the particle are equivalent to the relations

$$\ddot{X} = n^2 X, \quad \ddot{Y} = -n^2 Y, \quad \ddot{z} = n^2 z.$$

At time $t = 0$, the particle is at the origin with velocity

$$\mathbf{u} = u_x\mathbf{i} + u_y\mathbf{j} + u_z\mathbf{k}.$$

Show that subsequently its position vector is given by

$$\begin{aligned} n\mathbf{r} &= \tfrac{1}{2}\{(u_x + u_y)\sinh nt + (u_x - u_y)\sin nt\}\mathbf{i} \\ &\quad + \tfrac{1}{2}\{(u_x + u_y)\sinh nt - (u_x - u_y)\sin nt\}\mathbf{j} \\ &\quad + u_z \sinh nt\,\mathbf{k}. \end{aligned}$$

Verify directly from this result that the energy equation is indeed satisfied.

5.11 The potential of an electrostatic field satisfies the equation

$$\frac{\partial^2 V}{\partial x^2} + \frac{\partial^2 V}{\partial y^2} + \frac{\partial^2 V}{\partial z^2} = 0.$$

Assume that when $(x^2 + y^2) \ll 1$, the potential is of the form

$$V = V_0(z) + (x^2 + y^2)V_1(z) + (x^2 + y^2)^2 V_2(z) + \ldots$$

and show that

$$V_0'' + 4V_1 = 0, \quad V_1'' + 16V_2 = 0, \ldots.$$

A particle of unit mass is subject to such a potential. Show that, when second order terms are neglected, the equations of motion take the form

$$\ddot{x} = -2xV_1, \quad \ddot{y} = -2yV_1, \quad \ddot{z} = -V_0'$$

and, hence, that the speed v of the particle is given by the equation

$$\tfrac{1}{2}v^2 = C - V_0(z)$$

approximately, where C is a constant.

Show that $v^{1/2}x$ and $v^{1/2}y$ both satisfy the equation

$$\frac{d^2 q}{dz^2} + \frac{3v'^2}{4v^2} q = 0.$$

5.12 Two particles of masses $7m$ and $3m$ are fastened to the ends A, B respectively of a weightless rigid rod of length l, which is freely hinged at a point O of the rod such that $AO = l/3$. Prove that if the rod is just disturbed from its position of unstable equilibrium, the velocity with which A will pass through its position of stable equilibrium is $(4gl/57)^{1/2}$.

5.13 A particle is attached to one end of a light inelastic string of length l, the other end of which is tied to a fixed peg O. The particle is projected horizontally with speed u from the point at a depth l vertically below O. Show that if $2gl < u^2 < 5gl$ the string will not remain taut during the subsequent motion. Show: (i) that if $u^2 = (2 + 3^{1/2})gl$, the particle will subsequently strike the peg; and (ii) that if $u^2 = (7/2)gl$, the string will slacken when it has turned through an angle $2\pi/3$, and that when it again tightens the particle will be at the point from which it was originally projected.

5.14 Two particles, of masses m_1, m_2 $(m_2 > m_1)$, are connected by a light string of length πa, which is hung over a fixed, smooth, circular cylinder of radius a with its axis horizontal. Initially the string is held at rest on the upper half of the cylinder in a plane perpendicular to the axis. Show that if the string is released, and if m_1 remains in contact with the cylinder, the common speed of the masses when m_2 has fallen a distance $a\theta$ $(\theta < \pi)$ is

$$\left(\frac{2ga}{m_1 + m_2}(m_2\theta - m_1 \sin\theta)\right)^{1/2}.$$

Find the reaction between m_1 and the cylinder at this instant. Hence show that m_1 will certainly remain in contact with the cylinder initially if $m_2 < 3m_1$.

(Ans.: $m_1 g\{(3m_1 + m_2)\sin\theta - 2m_2\theta\}/(m_1 + m_2)$.)

5.15 A light string rests over two smooth pegs, which are at a distance $2a$ apart and at the same horizontal level. A smooth ring, of mass m_3, slides freely on the string and two particles, of masses m_1, m_2 respectively, are attached to the ends of the string.

Initially the system is held with that part of the string between the pegs horizontal and the ring midway between the pegs. The portions of the string beyond the pegs are vertical with the particles hanging at rest.

The system is then released from rest. Let y_1, y_2, y_3 denote, respectively, the distances which m_1, m_2, m_3 descend below their initial levels, so that $y_1 = y_2 = y_3 = 0$ initially.

Explain why m_3 descends vertically.

Write down the equations of motion for the three masses. Hence or otherwise show that:

(i) $y_1 + y_2 + 2(a^2 + y_3^2)^{1/2} = 2a$,
(ii) $m_1\ddot{y}_1 - m_2\ddot{y}_2 = (m_1 - m_2)g$,
(iii) $\frac{1}{2}(m_1\dot{y}_1^2 + m_2\dot{y}_2^2 + m_3\dot{y}_3^2) = g(m_1y_1 + m_2y_2 + m_3y_3)$.

Find y_1 and y_2 in terms of y_3 and t.

Show also that if m_3 comes to rest in the subsequent motion it will first do so after it has descended a distance

$$\frac{2ak}{1-k^2}, \quad \text{where } k = \frac{m_3(m_1+m_2)}{4m_1m_2}$$

and that this is possible only if $k < 1$.

5.16 Assume that the human heart works at an average rate of 3 W. Show that this is just sufficient to allow a person of mass 70 kg to climb from the bottom to the top of the Empire State Building, New York (380 m) in 24 h.

5.17 A waterfall of height h is fed by a stream of density ρ, breadth b and depth d approaching the top with speed u. Show that the rate at which work could be done by the waterfall, assuming no energy dissipation, is $bdu\rho(u^2 + 2gh)/2$.

In particular if $b = 40$ m, $d = 1$ m, $h = 30$ m, $u = 3$ m s^{-1} show that the maximum power that could be generated is $3{\cdot}6 \times 10^7$ W.

5.18 The engine of a car produces a power P when the car is running on the level at a speed v, where

$$P = P_0 n \frac{v}{u}\left(1 - n\frac{v}{u}\right).$$

Here P_0 and u are constants and n takes the values 3, 2, 1 respectively for first, second and third gears.

Write down the equation for motion on a level track with all resistance neglected. Hence show that the time taken to reach third gear on a level track after starting from rest cannot be less than

$$\left(\tfrac{1}{2}\log 3 - \tfrac{5}{36}\log 5 - \tfrac{1}{9}\log 2\right)\frac{mu^2}{P_0} = 0{\cdot}25\,\frac{mu^2}{P_0},$$

where m is the mass of the car.

5.19 A car of mass m travels on level ground. Its engine can deliver any power less than a certain constant P to the back wheels, the coefficient of friction between those wheels and the ground is μ, and the weight supported by them is W. Show that if all four wheels are assumed massless and their bearings frictionless, attempts to accelerate too rapidly may result in skidding if the speed is less than $P/\mu W$, and find the minimum time in which any speed greater than $P/\mu W$ can be attained from rest. If the car were going down a hill (with μ unaltered), would the maximum speed for skidding while accelerating be greater, less, or the same as before? Give reasons.

(Ans.: $(v^2 + P^2/\mu^2 W^2)m/2P$, greater.)

5.20 A train is drawn along a straight level track by an engine which exerts a constant force. The initial speed is u_1, and after an interval of time t the speed is u_2. All resisting forces are neglected. Show that the distance travelled is $\frac{1}{2}(u_1 + u_2)t$.

Now consider the problem with an engine working at constant power instead of constant force. Obtain the distance travelled assuming that u_1, u_2 and t remain the same. Deduce that the distance travelled in the second case is the greater.

Show that the engines are working at the same rate at time $t/2$.

6 Oscillations

6.1 SIMPLE HARMONIC OSCILLATIONS

In this chapter we shall be particularly concerned with the one-dimensional motion of a particle which oscillates to and fro about an equilibrium position. When the motion repeats itself after a definite time, it is called periodic. Of all the forces which can produce periodic motion, a restoring force proportional to the displacement of the particle is of particular interest. The resulting equation of motion has many applications in mechanics and indeed in several other branches of mathematical physics.

Let $x\mathbf{i}$ be the displacement of the particle from the origin, and let the force acting on the particle be $-n^2 x\mathbf{i}$ per unit mass. The equation of motion is then

$$\ddot{x} = -n^2 x. \tag{6.1.1}$$

Here $n > 0$ is a constant and the negative sign ensures that the force is a restoring force acting always towards the equilibrium value $x = 0$, which is a particular solution of (6.1.1). Such a force is conservative and can be derived from a potential energy $\frac{1}{2}n^2x^2$ per unit mass. Hence there is an energy equation of the form

$$\tfrac{1}{2}\dot{x}^2 + \tfrac{1}{2}n^2x^2 = \tfrac{1}{2}n^2a^2 \tag{6.1.2}$$

say, where the constant total energy has been written as $\frac{1}{2}n^2a^2$. It is easily verified that (6.1.1) follows from (6.1.2) by differentiation. Equation (6.1.2) is equivalent to

$$\frac{\mathrm{d}x}{\mathrm{d}t} = \pm n(a^2 - x^2)^{1/2}, \tag{6.1.3}$$

where the choice of sign will depend on whether x is increasing or decreasing with time. Finally (6.1.3) can be integrated to give

$$nt + \alpha = \pm\int \frac{\mathrm{d}x}{(a^2 - x^2)^{1/2}} = \pm\cos^{-1}\frac{x}{a}$$

or

$$x = a\cos(nt + \alpha). \tag{6.1.4}$$

Note that the motion is confined to $|x| \leqslant a$.

A more direct derivation of (6.1.4) is obtained by assuming that (6.1.1) has solutions of the form $x = Ae^{\lambda t}$, where A, λ are constants. It follows, by substitution in (6.1.1), that $\lambda^2 = -n^2$. This gives the two solutions $A_1 e^{int}$, $A_2 e^{-int}$.

which, for a linear differential equation, can be combined linearly to give the general solution

$$x = A_1 e^{int} + A_2 e^{-int}.$$

When this solution is applied to a physical problem, we are interested in a form which is real. Since the constants A_1 and A_2 are still at our disposal we can write

$$A_1 = \tfrac{1}{2}ae^{i\alpha}, \quad A_2 = \tfrac{1}{2}ae^{-i\alpha}$$

so that $x = a\cos(nt + \alpha)$ as in (6.1.4).

The solution (6.1.4) is determined by the values of the three parameters a, n, α called respectively the amplitude, frequency and phase. The motion is periodic, and repeats itself after a time $2\pi/n$, called the periodic time. In the above derivation a and α occur as constants of integration which will be determined in applications by some given initial conditions. Sometimes it is more convenient to put $A = a\cos\alpha, B = -a\sin\alpha$ and to write the solution in the form

$$x = A\cos nt + B\sin nt. \tag{6.1.5}$$

The motion described by the basic equation (6.1.1) with the corresponding solution (6.1.4) or (6.1.5) is called simple harmonic motion. It is a pure sinusoidal displacement in time, with a single frequency n. Note that a displacement such as

$$x = a_1\cos(nt + \alpha_1) + a_2\cos(2nt + \alpha_2),$$

which consists of two superimposed harmonics, is also periodic with period $2\pi/n$, but it involves the two frequencies n and $2n$. It does not arise from an equation of motion of the form (6.1.1) although, as we shall see, it may arise when two oscillating particles are coupled.

One of the simplest applications of (6.1.1) is to the motion of a particle suspended from an elastic string or spring, since the tension exerted by a string or spring is proportional to the extension. This empirical law was first discovered by Hooke and is known as Hooke's law. Note that an elastic string can only exert a force when extended beyond its natural length, whereas a spring also satisfies the law under compression. If the unstretched length is l, and the stretched length is $l + x$, the tension is given by the relation

$$T = \sigma x \tag{6.1.6}$$

where the constant of proportionality σ, called the stiffness, depends on the material but is independent of x. It also depends on l since a change in l will alter the extension x proportionately if the tension remains unaltered. Hence σ is inversely proportional to l and (6.1.6) is often written in the form

$$T = \lambda x/l \tag{6.1.7}$$

where, for a given material, λ is independent of l and is called the modulus of rigidity.

Hooke's law is also valid for a metal wire under tension provided that it is not stretched beyond its elastic limit, beyond which it flows plastically and is permanently deformed. The modulus of rigidity for a metal wire is independent of l but is proportional to the cross-section of the wire. For this reason the elasticity is usually specified by the ratio λ/A, called Young's modulus, which is a constant for a given material.

Example 6.1.1

A light spring has one end attached to a fixed point and a mass m suspended from the free end. The equilibrium extension of the spring is c and the equilibrium tension is T_0 as in Figure 6.1.1. Therefore

$$T_0 = \sigma c = mg.$$

If the spring is disturbed through a distance x from its equilibrium position, we have

$$m\ddot{x} = mg - T,$$

and

$$T = \sigma(c + x) = mg + \sigma x.$$

Hence

$$\ddot{x} = -\sigma x/m = -gx/c,$$

and the motion is oscillatory with a solution of the type (6.1.4). The periodic

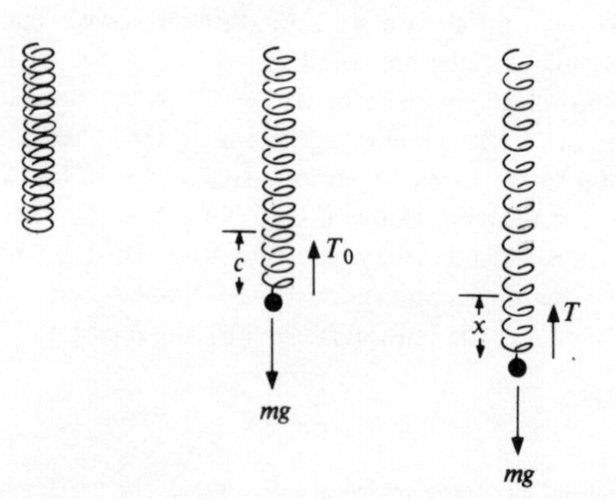

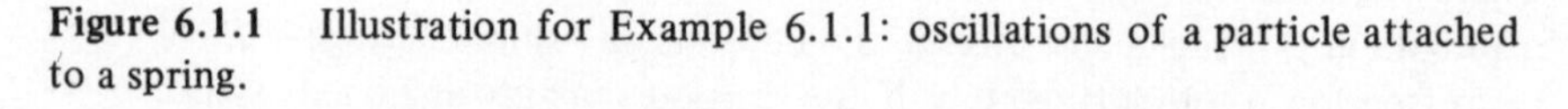

Figure 6.1.1 Illustration for Example 6.1.1: oscillations of a particle attached to a spring.

time is $2\pi(c/g)^{1/2}$. The amplitude and phase can be determined once the initial conditions have been prescribed.

Example 6.1.2

A light elastic string has one end attached to a fixed point and a mass m suspended from the free end. The equilibrium extension of the string is c. The mass is pulled down a further distance $2c$ and then released. Show that the period of the subsequent motion is $2(c/g)^{1/2}(3^{1/2} + 2\pi/3)$.

Initially the motion is simple harmonic, as in the previous example, with frequency $n = (g/c)^{1/2}$, amplitude $2c$ and zero phase. Hence $x = 2c \cos nt$, where x is measured from the equilibrium position (see Fig. 6.1.1). This motion persists until $x = -c$. This is when $\cos nt = -\frac{1}{2}$ and first arises when $nt = 2\pi/3$, so that

$$\dot{x} = -2cn \sin nt = -(3gc)^{1/2}.$$

The string then becomes slack and the particle continues to rise freely for a time $(3gc)^{1/2}/g = (3c/g)^{1/2}$. The time for a half period is therefore

$$\frac{2\pi}{3n} + \left(\frac{3c}{g}\right)^{1/2} = \left(\frac{c}{g}\right)^{1/2}\left(\frac{2\pi}{3} + 3^{1/2}\right).$$

It will take an equal time to return to its original position after which the motion will repeat itself.

Note that although the motion is periodic, and although the first part of the motion is governed by an equation of the form (6.1.1), it is not simple harmonic motion because for part of the time the particle is moving freely under gravity. If, however, the string is replaced by a light elastic spring of the same length and stiffness, the relation $x = 2c \cos nt$ holds throughout the motion.

Example 6.1.3

A particle of mass m is connected by a light spring of natural length l to a point O on a rough horizontal table. Let us consider the equation of motion when the particle is at a distance $l + x$ from O. Here the magnitude of the frictional force is μmg, but the direction will depend on the direction of motion of the particle.

Suppose first that x is increasing. The equation of motion is

$$m\ddot{x} = -T - \mu mg = -\frac{\lambda x}{l} - \mu mg,$$

or

$$\frac{d^2}{dt^2}\left(x + \frac{\mu mgl}{\lambda}\right) = -\frac{\lambda}{ml}\left(x + \frac{\mu mgl}{\lambda}\right), \tag{6.1.8}$$

where $T = \lambda x/l$ is the tension and $F = \mu mg$ is the frictional force (see Fig. 6.1.2). Equation (6.1.8) is similar to (6.1.1) with $n^2 = \lambda/ml$ and the variable changed to $(x + \mu mgl/\lambda)$. The motion is thus similar to a simple harmonic oscillation but

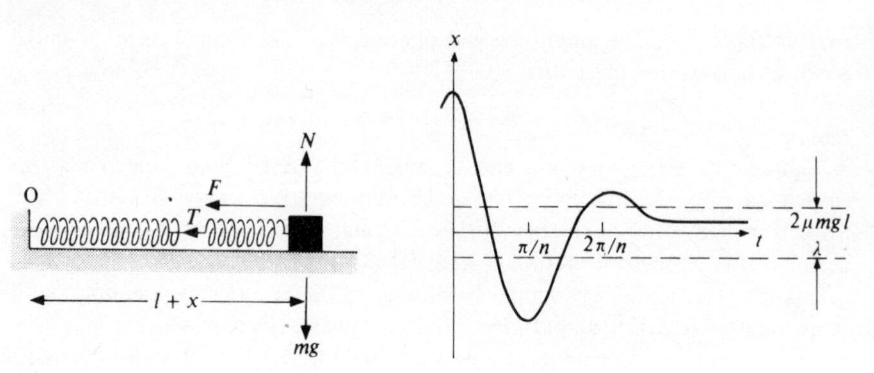

Figure 6.1.2 Illustration for Example 6.1.3: particle subject to a restoring force T and a frictional force F.

with symmetry about $x = -\mu mgl/\lambda$ instead of $x = 0$, as in the unresisted case.

When x is decreasing, however, the frictional force changes sign and the equation of motion becomes

$$m\ddot{x} = -\frac{\lambda x}{l} + \mu mg,$$

or

$$\frac{d^2}{dt^2}\left(x - \frac{\mu mgl}{\lambda}\right) = -\frac{\lambda}{ml}\left(x - \frac{\mu mgl}{\lambda}\right). \tag{6.1.9}$$

The centre of the motion thus switches to $x = \mu mgl/\lambda$. This switch of the centre of the motion each time the direction of motion changes decreases the effective amplitude of oscillation until eventually the particle comes to rest within a distance $\mu mgl/\lambda$ on either side of the origin $x = 0$. When this happens, the restoring force $-\lambda x/l$ is smaller in magnitude than the limiting frictional force μmg and the particle will remain permanently at rest (see Fig. 6.1.2). Note that the ultimate position of rest can be anywhere in the interval $-\mu mgl/\lambda \leqslant x \leqslant \mu mgl/\lambda$ depending on the initial conditions.

6.2 DAMPED OSCILLATIONS

In practice mechanical oscillations are damped by natural or artificially induced resistance. We consider here the effect of viscous damping proportional to the velocity. Let this be $-k\dot{x}\mathbf{i}$ per unit mass, so that (6.1.1) is modified to

$$\ddot{x} = -n^2 x - k\dot{x}. \tag{6.2.1}$$

This is still a linear differential equation with constant coefficients, and to solve it we look for solutions of the form

$$x = Ae^{-\lambda t}. \tag{6.2.2}$$

When this is substituted into (6.1.1) we get

$$\lambda^2 - k\lambda + n^2 = 0.$$

This equation has the two roots

$$\left.\begin{aligned}\lambda_1 &= \tfrac{1}{2}k + \tfrac{1}{2}(k^2 - 4n^2)^{1/2}\\ \lambda_2 &= \tfrac{1}{2}k - \tfrac{1}{2}(k^2 - 4n^2)^{1/2}\end{aligned}\right\} \qquad (6.2.3)$$

and the form of the solution will depend on the sign of $k^2 - 4n^2$.

If $k < 2n$, the general solution can be written

$$\begin{aligned} x &= \mathrm{e}^{-kt/2}[A_1 \exp\{-\mathrm{i}(n^2 - \tfrac{1}{4}k^2)^{1/2}t\} + A_2 \exp\{\mathrm{i}(n^2 - \tfrac{1}{4}k^2)^{1/2}t\}],\\ &= a\mathrm{e}^{-kt/2} \cos\{(n^2 - \tfrac{1}{4}k^2)^{1/2}t + \alpha\}, \end{aligned} \qquad (6.2.4)$$

where the arbitrary constants

$$A_1 = \tfrac{1}{2}a\mathrm{e}^{-\mathrm{i}\alpha}, \qquad A_2 = \tfrac{1}{2}a\mathrm{e}^{\mathrm{i}\alpha}$$

have been chosen to give a real form of the solution (6.2.4). In this case the damping is conveniently described as light. The motion is oscillatory, but the amplitude decays according to the factor $\mathrm{e}^{-kt/2}$ and so tends to zero as t tends to infinity. A typical example is a pendulum oscillating in air.

For $k > 2n$, the solution can be written directly as

$$x = A_1\mathrm{e}^{-\lambda_1 t} + A_2\mathrm{e}^{-\lambda_2 t}, \qquad (6.2.5)$$

where λ_1 and λ_2 ($<\lambda_1$) are now both real and positive. This solution represents exponential decay, without oscillation, to $x = 0$. If $A_1/A_2 < -1$ it will pass through $x = 0$ just once, otherwise it will decay monotonically to zero. Such heavy damping might arise from the motion of a pendulum in a very viscous medium.

This approach fails to give two independent solutions in the critical case $k = 2n$, when λ_1 and λ_2 are both equal to $\tfrac{1}{2}k$. We can see more clearly what happens in this case if equation (6.2.1) is rewritten in the equivalent form

$$\frac{\mathrm{d}^2}{\mathrm{d}t^2}(x\mathrm{e}^{kt/2}) = -(n^2 - \tfrac{1}{4}k^2)(x\mathrm{e}^{kt/2}). \qquad (6.2.6)$$

With $x\mathrm{e}^{kt/2}$ as the dependent variable, (6.2.6) is of the form (6.1.1) if $k < 2n$, or (3.3.10) if $k > 2n$. If $k = 2n$ it is easily integrated to give

$$x\mathrm{e}^{kt/2} = A_1 + A_2 t,$$

or

$$x = (A_1 + A_2 t)\mathrm{e}^{-kt/2}. \qquad (6.2.7)$$

Here again the solution tends to $x = 0$ without oscillation as t tends to infinity. The particle passes through the origin at most once, and then only if A_1 and A_2 are of opposite signs.

6.3 FORCED OSCILLATIONS

In some lightly damped systems, where it is desirable to maintain an oscillation, a compensating mechanism is required to overcome the resistance. The most familiar example is the spring in the escapement of a clock.

We assume that the compensating mechanism is represented by a force which is a prescribed function of the time, say $f(t)$ per unit mass. The equation of motion then becomes

$$\ddot{x} = -n^2 x - k\dot{x} + f(t). \tag{6.3.1}$$

The solution of equations such as (6.3.1) can be divided into two parts. The complementary function, which is the solution of the equation with $f(t)$ omitted, is exactly the solution discussed above in §6.2. The second part is the particular integral, which can be any particular solution of the full equation. The most general solution is the sum of the complementary function and the particular integral.

There are standard techniques for calculating particular integrals. Here we simply verify that a solution of the following form is possible. Let

$$x = \int_0^t X(t-\tau) f(\tau)\, d\tau \tag{6.3.2}$$

from which, by differentiation, we have

$$\dot{x} = X(0) f(t) + \int_0^t \dot{X}(t-\tau) f(\tau)\, d\tau, \tag{6.3.3}$$

$$\ddot{x} = X(0)\dot{f}(t) + \dot{X}(0) f(t) + \int_0^t \ddot{X}(t-\tau) f(\tau)\, d\tau. \tag{6.3.4}$$

If these expressions are substituted in (6.3.1), it will be seen that (6.3.2) is a particular integral provided that $X(t)$ is a solution of (6.2.1) which satisfies $X(0) = 0$, $\dot{X}(0) = 1$. For example if $n^2 > \frac{1}{4}k^2$, the appropriate solution for $X(t)$ is

$$X(t) = (n^2 - \tfrac{1}{4}k^2)^{-1/2} e^{-kt/2} \sin\{(n^2 - \tfrac{1}{4}k^2)^{1/2} t\}. \tag{6.3.5}$$

Equation (6.3.2) has been verified by a mathematical argument but it is worthwhile to consider the physical interpretation. Let x denote the displacement of a particle, of mass m, whose equation of motion reduces to (6.2.1). Let the particle suffer a blow of impulse $\bar{F}$ at $t = \tau$. Then there is a discontinuous change in the velocity of the particle, when $t = \tau$, of amount $\bar{F}/m$. The displacement is therefore modified by a term which is zero for $t \leqslant \tau$, which satisfies (6.2.1) and whose derivative is discontinuous by an amount $\bar{F}/m$ at $t = \tau$. For $t \geqslant \tau$ this modification is clearly $X(t - \tau)\bar{F}/m$ since this satisfies all the required conditions. A continuous force $mf(t)$ can be regarded as a series of impulses, of which that occurring in the interval of time $\delta\tau$ is $mf(\tau)\,\delta\tau$. The contribution to the displacement arising from $mf(\tau)\,\delta\tau$ is thus zero for $t \leqslant \tau$, and $X(t - \tau)f(\tau)\,\delta\tau$ for $t \geqslant \tau$. The contribution from the continuous force can then be deduced by summation over the several contributions to give in the limit the relation (6.3.2). Note that the upper limit will be t because only those impulses up to the current value of t will be operative. The lower limit is chosen so that the initial conditions can be satisfied by appropriate choice of the complementary function.

There are two special cases of interest, both of which are more readily dealt with directly rather than by the general formula (6.3.2). The first is when f is a constant f_0. By inspection a particular integral is $x = f_0/n^2$, and the general solution is the sum of f_0/n^2 and the complementary function discussed in §6.2. The effect of a constant force is to shift the point of equilibrium to $x = f_0/n^2$ and, since the complementary function always tends to zero as $t \to \infty$ whatever the initial conditions, the solution always tends eventually to the value of f_0/n^2. This is the situation which occurs in many instruments, such as speedometers, control mechanisms and instruments used for measuring electric currents. In all these instruments it is desirable that the reading should settle quickly to its correct value. If damping is absent or too small, the needle will oscillate or 'hunt' around the correct value without reaching a steady reading. Very heavy damping is equally undesirable because, as k becomes large, λ_2 tends to zero and hence one of the exponential factors in (6.2.5) tends only slowly to zero. The response of the needle is then sluggish. In practice a value of k approximating to that for critical damping is to be preferred (see also Exercise 6.15). Note, from Example 6.1.3, that solid friction is to be avoided as a damping mechanism because the instrument may settle down to the wrong reading.

The second example involves a periodic forcing term for which $f = f_0 \cos pt$. The simplest derivation of the particular integral here is from the equation

$$\ddot{z} + k\dot{z} + n^2 z = f_0 e^{ipt}, \tag{6.3.6}$$

where $z = x + iy$ is a complex function of t. Since an equation involving complex quantities can only be satisfied if the real and imaginary parts of the two sides of the equation are separately equal, it follows that if z is a solution of (6.3.6) then the real part x is a solution of (6.3.1) with $f = f_0 \cos pt$ (and y is also a solution of (6.3.1) with $f = f_0 \sin pt$).

A particular integral of (6.3.6) is $z = z_0 e^{ipt}$, where z_0 is a constant satisfying the equation

$$(-p^2 + kip + n^2)z_0 = f_0.$$

It follows that
$$z = x + \mathrm{i}y = \frac{f_0 e^{\mathrm{i}pt}}{n^2 - p^2 + \mathrm{i}kp}$$

and so
$$x = \frac{f_0 \cos(pt - \beta)}{\{(n^2 - p^2)^2 + k^2 p^2\}^{1/2}}, \tag{6.3.7}$$

where
$$\tan \beta = \frac{kp}{n^2 - p^2}. \tag{6.3.8}$$

This is called the forced oscillation to distinguish it from the free oscillation, represented by the complementary function, to which the particle can respond without the help of the forcing term. As in all resisted oscillations, the free, or transient, oscillations eventually decay and ultimately the motion is just that represented by (6.3.7). The frequency of this forced oscillation is that of the forcing term, but there is a change in amplitude and phase. When k is small there is a significant increase in amplitude for a narrow band of frequencies near the frequency n of free, unresisted oscillations, that is when $(n - p)$ is small. Indeed there is a singularity in the amplitude for $n = p$ and $k = 0$. This phenomenon is called resonance and can be a source of potential danger in mechanical structures. In fact in many applications involving spring supports and periodic forcing terms, it is desirable to include sufficient damping unless the frequency of natural vibrations differs substantially from the frequencies of the forced oscillations that might be encountered. One practical example is the vibrations of a motor with a resilient mounting. Equation (6.3.7) shows that the amplitude of the forced oscillation can be written

$$\frac{f_0}{\{n^4 + p^2(p^2 + k^2 - 2n^2)\}^{1/2}}.$$

This is certainly a decreasing function of p provided $k^2 \geqslant 2n^2$, and if p always exceeds a certain minimum value p_{m} then it is sufficient that $k^2 \geqslant 2n^2 - p_{\mathrm{m}}^2$ to ensure that the amplitude is a decreasing function of p.

When k is small the lag in phase of the forced oscillation relative to the forcing term is nearly equal to 0 or π according as $p \lessgtr n$, so that there is a change in phase by this amount in the transition through resonance. But the change is continuous provided $k \neq 0$, so that when $p = n$ for example, the maximum displacement follows the maximum force with an interval of a quarter period, since $\beta = \pi/2$.

The lag in phase is associated with the supply of energy from the forcing term, the average energy being zero when $\beta = 0$. Physically this means that there is no resistance and so no work required, on average, to maintain the oscillation when $\beta = 0$. To see this let us consider the solution when the free oscillations are negligible. The rate of working of the force $f_0 \cos pt$ is, by (6.3.7),

$$f_0 \cos pt \frac{dx}{dt} = \frac{-pf_0^2}{\{(n^2 - p^2)^2 + k^2p^2\}^{1/2}} \cos pt \sin(pt - \beta),$$

$$= \frac{pf_0^2}{2\{(n^2 - p^2)^2 + k^2p^2\}^{1/2}} \{\sin \beta - \sin(2pt - \beta)\}. \quad (6.3.9)$$

Hence the work done in one complete period $0 \leqslant t \leqslant 2\pi/p$ is

$$\int_0^{2\pi/p} f_0 \cos pt \frac{dx}{dt} dt = \frac{\pi f_0^2 \sin \beta}{\{(n^2 - p^2)^2 + k^2p^2\}^{1/2}} = \frac{\pi kpf_0^2}{(n^2 - p^2)^2 + k^2p^2}, \quad (6.3.10)$$

where (6.3.8) has been used. For small values of k this is also small, and approximately proportional to k, except near resonance where there is a sharp increase in the supply of energy. When $n = p$ this supply is actually proportional to k^{-1}, so giving rise to large amplitudes.

The case for which $n = p$ and $k = 0$, namely

$$\ddot{x} + n^2 x = f_0 \cos nt, \quad (6.3.11)$$

is physically an idealisation and mathematically rather special. For the sake of completeness we derive the particular integral, which can be evaluated using (6.3.2). With

$$f(t) = f_0 \cos nt \quad (6.3.12)$$

and, from (6.3.5) with $k = 0$,

$$X(t) = n^{-1} \sin nt, \quad (6.3.13)$$

the particular integral is

$$x = \frac{f_0}{n} \int_0^t \sin n(t - \tau) \cos n\tau \, d\tau$$

$$= \frac{f_0}{2n} \int_0^t \{\sin nt + \sin n(t - 2\tau)\} \, d\tau$$

$$= \frac{f_0}{2n} t \sin nt. \quad (6.3.14)$$

The solution is no longer periodic, but has an amplitude which grows linearly with time. This behaviour is typical of undamped resonance. In practice, however, there will invariably be some damping present and often there will be non-linear effects brought into play as the amplitude grows, thereby modifying the ultimate behaviour.

6.4 COUPLED OSCILLATIONS

There are many physical applications of the mathematics relating to the interaction of oscillating systems. We introduce the ideas with a discussion of a pair of equations which typify such interactions, namely

$$\left.\begin{aligned}\frac{d^2x_1}{dt^2} &= -k_{11}x_1 - k_{12}x_2\\ \frac{d^2x_2}{dt^2} &= -k_{21}x_1 - k_{22}x_2\end{aligned}\right\} \tag{6.4.1}$$

The terms involving the constants k_{12} and k_{21} are those arising from the interaction between the two systems. When they are absent, the equations become uncoupled and are then of a type already discussed.

There are standard techniques for the integration of equations such as (6.4.1), but it is instructive to see how the solution can be built up in the following way. Equations (6.4.1) can be combined linearly to give

$$\frac{d^2}{dt^2}(x_1 + \kappa x_2) = -(k_{11} + \kappa k_{21})\left(x_1 + \frac{k_{12} + \kappa k_{22}}{k_{11} + \kappa k_{21}}\, x_2\right), \tag{6.4.2}$$

and if the constant κ is chosen to satisfy the relation

$$\kappa = \frac{k_{12} + \kappa k_{22}}{k_{11} + \kappa k_{21}} \tag{6.4.3}$$

equation (6.4.2) simplifies to

$$\frac{d^2}{dt^2}(x_1 + \kappa x_2) = -(k_{11} + \kappa k_{21})(x_1 + \kappa x_2). \tag{6.4.4}$$

This equation for the single variable $x_1 + \kappa x_2$ is of a known form. Now equation (6.4.3) can be written as a quadratic equation for κ, namely

$$k_{21}\kappa^2 - (k_{22} - k_{11})\kappa - k_{12} = 0, \tag{6.4.5}$$

which has the two solutions κ_1, κ_2, where

$$2k_{21}\kappa_{1,2} = (k_{22} - k_{11}) \pm \{(k_{22} - k_{11})^2 + 4k_{12}k_{21}\}^{1/2}. \tag{6.4.6}$$

For these two values of κ, (6.4.4) yields a pair of equations for $x_1 + \kappa_1 x_2$ and $x_1 + \kappa_2 x_2$, which can be used as alternatives to the original pair (6.4.1). A discussion of these equations in the general case presents no particular difficulty. However, we shall concentrate on the case for which the frequencies n_1, n_2 given by

$$2n_{1,2}^2 = 2(k_{11} + \kappa_{1,2}k_{21}) = (k_{22} + k_{11}) \pm \{(k_{22} - k_{11})^2 + 4k_{12}k_{21}\}^{1/2} \tag{6.4.7}$$

are real, distinct and positive. We then have, from (6.4.4),

$$\left.\begin{aligned} x_1 + \kappa_1 x_2 &= a\cos(n_1 t + \alpha_1) \\ x_1 + \kappa_2 x_2 &= a'\cos(n_2 t + \alpha_2) \end{aligned}\right\} \tag{6.4.8}$$

say, and this pair of equations can be solved to give x_1 and x_2 as a linear combination of two harmonic oscillations.

Alternatively we can solve (6.4.1) by looking for solutions such that both x_1 and x_2 are proportional to $e^{\lambda t}$; such a procedure has more general applicability for linear equations with constant coefficients. If the solutions are known to be purely oscillatory, this fact can be recognised from the beginning by choosing $\lambda = in$, say, where n is real. The procedure is then equivalent to looking for possible solutions such that

$$\frac{d^2x_1}{dt^2} = -n^2x_1, \quad \frac{d^2x_2}{dt^2} = -n^2x_2. \tag{6.4.9}$$

When (6.4.9) are substituted in (6.4.1) we get

$$\left.\begin{aligned} (n^2 - k_{11})x_1 - k_{12}x_2 &= 0 \\ -k_{21}x_1 + (n^2 - k_{22})x_2 &= 0 \end{aligned}\right\} \tag{6.4.10}$$

or, equivalently,

$$\frac{x_1}{x_2} = \frac{k_{12}}{n^2 - k_{11}} = \frac{n^2 - k_{22}}{k_{21}}. \tag{6.4.11}$$

This relation yields a quadratic equation for n^2, whose roots are given by (6.4.7) without the intermediate calculation of κ. When n_1 and n_2 are known, the general solution can be written as a linear combination of the corresponding solutions of (6.4.9), say

$$\left.\begin{aligned} x_1 &= a_1\cos(n_1 t + \alpha_1) + a_2\cos(n_2 t + \alpha_2) \\ x_2 &= \frac{k_{21}a_1}{n_1^2 - k_{22}}\cos(n_1 t + \alpha_1) + \frac{k_{21}a_2}{n_2^2 - k_{22}}\cos(n_2 t + \alpha_2) \end{aligned}\right\} \tag{6.4.12}$$

where the relation for x_2 follows with the help of (6.4.11). That equations (6.4.12) are equivalent to (6.4.8) may be verified directly.

We note that the oscillation frequencies are the same for both the displacements x_1 and x_2, but the amplitudes are different. In general the actual oscillations represented by (6.4.12) will not be harmonic, indeed they will not even be periodic if n_1/n_2 is irrational. But if one of the arbitrary constants a_1 or a_2 happens to be zero, the resulting oscillation is harmonic with the same frequency for both coordinates. Such an oscillation is called a normal mode of vibration. Here there are two possible normal modes of vibration, represented by (6.4.12) with either $a_1 = 0$ or $a_2 = 0$, and the system can be made to vibrate in one or other normal mode by suitable choice of the initial conditions. In general, however, the motion will be given by a superposition of the two normal modes of vibration.

There is one special case for which the above arguments fail, namely when

$$(k_{22} - k_{11})^2 + 4k_{12}k_{21} = 0. \tag{6.4.13}$$

Equations (6.4.6) and (6.4.7) then yield only one value for $\kappa = (k_{22} - k_{11})/2k_{21}$ and one associated value for $n = \{(k_{22} + k_{11})/2\}^{1/2}$. This will yield only one mode of oscillation and is not sufficient information to obtain the general solution. Although we shall not use this case, a method of solution is indicated for mathematical completeness.

The above values of κ and n give rise to one relation of the form (6.4.8), say

$$x_1 + \kappa x_2 = a\cos(nt + \alpha_1). \tag{6.4.14}$$

If this is substituted in the second of equations (6.4.1), we obtain a differential equation for x_2, namely

$$\begin{aligned}\frac{d^2x_2}{dt^2} &= -k_{21}a\cos(nt + \alpha_1) - (k_{22} - \kappa k_{21})x_2 \\ &= -k_{21}a\cos(nt + \alpha_1) - n^2x_2.\end{aligned} \tag{6.4.15}$$

Comparison with (6.3.11) and (6.3.14) shows that (6.4.15) is mathematically equivalent to the equation for undamped resonance oscillations, with a solution of the form

$$x_2 = -\frac{k_{21}a}{2n}\,t\sin(nt + \alpha_1) + a'\cos(nt + \alpha_2). \tag{6.4.16}$$

The associated expression for x_1 then follows immediately from (6.4.14).

Example 6.4.1

Two particles m_1, m_2 are connected by a light elastic string, of natural length l, which passes over a smooth peg, as in Figure 6.4.1. The particles are released

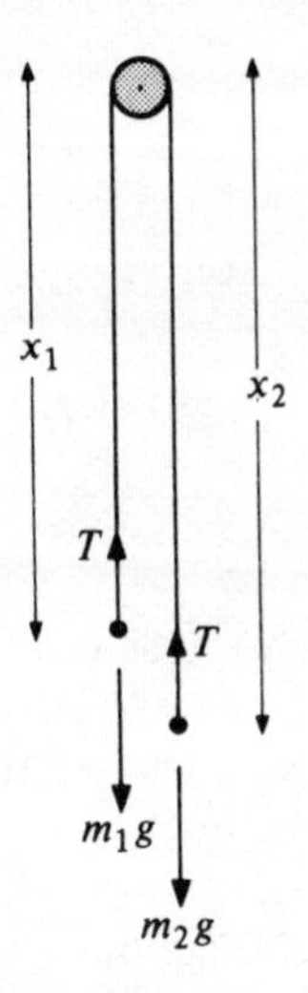

Figure 6.4.1 Illustration for Example 6.4.1.

from a rest position in which a length $l/2$ of the string is hanging vertically on either side of the peg.

To discuss the subsequent motion, let m_1, m_2 be at distances x_1, x_2 respectively from the peg. Then

$$m_1\ddot{x}_1 = m_1 g - T, \quad m_2\ddot{x}_2 = m_2 g - T, \quad T = \lambda(x_1 + x_2 - l)/l. \tag{6.4.17}$$

To solve these equations, we rewrite them in the form

$$\frac{d^2}{dt^2}(m_1 x_1 - m_2 x_2) = (m_1 - m_2)g, \tag{6.4.18}$$

$$\frac{d^2}{dt^2}\left(x_1 + x_2 - l - \frac{2glm_1m_2}{\lambda(m_1+m_2)}\right) = -\frac{\lambda(m_1+m_2)}{lm_1m_2}\left(x_1 + x_2 - l - \frac{2glm_1m_2}{\lambda(m_1+m_2)}\right). \tag{6.4.19}$$

Equation (6.4.18) is readily integrated. With the boundary conditions $x_1 = x_2 = l/2, \dot{x}_1 = \dot{x}_2 = 0$ at $t = 0$, the result is

$$m_1 x_1 - m_2 x_2 = \tfrac{1}{2}(m_1 - m_2)(gt^2 + l). \tag{6.4.20}$$

Equation (6.4.19) is just (6.1.1) with an appropriate change of variable and with

$$n^2 = \frac{\lambda(m_1 + m_2)}{lm_1m_2}.$$

The appropriate solution, which also satisfies the initial conditions, is

$$x_1 + x_2 = l + \frac{2glm_1m_2}{\lambda(m_1 + m_2)}(1 - \cos nt). \tag{6.4.21}$$

Equations (6.4.20) and (6.4.21) are sufficient to determine x_1 and x_2 separately as functions of t.

Example 6.4.2

Three light springs AB, BC, CD are joined end to end and stretched between two fixed points A, D on a smooth horizontal plane. Masses m_1, m_2 are fixed to the springs at B and C and the system is allowed to oscillate in the line AD (see Fig. 6.4.2).

Let the displacements of the particles B and C from their equilibrium positions be x_1 and x_2 respectively. Let the spring stiffnesses for AB, BC and CD be σ_1, σ and σ_2 respectively. In order to discuss the motions of B and C we require only the *changes* in the tensions of the springs, since the equilibrium values must cancel. For particle B, the net force is $-\sigma_1 x_1 - \sigma(x_1 - x_2)$ and for C is $-\sigma_2 x_2 + \sigma(x_1 - x_2)$. The equations of motion are therefore

$$m_1\ddot{x}_1 = -\sigma_1 x_1 - \sigma(x_1 - x_2), \quad m_2\ddot{x}_2 = -\sigma_2 x_2 + \sigma(x_1 - x_2). \tag{6.4.22}$$

These equations are solved by looking for solutions such that both B and C have oscillations of the same frequency, that is solutions for which, simultaneously,

$$\ddot{x}_1 = -n^2 x_1, \quad \ddot{x}_2 = -n^2 x_2. \tag{6.4.23}$$

Substitution of (6.4.23) in (6.4.22) gives the pair of equations

$$(m_1n^2 - \sigma_1 - \sigma)x_1 + \sigma x_2 = 0, \quad \sigma x_1 + (m_2n^2 - \sigma_2 - \sigma)x_2 = 0, \tag{6.4.24}$$

from which x_1 and x_2 can be eliminated to give

$$(m_1n^2 - \sigma_1 - \sigma)(m_2n^2 - \sigma_2 - \sigma) - \sigma^2 = 0. \tag{6.4.25}$$

This is a quadratic equation for n^2 which in general has two roots, say n_1^2 and n_2^2. For each value there is a possible solution, with the ratio x_1/x_2 fixed and

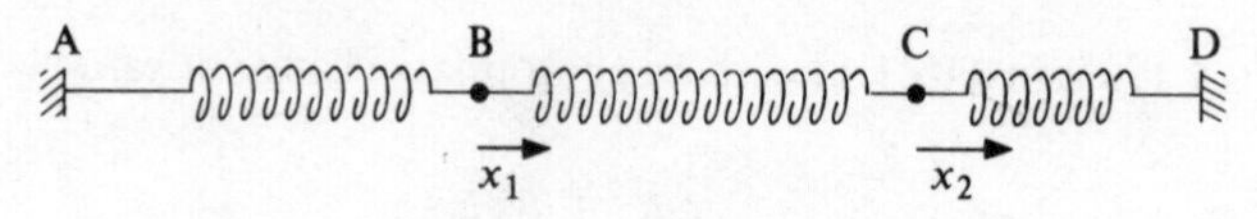

Figure 6.4.2 Illustration for Example 6.4.2.

given by (6.4.24). These solutions are the normal modes of oscillation. The most general solution of (6.4.22) is a linear combination of these two solutions, say

$$x_1 = a_1 \cos(n_1 t + \alpha_1) + a_2 \cos(n_2 t + \alpha_2) \tag{6.4.26}$$

$$x_2 = -\frac{m_1 n_1^2 - \sigma_1 - \sigma}{\sigma} a_1 \cos(n_1 t + \alpha_1) - \frac{m_1 n_2^2 - \sigma_1 - \sigma}{\sigma} a_2 \cos(n_2 t + \alpha_2), \tag{6.4.27}$$

where a_1, a_2, α_1, α_2 are arbitrary constants to be determined from the initial conditions. If the latter are adjusted so that a_1 or a_2 is zero then the system will oscillate in a normal mode with a definite frequency of oscillation. In general, however, this will not be the case, and the solution is a combination of the two modes of oscillation.

Consider the special case $m_1 = m_2$, $\sigma_1 = \sigma_2$ so that, from (6.4.25),

$$n_1^2 = (\sigma_1 + 2\sigma)/m_1, \quad n_2^2 = \sigma_1/m_1 \tag{6.4.28}$$

and

$$x_1 = a_1 \cos n_1 t + a_2 \cos n_2 t, \tag{6.4.29}$$

$$x_2 = -a_1 \cos n_1 t + a_2 \cos n_2 t. \tag{6.4.30}$$

Here the first normal mode is such that the two particles have the same amplitude a_1 and the same frequency n_1 but are π out of phase so that they always move in opposite directions. In the second slower normal mode the amplitude, frequency and phase are all identical. In this mode the spring BC exerts no influence. For arbitrary initial conditions the motion is a superposition of these two oscillations. Suppose for example that the initial conditions are

$$x_1 = 2a, \quad x_2 = \dot{x}_1 = \dot{x}_2 = 0 \qquad \text{when } t = 0.$$

Then the subsequent oscillations are given by (6.4.29), (6.4.30) with $a_1 = a_2 = a$. An alternative form of these equations is

$$x_1 = 2a \cos\left(\frac{n_1 - n_2}{2} t\right) \cos\left(\frac{n_1 + n_2}{2} t\right), \tag{6.4.31}$$

$$x_2 = 2a \sin\left(\frac{n_1 - n_2}{2} t\right) \sin\left(\frac{n_1 + n_2}{2} t\right). \tag{6.4.32}$$

Now $$\frac{n_1 + n_2}{2} = \frac{(\sigma_1 + 2\sigma)^{1/2} + \sigma_1^{1/2}}{2m_1^{1/2}}, \quad \frac{n_1 - n_2}{2} = \frac{(\sigma_1 + 2\sigma)^{1/2} - \sigma_1^{1/2}}{2m_1^{1/2}}$$

and if $\sigma = 0$, we see that the solution is $x_1 = 2a \cos(\sigma_1/m_1)^{1/2} t$, $x_2 = 0$ as expected.

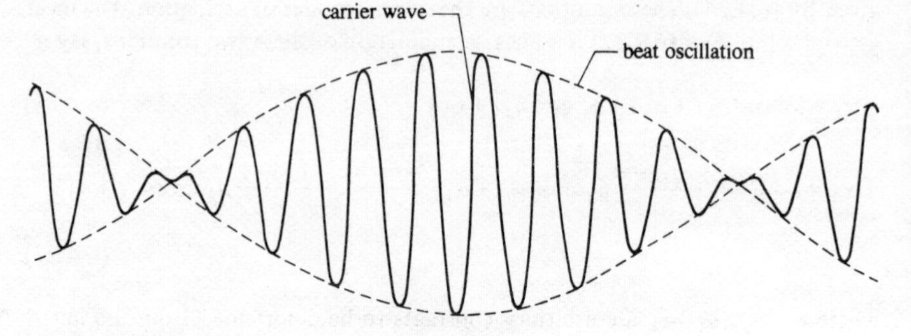

Figure 6.4.3 Superimposed oscillations of nearly equal frequency.

If $\sigma \ll \sigma_1$, the frequency $\frac{1}{2}(n_1 + n_2)$ approximates to $(\sigma_1/m_1)^{1/2}$. But note that there is still the slowly varying factor $\cos\frac{1}{2}(n_1 - n_2)t$, and provided that $(n_1 - n_2)$ is not zero this will eventually alter the solution for x_1 substantially. Furthermore x_2 will eventually acquire a substantial oscillation. This is an example of a situation in which a small effect can accumulate over a sufficient length of time to produce a significant modification of the solution. A vibration of the kind described by (6.4.31) or (6.4.32) is called a modulated oscillation. The slow oscillation superimposed on the modulated oscillation is called a beat oscillation and can be used in tuning musical instruments. The combined effect of two notes of approximately equal frequency has an amplitude which slowly rises and falls as shown in Figure 6.4.3. These surges in volume are easily detected by the ear and tuning is effected by arranging that the frequency of the beat oscillation tends to zero. This type of oscillation also has applications in the theory of electromagnetic waves. Radio waves are transmitted at frequencies of the order of 10^6 oscillations per second. These are called carrier waves. But the actual information is transmitted by modulating the carrier wave at relatively lower acoustic frequencies, not more than 10^4 oscillations per second. The receiving apparatus then separates the acoustic oscillation from the carrier wave for further use.

Example 6.4.3

The calculation of normal modes of oscillation can be extended to more complex systems. We consider here the motion of r equal particles fastened at

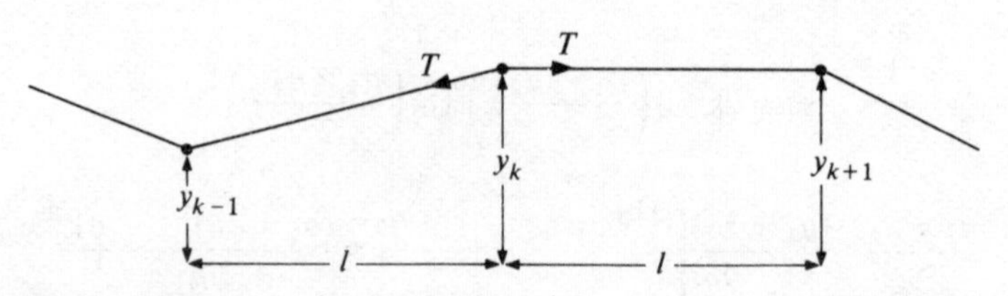

Figure 6.4.4 Illustration for Example 6.4.3: transverse oscillations of particles on a stretched string.

equal distances to a string of length $(r+1)l$. Both ends of the string are fixed and the only force influencing the motion of the particles is assumed to be the tension T in the string.

We consider small lateral displacements $y_1, y_2, \ldots, y_r$ of the particles. Then the change in the tension T from its equilibrium value T_0 will be small and we can write, correct to the first order,

$$m\ddot{y}_k = -T_0\left(\frac{y_k - y_{k-1}}{l} + \frac{y_k - y_{k+1}}{l}\right)$$

as the equation of motion of the kth particle. To find the normal modes we look for solutions for which

$$\ddot{y}_k = -n^2 y_k, \quad (k = 1, 2, \ldots, r).$$

Substitution in the equation of motion gives

$$y_{k+1} + (\sigma - 2)y_k + y_{k-1} = 0, \quad (k = 1, 2, \ldots, r),$$

where $\sigma = mn^2 l/T_0$.

We have now a set of r linear difference equations which are to be solved subject to the boundary conditions

$$y_0 = y_{r+1} = 0.$$

If we assume a solution of the form

$$y_k = a \sin(k\beta + \epsilon)\cos(nt + \alpha),$$

so that

$$\begin{aligned} y_{k+1} + y_{k-1} &= \{a \sin(k\beta + \epsilon + \beta) + a \sin(k\beta + \epsilon - \beta)\} \cos(nt + \alpha) \\ &= 2a \cos \beta \sin(k\beta + \epsilon) \cos(nt + \alpha), \end{aligned}$$

it is seen that the difference equation is satisfied provided that

$$\cos \beta = 1 - \tfrac{1}{2}\sigma.$$

To satisfy the boundary conditions, we must have $y_0 = y_{r+1} = 0$, or

$$a \sin \epsilon = 0, \quad a \sin\{(r+1)\beta + \epsilon\} = 0,$$

and so $\epsilon = 0$ and $\sin(r+1)\beta = 0$. Hence there are r normal oscillations corresponding to the values of β given by

$$\beta = \frac{k\pi}{r+1}, \quad (k = 1, 2, \ldots, r).$$

The corresponding frequencies n can be calculated from

$$mn^2 l/T_0 = \sigma = 2(1 - \cos\beta).$$

6.5 NON-LINEAR OSCILLATIONS

We consider a particle in one-dimensional motion, with displacement x, and subject to a conservative force having a potential $V(x)$ per unit mass. The energy equation can then be written

$$\tfrac{1}{2}\dot{x}^2 + V(x) = E, \tag{6.5.1}$$

where E is the total energy which can be determined from the initial conditions. Note that the acceleration is given by $-V'(x)$ and that, if the acceleration is zero at $x = x_0$, then $V'(x_0) = 0$. If the initial conditions are such that, in addition, $E = V(x_0)$, both the velocity and acceleration are zero at $x = x_0$ and the particle can rest in equilibrium. Otherwise the details of the motion are deducible from the relation

$$2^{1/2}t = \pm \int \frac{dx}{(E - V(x))^{1/2}}, \tag{6.5.2}$$

which is a direct consequence of (6.5.1). The ambiguity of sign is resolved by the fact that t must always increase. Hence the positive sign is appropriate when x is increasing, the negative sign when x is decreasing.

For certain force fields it is possible for the motion of the particle to be bounded. This possibility arises when $E - V(x)$ is of the form

$$E - V(x) = (x - b)(a - x)\psi(x), \tag{6.5.3}$$

where $b < a$ and $\psi(x)$ is strictly positive and single-valued in the range $b \leqslant x \leqslant a$. Then the motion is confined to this range, since only values of $E - V(x)$ which are non-negative are appropriate. Equation (6.5.1) now becomes

$$\tfrac{1}{2}\dot{x}^2 = (x - b)(a - x)\psi(x), \tag{6.5.4}$$

which shows that the velocity is zero at $x = b, x = a$ and, by differentiation, that the acceleration is positive at $x = b$ and negative at $x = a$. Accordingly the particle oscillates between b and a and the motion is a repetition of that which occurs in one complete oscillation. The periodic time T is the time for one complete oscillation from b to a and back again. From (6.5.2), with due attention paid to sign, T is given by

$$2^{1/2}T = \int_b^a \frac{\mathrm{d}x}{(E - V(x))^{1/2}} - \int_a^b \frac{\mathrm{d}x}{(E - V(x))^{1/2}} = \int_b^a \frac{2\mathrm{d}x}{(E - V(x))^{1/2}}. \tag{6.5.5}$$

Note that although the integrand is singular at $x = b$ and $x = a$, the integral is in fact convergent. For example if, as in equation (6.1.2), we have

$$\tfrac{1}{2}\dot{x}^2 = E - V(x) = \tfrac{1}{2}n^2(a^2 - x^2), \tag{6.5.6}$$

the periodic time is

$$T = \frac{2}{n}\int_{-a}^{a} \frac{\mathrm{d}x}{(a^2 - x^2)^{1/2}} = \frac{2\pi}{n}. \tag{6.5.7}$$

This is the special case of simple harmonic oscillations as described in §6.1, where T is independent of the amplitude a.

We now show that, in certain circumstances, a small perturbation about a position of equilibrium is approximately described by (6.5.6). This is important because in practice there are always small extraneous disturbances present and it is useful to know their effect on a state of equilibrium.

Let $x = x_0$ be a position of equilibrium, so that $V'(x_0) = 0$. We consider a motion for which $\dot{x} = 0$ when $x = x_0(1 + \epsilon)$ so that

$$\tfrac{1}{2}\dot{x}^2 = E - V(x) = V(x_0 + x_0\epsilon) - V(x),$$

or

$$\tfrac{1}{2}x_0^2\dot{\xi}^2 = V(x_0 + x_0\epsilon) - V(x_0 + x_0\xi) \tag{6.5.8}$$

if

$$x = x_0(1 + \xi). \tag{6.5.9}$$

We assume that $V(x)$ has an expansion in the neighbourhood of $x = x_0$ of the form

$$\begin{aligned} V(x_0 + x_0\xi) &= V(x_0) + x_0\xi V'(x_0) + \tfrac{1}{2}x_0^2\xi^2 V''(x_0) + \ldots \\ &= V(x_0) + \tfrac{1}{2}x_0^2\xi^2 V''(x_0) + \ldots \end{aligned} \tag{6.5.10}$$

since $V'(x_0) = 0$. If we also assume that $0 < \epsilon \ll 1$ and that $|\xi| \ll 1$, equation (6.5.8) gives, approximately,

$$\dot{\xi}^2 = V''(x_0)(\epsilon^2 - \xi^2). \tag{6.5.11}$$

Suppose first that $V''(x_0) > 0$ so that V has a minimum at $x = x_0$. It is clear by comparison with (6.5.6), which represents simple harmonic motion, that ξ is then confined to the range $|\xi| \leqslant \epsilon$ and that the perturbation is a simple harmonic oscillation given by

$$\xi = \epsilon \cos nt, \quad n^2 = V''(x_0). \tag{6.5.12}$$

The periodic time is

$$T = 2\pi(V''(x_0))^{-1/2}. \tag{6.5.13}$$

On the other hand, if $V''(x_0)$ is negative, $|\xi| \geqslant \epsilon$, the acceleration at $\xi = \epsilon$ is positive and the perturbation will grow according to the solution

$$\xi = \epsilon \cosh nt, \quad n^2 = -V''(x_0). \tag{6.5.14}$$

The perturbation is therefore unstable if $V''(x_0)$ is negative and approximates to a simple harmonic oscillation if $V''(x_0)$ is positive. If $V''(x_0) = 0$, the above approximate analysis is not sufficiently sensitive and needs to be refined. But it is still true that the perturbation is oscillatory if V has a minimum, for this is sufficient to ensure that, at least for sufficiently small ϵ, there is a finite interval for which the right-hand side of (6.5.8) remains positive and hence expressible in the form (6.5.4). It will, however, not be a simple harmonic oscillation.

Example 6.5.1 The simple Pendulum

We consider the motion of a particle of mass m attached to one end of a light rigid rod of length l, the other end being free to rotate about a fixed point (see Fig. 6.5.1). It is assumed that the particle moves in a vertical plane. When the angular displacement of the rod from the vertical position is θ, the equation of motion of the particle has a component perpendicular to the rod which is

$$ml\ddot{\theta} = -mg \sin \theta. \tag{6.5.15}$$

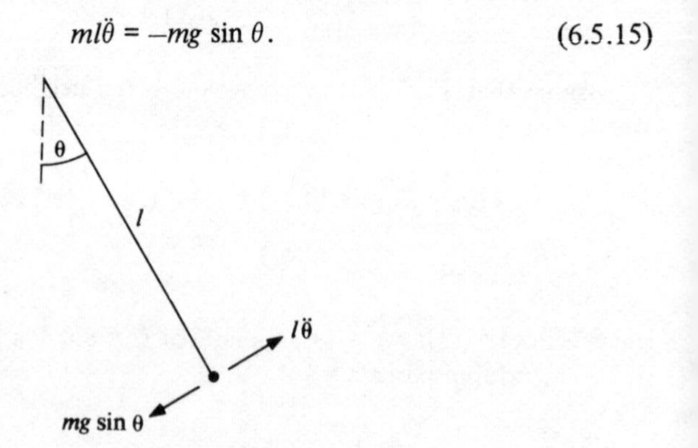

Figure 6.5.1 Illustration for Example 6.5.1: the simple pendulum.

The configuration is known as a simple pendulum. The motion is oscillatory but not simple harmonic. However, when the value of θ is small, $\sin\theta$ can be approximated by θ, and with this simplification (6.5.15) reduces to equation (6.1.1) for simple harmonic motion in θ with frequency $n = (g/l)^{1/2}$. The periodic time, $2\pi(l/g)^{1/2}$, is then independent of the amplitude within the approximation considered which is a useful property when the pendulum is used for time keeping. If $l = g/\pi^2 = 0{\cdot}994$ m, the pendulum will make one complete swing, or half an oscillation, in one second.

For complete accuracy, however, the full equation must be integrated. One integration is immediate, namely

$$\tfrac{1}{2}\dot{\theta}^2 = \frac{g}{l}\cos\theta + E,$$

which is the energy equation. We write this in the alternative form

$$\dot{\theta}^2 = 4n^2(a^2 - \sin^2\tfrac{1}{2}\theta), \tag{6.5.16}$$

where

$$n^2 = g/l, \quad a^2 = \tfrac{1}{2}(1 + El/g).$$

One further integration along the lines suggested by (6.5.2) gives θ as a function of t. The result is expressible in terms of elliptic functions which are tabulated in the literature. The following discussion is restricted to the calculation of the periodic time for non-linear oscillations.

Two types of oscillation are possible. If $0 < a < 1$, say $a = \sin\tfrac{1}{2}\alpha$, then $\dot{\theta} = 0$ when $\theta = \pm\alpha$ and α represents the maximum angular displacement of the pendulum, which will oscillate between $\theta = \pm\alpha$. The periodic time, T, is given by

$$T = \frac{1}{n}\int_{-\alpha}^{\alpha}\frac{d\theta}{(\sin^2\tfrac{1}{2}\alpha - \sin^2\tfrac{1}{2}\theta)^{1/2}}\,. \tag{6.5.17}$$

The integral is called an elliptic integral. The standard form is obtained with the transformation

$$\sin\tfrac{1}{2}\theta = \sin\tfrac{1}{2}\alpha\sin\phi = a\sin\phi, \tag{6.5.18}$$

and if we write $T_0 = 2\pi/n$ as the periodic time for small oscillations, equation (6.5.17) can be written

$$\frac{T}{T_0} = \frac{2}{\pi}\int_0^{\pi/2}\frac{d\phi}{(1 - a^2\sin^2\phi)^{1/2}} = \frac{2K(a)}{\pi}, \tag{6.5.19}$$

where $K(a)$ is called the complete elliptic integral of the first kind. It can be found tabulated in all the more comprehensive tables of functions. Note that

$$K(a) \to \pi/2 \quad \text{as } a \to 0$$

and the periodic time $T = T_0$ for small oscillations is recovered. A better approximation for small a is obtained by writing

$$K(a) = \int_0^{\pi/2} \frac{d\phi}{(1 - a^2 \sin^2 \phi)^{1/2}}$$

$$= \int_0^{\pi/2} (1 + \tfrac{1}{2}a^2 \sin^2 \phi + \dots)\, d\phi = \tfrac{1}{2}\pi(1 + \tfrac{1}{4}a^2 + \dots), \qquad (6.5.20)$$

so that, approximately,

$$T = T_0 \left(1 + \frac{a^2}{4}\right) = T_0 \left(1 + \frac{\alpha^2}{16}\right). \qquad (6.5.21)$$

When $a > 1$, $\dot{\theta}$ never becomes zero and the pendulum swings in complete circles rather than to and fro. Here the periodic time is the time for θ to increase by 2π, and so

$$\frac{T}{T_0} = \frac{1}{4\pi} \int_0^{2\pi} \frac{d\theta}{(a^2 - \sin^2 \frac{1}{2}\theta)^{1/2}} = \frac{1}{\pi a} \int_0^{\pi/2} \frac{d\phi}{(1 - a^{-2} \sin^2 \phi)^{1/2}} = \frac{K(a^{-1})}{\pi a}, \qquad (6.5.22)$$

where the substitution $\theta/2 = \phi$ has been used. Note that

$$\frac{T}{T_0} \sim \frac{1}{2a} \quad \text{as } a \to \infty.$$

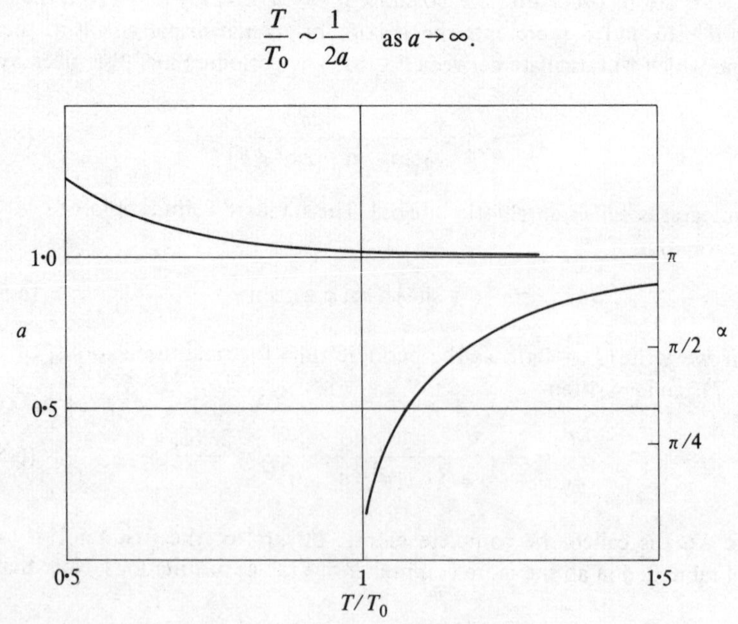

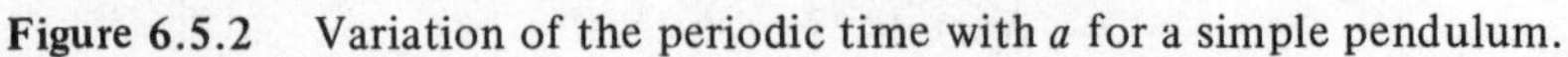

Figure 6.5.2 Variation of the periodic time with a for a simple pendulum.

Figure 6.5.2 shows the variation of T/T_0 with a, produced from tables of the function K. When $\alpha = \pi/4 (a = 0{\cdot}383)$, for example, the periodic time is $1{\cdot}04T_0$ so that there is a 4% error in using the small amplitude approximation T_0. The periodic time has increased to $1{\cdot}17T_0$ at $\alpha = \pi/2 (a = 0{\cdot}707)$ and tends to infinity as $\alpha \to \pi (a \to 1)$. When $a = 1{\cdot}016$ the periodic time is again T_0, and for $a > 5$ the periodic time is $(2a)^{-1}$ with an error of less that 1%.

Example 6.5.2

A bead B, of mass m, is free to slide on a fixed horizontal smooth wire. It is constrained by two springs AB, CB, each of natural length l and modulus λ, which are attached to the bead and to fixed points A and C (see Fig. 6.5.3). The bead can rest in equilibrium with AB = BC and ABC perpendicular to the wire. Let us consider the possible oscillations of the bead on the wire.

We first note that there is an energy equation since the tension in a spring is conservative. If the tension is $\lambda y/l$ when the spring is stretched a distance y, the potential energy is

$$\int_0^y \frac{\lambda y}{l}\,\mathrm{d}y = \frac{\lambda y^2}{2l}. \tag{6.5.23}$$

Let AC = $2b$, and let x denote the displacement of the bead from AC. Then each spring is stretched a distance $(b^2 + x^2)^{1/2} - l$ and the total potential energy is, from (6.5.23),

$$V(x) = \lambda\{(b^2 + x^2)^{1/2} - l\}^2/ml \tag{6.5.24}$$

per unit mass of the bead. The energy equation is therefore

$$\tfrac{1}{2}\dot{x}^2 = E - V(x) = E - \frac{\lambda}{ml}\{(b^2 + x^2)^{1/2} - l\}^2. \tag{6.5.25}$$

Let us consider the equilibrium positions and their stability. From (6.5.24)

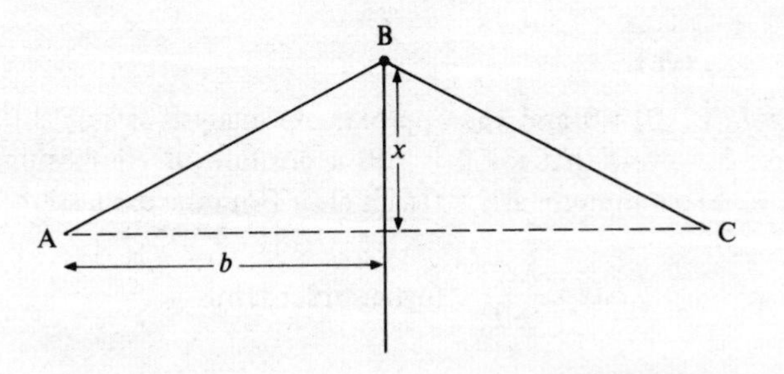

Figure 6.5.3　Illustration for Example 6.5.2.

we have that

$$V'(x) = \frac{2\lambda x}{ml}\left(1 - \frac{l}{(b^2 + x^2)^{1/2}}\right), \tag{6.5.26}$$

$$V''(x) = \frac{2\lambda}{ml}\left(1 - \frac{lb^2}{(b^2 + x^2)^{3/2}}\right). \tag{6.5.27}$$

Equilibrium is possible when $V'(x) = 0$, that is, when $x = 0$ or $x^2 = (l^2 - b^2)$. The latter condition is possible only when $l > b$ and the spring can sustain compression. Now

$$V''(0) = \frac{2\lambda}{ml}\left(1 - \frac{l}{b}\right) \tag{6.5.28}$$

and so a perturbation about $x = 0$ is oscillatory if $l < b$, unstable if $l > b$. Also

$$V''((l^2 - b^2)^{1/2}) = \frac{2\lambda}{ml}\left(1 - \frac{b^2}{l^2}\right), \tag{6.5.29}$$

and so a perturbation about $x = (l^2 - b^2)^{1/2}$ is oscillatory whenever equilibrium is possible in that position.

The details of the motion can be obtained from a further integration of (6.5.25), but for small displacements about the position of equilibrium the approximate period will be, from (6.5.13),

$$2\pi\left(\frac{mbl}{2\lambda(b - l)}\right)^{1/2}$$

about $x = 0$ and

$$2\pi\left(\frac{ml^3}{2\lambda(l^2 - b^2)}\right)^{1/2}$$

about $x = (l^2 - b^2)^{1/2}$.

When $b = l$, $V''(0) = 0$ and the approximate analysis is not sufficiently accurate. Note, however, that $x = 0$ is still a position of equilibrium. It is also stable since V has a minimum at $x = 0$ as is clear from the expression

$$V = \frac{\lambda x^4}{4ml^3} + \text{higher order terms} \tag{6.5.30}$$

when $x \ll l$. The energy equation (6.5.25) gives approximately

$$\tfrac{1}{2}\dot{x}^2 = \frac{\lambda}{4ml^3}(a^4 - x^4) \tag{6.5.31}$$

provided $\dot{x} = 0$ when $x = a$ and $a \ll l$. The periodic time is

$$2\left(\frac{2ml^3}{\lambda}\right)^{1/2}\int_{-a}^{a}\frac{\mathrm{d}x}{(a^4 - x^4)^{1/2}} = \frac{4}{a}\left(\frac{ml^3}{\lambda}\right)^{1/2}\int_0^{\pi/2}\frac{\mathrm{d}\phi}{(1 - \frac{1}{2}\sin^2\phi)^{1/2}},$$

$$= \frac{4}{a}\left(\frac{ml^3}{\lambda}\right)^{1/2} K(2^{-1/2}), \tag{6.5.32}$$

where the substitution $x = a\cos\phi$ has been used. The function $K(2^{-1/2})$ is the complete elliptic integral with argument $2^{-1/2}$ (see equation (6.5.19)) and has the value 1·854. Hence the periodic time is

$$\frac{7{\cdot}42}{a}\left(\frac{ml^3}{\lambda}\right)^{1/2}.$$

Unlike the simple harmonic case, the periodic time is now inversely proportional to the amplitude.

EXERCISES

6.1 Verify that $x = a\sin^m nt$ describes simple harmonic motion, about a suitable origin, for $m = 1, 2$ but not for $m \geqslant 3$.

6.2 A mass is suspended at the lower end of a light, perfectly elastic vertical string whose upper end is fixed. The extension of the spring is 0·06 m when the mass is in its equilibrium position C. The mass is pulled up 0·02 m above C and set in motion with an initial upward speed of 0·4 m s^{-1}. Find the potential energy in any position and hence use energy considerations to determine:

(i) the highest position of the mass,
(ii) the speed of the mass 0·02 m below C,
(iii) the position when the speed is a maximum.

Explain why the last two results can be deduced without detailed calculation.
(Ans.: 0·037 m above C, 0·4 m s^{-1}, C.)

6.3 A particle P is constrained to move along the x-axis under the several forces $k_r\overline{\mathrm{PP}_r}$, $r = 1, 2, \ldots, n$, where k_r is a constant scalar and P_r is a point on the x-axis.

Show that the motion is simple harmonic about an origin C such that $\Sigma k_r \overline{\mathrm{CP}}_r = \mathbf{0}$ provided that $\Sigma k_r > 0$. Discuss the motion when $\Sigma k_r < 0$ and when $\Sigma k_r = 0$.

6.4 A particle of mass m slides on a smooth plane, which is inclined at an angle α to the horizontal. The particle is attached to a fixed point on the plane by a spring of stiffness σ. Show that the extension of the spring in equilibrium is $\sigma^{-1} mg \sin \alpha$ and that the frequency of oscillations about the equilibrium position is independent of α.

Assume that the particle oscillates along a line of greatest slope.

6.5 A particle executes simple harmonic oscillations with frequency n.

(i) The displacements of the particle are x_1, x_2, $\bar{x}$ at times t_1, t_2, $\frac{1}{2}(t_1 + t_2)$ respectively. Show that

$$\frac{x_2 - x_1}{\dot{x}_1 - \dot{x}_2} = \frac{\dot{\bar{x}}}{n^2 \bar{x}}, \qquad \frac{x_1 + x_2}{\dot{x}_1 + \dot{x}_2} = \frac{\bar{x}}{\dot{\bar{x}}}.$$

(ii) At a given instant the displacement of the particle from the equilibrium position O is x, and it is moving away from O with speed v. Show that it next reaches O after a time

$$\frac{1}{n}\left(\pi - \tan^{-1} \frac{nx}{v}\right).$$

(iii) At the instant that the particle is moving away from O with speed v, it is given an impulse which instantaneously changes the speed from v to $v + \delta v$, where $\delta v \ll v$. Show that the time to reach O alters by $\delta v(n^2 a^2 - v^2)^{1/2}/a^2 n^3$ approximately, where a is the amplitude.

6.6 Two light springs, of the same natural length l and of stiffnesses σ_1, σ_2, can be combined in parallel to form a combination also having the natural length l. Show that a particle of mass m, when suspended from such a combination, oscillates with a periodic time $2\pi\{m/(\sigma_1 + \sigma_2)\}^{1/2}$.

Show that if the springs are combined in series to form a combination having the natural length $2l$, the particle will then oscillate with a periodic time $2\pi\{m(\sigma_1^{-1} + \sigma_2^{-1})\}^{1/2}$.

Which is the longer period?

6.7 A particle of mass m is attached to one end of an elastic string, of natural length l, whose other end is fixed at a point O. The particle is allowed to fall from rest at O and the maximum depth of the particle below O is written in the form $l \cot^2 \frac{1}{2}\theta$, where θ is a positive acute angle. Show that the particle attains this depth in a time $(2l/g)^{1/2}\{1 + (\pi - \theta)\cot\theta\}$. Show that the modulus of elasticity of the string is $\frac{1}{2}mg \tan^2 \theta$.

6.8 A particle is attached to the mid point of a light elastic string of natural length l. The ends of the string are attached to fixed points A and B, A being at a height $2l$ vertically above B, and in equilibrium the particle rests at a depth $5l/4$ below A. The particle is projected vertically downwards from this position with speed $(gl)^{1/2}$.

Prove that the lower string slackens after a time $\pi(l/g)^{1/2}/12$, and that the particle comes to rest after a further time $\beta(l/2g)^{1/2}$, where β is the acute angle defined by $\tan \beta = (3/2)^{1/2}$.

6.9 A light spring, of stiffness σ_1, is fixed at one end and a smooth pulley of negligible mass is suspended from the other. Over the pulley passes a light elastic string of stiffness σ_2. One end of the string is fixed to the ground, from the other end is suspended a particle of mass m. In equilibrium the spring and the two portions of the string are vertical. Show that, provided the string remains taut and vertical, the particle can perform simple harmonic oscillations with period

$$2\pi\{m(4\sigma_2 + \sigma_1)/(\sigma_1\sigma_2)\}^{1/2}.$$

Deduce the period of oscillations in the two special cases:

(i) when the spring is replaced by a rigid support,
(ii) when the string is inextensible.

6.10 A particle, of mass m, moves along the x-axis and is subject to a restoring force equal to $-mn^2x$. It is also subject to a constant frictional resistance, of magnitude mn^2a, but which acts only when x is positive. The particle is released from rest when $x = -b(b > 0)$. Show that it next comes to rest when $x = (a^2 + b^2)^{1/2} - a$ and that it will then stay permanently at rest if $b < 3^{1/2}a$.

6.11 The displacement of a particle is given by

$$x = a\,\mathrm{e}^{-kt/2}\cos pt.$$

Show that in each half oscillation:

(i) the displacement is a maximum (in magnitude) when $(pt + \alpha)/\pi$ is an integer, where $\tan\alpha = k/2p$;
(ii) the speed is a maximum at a time $2\alpha/p$ before the displacement is zero;
(iii) the acceleration is a maximum (in magnitude) at a time $2\alpha/p$ before the displacement is a maximum;
(iv) the time from zero displacement to the next maximum displacement is $(\pi - 2\alpha)/2p$, and the time for the return to zero displacement is $(\pi + 2\alpha)/2p$;
(v) the maximum displacement is reduced in the ratio $\mathrm{e}^{-\pi k/2p}$.

6.12 Show that, if x satisfies the equation

$$\ddot{x} + k\dot{x} + n^2(x - x_0) = 0$$

and if x_1, x_2, x_3 are consecutive values of x for which $\dot{x} = 0$, then

$$x_0 = \frac{x_1x_3 - x_2^2}{x_1 + x_3 - 2x_2},$$

$$n = \frac{k}{2}\left\{1 + \left(\frac{\pi}{\log\{(x_1 - x_2)/(x_3 - x_2)\}}\right)^2\right\}^{1/2}.$$

Show that, if the resistance is small,

$$x_0 = \tfrac{1}{4}(x_1 + 2x_2 + x_3)$$

approximately.

6.13 A spring balance consists of a pan of mass m supported by a spring of stiffness σ. Its motion is subject to a resistance equal to K times its speed. A particle of mass m_p is gently placed on the pan, which is initially at rest. Show that when the pan comes to rest again for the first time, its displacement from its initial position is

$$\frac{m_\text{p} g}{\sigma}\left\{1 + \exp\left(-\frac{K\pi}{(4\sigma(m + m_\text{p}) - K^2)^{1/2}}\right)\right\},$$

provided that $K^2 < 4\sigma(m + m_\text{p})$.

6.14 A simple pendulum, of mass m and length l, is completely immersed in a viscous liquid and the resistance to its motion is mk times the speed of the bob. Show that the angular displacement θ of the pendulum from the position of stable equilibrium satisfies the equation

$$\ddot{\theta} + k\dot{\theta} + n^2 \sin\theta = 0,$$

where $n^2 = |1 - s|g/l$, and s is the ratio of the density of the liquid to that of the bob.

Now consider a small perturbation from the position of stable equilibrium.

(i) Show that the motion is not oscillatory if

$$1 - \frac{k^2 l}{4g} \leqslant s \leqslant 1 + \frac{k^2 l}{4g}.$$

(ii) Show that if the motion is oscillatory and $k \ll n$, to the first order the resistance affects only the amplitude and the buoyancy affects only the frequency.

(When a body is immersed in a liquid it experiences a buoyancy force which is equal and opposite to the weight of the liquid displaced by the body.)

6.15 A convenient measure of the spread of the values of $x(t)$ $(0 < t < \infty)$ about the value $x = 0$ is

$$\int_0^\infty x^2 \, dt$$

in the sense that if the integral is small, $|x(t)|$ must itself be small most of the time.

The motion of a particle is governed by the equation

$$\ddot{x} + k\dot{x} + n^2x = 0,$$

and initially $x = a, \dot{x} = u$. Show that

$$\int_0^\infty x^2 \, dt = \frac{1}{2kn^2} \{(u + ka)^2 + n^2a^2\}.$$

Show that, as a function of k, this is a minimum when

$$k^2 = n^2 + u^2/a^2.$$

(Hint. To evaluate the integral multiply the original equation by $\dot{x} + kx$ and integrate.)

6.16 For time $t \leqslant 0$ a particle hangs in equilibrium at the lower end of a light elastic string, which is extended a distance a beyond its natural length. For $t > 0$ the upper end of the string moves vertically, its downward displacement being $b \sin pt$ at time t. Show that, provided the string remains taut, the downward displacement x of the particle satisfies the equation

$$\ddot{x} + n^2x = n^2b \sin pt,$$

where $n^2 = g/a$.

Show that the string remains taut throughout the motion if

$$\frac{pb}{n^2 - p^2}(n \sin nt - p \sin pt) < a$$

for all values of t. Show that this is certainly true if

$$pb < |n - p| a.$$

6.17 A simple pendulum consists of a light inelastic string, of length l, with a particle, of mass m, attached to its lower end. The upper end of the string is made to move in a horizontal straight line so that its distance from a fixed point in the line is equal to $\xi(t)$ at time t.

Give the full equations of motion of the particle in terms of ξ and θ and their derivatives, where θ is the inclination of the string to the vertical.

Suppose that $\xi = b \sin pt$, where $b \ll l$, $p < n$, and that the string is vertical, with zero angular speed, when $t = 0$. Show that, provided θ remains small in the ensuing motion,

$$\theta = \frac{bp^2}{nl(n^2 - p^2)}(n \sin pt - p \sin nt)$$

approximately, where $n^2 = g/l$.

Show that the next time the angular speed vanishes the value of t is $2\pi/(n+p)$.

6.18 Reciprocating or rotary engines are frequently mounted on their foundations with rubber supports. Let these supports be modelled as a combination of a restoring force proportional to the displacement and a damping force proportional to the velocity. Because of its vibration, the engine is subject to a periodic force with frequency equal to that of its periodic motion. Thus the equation of motion, for vertical oscillations about equilibrium, may be written

$$\ddot{x} = f_0 \cos pt - (k\dot{x} + n^2 x),$$

where $f_0 \cos pt$ is the periodic force per unit mass and $(k\dot{x} + n^2 x)$ is the force per unit mass which arises from the supports and hence, by Newton's third law, is the force transmitted to the foundation. Show that when the transient oscillations are negligible, the ratio of the amplitudes of these two forces is

$$\left(\frac{n^4 + k^2 p^2}{(n^2 - p^2)^2 + k^2 p^2}\right)^{1/2}$$

Hence show that the amplitude of the force transmitted to the foundations is always greater than that of the exciting force if $p < 2^{1/2} n$.

Show that the ratio is less than $\frac{1}{2}$ if either

(i) k^2 is negligible and $p > 3^{1/2} n$, or
(ii) k is equal to half the value for critical damping and $p > 2.35n$.

6.19 A particle of mass m experiences a restoring force of magnitude $n^2 x$ per unit mass and a resistance of magnitude $k\dot{x}$ per unit mass, where x is the displacement. Given that $x = 0$, $\dot{x} = u$ when $t = 0$, and that $k < 2n$, show that the energy dissipated by the resistance in the interval $0 \leqslant t \leqslant 2\pi/(4n^2 - k^2)^{1/2}$ is

$$\tfrac{1}{2} m u^2 [1 - \exp\{-2\pi k/(4n^2 - k^2)^{1/2}\}].$$

Deduce that if the kinetic energy of the particle is increased by this amount each time $x = 0$, the motion will be periodic. If the energy is given by a regular impulse, show that it should be of magnitude

$$mu[1 - \exp\{-\pi k/(4n^2 - k^2)^{1/2}\}].$$

6.20 Verify that the equation

$$\ddot{x} + n^2 x = f_0 \cos nt$$

has the particular integral

$$\frac{f_0 t}{2n} \sin nt.$$

Hence obtain the solution which satisfies $x = a, \dot{x} = u$ when $t = 0$.

Now obtain the solution of the equation

$$\ddot{x} + k\dot{x} + n^2 x = f_0 \cos nt$$

which satisfies the same boundary conditions. Show that the two solutions agree in the limit $k \to 0$.

Sketch the solution when $a = u = k = 0$.

(Ans.: $kx = n^{-1} f_0 \sin nt + \mathrm{e}^{-kt/2} \{ka \cos \omega t + \omega^{-1}(ku + \frac{1}{2}k^2 a - f_0)\sin \omega t\}$, where $\omega = (n^2 - k^2/4)^{1/2}$.)

6.21 Consider forced oscillations governed by the equation

$$\ddot{x} + k\dot{x} + n^2 x = f_0 \cos pt,$$

when free oscillations are negligible. Verify from the solution that the energy dissipated by the resisting force in one complete period is equal to the work done by the forcing term as given by equation (6.3.10). Why does the instantaneous rate of working of the forcing term not equal the rate of dissipation of energy?

6.22 From the equation

$$\ddot{x} + k\dot{x} + n^2 x = f(t),$$

show directly, without appeal to the solution, that

$$\int_{t_1}^{t_2} \{f(t)\dot{x} - k\dot{x}^2\}\mathrm{d}t = [\tfrac{1}{2}\dot{x}^2 + \tfrac{1}{2}n^2 x^2]_{t_1}^{t_2}.$$

This result also shows that the work done by the forcing term is, on average, equal to the dissipated energy once the motion has become periodic, since then the right-hand side is zero if $t_2 - t_1$ is the periodic time.

6.23 An oscillator satisfies the equation

$$\ddot{x} + n^2 x = f_0 \,\mathrm{e}^{-qt}.$$

Show that the solution which satisfies $x = \dot{x} = 0$ when $t = 0$ is

$$x = \frac{f_0}{n^2 + q^2}\left(\mathrm{e}^{-qt} + \frac{\sin(nt - \alpha)}{\sin \alpha}\right),$$

where $\tan \alpha = n/q$.

6.24 An oscillator satisfies the equation

$$\ddot{x} + n^2 x = f(t).$$

Verify that the solution which satisfies $x = \dot{x} = 0$ when $t = 0$ is

$$x = \frac{1}{n}\int_0^t f(\tau)\sin n(t-\tau)\mathrm{d}\tau.$$

Let

$$f(t) = \begin{cases} \tfrac{1}{2}f_0 & 0 \leqslant t \leqslant t_0, \\ 0 & t > t_0. \end{cases}$$

Show that the solution is then

$$x = \begin{cases} \dfrac{f_0}{2n^2}(1-\cos nt), & 0 \leqslant t \leqslant t_0, \\ \dfrac{f_0}{n^2}\sin \tfrac{1}{2}nt_0 \sin n(t-\tfrac{1}{2}t_0), & t \geqslant t_0. \end{cases}$$

Sketch the solution for $t_0 = \pi/n$ and for $t_0 = 2\pi/n$.

For $t_0 = 2\alpha/n$, where $\tan\alpha = n/q$, show that this solution is the same as the asymptotic solution, as $t \to \infty$, of Exercise 6.23. In particular note the special case $q \ll n$, when t_0 is approximately π/n and the asymptotic oscillation is approximately

$$x = -\frac{f_0 \cos nt}{n^2}.$$

The implication of this result is that a force which is approximately constant and equal to f_0 for an interval of time of order q^{-1} produces the same asymptotic oscillation as a force $\tfrac{1}{2}f_0$ acting over the much shorter interval of time of order n^{-1}. What is the reason for this?

6.25 A light elastic string lies unstretched in a straight line on a horizontal plane. A particle of mass m, which is initially at rest, is attached to one end of the string and, for time $t > 0$, the other end moves with a velocity $V(t)$ away from the particle under the action of an appropriate force $F(t)$. The motion of the particle is resisted by a force $2m\omega v$, where v is its speed and ω is constant. Show that if the extension of the string at time t is x, and if the tension in the string is then $5m\omega^2 x$,

$$\ddot{x} + 2\omega\dot{x} + 5\omega^2 x = \dot{V}(t) + 2\omega V(t).$$

Verify that a solution can be written in the form

$$x = \int_0^t X(t-\tau)V(\tau)\mathrm{d}\tau,$$

where $X(t)$ satisfies the equation

$$\ddot{X} + 2\omega\dot{X} + 5\omega^2 X = 0$$

and the initial conditions $X(0) = 1$, $\dot{X}(0) = 0$.

Show that the above solution also satisfies the correct boundary conditions on x.

Find $X(t)$ and hence show that, if $V(t) = e^{-\omega t}$,

$$F(t) = \tfrac{5}{2}m\omega \sin \omega t\{\sin \omega t + 2 \cos \omega t\}e^{-\omega t}.$$

6.26 A particle moves in the (x, y) plane and is subject to a force $\omega^2(x - y)\mathbf{j}$ per unit mass. It is initially at the origin moving with velocity $u\mathbf{i}$, where ω and u are constants.

Show that the path of the particle in the subsequent motion is

$$y = x - \frac{u}{\omega}\sin\left(\left|\frac{\omega x}{u}\right|\right).$$

Show that the work done by the force field in the time interval $0 \leqslant t \leqslant T$ is $2u^2 \sin^4(\omega T/2)$ per unit mass.

6.27 Two particles A, B, of equal mass m, rest on a smooth horizontal table and are connected by a light spring of natural length l and modulus λ. Initially AB $= l$, A is at rest and B is given a velocity u in the direction AB.

Let x and y be the distances moved by A and B respectively, after time t. Write down the equations of motion and show that

$$\ddot{x} + \ddot{y} = 0, \quad \ddot{x} - \ddot{y} = -n^2(x - y),$$

where $n^2 = 2\lambda/ml$.

Calculate x and y as functions of t.

(Ans.: $2x = ut + un^{-1} \sin nt$, $2y = ut - un^{-1} \sin nt$.)

6.28 A light spring, of modulus λ and natural length $2l$, has one end fixed to a point O and has particles of masses m_1, m_2 suspended respectively from its mid-point and from its other end. Show that in equilibrium the particles are at depths

$$l + (m_1 + m_2)gl/\lambda \quad \text{and} \quad 2l + (m_1 + 2m_2)gl/\lambda,$$

respectively, below O.

Find the equations of motion for vertical oscillations of the particles about the equilibrium positions and show that the frequencies of the normal modes of vibration are given by the roots of the equation

$$n^4 - \frac{\lambda(m_1 + 2m_2)}{lm_1m_2}n^2 + \frac{\lambda^2}{l^2m_1m_2} = 0.$$

6.29 Two simple pendulums A_1B_1 and A_2B_2, each of length l and mass m, are suspended from two points A_1 and A_2 at the same level and at a distance a apart. The bobs B_1 and B_2 are connected by a light spring of modulus λ, whose unstretched length is a. Show that there are particular oscillations for which:

(i) the two pendulums are exactly in phase and oscillate with a periodic time which is identical to that for either pendulum oscillating independently with the spring removed;

(ii) the two pendulums are exactly out of phase, so that their angular displacements are θ and $-\theta$ respectively, and such that the energy equation can be written in the form

$$\dot{\theta}^2 + \frac{2g}{l}(1 - \cos\theta) + \frac{2\lambda}{ma}\sin^2\theta = \omega^2,$$

where ω is the value of $\dot{\theta}$ when $\theta = 0$. Show that the frequency of oscillation when θ is small is $(g/l + 2\lambda/ma)^{1/2}$.

The system is held at rest initially with A_2B_2 vertical and A_1B_1 displaced, in the vertical plane through A_1A_2, by a small angle α from the vertical. The system is released from rest at time $t = 0$. On a linearised theory, show that the angular displacements θ_1 and θ_2 of the two pendulums satisfy the equations

$$\ddot{\theta}_1 = -(g/l)\theta_1 + (\lambda/ma)(\theta_2 - \theta_1)$$
$$\ddot{\theta}_2 = -(g/l)\theta_2 - (\lambda/ma)(\theta_2 - \theta_1)$$

and hence that θ_1 and θ_2 can be represented as a linear combination of the oscillations described by (i) and (ii) above. Is this also true of the exact non-linear oscillations?

Show that

$$\theta_1 = \alpha\cos\sigma_1 t\cos\sigma_2 t, \quad \theta_2 = \alpha\sin\sigma_1 t\sin\sigma_2 t,$$

where σ_1 and σ_2 are constants which should be defined.

Discuss briefly the case when $\lambda l/mag$ is small.

(Ans.: $\sigma_1 = (g/l + 2\lambda/ma)^{1/2} + (g/l)^{1/2}$, $\sigma_2 = (g/l + 2\lambda/ma)^{1/2} - (g/l)^{1/2}$.)

6.30 Two particles, of masses m_1, m_2 $(m_1 > m_2)$, are attached to the ends of a string of length $2l$, which passes over a smooth peg at a height l above a smooth plane. The plane is inclined at an angle α to the vertical and the particles are held at rest on it at the point vertically below the peg. Prove that, if the particles are released, m_2 will oscillate through a vertical distance $2lm_1(m_1 - m_2)(m_2^2\sec^2\alpha - m_1^2)$ provided that $\tan^2\alpha > (3m_1 + m_2)(m_1 - m_2)/m_2^2$.

6.31 A light spring is designed to exert a restoring force equal to $n^2x + \nu x^3$ when extended a distance x. When a mass m is suspended from the spring the equilibrium extension is a.

(i) Show that if $\nu > 0$ any motion is oscillatory, about the equilibrium position $x = a$, for all values of the total energy E. Show that if the mass oscillates with a small amplitude, such that $|x - a| \ll a$, the oscillations are approximately simple harmonic with frequency

$$\left(\frac{g(n^2 + 3\nu a^2)}{a(n^2 + \nu a^2)}\right)^{1/2}.$$

(ii) Show that if $n^2 + 3\nu a^2 = 0$, there are two positions of equilibrium, at $x = a$ and $x = -2a$, both of which are unstable.

6.32 The one-dimensional motion of a particle is governed by the equation

$$\ddot{x} = -xn^2(x^2)$$

where $n^2(x^2)$ is some monotonic function of x^2 which is non-zero for $0 \leqslant x^2 \leqslant a^2$. Show that if $\dot{x} = 0$ when $x = a$, the motion is periodic with periodic time

$$T = 4\int_0^1 \left(\int_{\xi^2}^1 n^2(a^2\eta)\,\mathrm{d}\eta\right)^{-1/2} \mathrm{d}\xi.$$

Deduce that the periodic time increases or decreases with amplitude according as $n^2(x^2)$ decreases or increases with x^2.

6.33 A small bead is free to slide on a smooth wire whose shape is defined by the equation $y = f(x)$, where y is vertically upwards and x is horizontal. Show that there is an energy equation of the form

$$\tfrac{1}{2}\dot{x}^2\left\{1 + \left(\frac{\mathrm{d}f}{\mathrm{d}x}\right)^2\right\} + gf(x) = \text{constant}.$$

Let

$$f(x) = l\,|x|^\alpha/a^\alpha,$$

where l is positive, $\alpha > 1$ and a is the value of $|x|$ for which $\dot{x} = 0$. Show that the bead can perform oscillations with periodic time

$$T = \left(\frac{8l}{g}\right)^{1/2} F(\alpha, a/l)$$

where

$$F(\alpha, \beta) = \int_0^1 \left(\frac{\beta^2 + \alpha^2\xi^{2\alpha-2}}{1 - \xi^\alpha}\right)^{1/2} \mathrm{d}\xi.$$

Deduce the following properties of $F(\alpha, \beta)$:

(i) $F(1, \beta) = 2(1 + \beta^2)^{1/2}$,
(ii) $F(\alpha, 0) = 2$,

(iii) $F(\alpha, \beta) = \beta G(\alpha) + 0(\beta^{-1})$ for $\beta \gg 1$, where $G(\alpha)$ decreases monotonically from 2 to 1 as α increases from 1 to ∞,

(iv) $F(\alpha, \beta) \to 2 + \beta$ as $\alpha \to \infty$,

(v) $F(2, \beta) = 2yE(y^{-1})$,
where $y = (1 + \beta^2/4)^{1/2}$ and E is the complete elliptic integral of the second kind, namely

$$E(k) = \int_0^{\pi/2} (1 - k^2 \sin^2 \theta)^{1/2} d\theta.$$

Verify that the empirical approximation

$$\alpha^{-1}\{2(\alpha^2 + \beta^2)^{1/2} + \beta(\alpha - 1)\}$$

to $F(\alpha, \beta)$ has the properties (i)–(iv) with $G(\alpha)$ approximated by $(\alpha + 1)/\alpha$. Show that the error in $G(2)$ is less than 5% and the error in $F(2, 1)$ is less than 4%. ($E(2/5^{1/2}) = 1.18$.)

6.34 The one-dimensional equation of motion of a particle is

$$\ddot{x} = \frac{h^2}{x^3} + f(x),$$

where h is a positive constant.

Show that $x = a$ is a position of unstable equilibrium provided that

$$f(a) = \frac{h^2}{a^3}, \qquad -\frac{af'(a)}{f(a)} = n > 3.$$

Show that, if $n < 3$, the particle can perform small oscillations about $x = a$ with frequency $h(3 - n)^{1/2}/a^2$.

Sketch the potential energy curve when $f(x) = -\mu/x^2$ $(\mu > 0)$. Hence discuss the motion of the particle qualitatively. State the condition to be satisfied by the total energy E for oscillatory solutions to be possible, and calculate the periodic time of these oscillations.
(Ans.: $-\mu^2/2h^2 < E < 0$, $2\pi\mu/(-2E)^{3/2}$.)

6.35 A jet of water issues vertically upwards from a nozzle. The mass of water issued per unit time is km and the initial speed is u. The jet strikes a ball of mass m and the vertical velocity of the water relative to the ball is destroyed on impact.

Show that the momentum destroyed per unit time is

$$km(v - \dot{x})^2/v,$$

where $v = (u^2 - 2gx)^{1/2}$ is the speed of the jet at a height x above the nozzle.

Hence show that the equation for vertical motion of the ball is

$$\ddot{x} = k(v - \dot{x})^2/v - g.$$

Show that the ball can rest in equilibrium at a height

$$x = \frac{u^2}{2g} - \frac{g}{2k^2}$$

and that a perturbation about the equilibrium position is critically damped.

7 Central Forces

7.1 BASIC EQUATIONS

A force field with the property that the force at every point acts along the position vector of that point relative to an origin O is called a central force field. The motion of a particle subject to a central force is of particular interest because of its wide application to problems as diverse as the motion of the planets and the motion of electrons. The central force most commonly found in nature is one which varies inversely as the square of the distance from the centre of force O. We shall begin, however, by stating the equations for an arbitrary central force.

Let the position vector of the particle relative to O be $\mathbf{r}$, and let the force acting on the particle be $f(\mathbf{r})\mathbf{r}/r$ per unit mass. The equation of motion is then

$$\ddot{\mathbf{r}} = \frac{f(\mathbf{r})\mathbf{r}}{r}. \tag{7.1.1}$$

It follows immediately that $\mathbf{r} \times \ddot{\mathbf{r}} = \mathbf{0}$, from which we have

$$\mathbf{r} \times \ddot{\mathbf{r}} = \frac{\mathrm{d}}{\mathrm{d}t}(\mathbf{r} \times \dot{\mathbf{r}}) = \mathbf{0} \tag{7.1.2}$$

or

$$\mathbf{r} \times \dot{\mathbf{r}} = h\mathbf{k} \tag{7.1.3}$$

say, where h is a constant and $\mathbf{k}$ is a unit vector in a fixed direction. Because $\mathbf{r}$ and $\dot{\mathbf{r}}$ are always perpendicular to the fixed direction of $\mathbf{k}$, the motion must take place in a fixed plane defined by the origin and the initial directions of $\mathbf{r}$ and $\dot{\mathbf{r}}$.

For a particle with position vector $\mathbf{r}$ and momentum $m\dot{\mathbf{r}}$, the vector product $\mathbf{r} \times m\dot{\mathbf{r}}$ is called the moment of the momentum, or the angular momentum, of the particle. Equation (7.1.3) shows that the angular momentum $mh\mathbf{k}$ of a particle subject to a central force is conserved in both magnitude and direction. Because the direction is fixed it is not usual in this context to emphasise the vector character, and henceforth the constant h will be referred to as the angular momentum (per unit mass).

If the position of the particle is defined by polar coordinates (r, θ) in the plane of the motion, equation (7.1.1) has components along and perpendicular to the radius vector given by

$$\ddot{r} - r\dot{\theta}^2 = f(\mathbf{r}), \tag{7.1.4}$$

$$r\ddot{\theta} + 2\dot{r}\dot{\theta} = \frac{1}{r}\frac{d}{dt}(r^2\dot{\theta}) = 0. \tag{7.1.5}$$

Equation (7.1.5) implies on integration that

$$r^2\dot{\theta} = h, \tag{7.1.6}$$

where h is just the angular momentum as previously defined in equation (7.1.3). Equation (7.1.6) is in fact deducible from (7.1.3) with $\mathbf{r}$ and $\dot{\mathbf{r}}$ written in terms of their components in polar coordinates, namely $(r, 0)$ and $(\dot{r}, r\dot{\theta})$.

Example 7.1.1
We show that the radius vector to the particle from the centre of force sweeps out equal areas in equal times.

This result follows from the fact that if in time δt θ increases to $\theta + \delta\theta$, the area of the elementary triangle swept out by the radius vector is $\frac{1}{2}r^2\,\delta\theta$ correct to the first order (see Fig. 7.1.1). Accordingly, if we denote this area by δA, we have that

$$\delta A = \tfrac{1}{2}r^2\,\delta\theta(1 + \epsilon),$$

where $|\,\epsilon\,| \to 0$ as $\delta t \to 0$. It follows that

$$dA/dt = \lim_{\delta t \to 0} \delta A/\delta t = \tfrac{1}{2}r^2\,d\theta/dt = \tfrac{1}{2}h, \tag{7.1.7}$$

and so the rate of sweep of area is constant and equal to $\frac{1}{2}h$. This result is a direct consequence of the conservation of angular momentum.

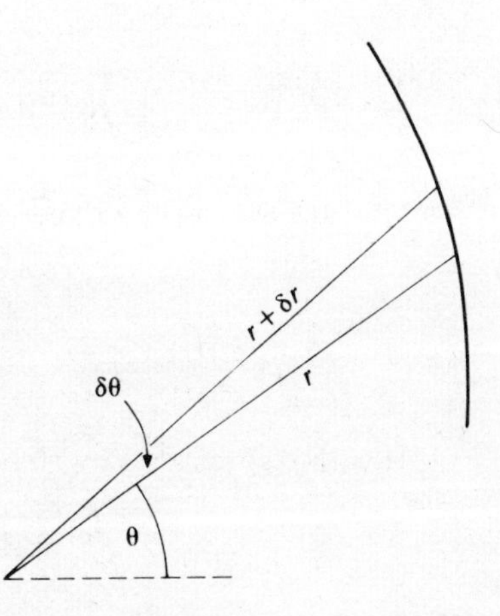

Figure 7.1.1 Illustration for Example 7.1.1: elementary area swept out by radius vector.

7.2 CONSERVATIVE FORCES

We have already seen in (7.1.6) that the angular momentum of a particle moving in a central force field is constant. When the force field is conservative, the total energy is also constant. These two conservation equations are sufficient to define the motion completely, when combined with appropriate initial conditions. We have also seen that a central force whose magnitude depends only on the distance r is conservative (see Example 1.10.1 but note that the potential energy is now $-\phi$). This covers most cases of physical interest and so we shall investigate the motion of a particle in such a force field in more detail.

The force per unit mass can now be written as $f(r)\mathbf{r}/r$, and the associated potential energy per unit mass is

$$V(r) = -\int^{\mathbf{r}} \frac{f(r)\mathbf{r} \,.\, \mathrm{d}\mathbf{r}}{r} = -\int^{r} f(r)\,\mathrm{d}r, \tag{7.2.1}$$

where the lower limit of the integral can be chosen to be any convenient level of zero potential energy and

$$\mathbf{r} \,.\, \mathrm{d}\mathbf{r} = \mathrm{d}(\tfrac{1}{2}\mathbf{r} \,.\, \mathbf{r}) = \mathrm{d}(\tfrac{1}{2}r^2) = r\,\mathrm{d}r$$

has been used. The energy equation is

$$\tfrac{1}{2}v^2 + V(r) = E, \tag{7.2.2}$$

where E is the total energy per unit mass. Since the speed of the particle in polar coordinates is such that $v^2 = \dot{r}^2 + r^2\dot{\theta}^2$ we have, with the help of (7.1.6), (7.2.1) and (7.2.2),

$$\tfrac{1}{2}\dot{r}^2 + U(r) = E, \tag{7.2.3}$$

where

$$U(r) = \frac{h^2}{2r^2} + V(r) = \frac{h^2}{2r^2} - \int^{r} f(r)\,\mathrm{d}r. \tag{7.2.4}$$

It is a straightforward matter to check that differentiation of (7.2.3) with respect to the time gives

$$\ddot{r} = \frac{h^2}{r^3} + f(r), \tag{7.2.5}$$

which is equivalent to (7.1.4) with $f(\mathbf{r})$ replaced by $f(r)$ and with $\dot{\theta}$ eliminated with the help of (7.1.6).

Since

$$\frac{\mathrm{d}r}{\mathrm{d}t} = \dot{\theta}\,\frac{\mathrm{d}r}{\mathrm{d}\theta} = \frac{h}{r^2}\,\frac{\mathrm{d}r}{\mathrm{d}\theta} \tag{7.2.6}$$

we can eliminate t in favour of θ as the independent variable in (7.2.3), provided that the trajectory is not one-dimensional with θ = constant and $h = 0$. This gives

$$\frac{h^2}{2r^4}\left(\frac{\mathrm{d}r}{\mathrm{d}\theta}\right)^2 + U(r) = E. \tag{7.2.7}$$

In principle, equation (7.2.7) can be integrated to give directly the equation of the trajectory in polar coordinates.

Comparison of (7.2.3), with (6.5.1) shows that it is analogous to the energy equation for the one-dimensional motion of a particle subject to a conservative force, with the radial coordinate r replacing the one-dimensional coordinate x, and with an effective potential energy

$$U(r) = \frac{h^2}{2r^2} + V(r).$$

Note that the contribution $h^2/2r^2$ to $U(r)$ arises from an effective force h^2/r^3. This is simply the term $r\dot{\theta}^2$, with $\dot{\theta}$ replaced by h/r^2, transferred to the right-hand side of equation (7.1.4) and treated as a contribution to the force instead of to the acceleration, as in (7.2.5). When it is interpreted in this fashion the term $r\dot{\theta}^2$ is called a centrifugal force. The motivation is that $\ddot{r}$ can then be regarded as the effective acceleration, in direct analogy with a one-dimensional problem, and the arguments used to investigate the associated energy equation (6.5.1) in the one-dimensional problem are also applicable to (7.2.3).

The motion is restricted to values of r for which $U(r) \leqslant E$, since $\dot{r}^2$ cannot be negative. When $U(r) = E$, $\dot{r}$ (and $\mathrm{d}r/\mathrm{d}\theta$) is zero so that r has a stationary value and the particle is moving perpendicular to the radius vector. A point on the trajectory of the particle for which r has a stationary value is called an apse. The direction of $\mathbf{r}$ there is called the apse line and r is the apsidal distance.

When $U(r)$ is such that

$$\tfrac{1}{2}\dot{r}^2 = E - U(r) = (r-b)(a-r)\psi(r), \tag{7.2.8}$$

where $b < a$ and $\psi(r)$ is strictly positive and single valued in the range $b \leqslant r \leqslant a$, there is a solution in which r is confined to the range $b \leqslant r \leqslant a$ (see equation (6.5.3)). There are just two apsidal distances, namely $r = b, r = a$, and r oscillates between these two values. The trajectory of the particle is in fact deducible from its behaviour in one of the intervals $b \leqslant r \leqslant a$. This is illustrated in Figure 7.2.1, where B_1, A_1, B_2, A_2, represent adjacent apses.

The situation can also be discussed using a more physical approach. Since the force is always directed towards the origin, and is always the same at the same distance, it follows that if at an apse the velocity of a particle were reversed, it would retrace its previous path. It also follows that the path is symmetrical about an apse line. Thus if adjacent apses on either side of OA_1 are B_1 and B_2, we must have $OB_1 = OB_2$. Similarly if A_2 is the next apse beyond B_2, then

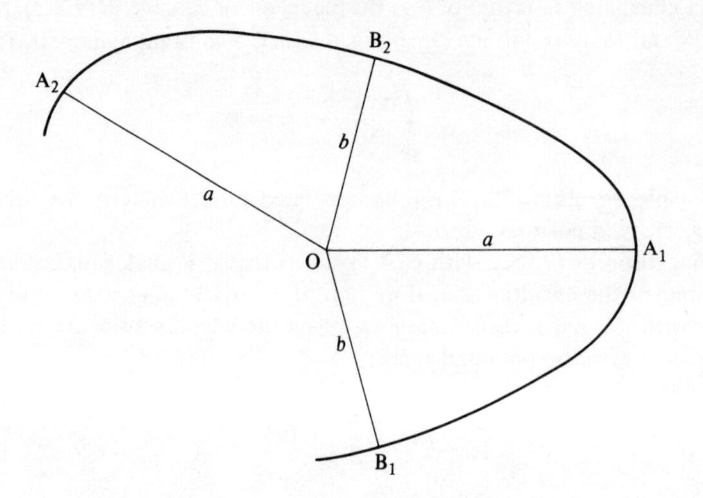

Figure 7.2.1 Apses and apse lines.

$OA_1 = OA_2$, and the trajectory can be built up by continual reflection about an apse line.

The angle between two adjacent apse lines is called the apsidal angle and is given by the expression

$$\frac{h}{2^{1/2}} \int_b^a \frac{dr}{r^2(E - U)^{1/2}},$$

which follows from (7.2.7).

By way of illustration, Figure 7.2.2 shows the variation with r of

$$U(r) = \frac{h^2}{2r^2} - \frac{\mu}{r}.$$

This is the effective potential for the inverse square law, with $f(r) = -\mu/r^2$. For the energy level E_0 and $\mu > 0$ in Figure 7.2.2, the motion is circular with radius r_0. This is the minimum energy level for which motion is possible. For $E = E_1$ and $\mu > 0$, the trajectory is bounded and r lies between $r = r_1$ and $r = r_1'$. In fact we shall see later that the apsidal angle is π and that there are just two apses. For $E = E_2$, the motion is restricted to $r \geqslant r_2$ $(\mu \geqslant 0)$ or $r \geqslant r_2'$ $(\mu < 0)$. There is only one apse and the trajectory is unbounded.

The inverse square law is appropriate for the determination of the motion of the planets round the sun. The apses are then called the perihelion, where r is a minimum, and the aphelion, where r is a maximum. The corresponding points for orbits relative to the earth as the centre of force are called the perigee and the apogee.

7.3 CIRCULAR ORBITS AND STABILITY

We continue the analogy with the one-dimensional problem with a discussion of the possibility of circular orbits and their stability.

In the one-dimensional motion of a particle subject to a conservative force, a point of equilibrium, where x is constant, corresponds to a stationary value of the potential. In the present context, the constant values of r for which $U(r)$ has a stationary value are the radii of possible circular orbits. For example

$$U(r) = \frac{h^2}{2r^2} - \frac{\mu}{r}$$

has a stationary value when $r = r_0 = h^2/\mu$. If

$$E = E_0 = U(r_0) = -\mu^2/2h^2,$$

then a circular orbit is possible with radius r_0 (see Fig. 7.2.2).

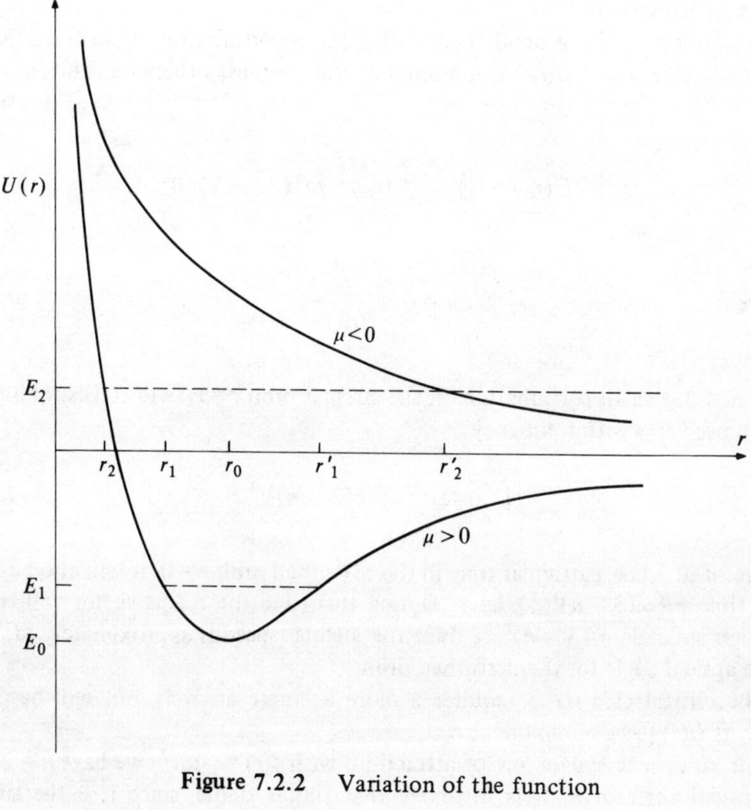

Figure 7.2.2 Variation of the function

$$U(r) = \frac{h^2}{2r^2} - \frac{\mu}{r}.$$

In the more general case, with $U(r)$ given by (7.2.4), a circular orbit with radius r_0 is possible if

$$U'(r_0) = -\frac{h^2}{r_0^3} - f(r_0) = 0. \tag{7.3.1}$$

The result also follows directly from the original equations of motion. In the first place equation (7.1.6) shows that

$$\dot{\theta} = h/r_0^2 = \omega$$

say, where ω is the constant angular speed with which the orbit is traversed. Then (7.1.4), with $\ddot{r} = 0$, gives

$$f(r_0) = -r_0\omega^2 = -h^2/r_0^3 \tag{7.3.2}$$

in agreement with (7.3.1). It follows that $f(r_0)$ must be negative so that the force is one of attraction.

According to the general theory of §6.5, a perturbation of such a solution will be oscillatory if $U(r_0)$ is a minimum and unstable otherwise. Thus to avoid instability we must have

$$U''(r_0) = \frac{3h^2}{r_0^4} - f'(r_0) = \omega^2(3-n) > 0, \tag{7.3.3}$$

where

$$n = -\frac{r_0 f'(r_0)}{f(r_0)}. \tag{7.3.4}$$

If $n < 3$, a small perturbation on the circular orbit $r = r_0$ will consist of a term which oscillates with frequency

$$(U''(r_0))^{1/2} = \omega(3-n)^{1/2}. \tag{7.3.5}$$

Hence, if at some particular time in the perturbed orbit $\dot{r} = 0$, it will also be zero at a time $\pi/\{\omega(3-n)^{1/2}\}$ later. During this time the radius vector will travel through an angle $\pi/(3-n)^{1/2}$, since the angular speed is approximately ω. This is the apsidal angle for the perturbed orbit.

The critical case $n = 3$ requires a more accurate analysis, but will be oscillatory if $U(r)$ has a minimum at r_0.

For an inverse square law of attraction, with $f(r) = -\mu/r^2$, we have $n = 2$ and an apsidal angle of π. It is fortunate that this is stable, since it is the law of attraction between the sun and the earth, whose orbit round the sun can be regarded as a small perturbation of a circular orbit.

Example 7.3.1

A particle of mass m_1 moves on a smooth horizontal table. It is attached to a light, inextensible string which passes through a small hole in the table and carries a mass m_2 hanging vertically.

Let r be the distance of m_1 from the hole, and let θ be the angular displacement of the horizontal part of the string. During the motion of the system, there is no work done by the tension in the string (see Example 5.3.1). The only force which does work on the system is the gravitational force acting on m_2, and this is conservative. The two conservation laws for the system as a whole are

$$r^2\dot{\theta} = h, \quad \tfrac{1}{2}m_1(\dot{r}^2 + r^2\dot{\theta}^2) + \tfrac{1}{2}m_2\dot{r}^2 + m_2 gr = \text{constant}.$$

where $m_1 h$ is the constant angular momentum of m_1. Elimination of $\dot{\theta}$ leads to the equation

$$\tfrac{1}{2}\dot{r}^2 + U(r) = E, \tag{7.3.6}$$

where
$$U(r) = \frac{1}{2}\frac{m_1}{m_1 + m_2}\frac{h^2}{r^2} + \frac{m_2}{m_1 + m_2}gr. \tag{7.3.7}$$

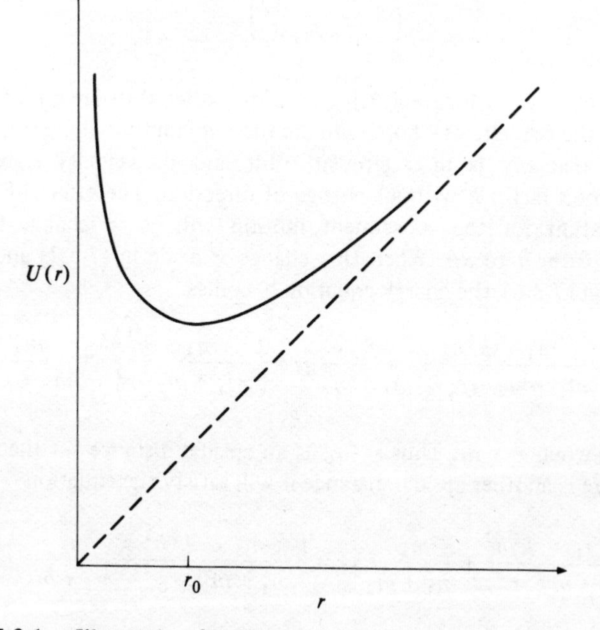

Figure 7.3.1 Illustration for Example 7.3.1: variation of the function

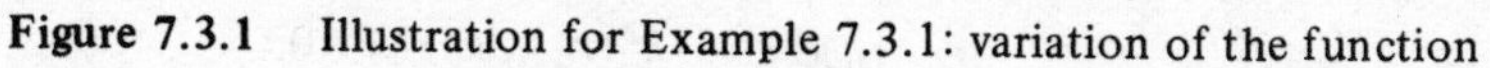

$$U(r) = \tfrac{1}{2}\,\frac{m_1}{m_1 + m_2}\frac{h^2}{r^2} + \frac{m_2}{m_1 + m_2}gr.$$

Figure 7.3.1 shows the variation of $U(r)$ with r. There is one stationary value, when $r = r_0$, given by

$$U'(r_0) = -\frac{m_1}{m_1 + m_2}\frac{h^2}{r_0^3} + \frac{m_2}{m_1 + m_2} g = 0.$$

It follows that for the particle m_1 to move in a circular orbit of a given radius r_0, it must be given an angular momentum h which satisfies the relation

$$h^2 = m_2 g r_0^3 / m_1, \qquad (7.3.8)$$

which is equivalent to a constant angular speed

$$\omega = \frac{h}{r_0^2} = \left(\frac{m_2 g}{m_1 r_0}\right)^{1/2} \qquad (7.3.9)$$

Since

$$U''(r_0) = \frac{3m_1}{m_1 + m_2}\frac{h^2}{r_0^4} = \frac{3m_2}{m_1 + m_2}\frac{g}{r_0}$$

is essentially positive, a perturbation of this circular orbit will consist of a superimposed oscillation with frequency

$$\left(\frac{3m_2}{m_1 + m_2}\frac{g}{r_0}\right)^{1/2}.$$

In fact it is clear from Figure 7.3.1 that all possible orbits are bounded and will lie between the two circles whose radii are the two appropriate apsidal distances.

Suppose that m_1 is in a circular orbit and its velocity is increased in magnitude by a factor k without change of direction. The effect of this on the energy equation for the subsequent motion will be to change the angular momentum from h to kh. When this change is made in (7.3.7) and the result substituted in (7.3.6), the energy equation becomes

$$\frac{1}{2}\dot{r}^2 + \frac{1}{2}\frac{m_1}{m_1 + m_2}\frac{k^2 h^2}{r^2} + \frac{m_2}{m_1 + m_2} gr = \frac{1}{2}\frac{m_1}{m_1 + m_2}\frac{k^2 h^2}{r_0^2} + \frac{m_2}{m_1 + m_2} g r_0,$$

since $\dot{r} = 0$ when $r = r_0$. Thus $r = r_0$ is an apsidal distance for the subsequent orbit. If there is another apsidal distance it will satisfy the equation

$$\frac{1}{2}\frac{m_1}{m_1 + m_2}\frac{k^2 h^2}{r^2} + \frac{m_2}{m_1 + m_2} gr = \frac{1}{2}\frac{m_1}{m_1 + m_2}\frac{k^2 h^2}{r_0^2} + \frac{m_2}{m_1 + m_2} g r_0$$

or

$$(r - r_0)(2r^2 - k^2 r r_0 - k^2 r_0^2) = 0, \qquad (7.3.10)$$

where the relation (7.3.8) has been used.

Apart from the root $r = r_0$, equation (7.3.10) has one other positive solution for r given by

$$r = \tfrac{1}{4}k\{k + (8 + k^2)^{1/2}\}r_0,$$

and this is the second apsidal distance.

7.4 ELLIPTIC HARMONIC MOTION

When $f(r) = -n^2 r$, where n is a positive constant, the equation of motion (7.1.1) becomes

$$\ddot{\mathbf{r}} = -n^2 \mathbf{r}. \tag{7.4.1}$$

Each component of this equation is of the form (6.1.1) with a corresponding solution of the form (6.1.5). When the solutions for the components are recombined into a single vector, the result is a solution of (7.4.1) of the form

$$\mathbf{r} = \mathbf{a}\cos nt + \mathbf{b}\sin nt, \tag{7.4.2}$$

where **a** and **b** are constant vectors. In terms of **a** and **b**, the energy and angular momentum are, from (7.2.3) and (7.1.3),

$$E = \tfrac{1}{2}\dot{\mathbf{r}}\,.\,\dot{\mathbf{r}} + \tfrac{1}{2}n^2\mathbf{r}\,.\,\mathbf{r} = \tfrac{1}{2}n^2(a^2 + b^2), \tag{7.4.3}$$

$$h = |\mathbf{r} \times \dot{\mathbf{r}}| = n\,|\mathbf{a} \times \mathbf{b}|. \tag{7.4.4}$$

We proceed to show that the particle describes an elliptic orbit. Let r_m be the maximum value of r, and let the origin of time be chosen so that $\mathbf{r} = \mathbf{r}_m$ when $t = 0$. Define the cartesian axes Ox in the direction $\mathbf{r}_m$ and Oy perpendicular to $\mathbf{r}_m$ in the initial direction of $\dot{\mathbf{r}}$. Because of the choice of axes these initial conditions are that

$$\mathbf{r} = (x, 0, 0), \quad \dot{\mathbf{r}} = (0, \dot{y}, 0) \tag{7.4.5}$$

when $t = 0$. Thus

$$\mathbf{a} = (a, 0, 0), \quad \mathbf{b} = (0, b, 0) \tag{7.4.6}$$

and so

$$x = a\cos nt, \quad y = b\sin nt. \tag{7.4.7}$$

Elimination of t from relations (7.4.7) gives the equation

$$\frac{x^2}{a^2} + \frac{y^2}{b^2} = 1, \tag{7.4.8}$$

which represents an ellipse with semi-axes of lengths a, b and centre at the origin. The ellipse is traversed in the periodic time $2\pi/n$.

Example 7.4.1. The Spherical Pendulum

In Example 6.5.1 we considered the simple pendulum whose motion is constrained to lie in a vertical plane. With a spherical pendulum, the only constraint is that the position vector $\mathbf{l}$ of the bob relative to the point of suspension has constant magnitude; in other words the bob moves on the surface of a sphere. The equation of motion is

$$m\frac{\mathrm{d}^2\mathbf{l}}{\mathrm{d}t^2} = m\mathbf{g} - \frac{T\mathbf{l}}{l}, \tag{7.4.9}$$

where T is the tension (see Fig. 7.4.1).

Let $\mathbf{l}$ have components $\mathbf{z}$ vertically and $\mathbf{r}$ horizontally. The components of (7.4.9) then give

$$\frac{\mathrm{d}^2 z}{\mathrm{d}t^2} = g - \frac{Tz}{ml} \tag{7.4.10}$$

$$\frac{\mathrm{d}^2\mathbf{r}}{\mathrm{d}t^2} = -\frac{T\mathbf{r}}{ml}. \tag{7.4.11}$$

If θ is the inclination of the pendulum to the vertical, we have $z = l\cos\theta$, which is constant to first order if θ is small. Thus for small oscillations (7.4.10) gives

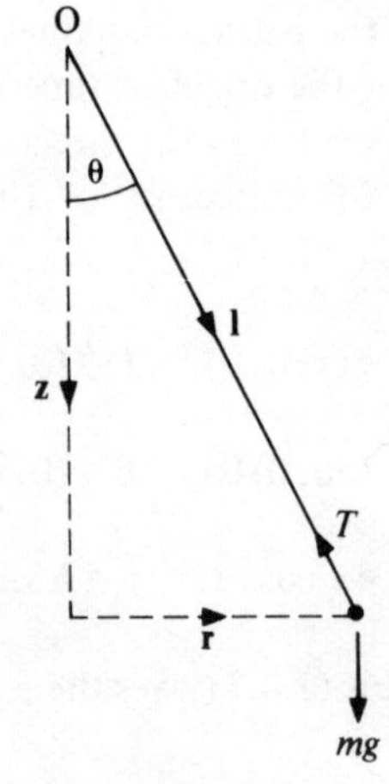

Figure 7.4.1 Illustration for Example 7.4.1: notation for the spherical pendulum.

$T = mg$ and then (7.4.11) becomes

$$\frac{d^2\mathbf{r}}{dt^2} = -\frac{g\mathbf{r}}{l}$$

correct to first order. This represents elliptic harmonic motion with period $2\pi(l/g)^{1/2}$.

7.5 THE INVERSE SQUARE LAW

The theory of motion under a central force was originally developed by Newton as a consequence of his law of gravitation, which states that the force on a particle P_1, of mass m_1 and position vector $\mathbf{r}_1$, arising from the attraction between P_1 and a second particle P_2, of mass m_2 and position vector $\mathbf{r}_2$, is

$$\frac{Gm_1m_2(\mathbf{r}_2 - \mathbf{r}_1)}{|\mathbf{r}_2 - \mathbf{r}_1|^3},$$

where $$G = 6{\cdot}67 \times 10^{-11}\ \text{m}^3\ \text{kg}^{-1}\ \text{s}^{-2}$$

and is called the gravitational constant. There is an equal and opposite force acting on P_2.

This is the source of the earth's gravitational attraction, but has a much wider application including, for example, the motion of the planetary system. Since G is small in magnitude, the force is appreciable only when large masses, such as the stars and planets, or small distances are involved. Although the law is stated here for the attraction between particles, it can be shown that spherically symmetrical distributions of mass attract in the same way as particles of similar mass at their centres. This is a good approximation for the planets.

One of Newton's earliest calculations in this connection was concerned with the attractive force of the earth for the moon. Let r_e be the radius of the earth and let d be the distance of the centre of the moon from the centre of the earth. According to the inverse square law, the force of attraction at the moon is $g(r_e/d)^2$ per unit mass, where g is the force per unit mass at the surface of the earth. From equation (7.3.2), with $f(d) = -g(r_e/d)^2$, it can be seen that the moon will describe a circle of radius d around the earth if

$$g\,(r_e/d)^2 = d\omega^2,$$

where ω is the angular speed of the moon in its orbit and is equal to $2\pi/T$, where T is the periodic time. Hence

$$\left(\frac{r_e}{d}\right)^3 = \frac{4\pi r_e}{gT^2}.$$

The quantity $\sin^{-1}(r/d)$ is called the moon's horizontal parallax and is known from observation, as indeed are all the other quantities in the above relation. This calculation provided one of the first tests of the law of gravitation. Subsequent work in astronomy and in the dynamics of satellites has abundantly confirmed the accuracy of the law in these fields.

The inverse square law also governs the attraction between electrically and magnetically charged particles, and can be either an attraction or a repulsion depending on the sign of the charges. One of the important applications here is to the motion of an electron in an electromagnetic field, for example in its orbit around the nucleus of an atom. Strictly speaking, the latter problem requires the theory of quantum mechanics for its complete formulation, although Bohr's original calculation of the energy levels of the hydrogen atom was based substantially on the classical theory.

The polar equation of the orbit for a general law of force may be obtained from the integral of equation (7.2.7). It so happens that there is an elementary integral for the inverse square law. We put $f(r) = -\mu/r^2$, so that μ is positive for an attractive force, and for the moment we restrict the analysis to $\mu > 0$. It is convenient to take the zero level for the potential energy at infinity and define

$$V(r) = \int_r^\infty f(r)\,\mathrm{d}r = \int_r^\infty -\frac{\mu}{r^2}\,\mathrm{d}r = -\frac{\mu}{r}. \tag{7.5.1}$$

Equation (7.2.7) then becomes, with the help of (7.2.4) and (7.5.1),

$$\frac{h^2}{2r^4}\left(\frac{\mathrm{d}r}{\mathrm{d}\theta}\right)^2 + \frac{h^2}{2r^2} - \frac{\mu}{r} = E. \tag{7.5.2}$$

Equation (7.5.2) is simplified by the substitution $u = 1/r$, which gives

$$\frac{\mathrm{d}u}{\mathrm{d}\theta} = -\frac{1}{r^2}\frac{\mathrm{d}r}{\mathrm{d}\theta}$$

so that equation (7.5.2) can be written

$$\tfrac{1}{2}h^2(\mathrm{d}u/\mathrm{d}\theta)^2 + \tfrac{1}{2}h^2u^2 - \mu u = E$$

or, equivalently,

$$\left\{\frac{\mathrm{d}}{\mathrm{d}\theta}\left(\frac{h^2u}{\mu} - 1\right)\right\}^2 + \left(\frac{h^2u}{\mu} - 1\right)^2 = 1 + \frac{2h^2E}{\mu^2}. \tag{7.5.3}$$

Equation (7.5.3) can now be compared with the energy equation (6.1.2) for simple harmonic motion. In fact the transformations

$$\theta = t, \quad \frac{h^2 u}{\mu} - 1 = x \tag{7.5.4}$$

change (7.5.3) into (6.1.2) with $n = 1$ and amplitude $(1 + 2h^2E/\mu^2)^{1/2}$. The solution of (7.5.3) is therefore deducible from (6.1.4) as

$$\frac{h^2 u}{\mu} = \frac{h^2}{\mu r} = 1 + \left(1 + \frac{2h^2 E}{\mu^2}\right)^{1/2} \cos(\theta + \alpha). \tag{7.5.5}$$

Equation (7.5.5) is the polar equation of a conic section (ellipse, parabola, hyperbola) with a focus as the origin of coordinates, and we proceed to classify the various possibilities.

Elliptic Orbit $(-\mu^2/2h^2 \leqslant E < 0, \mu > 0)$

Although the kinetic energy is essentially non-negative, it follows from (7.5.1) that the potential energy is everywhere negative for an inverse square law of attraction. Therefore it is possible for the total energy E to be negative when $\mu > 0$. If, however, equation (7.5.3) is to have a physical interpretation it is clear that $1 + 2h^2E/\mu^2$ cannot be negative. Equation (7.5.3), or more directly (7.5.5), also shows that the special case $E = -\mu^2/2h^2$ corresponds to a circular orbit with radius $r = h^2/\mu$. This is the energy level denoted by E_0 in Figure 7.2.2.

More generally, for an energy level such as E_1 in Figure 7.2.2, equation (7.5.5) is usually written in the form

$$l/r = 1 + e \cos\theta, \tag{7.5.6}$$

where

$$l = h^2/\mu, \tag{7.5.7}$$

$$e = \left(1 + \frac{2h^2 E}{\mu^2}\right)^{1/2} < 1 \tag{7.5.8}$$

and the polar axis is chosen so that $dr/d\theta$ is a minimum there, hence $\alpha = 0$.

An alternative form of the equation can be obtained using cartesian coordinates $x = r\cos\theta, y = r\sin\theta$, so that (7.5.6) gives

$$r^2 = x^2 + y^2 = (l - ex)^2.$$

When the terms in this equation are appropriately rearranged, the result is

$$\frac{(x + ae)^2}{a^2} + \frac{y^2}{b^2} = 1, \tag{7.5.9}$$

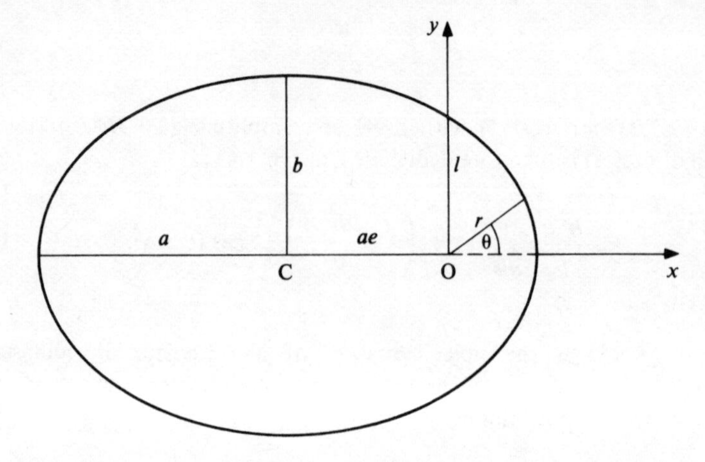

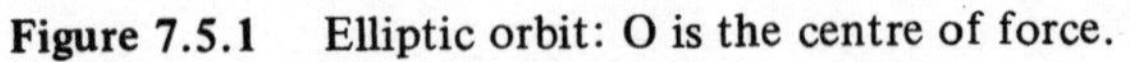

Figure 7.5.1 Elliptic orbit: O is the centre of force.

where
$$a = \frac{l}{1 - e^2} = -\frac{\mu}{2E}, \tag{7.5.10}$$

$$b = \frac{l}{(1 - e^2)^{1/2}} = a(1 - e^2)^{1/2} = \frac{h}{(-2E)^{1/2}}. \tag{7.5.11}$$

Equation (7.5.9) is the cartesian equation for an ellipse with centre at $x = -ae$, $y = 0$, semi-major axis a, semi-minor axis b, eccentricity e and semi-latus rectum l (see Fig. 7.5.1).

For an ellipse, the energy equation in its original form

$$\tfrac{1}{2}v^2 + V(r) = \tfrac{1}{2}v^2 - \mu/r = E, \tag{7.5.12}$$

simplifies to
$$v^2 = \mu\left(\frac{2}{r} - \frac{1}{a}\right) \tag{7.5.13}$$

with the substitution for $V(r)$ and E from relations (7.5.1) and (7.5.10) respectively. This is a useful relation giving the speed in terms of the distance from the centre. It shows that the speed is always less than $(2\mu/r)^{1/2}$ in an elliptic orbit. For the circular orbit $r = a$ the result

$$v^2 = \mu/a \tag{7.5.14}$$

is recovered. It is equivalent to (7.3.2) with the substitutions $f(a) = -\mu/a^2$ and $\omega = v/a$.

Equation (7.1.7) offers a convenient way of calculating the time to traverse any part of the orbit, a calculation which is of interest in astronomy. If A is the area traversed by the radius vector, the time of traverse is $2A/h$. In particular the periodic time T of a particle in an elliptic orbit is the time taken to traverse the area of the ellipse, which is πab, at the uniform rate $\frac{1}{2}h$. Hence, with the help of (7.5.10) and (7.5.11),

$$T = \frac{2\pi ab}{h} = \frac{2\pi\mu}{(-2E)^{3/2}} = \frac{2\pi a^{3/2}}{\mu^{1/2}}. \tag{7.5.15}$$

For a given central force, the periodic time in an elliptic orbit varies as the 3/2 power of the major axis. This is a result which has been checked with considerable accuracy for the planetary system.

At the surface of the earth $\mu/r_e^2 = g$, where r_e is the radius of the earth. It follows that for a particle to travel in a circular orbit just above the earth's surface requires a speed slightly less than

$$v = (\mu/r_e)^{1/2} = (gr_e)^{1/2},$$

if resistance is neglected. With

$$g = 9{\cdot}81 \text{ m s}^{-2}, \quad r_e = 6{\cdot}37 \times 10^6 \text{ m}$$

this gives

$$v = 7{\cdot}9 \times 10^3 \text{ m s}^{-1}.$$

The periodic time is

$$T = 2\pi(r_e/g)^{1/2} = 5063 \text{ s} = 84 \text{ min}.$$

The periodic time of artificial satellites in orbit around the earth is larger than this because they must be at a height greater than 10^5 m above the earth's surface in order to escape the resisting effect of the earth's atmosphere. A satellite in a circular orbit of radius $a = 7{\cdot}37 \times 10^6$ m for example is at a height of 10^6 m and well beyond the atmosphere. The periodic time is

$$T = \frac{2\pi a^{3/2}}{\mu^{1/2}} = \frac{2\pi a^{3/2}}{g^{1/2} r_e} = 105 \text{ min}.$$

Parabolic Orbit ($E = 0, \mu > 0$)

For a parabolic orbit, $\mu > 0$ and $e = 1$. Equation (7.5.5) gives, for the polar equation of the orbit,

$$l/r = 1 + \cos\theta. \tag{7.5.16}$$

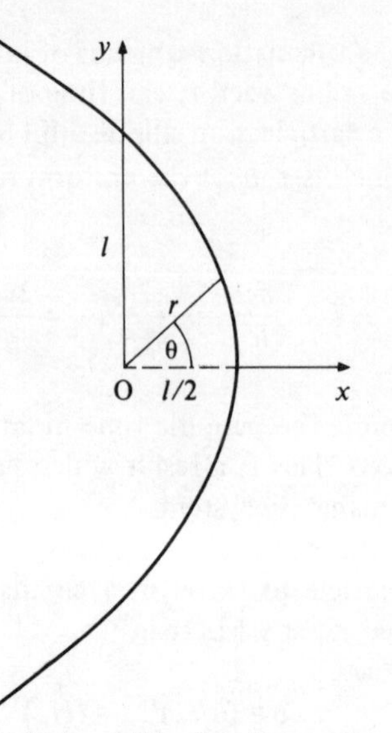

Figure 7.5.2 Parabolic orbit: O is the centre of force.

In cartesian coordinates this becomes

$$y^2 = 2l(\tfrac{1}{2}l - x), \tag{7.5.17}$$

which represents a parabola with latus rectum l and vertex at the point $x = l/2$, $y = 0$ (see Fig. 7.5.2).

Equation (7.5.12) simplifies to

$$v^2 = 2\mu/r. \tag{7.5.18}$$

In this orbit a particle will eventually recede to infinity, but with its kinetic energy and potential energy both tending to zero. Thus for a given initial value of r, $(2\mu/r)^{1/2}$ represents the minimum initial value of the speed required if the particle is to escape to infinity. This is just $2^{1/2}$ times the speed required for a circular orbit at the same radial distance.

From the surface of the earth the escape speed is

$$v = (2\mu/r_e)^{1/2} = (2gr_e)^{1/2} = 11{\cdot}2 \times 10^3 \text{ m s}^{-1}.$$

It is independent of the angle of projection. For a starting speed greater than this, the total energy will be positive, and so the kinetic energy when the particle reaches infinity will also be positive. The orbit in this case is a hyperbola whose properties are described below.

Hyperbolic Orbit ($E > 0, \mu \gtrless 0$)

When $E > 0$, so that $e = (1 + 2h^2E/\mu^2) > 1$, the equation

$$l/r = 1 + e \cos \theta$$

represents a hyperbola. We also note that when $\mu < 0$, so that the force is one of repulsion, we must have $E > 0$ because the kinetic energy $\frac{1}{2}mv^2$ cannot be negative and the potential energy $V = -\mu/r$ is now positive. It is convenient to treat the cases $E > 0$, $\mu \gtrless 0$ together.

For $\mu < 0$ we have an inverse square law of repulsion. We write the energy equation (7.5.2) in the form

$$\frac{1}{2}\frac{h^2}{r^4}\left(\frac{\mathrm{d}r}{\mathrm{d}\theta}\right)^2 + \frac{1}{2}\frac{h^2}{r^2} + \frac{|\mu|}{r} = E,$$

and the associated equation (7.5.3) becomes

$$\left\{\frac{\mathrm{d}}{\mathrm{d}\theta}\left(\frac{h^2u}{|\mu|} + 1\right)\right\}^2 + \left(\frac{h^2u}{|\mu|} + 1\right)^2 = 1 + \frac{2h^2E}{\mu^2}, \tag{7.5.19}$$

where $u = 1/r$. If the definition of l in (7.5.7) is generalised to

$$l = h^2/|\mu|, \tag{7.5.20}$$

the solution of (7.5.19) is

$$l/r = -1 + e \cos \theta$$

with a suitable choice of polar axis. The orbits are therefore given by

$$l/r = \pm 1 + e \cos \theta \tag{7.5.21}$$

with the positive sign for $\mu > 0$ and the negative sign for $\mu < 0$.

Relations (7.5.21) represent the two branches of the same hyperbola. In terms of cartesian coordinates $x = r \cos \theta$, $y = r \sin \theta$, (7.5.21) becomes

$$\frac{(x - ae)^2}{a^2} - \frac{y^2}{b^2} = 1, \tag{7.5.22}$$

where

$$a = \frac{l}{e^2 - 1} = \frac{|\mu|}{2E}, \tag{7.5.23}$$

$$b = \frac{l}{(e^2 - 1)^{1/2}} = a(e^2 - 1)^{1/2} = \frac{h}{(2E)^{1/2}}. \tag{7.5.24}$$

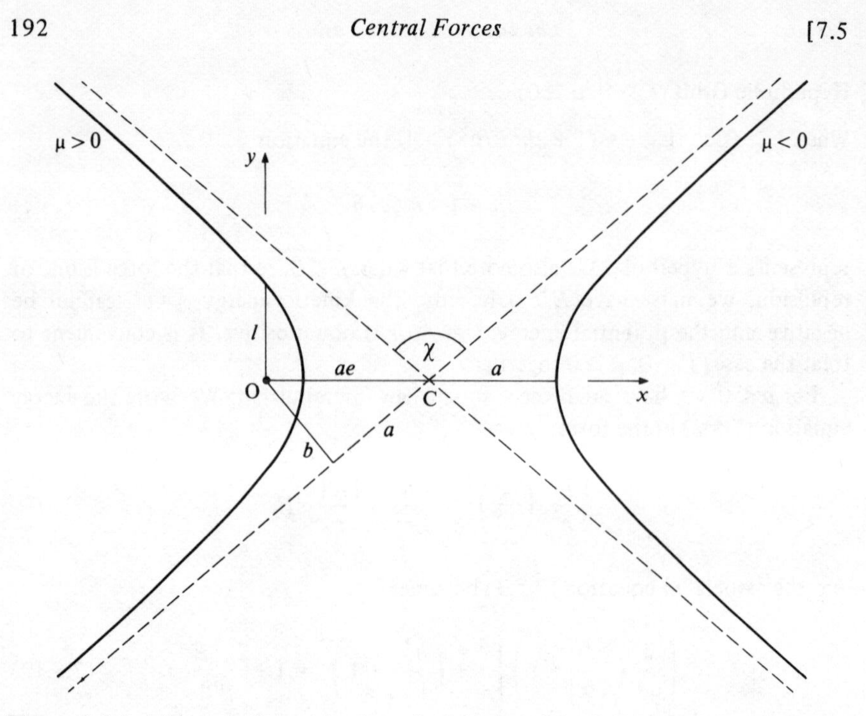

Figure 7.5.3 Hyperbolic orbits for $\mu > 0$ and $\mu < 0$: O is the centre of force.

Equation (7.5.22) represents a hyperbola with its centre C at $x = ae, y = 0$ and asymptotes $y = \pm b(x - ae)/a$ (see Fig. 7.5.3). For the hyperbolic orbit the energy equation (7.5.12) gives, with the help of (7.5.23),

$$v^2 = 2\frac{\mu}{r} + \frac{|\mu|}{a}. \tag{7.5.25}$$

It is clear from Figure 7.5.3 that a particle coming in from infinity along one asymptote and receding to infinity along the other is ultimately deflected through an angle $2 \tan^{-1} a/b$. In terms of the speed at infinity, which is

$$v_\infty^2 = |\mu|/a \tag{7.5.26}$$

from (7.5.25), this angle of deflection is

$$\chi = 2 \tan^{-1}(|\mu|/bv_\infty^2). \tag{7.5.27}$$

Relation (7.5.27) is important in the study of the scattering of interacting atomic particles, where it is not possible to follow the trajectory of an individual particle. What can be measured is the number of particles that are scattered between angles χ and $\chi + d\chi$. In such experiments an initially uniform parallel beam of particles impinges on the scattering centre and the particles are scat-

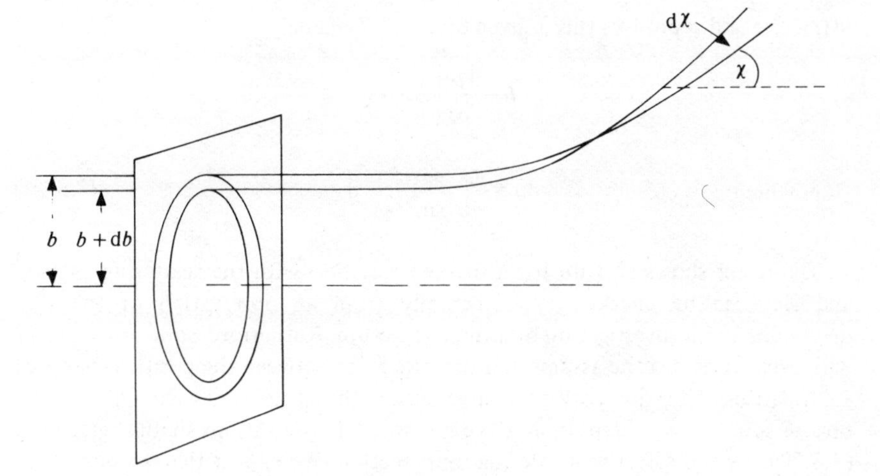

Figure 7.5.4 Scattering of neighbouring particles in a uniform incident beam.

tered through different angles χ depending on the distance b of their initial line of approach from the scattering centre (see Fig. 7.5.4); b is called the impact parameter. The number dN of particles scattered between angles χ and $\chi + d\chi$ will be proportional to the flux n of particles in the incident uniform beam, that is the number of particles crossing unit area normal to the beam in unit time. Hence we write $dN = n d\sigma$, where $d\sigma$ has the dimensions of area and is called the effective scattering cross-section. It is determined by the nature of the scattering process and is an important characteristic of that process. We assume that the scattering process is symmetrical about an axis through the scattering centre parallel to the incoming beam, as is the case for a fixed central force field. We also assume that the particles which are scattered through angles between χ and $\chi + d\chi$ are those which originated with impact parameters between b and $b + db$ (see Fig. 7.5.4). Since the flux of particles across a ring of thickness $|db|$ normal to the incident beam is $n 2\pi b |db|$, it follows that

$$d\sigma = 2\pi b |db|,$$

where the modulus sign is used to ensure that $d\sigma$ is positive even when db is negative, as it is in Figure 7.5.4. In order to obtain the dependence of σ on χ, we regard $d\chi$ as a positive increment in χ and write

$$d\sigma = 2\pi b \left| \frac{db}{d\chi} \right| d\chi. \qquad (7.5.28)$$

This is the desired relation in general terms, and is valid for any law of scatter. The precise form of (7.5.28) depends on the details of the scattering process, which determine the variation of b with χ. For the inverse square law (both attractive and repulsive) this is given by (7.5.27) namely

$$b = \frac{|\mu|}{v_\infty^2} \cot \tfrac{1}{2}\chi$$

from which

$$d\sigma = \frac{\pi\mu^2}{v_\infty^4} \frac{\cos \frac{1}{2}\chi}{\sin^3 \frac{1}{2}\chi} d\chi. \tag{7.5.29}$$

The result shows that $d\sigma$ has a marked variation with the scattering angle χ, and this can be checked experimentally from an observation of the flux $dN = n d\sigma$. In his investigation of atomic structure, Rutherford bombarded atoms with α-particles. On the assumption that the force between the positive charge of an α-particle and the positive charge inside the atom varies according to an inverse square law of repulsion, the experimental observations should agree with (7.5.29) for the effective scattering cross-section. We expect the assumption to be valid provided that the charges can be treated as point charges. But if positive charge were distributed throughout the volume of the atom, a deviation from (7.5.29) would be expected for those α-particles which penetrated the atom. Rutherford obtained agreement between theory and experiment provided that the distance of closest approach was larger than 10^{-14} m and so concluded that the positive charge was concentrated within such a radius. This was the origin of the nuclear theory of the atom.

The above calculation leading to (7.5.29) assumes a fixed centre of force and so is applicable only when the nucleus of the atom is much heavier than the α-particle. The modification required when the masses are comparable will be discussed in Example 8.3.4. The α-particles will also be deflected by the negative electrons, but the mass of the electron is relatively very small and the deflection is consequently insignificant (see Example 8.3.4).

Finally it should be noted that the interaction of an α-particle with an atomic nucleus is strictly a problem of quantum mechanics, where the concept of a definite particle with a definite trajectory is not appropriate. On the other hand, the concept of a scattering cross-section for a scattered beam of particles is appropriate in quantum mechanics, and (7.5.29) does in fact remain valid.

Summary

For reference purposes we end this section with a summary of the formulae involved in the study of orbits governed by the inverse square law. The numbering of the equations agrees with that used in the body of the chapter.

The eccentricity e and the semi-latus rectum l are

$$e = \left(1 + \frac{2h^2 E}{\mu^2}\right)^{1/2}, \tag{7.5.8}$$

$$l = h^2/|\mu|. \tag{7.5.20}$$

The polar equation of the orbit is

$$l/r = \pm 1 + e\cos\theta, \tag{7.5.21}$$

with the positive sign for $\mu > 0$ and the negative sign for $\mu < 0$.

Elliptic Orbit

$$\frac{(x+ae)^2}{a^2} + \frac{y^2}{b^2} = 1, \tag{7.5.9}$$

$$\tfrac{1}{2}v^2 - \mu/r = E = -\mu/2a, \tag{7.5.12}$$

$$T = \frac{2\pi ab}{h} = \frac{2\pi\mu}{(-2E)^{3/2}} = \frac{2\pi a^{3/2}}{\mu^{1/2}}, \tag{7.5.15}$$

$$a = \frac{l}{1-e^2} = -\frac{\mu}{2E}, \tag{7.5.10}$$

$$b = \frac{l}{(1-e^2)^{1/2}} = a(1-e^2)^{1/2} = \frac{h}{(-2E)^{1/2}}. \tag{7.5.11}$$

Parabolic Orbit

$$y^2 = 2l(\tfrac{1}{2}l - x), \tag{7.5.17}$$

$$\tfrac{1}{2}v^2 - \mu/r = E = 0. \tag{7.5.18}$$

Hyperbolic Orbit

$$\frac{(x-ae)^2}{a^2} - \frac{y^2}{b^2} = 1, \tag{7.5.22}$$

$$\tfrac{1}{2}v^2 - \mu/r = E = |\mu|/2a, \tag{7.5.25}$$

$$a = \frac{l}{e^2-1} = \frac{|\mu|}{2E}, \tag{7.5.23}$$

$$b = \frac{l}{(e^2-1)^{1/2}} = a(e^2-1)^{1/2} = \frac{h}{(2E)^{1/2}}. \tag{7.5.24}$$

Example 7.5.1

In the following example the earth is regarded as a sphere of radius r_e and the rotation of the earth about its axis is neglected. The resisting effect of the atmosphere is also neglected.

(i) A particle is projected from the surface of the earth with speed v_e at an angle β to the vertical. Show that the maximum range is $2r_e\alpha_m$, where $\sin\alpha_m = K/(2-K)$ and $K = v_e^2/gr_e < 1$. Show that the maximum range is attained when $\tan\beta = (1-K)^{-1/2}$.

(ii) If at its maximum height the speed is increased so as to put the particle into circular orbit, what is the additional speed required?

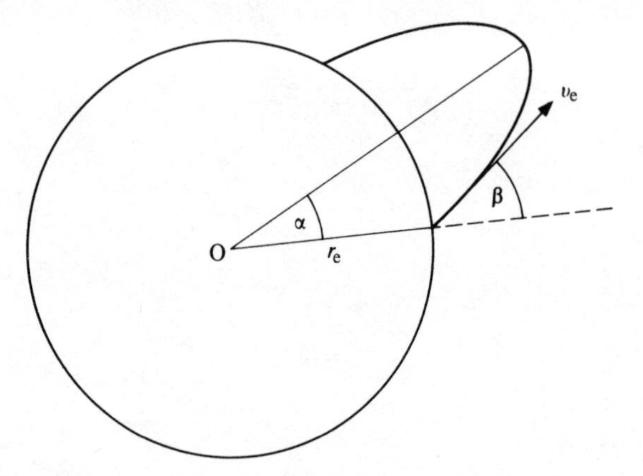

Figure 7.5.5 Illustration for Example 7.5.1: trajectory of a particle fired from the surface of the earth.

(i) The trajectory of the particle is an ellipse. The centre of the earth is the centre of attraction O. The maximum height is achieved at the apse which is furthest from O (the apogee, see Fig. 7.5.5).

It is convenient to take the polar axis along the radius vector from the centre of the earth to the apogee. The equation of the trajectory is then

$$l/r = 1 - e\cos\theta.$$

Since the orbit is symmetrical about the apogee, the range is $2r_e\alpha$, where $\theta = \alpha$ when $r = r_e$ (see Fig. 7.5.5). The maximum range is $2r_e\alpha_m$, where α_m is the maximum value of α as a function of β. The equation of the trajectory gives

$$l/r_e = 1 - e\cos\alpha, \tag{7.5.30}$$

and to find the maximum value of α we must determine the parameters l and e as functions of β.

We first determine the constants μ, h, E which characterise the motion. Since the law of attraction is of the form μ/r^2 per unit mass, we have

$\mu/r_e^2 = g$, the gravitational acceleration, and this serves to determine $\mu = gr_e^2$. The angular momentum h and the total energy E are determined by the initial conditions, which give

$$h = |\mathbf{r} \times \dot{\mathbf{r}}| = r_e v_e \sin\beta, \quad E = \tfrac{1}{2}v_e^2 - \mu/r_e = \tfrac{1}{2}v_e^2 - gr_e.$$

Equations (7.5.7) and (7.5.8) then give

$$l = \frac{h^2}{\mu} = \frac{v_e^2 \sin^2\beta}{g} = r_e K \sin^2\beta,$$

$$e^2 = 1 + \frac{2h^2 E}{\mu^2} = 1 - K(2 - K)\sin^2\beta.$$

These relations are now substituted in (7.5.30). The result is

$$\cos\alpha = \frac{1 - K\sin^2\beta}{\{1 - K(2 - K)\sin^2\beta\}^{1/2}},$$

whence

$$\tan\alpha = \frac{K\tan\beta}{1 + (1 - K)\tan^2\beta} = \frac{K}{(1 - K)^{1/2}}\frac{\xi}{1 + \xi^2},$$

where $\xi = (1 - K)^{1/2}\tan\beta$. Now the maximum value of $\xi/(1 + \xi^2)$ is $\frac{1}{2}$, attained when $\xi = 1$. It follows that

$$\tan\alpha_m = \frac{K}{2(1 - K)^{1/2}} \quad \text{or} \quad \sin\alpha_m = \frac{K}{2 - K}$$

and that the maximum range is attained when $(1 - K)^{1/2}\tan\beta = 1$.

The above discussion is appropriate for $0 < K < 1$. If $1 < K < 2$ all points on the earth's surface are attainable, since α can take all values between 0 and $\pi/2$. The value $\alpha = \pi/2$ is attained when $\tan\beta = \pm(K - 1)^{-1/2}$. The degenerate case $K = 1$, $\beta = \pi/2$ is the condition for a circular orbit of radius r_e. If $K \geqslant 2$ the particle will not move in a bounded trajectory but will escape from the earth's gravitational attraction, for then $v_e^2 > \mu r_e$.

(ii) Now let us consider the problem of putting the particle into circular orbit upon reaching the apogee. The radial coordinate r_a there is given by

$$r_a = l/(1 - e)$$

from the equation of the trajectory with $\theta = 0$. The speed v_a is given by

$$\tfrac{1}{2}v_a^2 - \mu/r_a = \tfrac{1}{2}v_e^2 - \mu/r_e,$$

or

$$v_a^2 = v_e^2 - 2gr_e\left(1 - \frac{1-e}{K\sin^2\beta}\right).$$

The speed v_0 required for a circular orbit of radius r_a is given by

$$v_0^2 = \frac{\mu}{r_a} = \frac{gr_e(1-e)}{K\sin^2\beta}.$$

Since e has already been found to be $\{1 - K(2-K)\sin^2\beta\}^{1/2}$, both v_a and v_0 can be determined in terms of the initial conditions. The difference $v_0 - v_a$ is the additional speed required to boost the particle into circular orbit.

Example 7.5.2

Two satellites, S_1 and S_2, are describing circular orbits in the same plane, of radii r_1 and r_2 ($>r_1$), around a fixed centre of force O. Let us consider the manoeuvre required to effect a rendezvous by boosting the velocity of S_1. The most effective way to do this is to increase the velocity of S_1 without change of direction so as to put S_1 into an elliptic orbit whose apsidal distances are r_1 and r_2 (see Fig. 7.5.6).

The orbit required will have the equation

$$l/r = 1 + e\cos\theta$$

where

$$l/r_1 = 1 + e, \quad l/r_2 = 1 - e$$

or

$$l = \frac{2r_1r_2}{r_1 + r_2}, \quad e = \frac{r_2 - r_1}{r_2 + r_1}.$$

This defines the orbit. The speed in the elliptic orbit at $\theta = 0$ is given by

$$v_e^2 = \mu\left(\frac{2}{r_1} - \frac{1}{a}\right) = \mu\left(\frac{2}{r_1} - \frac{2}{r_1 + r_2}\right) = \frac{2\mu r_2}{r_1(r_1 + r_2)}.$$

Hence the increase in speed for the required elliptic orbit is

$$v_e - \left(\frac{\mu}{r_1}\right)^{1/2} = \left(\frac{\mu}{r_1}\right)^{1/2}\left\{\left(\frac{2r_2}{r_1 + r_2}\right)^{1/2} - 1\right\}.$$

The time taken to arrive at the orbit of S_2 is one half of the periodic time in

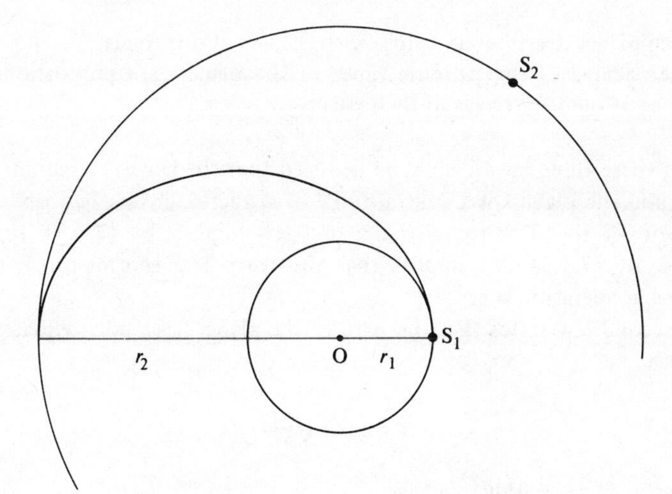

Figure 7.5.6 Illustration for Example 7.5.2: rendezvous of two satellites in circular orbit.

elliptic orbit, that is

$$\frac{\pi a^{3/2}}{\mu^{1/2}} = \frac{\pi}{\mu^{1/2}} \left(\frac{r_1 + r_2}{2}\right)^{3/2}.$$

Since S_2 has angular speed $(\mu/r_2^3)^{1/2}$, in this time it will have travelled an angular distance

$$\frac{\pi}{\mu^{1/2}} \left(\frac{r_1 + r_2}{2}\right)^{3/2} \left(\frac{\mu}{r_2^3}\right)^{1/2} = \pi \left(\frac{r_1 + r_2}{2r_2}\right)^{3/2}.$$

In order to effect a rendezvous, S_1 should be fired into orbit when S_2 is in advance of S_1 by an angular distance

$$\pi \left\{1 - \left(\frac{r_1 + r_2}{2r_2}\right)^{3/2}\right\}.$$

7.6 KEPLER'S LAWS

At the beginning of the seventeenth century, and sixty years before Newton postulated his theory of gravitation, Kepler had enunciated the following three laws, which he deduced empirically from the extensive series of observations of Tycho Brahe:

(i) The radius vector drawn from the sun to a planet describes equal areas in equal times.

(ii) Each planet describes an ellipse with the sun at one focus.
(iii) The squares of the periodic times of the planets are proportional to the cubes of the major axes of their paths.

From these three laws it may be deduced that the law of attraction between the sun and the planets is a central force of attraction inversely proportional to the square of the distance. For the first law implies, by (7.1.7), that $r^2\dot{\theta}$ is constant. By (7.1.5) this implies that the force is a central force, since the transverse acceleration is zero.

The second law states that the path of the planet has a polar equation of the form

$$l/r = 1 + e \cos \theta .$$

Hence, since $\mathrm{d}/\mathrm{d}t = \dot{\theta}\, \mathrm{d}/\mathrm{d}\theta$, we have

$$\ddot{r} = \dot{\theta} \frac{\mathrm{d}}{\mathrm{d}\theta}\left(\dot{\theta}\frac{\mathrm{d}r}{\mathrm{d}\theta}\right) = \frac{h}{r^2}\frac{\mathrm{d}}{\mathrm{d}\theta}\left(\frac{h}{r^2}\frac{\mathrm{d}r}{\mathrm{d}\theta}\right) = \frac{h}{r^2}\frac{\mathrm{d}}{\mathrm{d}\theta}\left(\frac{he\sin\theta}{l}\right)$$

$$= \frac{h^2 e \cos\theta}{lr^2} = \frac{h^2}{lr^2}\left(\frac{l}{r} - 1\right) .$$

Substitution in (7.1.4) gives

$$f(r) = -h^2/lr^2 = -\mu/r^2 .$$

At this stage we have deduced that for each planet the force varies inversely as the square of the distance, but so far we have only shown that μ is a constant for a particular planet. The third law now shows that μ is the same constant for all planets since, by (7.5.15),

$$\mu = 4\pi^2 a^3/T^2$$

and, by the third law, a^3/T^2 is the same for all planets.

In point of fact, accurate observation indicates a small discrepancy in Kepler's third law. This has a simple theoretical explanation in that gravity is both a mutual and a universal force, and the model of a planet subjected to a single force arising from the sun as a fixed centre is oversimplified. Consider the mutual force between the sun, of mass m_s, and a planet, of mass m_p. The law of gravitation implies that the planet is subjected to an acceleration Gm_s/r^2. However, since the force between sun and planet is a mutual one, the sun is subjected to an acceleration Gm_p/r^2. Thus the acceleration of the planet relative to the sun is $G(m_s + m_p)/r^2$. The constant of proportionality, μ, is therefore effectively $G(m_s + m_p)$ instead of Gm_s, and is hence in error by a factor $(1 + m_p/m_s)$ when calculated on the assumption of a fixed centre of force. One effect of this correction is to change the periodic time to

$$T = \frac{2\pi a^{3/2}}{\mu^{1/2}(1 + m_p/m_s)^{1/2}}$$

and the constant of proportionality is now different for different planets. The ratio of m_p/m_s is, however, very small as can be seen by reference to Table 7.6.1, which summarises some of the more important data concerning the planets. Its largest value, for Jupiter, is 10^{-3}. For the earth it is 3×10^{-6}.

Of the information summarised in Table 7.6.1, the periods of revolution and periods of rotation are obtained by telescopic and radar observation. The length of the orbital axis is then deduced from Kepler's third law, which states that

$$T_1^2/T_2^2 = a_1^3/a_2^3.$$

The eccentricity of the earth's orbit can be deduced from the fact that the distance to the sun is inversely proportional to the sun's apparent diameter.

Table 7.6.1 Planetary data.

		Mass	*Mean radius*	*Sidereal period of axial rotation*	*Sidereal period of revolution about the sun*	*Semi-major axis of orbit*	*Eccentricity*
		⊕ = 1	⊕ = 1	days	years	⊕ = 1	
Sun	☉	332 x 10^3	109·20	25·37	—	—	—
Mercury	☿	0·055	0·38	59 ± 5	0·241	0·387	0·206
Venus	♀	0·815	0·96	247 ± 5	0·615	0·723	0·007
Earth	⊕	1	1	0·997	1	1	0·017
Mars	♂	0·108	0·53	1·03	1·88	1·524	0·093
Jupiter	♃	318·4	10·97	0·41	11·86	5·202	0·048
Saturn	♄	95·3	9·06	0·43	29·46	9·539	0·056
Uranus	♅	14·5	3·62	0·45	84·01	19·182	0·047
Neptune	♆	17·2	3·52	0·66	164·8	30·057	0·009
Pluto	♇	0·14	0·47	0·28	250·6	39·520	0·250
		10^{24} kg	km	days	days	10^6 km	
Earth	⊕	5·976	6370	0·997	365·26	149·6	0·017
Moon	☾	0·0735	1738	27·32	27·32	0·384	0·055

Notes

1. The earth is used as the unit, but note that the sidereal period of axial rotation is relative to the fixed stars. For the earth this is 365/366 times the apparent period relative to the sun (1 day). The difference arises from the motion of the earth around the sun.
2. The last two rows give, for comparison, data for the earth and moon in metric units. Data for the moon refer to its orbit about the earth.

From the geometry of the ellipse (Fig. 7.5.1) the maximum and minimum distances to the sun are in the ratio $(1+e)/(1-e)$. This is equated to the ratio of the maximum and minimum apparent diameters of the sun. The same principle holds for the other planets, but the calculations are more elaborate.

If a planet has a satellite, the mass of the planet relative to that of the sun can be calculated from observations of the period of revolution of the planet round the sun, and the satellite round the planet. For example, if the suffix m refers to the orbit of the moon round the earth (of mass m_e) and the suffix e refers to the orbit of the earth round the sun (of mass m_s) we have, from (7.5.15) with $\mu = Gm$,

$$T_m^2 = \frac{4\pi^2 a_m^3}{Gm_e}, \quad T_e \doteq \frac{4\pi^2 a_e^3}{Gm_s}$$

and so

$$\frac{m_e}{m_s} = \frac{a_m^3 T_e^2}{a_e^3 T_m^2}.$$

If a_m/a_e and T_m/T_e are known from observation, this relation will serve to determine the ratio m_e/m_s.

To determine the mass of a planet in terrestrial units requires a knowledge of G, and this in turn is only determinable from an experiment involving known masses. The most recent of such experiments gives the value $6{\cdot}67 \times 10^{-11}\ \mathrm{m^3\ kg^{-1}\ s^{-2}}$.

7.7 THE EFFECT OF RESISTANCE

In the discussion so far the effect of resistance on orbits has been neglected. This is quite legitimate in the case of a planet moving round the sun. However air resistance will have a modifying effect near the surface of the earth, for example. Compared with the radius of the earth, the surrounding atmosphere is only a thin layer. At an altitude of 100 km above sea level the density of the atmosphere is reduced by a factor of 10^{-8} compared with its value at sea level. Weather satellites, which are in circular orbits higher than this, are moving in a vacuum for all practical purposes and resistance is negligible.

As a first example of the effect of resistance, we consider a satellite whose elliptic orbit round the earth lies outside the earth's atmosphere except for a small portion near the point of closest approach (the perigee) as in Figure 7.7.1. Since the resistive effect acts over a relatively short interval, we idealise the resistive force into an impulse, applied at the perigee along the line of flight. This will not affect the position of the perigee, since the impulse will not alter the radial distance r_p and the velocity of the satellite will still be perpendicular to the radius vector after the impulse. But the speed v_p at the perigee will be changed by a small amount δv_p. There will be a corresponding change δa in a, the semi-major axis of the orbit. Thus, by (7.5.13),

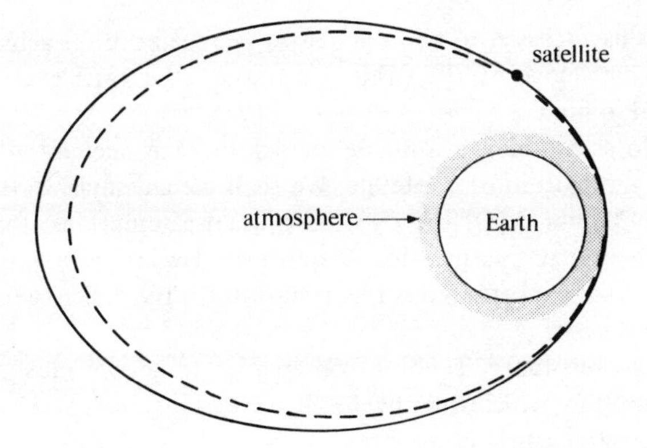

Figure 7.7.1 Satellite in orbit round the earth.

$$v_p^2 = \mu\left(\frac{2}{r_p} - \frac{1}{a}\right) \tag{7.7.1}$$

before the impulse, and

$$(v_p + \delta v_p)^2 = \mu\left(\frac{2}{r_p} - \frac{1}{a + \delta a}\right) \tag{7.7.2}$$

after the impulse. Subtraction of (7.7.1) from (7.7.2) and neglect of second and higher powers of δv_p and δa gives

$$2v_p\delta v_p = \frac{\mu}{a^2}\,\delta \dot{a}. \tag{7.7.3}$$

Since δv_p is negative, so is δa and the semi-major axis of the orbit is decreased. The periodic time, since it is proportional to $a^{3/2}$, is also decreased. Also, for the elliptic orbit,

$$r_p = a(1 - e) = (a + \delta a)(1 - e - \delta e). \tag{7.7.4}$$

Hence there is a small change δe in the eccentricity given, to the first order, by the relation

$$\frac{\delta a}{a} = \frac{\delta e}{1 - e}. \tag{7.7.5}$$

Thus e also decreases with a and v_p, and each time the satellite passes through the earth's atmosphere the effect, according to the above model, is to leave the minimum distance r_p unchanged, but reduce the maximum distance $a(1 + e)$ from the origin. The orbit will therefore tend to become circular. This model, however, will have ceased to be valid before a circular orbit is achieved since the assumption that most of the orbit lies outside the earth's atmosphere will eventually fail to hold.

We therefore consider a suitable model to take account of a persistent resistance to the motion of a satellite. We shall use an empirical law of the type discussed in Chapter 4. Of the two more usual assumptions, that of a linear variation with velocity, rather than a quadratic law, is more appropriate for a highly rarefied atmosphere. Since this is also much the easier to handle we shall proceed on this basis.

Because the resistance is not a conservative force, we must return to the equations of motion, which will now be augmented by a force $-k\dot{\mathbf{r}}$ per unit mass, where k is a positive constant, to give

$$\ddot{r} - r\dot{\theta}^2 = -\frac{\mu}{r^2} - k\dot{r}, \tag{7.7.6}$$

$$\frac{1}{r}\frac{\mathrm{d}}{\mathrm{d}t}(r^2\dot{\theta}) = -kr\dot{\theta}. \tag{7.7.7}$$

Equation (7.7.7) is easily integrated to give

$$r^2\dot{\theta} = h\mathrm{e}^{-kt}, \tag{7.7.8}$$

where h is a positive constant. This relation shows that the angular momentum is no longer constant but is subject to an exponential decay because of the resistance. This decay will be slow if k is small but persistent and cumulative.

Equation (7.7.6) is not so readily integrated and to proceed further requires a preliminary transformation, the reason for which is not immediately apparent. It should be judged by the simplicity achieved in the final results. We first note the following relation:

$$\mathrm{e}^{kt}\frac{\mathrm{d}}{\mathrm{d}t}\left(\mathrm{e}^{-3kt}\frac{\mathrm{d}}{\mathrm{d}t}(\mathrm{e}^{2kt}r)\right) = \mathrm{e}^{kt}\frac{\mathrm{d}}{\mathrm{d}t}\{\mathrm{e}^{-kt}(\dot{r}+2kr)\} = \ddot{r} + k\dot{r} - 2k^2 r. \tag{7.7.9}$$

This relation is substituted in (7.7.6) and $\dot{\theta}$ is eliminated with the help of (7.7.8). The result is

$$\mathrm{e}^{kt}\frac{\mathrm{d}}{\mathrm{d}t}\left(\mathrm{e}^{-3kt}\frac{\mathrm{d}}{\mathrm{d}t}(\mathrm{e}^{2kt}r)\right) - \frac{h^2\mathrm{e}^{-2kt}}{r^3} + \frac{\mu}{r^2} = -2k^2 r. \tag{7.7.10}$$

Next we make the transformation

$$r = l\rho\mathrm{e}^{-2kt}, \tag{7.7.11}$$

where l is a constant with the dimensions of a length and serves to give the order of magnitude of r in the orbit of the satellite. We then have a non-dimensional variable ρ of order unity satisfying the equation

$$e^{-3kt}\frac{d}{dt}\left(e^{-3kt}\frac{d\rho}{dt}\right) - \frac{h^2}{l^4\rho^3} + \frac{\mu}{l^3\rho^2} = -2k^2\rho e^{-6kt}. \qquad (7.7.12)$$

Finally we replace t by θ as the independent variable, so that

$$\frac{d\rho}{dt} = \frac{d\theta}{dt}\frac{d\rho}{d\theta} = \frac{h}{l^2\rho^2}e^{3kt}\frac{d\rho}{d\theta} = -\frac{he^{3kt}}{l^2}\frac{d}{d\theta}\left(\frac{1}{\rho}\right), \qquad (7.7.13)$$

$$e^{-3kt}\frac{d}{dt}\left(e^{-3kt}\frac{d\rho}{dt}\right) = e^{-3kt}\frac{d\theta}{dt}\frac{d}{d\theta}\left\{-\frac{h}{l^2}\frac{d}{d\theta}\left(\frac{1}{\rho}\right)\right\} = -\frac{h^2}{l^4\rho^2}\frac{d^2}{d\theta^2}\left(\frac{1}{\rho}\right). \qquad (7.7.14)$$

When (7.7.14) is substituted in (7.7.12), the resulting equation can be written in the form

$$\frac{d^2}{d\theta^2}\left(\frac{1}{\rho} - \frac{\mu l}{h^2}\right) + \left(\frac{1}{\rho} - \frac{\mu l}{h^2}\right) = +\frac{2k^2 l^4}{h^2}\rho^3 e^{-6kt}. \qquad (7.7.15)$$

This equation would be of the simple form already discussed in §6.1 if it were not for the awkward term on the right-hand side. We now consider the conditions under which it would be appropriate to neglect this term, and this requires an estimate of its magnitude. For this purpose it is convenient to write h, which from (7.7.8) is the angular momentum apart from the decay factor, as $l^2\omega$. Here l, as already introduced, is a typical measure of the radius vector in the orbit. Then ω is a typical measure of the angular speed of the satellite in the orbit. The factor $k^2 l^4/h^2$ which appears on the right-hand side of (7.7.15) becomes, with the suggested substitution, k^2/ω^2. This is a non-dimensional quantity since both ω^{-1} and k^{-1} have the dimensions of time. But whereas ω^{-1} is a typical measure of the time associated with the motion of the satellite in its orbit, k^{-1} is an appropriate time scale with which to measure the effects of resistive decay. For example, equation (7.7.8) shows that significant changes in the angular momentum, through the factor e^{-kt}, occur when the time changes by an amount of the order of k^{-1}. In a typical practical situation the ratio k/ω will be very small, and neglect of the right-hand side of equation (7.7.15) amounts to the realistic assumption that k^2/ω^2 will be negligibly small. With this simplification the solution is immediate and can be written

$$\frac{1}{\rho} = \frac{\mu l}{h^2} + \epsilon\cos(\theta + \alpha), \qquad (7.7.16)$$

where ϵ and α are constants of integration. By analogy with (7.5.7) we now choose l to be h^2/μ, since its precise definition is still at our disposal. Also $\alpha = 0$ if the polar axis is chosen appropriately so that finally we have

$$l e^{-2kt}/r = 1 + \epsilon \cos\theta. \tag{7.7.17}$$

In this equation ϵ is used in preference to e on the right-hand side simply to avoid confusion with the exponential factor e^{-2kt} on the left-hand side. With this in mind the equation should be compared with (7.5.6). For small resistance it shows that the orbit is still approximately a conic, but that the latus rectum is slowly decreasing with time. For a bounded orbit approximating to an ellipse, it means that the orbit is not closed, but spirals slowly inward towards the centre of force.

Equation (7.7.17) is considerably simpler than one might have expected, though it can be argued that it is not self-contained because of the appearance of all three variables r, θ, t. To complete the solution satisfactorily, a relation between θ and t is required. This can be obtained from (7.7.8) and (7.7.17), which combine to give

$$\dot{\theta} = \frac{h}{l^2}(1 + \epsilon\cos\theta)^2 e^{3kt}, \tag{7.7.18}$$

or, on integration,

$$e^{3kt} = 1 + \frac{3k}{\omega} I(\theta), \tag{7.7.19}$$

where

$$I(\theta) = \int_0^\theta \frac{d\theta}{(1 + \epsilon\cos\theta)^2}, \tag{7.7.20}$$

$\omega = h/l^2$ and it has been assumed that $t = 0$ when $\theta = 0$. Substitution of (7.7.19) in (7.7.17) then shows that the equation of the orbit is

$$l/r = (1 + \epsilon\cos\theta)\left(1 + \frac{3k}{\omega} I(\theta)\right)^{2/3}. \tag{7.7.21}$$

The integral $I(\theta)$ is expressible in closed form* though at the cost of the desirable simplicity which so far has given qualitative insight merely by inspection of the results. If we assume that the orbit is nearly circular, again a property often met in applications, then the analysis simplifies, for $\epsilon \ll 1$ and $I = \theta + O(\epsilon)$. Since k/ω is also assumed small, the error incurred when I is replaced by θ in (7.7.21) is a second order term and hence negligible on a linear theory. The

$$^*\int_0^\theta \frac{d\theta}{(1+\epsilon\cos\theta)^2} = \frac{2}{(1-\epsilon^2)^{3/2}} \tan^{-1}\left\{\left(\frac{1-\epsilon}{1+\epsilon}\right)^{1/2} \tan\frac{\theta}{2}\right\} - \frac{\epsilon\sin\theta}{(1-\epsilon^2)(1+\epsilon\cos\theta)} \quad \text{if } \epsilon < 1.$$

equation for the orbit is then

$$\frac{l}{r} = (1 + \epsilon \cos \theta)\left(1 + \frac{3k}{\omega}\theta\right)^{2/3}. \qquad (7.7.22)$$

Further simplification on the grounds that k/ω is small is not possible. The required condition is strictly $k\theta/\omega \ll 1$, and this depends on the magnitude of θ as well as that of k/ω.

EXERCISES

7.1 Extend Table 7.6.1 by calculating the gravitational acceleration at the surfaces of the planets.

Because of the acceleration arising from the rotation of the planets about their axes, there is a correction to the gravitational acceleration. Show that on the earth's equator the correction amounts to 0·35% of the gravitational acceleration. Show that for Saturn, where the correction is largest, the corresponding figure is 15%.

7.2 Show that the minimum distance of Pluto from the sun is less than the minimum distance of Neptune.

7.3 Before the discovery of Kepler's laws it was thought that a planet travelled in a circle of radius a and centre O, the sun S being a distance $\epsilon a (\ll a)$ from O. In addition a point R was defined by $\overline{RO} = \overline{OS}$, and the radius vector from R to the planet was supposed to rotate about R with constant angular velocity ω. Find the equation of this orbit in polar coordinates with origin at S and polar line along ROS, ignoring terms of order ϵ^2 or smaller. Show that it approximates to the polar equation of an ellipse with semi-latus rectum a and eccentricity ϵ. Show also that the angular momentum about S is constant to this order.

7.4 The orbit of a communications satellite is to be chosen so that the satellite appears to be stationary when viewed from the earth. Show that the appropriate orbit is circular in the earth's equatorial plane with a radius

$$\left(\frac{gT^2}{4\pi^2 r_e}\right)^{1/3} r_e = 6{\cdot}6\, r_e,$$

where r_e is the radius of the earth, which is assumed to be spherical, T is the periodic time of the earth about its own axis and g is the gravitational acceleration at the earth's surface.

7.5 A particle describes an elliptic orbit under an inverse square law of force directed to one focus. Show that the times taken to traverse the two parts of the orbit separated by the minor axis are in the ratio $\pi - 2e : \pi + 2e$.

For the earth in its orbit round the sun, show that the difference in these times is about four days.

7.6 A satellite is in circular orbit under an inverse square law of attraction, and the periodic time for one revolution is T. Show that if the satellite were suddenly stopped in its orbit, it would fall into the centre of force in a time $2^{1/2}T/8$.

7.7 Two uniform solid metal spheres, each of diameter a and density ρ, are placed at rest with their centres a distance $2a$ apart. Show that if they are subject only to their own mutual attraction, they will come into contact in a time $(1 + \pi/2)(3/G\pi\rho)^{1/2}$, where G is the gravitational constant.

Show that for iron spheres, of density $7{\cdot}9 \times 10^3$ kg m^{-3}, the time is about an hour.

It may be assumed that the gravitational force of attraction between the spheres is equivalent to that between particles of the same mass situated at their centres.

7.8 A particle describes a circle of radius a under the influence of a central force varying inversely as the square of the distance. Show that if the direction of motion is suddenly changed, without change of speed, the new orbit is an ellipse with semi-major axis a and the same periodic time as the circular orbit.

7.9 A particle describes an ellipse of semi-major axis a under the action of a force μ/r^2 per unit mass towards a fixed focus. When the particle is at one end of the minor axis it collides and coalesces with a second particle which is initially at rest. Prove that the new orbit is an ellipse and that, if the periodic times are T_1 and T_2 before and after the impact, $1 < T_1/T_2 < 2^{3/2}$.

7.10 A particle describes an ellipse under the central force $-\mu\mathbf{r}/r^2$ per unit mass. The minimum and maximum speeds are u, U, respectively.

Show that the eccentricity of the orbit and the speed of the particle are given by $e = (U - u)/(U + u)$, $v^2 = 2\mu/r - uU$.

7.11 Two satellites are travelling in the same direction in the same circular orbit of radius r_0 and initially are some distance apart. One of them, in an attempt to reach the other, increases its speed without change of direction by an amount u. Given that $k = u(r_0/\mu)^{1/2}$, where μ/r^2 is the force of attraction per unit mass at a distance r, and that $k^2 + 2k < 1$, show that the new orbit is an ellipse whose major axis is $r_0(1 - 2k - k^2)$ and whose eccentricity is $(2k + k^2)$.

Discuss the possibility of effecting a rendezvous by this means.

7.12 A particle moves in a bounded orbit under an inverse square law of force. Prove that the time average of the kinetic energy is half the magnitude of the time average of the potential energy.

7.13 A comet travels in a parabolic orbit round the sun. Its least distance from the sun is $kr_e (k < 1)$, where r_e is the radius of the earth's orbit. Show that the time during which the comet's distance from the sun is less than r_e is

$$2^{1/2}(1 - k)^{1/2}(1 + 2k)/3\pi \text{ years.}$$

Show that the time is a maximum when $k = \frac{1}{2}$, and that it is then $2/3\pi$ years.

7.14 A satellite moves in a circular orbit of radius $4r_m$ round the moon, which is assumed to be a uniform sphere of radius r_m. The gravitational force at the surface of the moon is g_m per unit mass. The satellite is to be intercepted by a rocket fired horizontally from the moon's surface with initial speed $(2gr_m)^{1/2}$ and launched from a point in the plane of the satellite's orbit. Show that the rocket should be fired at a time

$15{\cdot}85(r_m/g_m)^{1/2}$ after the satellite has passed vertically over the rocket launching site.

7.15 (i) A particle is subject to a central force $-\mu\mathbf{r}/r^3$ per unit mass, where $\mathbf{r}$ is the position vector. It is projected at right angles to the radius vector with speed u at a distance c from the centre of force. Show that

$$\dot{r}^2 = u^2\left(1 - \frac{c^2}{r^2}\right) + 2\mu\left(\frac{1}{r} - \frac{1}{c}\right).$$

(ii) A particle is at rest on the surface of a smooth sphere of radius c. It is subject to a central force $-\mu\mathbf{r}/r^3$ $(\mu > 0)$ per unit mass towards the centre of the sphere. There are no other forces apart from the reaction between the particle and the sphere. The particle is given a velocity u at right angles to the radius vector.

(a) For $u^2 \leqslant \mu/c$ describe the motion of the particle and give the periodic time.
(b) For $\mu/c < u^2 < 2\mu/c$, find the greatest and least distances from the centre of force attained by the particle in its orbit. Deduce the periodic time.
(c) For $u^2 \geqslant 2\mu/c$ what is the asymptotic value of the speed far from the centre?

(Ans.: $2\pi c/u$, $2\pi\mu(2\mu/c - u^2)^{-3/2}$, $(u^2 - 2\mu/c)^{1/2}$.)

7.16 A particle is subject to a central force $-\mu\mathbf{r}/r^3\,(\mu > 0)$ per unit mass.

(i) The particle is projected at right angles to the radius vector from a point at a distance c from the centre of force with speed nv_0, where v_0 is the speed which would be required for a circular orbit. Show that the path of the particle turns through an angle ϕ in the subsequent motion, where $\sin\phi = (n^2 - 1)^{-1}$, provided $n^2 > 2$.

(ii) The particle is projected from infinity with a velocity of magnitude nv_0, where v_0 is the speed required for a circular orbit of radius c, and the velocity is initially directed along a line whose perpendicular distance from the centre of force is c. Show that the path of the particle is turned through an angle 2ϕ, where $\tan\phi = n^{-2}$.

7.17 A uniform spherical star, of mass m_s and radius r_s, is moving with constant velocity $\mathbf{u}$ through a large cloud of particles. Show that all particles ahead of the star and within a cylinder of radius

$$r_s(1 + 2Gm_s/r_s u^2)^{1/2},$$

where G is the gravitational constant, will eventually collide with the star.

7.18 A particle is subject to an inverse square law central force of attraction. It is projected from infinity with speed v_∞ along a line whose perpendicular distance from the centre of force is b, and is ultimately deflected through an angle $\pi/2$.

Let $O(x, y, z)$ be a system of cartesian axes such that the particle starts from infinity along the positive x-axis and ultimately recedes to infinity along the positive y-axis.

Show that the centre of force is at the point $(b, b, 0)$.

A particle is projected from infinity with speed v_∞ along a line, parallel to the x-axis, whose perpendicular distance from the centre of force is c. Show that the orbit has asymptotes which intersect on the plane $x = 0$ and that the particle is ultimately deflected through an angle $2 \tan^{-1} c/b$.

Particles are projected from infinity with speed v_∞ along lines parallel to the x-axis within the cylinder $(y - b)^2 + z^2 = c^2$. Show that, if $c \ll b$, all such particles ultimately recede to infinity within a circular cone of semi-angle $2c/b$, whose axis is the line $y = b$, $z = 0$ and whose vertex is the point $(b/2, b, 0)$.

7.19 (i) Prove that

$$\frac{\mathrm{d}}{\mathrm{d}t}\left(\frac{\mathbf{r}}{r}\right) = \frac{1}{r^3}\left(\mathbf{r} \times \frac{\mathrm{d}\mathbf{r}}{\mathrm{d}t}\right) \times \mathbf{r}.$$

(ii) A particle P, whose position vector is $\mathbf{r}$, is subject to a central force. State the equation of motion in vector form. Show that $\mathbf{r} \times \dot{\mathbf{r}}$ is a constant vector $\mathbf{h}$, and deduce that the motion of P is confined to a plane through the origin perpendicular to $\mathbf{h}$.

Show that if the central force is of the form $-\mu\mathbf{r}/r^3$ per unit mass,

$$\mu \frac{\mathrm{d}}{\mathrm{d}t}\left(\frac{\mathbf{r}}{r}\right) = \frac{\mathrm{d}^2\mathbf{r}}{\mathrm{d}t^2} \times \mathbf{h}.$$

Hence show that

$$\mu(\mathbf{a} \cdot \mathbf{r} + r) = h^2$$

for some constant vector $\mathbf{a}$.

7.20 A useful formula, particularly when the central force $mf(r)$ corresponding to a given path is required, is

$$\frac{\mathrm{d}^2 u}{\mathrm{d}\theta^2} + u = -\frac{f(1/u)}{h^2 u^2}, \quad u = \frac{1}{r}.$$

Derive this formula.

A particle of mass m is describing the curve $r \cos 2\theta = a$ with angular momentum mh under a central force to the origin.

Show that the force is one of repulsion with magnitude $3mh^2/r^3$.

As the particle passes through the point $r = a, \theta = 0$, it receives an impulse which doubles its speed. Show that the equation of the new orbit is

$$a/r = \cos(7^{1/2}\theta/2).$$

7.21 A particle, which moves under the influence of a central force, describes the curve $r(\theta + 1)^2 = a$, where a is a constant and (r, θ) are its polar coordinates. What is the law of force?

When the particle is in the position $\theta = 0$ it receives an impulse which reduces its radial velocity to zero and doubles its transverse velocity. Show that its subsequent path is given by

$$3a/2r = 1 + \tfrac{1}{2}\cos(3^{1/2}\theta/2).$$

7.22 A particle is subject to a central force with potential energy $-\mu/4r^4$ per unit mass. The angular momentum is h per unit mass.

(i) Show that if the total energy is zero, the orbit consists of part of a pair of touching circles, each of radius $(\mu/8h^2)^{1/2}$.
(ii) Show that if the total energy is $h^4/4\mu$, the orbit is part of the curves

$$r = b\tanh(\theta/2^{1/2}), \quad r = b\coth(\theta/2^{1/2}),$$

where $b = (\mu/h^2)^{1/2}$.

7.23 A particle of mass m is subject to the central force

$$-m\left(\frac{\mu}{r^3} - \frac{\nu}{r^4}\right)\mathbf{r},$$

where μ and ν are positive constants. It is projected with speed v_0 from a point at a distance a from the centre of force, at right angles to the radius vector. Find the equation of the orbit and show that if $v_0^2 < 2\mu/a - \nu/a^2$ then the orbit is bounded. What is the additional condition for the orbit to be closed?

Prove that when the particle is next moving at right angles to the radius vector, the latter has turned through an angle $\pi a v_0/(\nu + a^2 v_0^2)^{1/2}$.

7.24 A particle of mass m is subject to the central force

$$-m\left(\frac{\mu}{r^3} + \frac{\nu}{r^5}\right)\mathbf{r}$$

with $\mu > 0$, $\nu > 0$. The angular momentum is mh.

Show that two circular orbits are possible provided that $K = 4\mu\nu/h^4 < 1$. Show that one of them is unstable and that an orbit which is a small perturbation of the other has an apsidal angle equal to $\pi/(1 - K)^{1/4}$.

Because of the oblateness of the earth, its gravitational attraction in the equatorial plane is of the above form with $\nu \ll \mu r_e^2$. Show that a nearly circular orbit can be regarded approximately as an ellipse whose axes rotate with angular speed $\nu\omega/\mu a^2$, where ω and a are respectively the mean angular speed and mean radial distance of the particle.

7.25 For the central force

$$-m\left(\frac{\mu}{r^3} + \frac{\nu}{r^5}\right)\mathbf{r},$$

the differential equation for the trajectory can be written in the form

$$\frac{d^2u}{d\theta^2} + u - \frac{\mu}{h^2} = \frac{\nu u^2}{h^2}, \quad u = \frac{1}{r},$$

(see Exercise 7.20). On the assumption that the term on the right-hand side of the first equation is small compared with the individual terms on the left-hand side, show that a solution of the form

$$lu = 1 + e \cos n\theta + \epsilon \cos 2n\theta, \quad \epsilon \ll 1,$$

is consistent with the equation provided that all terms of the second order of smallness are neglected. Substitute the assumed form for u into the equation and compare like terms to obtain the following relations:

$$1 - \frac{\mu l}{h^2} = \frac{\nu}{h^2 l}(1 + \tfrac{1}{2}e^2),$$

$$1 - n^2 = 2\nu/(h^2 l),$$

$$\epsilon(1 - 4n^2) = \nu e^2/(2h^2 l).$$

When $\mu\nu/h^4 \ll 1$, show that these relations can be solved to give, approximately,

$$l = \frac{h^2}{\mu} - \frac{\mu\nu}{h^4}(1 + \tfrac{1}{2}e^2),$$

$$n = 1 - \mu\nu/h^4,$$

$$\epsilon = -\mu\nu e^2/(6h^4).$$

For $e < 1$, show that the most significant terms describe an ellipse whose axes rotate slowly with the angular speed $\omega\mu\nu/h^4$, where $2\pi/\omega$ is the time to traverse the orbit.

For $e \ll 1$, show that the resulting nearly circular orbit agrees with that of Exercise 7.24.

Note, however, that the result is not now confined to nearly circular orbits. A similar modification of the inverse square law arises from the theory of general relativity, with $\nu = 3h^2\mu/c^2$, where c is the speed of light. It is used to explain the slow rotation of the orbit of the planet Mercury.

7.26 A particle moves under the influence of a central force $f(r)\mathbf{r}/r$ per unit mass, where $\mathbf{r}$ is the position vector of the particle.

When the orbit has two apsidal distances a, b, show that the apsidal speeds are in the ratio b/a.

Show also that the speed v at a distance r from the origin is given by

$$\tfrac{1}{2}(b^2 - a^2)v^2 = a^2 \int_r^a f(r)\,dr + b^2 \int_b^r f(r)\,dr.$$

7.27 It has been suggested that the gravitational constant may in fact vary with the time. Consider the equation

$$\frac{\mathrm{d}^2\mathbf{r}}{\mathrm{d}t^2} = -\mu\frac{\mathbf{r}}{r^3}$$

in which μ varies with the time. Show that angular momentum is conserved but that there is no equation of conservation of energy.

Suppose that the particle traverses a bounded orbit, and that the unit of time is adjusted so that the period of revolution is of order unity. Suppose also that μ is a slowly varying function on this time scale, so that

$$\mu = \mu(\epsilon t), \quad 0 < \epsilon \ll 1.$$

Show that with the transformations

$$\mathbf{s} = R(\epsilon t)\mathbf{r}, \quad \tau = \int_0^t T(\epsilon t)\,\mathrm{d}t,$$

the equation of motion becomes

$$\frac{T^2}{R}\frac{\mathrm{d}^2\mathbf{s}}{\mathrm{d}\tau^2} + \frac{\epsilon}{R^2}(RT' - 2TR')\frac{\mathrm{d}\mathbf{s}}{\mathrm{d}\tau} + 2\frac{\epsilon^2}{R^3}(R'^2 - RR'')\mathbf{s} = -\mu\frac{R^2\mathbf{s}}{s^3}$$

and that this equation simplifies to

$$\mathrm{d}^2\mathbf{s}/\mathrm{d}\tau^2 = -\mathbf{s}/s^3$$

provided that

$$T^{1/2} = R = \mu,$$

and that either terms of order ϵ^2 are neglected or $\mu = \mu_0 e^{-\epsilon t}$.

Is it now possible to define conservation equations in terms of $\mathbf{s}$?

7.28 A particle P rests on a smooth horizontal table and is connected to a string which passes through a small hole O in the table. Initially the string is taut, with OP $= a$, and the particle is suddenly given a horizontal velocity $a\omega$ perpendicular to the string. During the subsequent motion the particle is subject to a resistance equal to $-k$ times the velocity per unit mass, and the length of the string is made to vary in a prescribed manner

Show that the angular speed of the string is $a^2\omega \mathrm{e}^{-kt}/r^2$, where $r =$ OP and t is the time.

For $r = a\mathrm{e}^{nt}$, verify that the tension in the string is initially positive provided that $\omega^2 > kn + n^2$, and that it will become zero after a time

$$\frac{1}{2(k+2n)}\log\left(\frac{\omega^2}{kn+n^2}\right).$$

7.29 A particle of mass m describes a nearly circular orbit under a central attractive force $m\mu/r^2$ and a force $m\nu x$ in the x direction.

Prove that if $\nu \ll \omega^2$ the polar equation of the orbit is approximately

$$r = a - \frac{\nu a}{6\omega^2}(1 + \cos 2\theta),$$

where ω is the mean angular speed in the orbit and $a^3 = \mu/\omega^2$.

7.30 A particle traverses the ellipse $l/r = 1 + e \cos\theta$ under an inverse square law of force. The apses A_1 and A_2 are such that the apsidal distances are given by

$$r_1 = \frac{l}{1+e}, \quad r_2 = \frac{l}{1-e}.$$

As the particle passes through A_1 it receives an impulse which increases the magnitude v_1 of the velocity by a small increment δv_1 without changing its direction. Similarly as the particle passes through A_2, it received an impulse which increases v_2 by δv_2.

Show that

$$\frac{\delta r_1}{r_1} = \frac{4\delta v_2}{(1+e)v_2}, \quad \frac{\delta r_2}{r_2} = \frac{4\delta v_1}{(1-e)v_1}.$$

Show that if $v_1 \delta v_1 + v_2 \delta v_2 = 0$ then

$$\delta r_1 + \delta r_2 = 0.$$

8 Systems of Particles

8.1 MASS CENTRE

In the study of the motion of several particles, it is sometimes more convenient to consider the motion of the system as a whole rather than the details of the motion of each separate particle. For this purpose it is useful to introduce the concept of the mass centre.

Let a typical particle of the system be situated at the point P_i $(i = 1, 2, \ldots, n)$ and let its mass be m_i. A point C such that

$$\sum_i m_i \overline{\mathrm{CP}}_i = \mathbf{0} \tag{8.1.1}$$

is called the mass centre of the system of particles (see Fig. 8.1.1). If C′ is also a point such that

$$\sum_i m_i \overline{\mathrm{C'P}}_i = \mathbf{0}$$

then, by subtraction,

$$\sum_i m_i(\overline{\mathrm{CP}}_i - \overline{\mathrm{C'P}}_i) = \sum_i m_i \overline{\mathrm{CC'}} = \left(\sum_i m_i\right)\overline{\mathrm{CC'}} = \mathbf{0}.$$

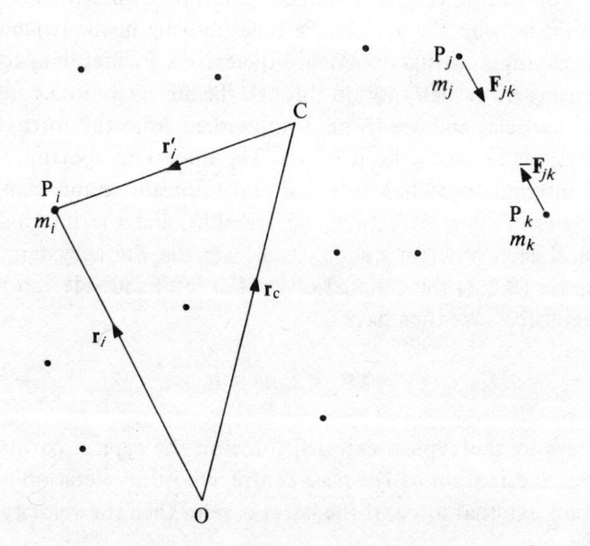

Figure 8.1.1 Illustration of the notation for systems of particles.

Hence C′ coincides with C and the mass centre is unique.

For a given origin O, (8.1.1) implies that

$$\sum_i m_i(\overline{\mathrm{CO}} + \overline{\mathrm{OP}_i}) = \mathbf{0},$$

or

$$\overline{\mathrm{OC}} = \frac{\Sigma_i m_i \overline{\mathrm{OP}}_i}{\Sigma_i m_i}. \tag{8.1.2}$$

This gives the position vector of the mass centre relative to O.

8.2 EQUATIONS OF MOTION

For any particle of mass m, position vector $\mathbf{r}$, acceleration $\mathbf{a}$ and subject to a force $\mathbf{F}$, we have the equation of motion

$$\mathbf{F} = m\,\mathrm{d}^2\mathbf{r}/\mathrm{d}t^2 = m\mathbf{a}. \tag{8.2.1}$$

Hence, by summation over a system of particles, we have

$$\Sigma\mathbf{F} = \Sigma m\frac{\mathrm{d}^2\mathbf{r}}{\mathrm{d}t^2} = \frac{\mathrm{d}^2}{\mathrm{d}t^2}\Sigma m\mathbf{r} = \frac{\mathrm{d}^2}{\mathrm{d}t^2}m_\mathrm{c}\mathbf{r}_\mathrm{c} = m_\mathrm{c}\frac{\mathrm{d}^2\mathbf{r}_\mathrm{c}}{\mathrm{d}t^2} = m_\mathrm{c}\mathbf{a}_\mathrm{c}, \tag{8.2.2}$$

where $m_\mathrm{c} = \Sigma m$ and $\mathbf{r}_\mathrm{c}$, $\mathbf{a}_\mathrm{c}$ are the position vector and acceleration of the mass centre. In (8.2.2) the suffix i has been omitted for economy of notation.

In the summation over the forces acting on the system we can distinguish between two contributions. The first contribution consists of all those forces originating from outside the system. Particles moving in the neighbourhood of the earth, for example, would experience the earth's gravitational attraction and a resistance arising from their motion through the air. Such forces are external to the system of particles and are to be distinguished from the forces arising from mutual interactions between the particles. The important dynamical distinction is that each internal force has an equal and opposite counterpart within the system, by Newton's law of action and reaction, and the internal forces will therefore cancel each other in a summation over the whole system of particles. Thus in equation (8.2.2) the summation on the left-hand side can be restricted to the external forces. We then have

$$\Sigma\mathbf{F} = \Sigma\mathbf{F}_\mathrm{e} = \Sigma m\mathbf{a} = m_\mathrm{c}\mathbf{a}_\mathrm{c}, \tag{8.2.3}$$

where $\mathbf{F}_\mathrm{e}$ refers to the typical external force on the typical particle. Equation (8.2.3) governs the motion of the mass centre, whose acceleration is determined by the resultant external force. If the latter is zero, then the velocity of the mass centre is constant.

Equation (8.2.3) can be integrated to give

$$\Sigma \int \mathbf{F} dt = \Sigma \int \mathbf{F}_e dt = [\Sigma m\mathbf{v}] = [m_c \mathbf{v}_c], \tag{8.2.4}$$

where $\mathbf{v}_c$ is the velocity of the mass centre and $[m_c\mathbf{v}_c]$ denotes the change in $m_c\mathbf{v}_c$ over the interval of integration. When the interval of time tends to zero, the limit of the integral

$$\int \mathbf{F}_e dt$$

is an impulse if it remains finite. The resultant external impulse is then equal to the instantaneous change in the linear momentum of the system. It should be emphasised that any external force which remains finite does not contribute to the impulse.

The work equation is not quite so straightforward. For a single particle we still have, as in (5.2.1),

$$\int \mathbf{F} \,.\, d\mathbf{r} = \int m \frac{d\mathbf{v}}{dt} \,.\, d\mathbf{r} = \int \frac{d}{dt}(\tfrac{1}{2}mv^2)\, dt = [\tfrac{1}{2}mv^2], \tag{8.2.5}$$

and so, by summation over the system

$$\Sigma \int \mathbf{F} \,.\, d\mathbf{r} = [\Sigma \tfrac{1}{2}mv^2]. \tag{8.2.6}$$

However, the internal forces may do work during the motion of the system, and this must be included in the summation on the left-hand side of (8.2.6). For example the work done by the forces acting on two particles connected by a spring is equal to $\frac{1}{2}$ x tension x extension.

Let m_j, m_k be the masses of two particles with position vectors $\mathbf{r}_j$ and $\mathbf{r}_k$ respectively. Let there be mutual forces $\pm\mathbf{F}_{jk}$ respectively on the two particles arising from their interaction. The work done by these forces is

$$\int \mathbf{F}_{jk} \,.\, (d\mathbf{r}_j - d\mathbf{r}_k) = \int \mathbf{F}_{jk} \,.\, d(\mathbf{r}_j - \mathbf{r}_k) \tag{8.2.7}$$

and so depends on the relative motion of m_j and m_k. Note, however, that if $\mathbf{F}_{jk}$ is parallel to $\mathbf{r}_k - \mathbf{r}_j$, and if $|\mathbf{r}_k - \mathbf{r}_j|$ is constant so that the particles move as a rigid body, no work is done by $\mathbf{F}_{jk}$ because it is perpendicular to the direction of relative motion. In this important special case the work done is confined to that of the external forces, and if they are conservative it can be calculated from the potential energy in the usual way.

There is an alternative expression for the kinetic energy $\Sigma\frac{1}{2}mv^2$, which is derived as follows. We denote by $\mathbf{r}'$ and $\mathbf{v}' = d\mathbf{r}'/dt$ the position vector and velocity of a typical particle relative to the mass centre C, so that $\mathbf{r} = \mathbf{r}_c + \mathbf{r}'$ and by differentiation, $\mathbf{v} = \mathbf{v}_c + \mathbf{v}'$ (see Fig. 8.1.1). We note from (8.1.1) that

$\Sigma m\mathbf{r}' = \mathbf{0}$ by definition of the mass centre. Hence, by differentiation, $\Sigma m\mathbf{v}' = \mathbf{0}$. The kinetic energy is therefore

$$\Sigma\tfrac{1}{2}mv^2 = \Sigma\tfrac{1}{2}m(\mathbf{v}_c + \mathbf{v}') \cdot (\mathbf{v}_c + \mathbf{v}') = \Sigma\tfrac{1}{2}mv_c^2 + \Sigma\tfrac{1}{2}mv'^2 + \mathbf{v}_c \cdot (\Sigma m\mathbf{v}')$$

$$= \tfrac{1}{2}m_c v_c^2 + \Sigma\tfrac{1}{2}mv'^2 \tag{8.2.8}$$

since $\Sigma m = m_c$ and $\Sigma m\mathbf{v}' = \mathbf{0}$. The first term of (8.2.8) is equal to the kinetic energy of the whole mass moving with the velocity of the mass centre. The second term is the kinetic energy of the motion relative to the mass centre.

Example 8.2.1
In this example we have a small ring, of mass m_1, which is connected by a string, of length l, to a bob of mass m_2. The ring is free to slide along a smooth, fixed, horizontal rail, as in Figure 8.2.1. We shall investigate the oscillations of this system in a vertical plane.

Let x be the displacement of the ring from some suitable origin, and let θ be the inclination of the string to the vertical. The ring then has a horizontal velocity $\dot{x}$. The bob also has the horizontal velocity $\dot{x}$ of the ring; in addition it has a velocity relative to the ring equal to $l\dot{\theta}$ perpendicular to the string; see Figure 8.2.1.

Of the forces acting on the system consisting of the ring and the bob, the tension in the string is an internal force acting mutually on the two masses. It does not contribute to the resultant force acting on the system. Moreover, provided the string remains taut and inextensible, the tension is always in a direction perpendicular to the relative displacement of the masses. It will therefore do no work during the motion. The external forces are the gravi-

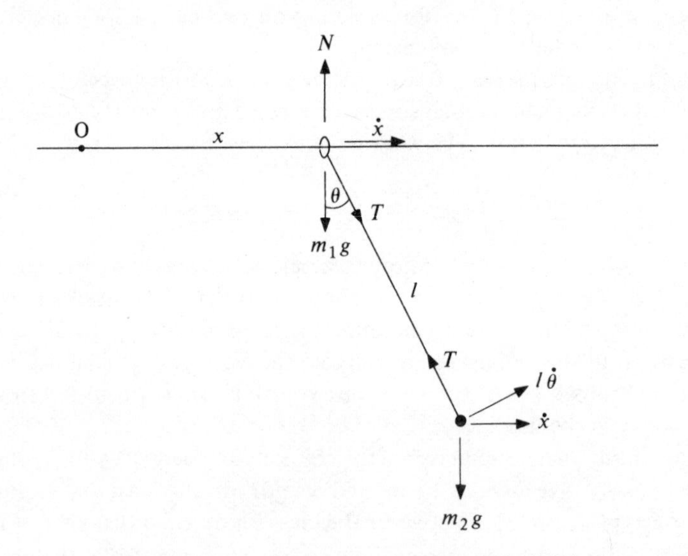

Figure 8.2.1 Illustration for Example 8.2.1.

tational forces and the normal reaction N between the ring and the wire. These are all vertical. Also N and $m_1 g$ act in a direction perpendicular to the displacement of the ring and so do no work during the motion. The gravitational force $m_2 g$ is conservative and so is derivable from a potential.

Since there is no resultant horizontal force, it follows from (8.2.4) that the horizontal component of the momentum is constant. We suppose that the system is started from rest; the horizontal component of momentum is then zero and we have

$$m_1 \dot{x} + m_2 (\dot{x} + l\dot{\theta} \cos \theta) = 0. \tag{8.2.9}$$

Since the only force which does work on the system is $m_2 g$, and since this force has the potential energy $-m_2 gl \cos \theta$, there is an energy equation of the form

$$\tfrac{1}{2} m_1 \dot{x}^2 + \tfrac{1}{2} m_2 \{(\dot{x} + l\dot{\theta} \cos \theta)^2 + l^2 \dot{\theta}^2 \sin^2 \theta\} - m_2 gl \cos \theta = E, \tag{8.2.10}$$

where E is a constant. Between (8.2.9) and (8.2.10), $\dot{x}$ can be eliminated to give an equation involving only θ. After simplification this equation can be written

$$\frac{1}{2} \frac{m_2}{m_1 + m_2} (m_1 + m_2 \sin^2 \theta) l^2 \dot{\theta}^2 - m_2 gl \cos \theta = E. \tag{8.2.11}$$

For a given value of E, which would be determined by some initial conditions, equation (8.2.11) is sufficient to determine θ as a function of the time. In particular, for small oscillations the equation becomes, approximately,

$$\dot{\theta}^2 + \frac{(m_1 + m_2) g \theta^2}{m_1 l} = \text{constant}.$$

This equation governs simple harmonic oscillations with period $2\pi/n$, where $n^2 = (m_1 + m_2) g / m_1 l$.

8.3 COLLISION PROBLEMS

The word collision is often used to denote the actual physical impact of two bodies, the effect of which is to produce impulsive changes in velocity. These changes take place in a very short time interval, outside which the velocities of the bodies are constant if no other forces are involved. In this discussion it is convenient to widen the definition and describe a collision between two particles as an interaction between them such that the initial and final velocities are constant. A collision can then take place without actual contact of the particles; an example we shall eventually discuss is the motion of two particles which are a large distance apart initially and which attract or repel each other with a force inversely proportional to the square of the distance apart.

To calculate the effect of a collision completely it is necessary to know the law of interaction, but some information about the initial and final states of the particles can be obtained simply from an application of the laws of conservation of energy and momentum. Since these laws are also applicable in quantum mechanics, the results deduced from them have wide validity. Indeed in modern physics conservation laws are regarded as basic postulates, but in their application we have to bear in mind that microscopic particles such as atoms and molecules possess internal energies associated with the motion of their constituent parts, and they are able to absorb or release energy when they collide. Macroscopic bodies which collide impulsively always absorb energy, which is ultimately dissipated into heat by frictional forces, so that there is always a loss of kinetic energy in this case. Finally, for particles moving with speeds comparable to the speed of light, there is a relativistic correction to the conservation laws.

We describe a perfectly elastic collision as one in which the net work done by the forces during the collision is zero, and the net change in the internal energy of the particles is zero. Since the energy is wholly kinetic in the initial and final states, in a perfectly elastic collision there is conservation of kinetic energy as well as momentum.

Example 8.3.1

We consider the spontaneous disintegration of a particle into two parts.

Let $\mathbf{u}$ be the velocity of the particle before disintegration. Let $m_1, m_2, \mathbf{v}_1, \mathbf{v}_2$ be the masses and velocities of the two parts after disintegration (see Fig. 8.3.1). From conservation of momentum we have

$$(m_1 + m_2)\mathbf{u} = m_1\mathbf{v}_1 + m_2\mathbf{v}_2.$$

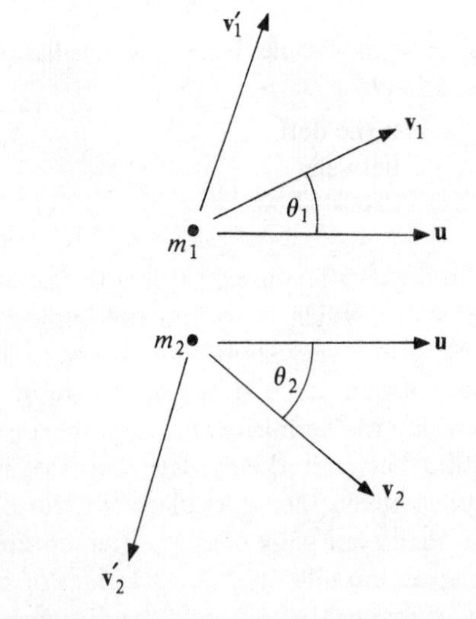

Figure 8.3.1 Illustration for Example 8.3.1: disintegration of a particle.

Since $\mathbf{v}_c = (m_1\mathbf{v}_1 + m_2\mathbf{v}_2)/(m_1 + m_2)$ is the velocity of the mass centre after disintegration, conservation of momentum shows that $\mathbf{v}_c$ remains unchanged and equal to $\mathbf{u}$. Also, since

$$m_1\mathbf{v}_1' + m_2\mathbf{v}_2' = \mathbf{0},$$

where

$$\mathbf{v}_1' = \mathbf{v}_1 - \mathbf{u}, \quad \mathbf{v}_2' = \mathbf{v}_2 - \mathbf{u}$$

are the velocities relative to the mass centre, it follows that the momenta of the two parts relative to the mass centre are equal in magnitude and have opposite directions.

The total energy in this problem can be calculated from the sum of the kinetic energy arising from the motion of the mass centre, the kinetic energy of the motion relative to the mass centre, and the internal energy which we denote by E_i. Conservation of energy then gives

$$\tfrac{1}{2}(m_1 + m_2)u^2 + E_i = \tfrac{1}{2}(m_1 + m_2)v_c^2 + \tfrac{1}{2}m_1 v_1'^2 + \tfrac{1}{2}m_2 v_2'^2 + E_{i1} + E_{i2}.$$

Since $v_c = u$ and $v_2' = m_1 v_1'/m_2$, this relation simplifies to

$$E_i - E_{i1} - E_{i2} = \frac{m_1(m_1 + m_2)}{2m_2} v_1'^2$$

which serves to determine v_1', and hence v_2', in terms of the internal energy released. The angles through which the particles are scattered, say θ_1 and θ_2, are not deducible without further information. If $v_1' > u$, any value is possible for θ_1, but if $v_1' < u$, there is a maximum value given by $\sin^{-1}(v_1'/u)$.

Example 8.3.2

We now investigate the collision of two particles.

Let the masses of the particles be m_1, m_2. Let the velocities be $\mathbf{u}_1, \mathbf{u}_2$ before collision and $\mathbf{v}_1, \mathbf{v}_2$ after collision. From conservation of momentum, we have

$$m_1\mathbf{u}_1 + m_2\mathbf{u}_2 = m_1\mathbf{v}_1 + m_2\mathbf{v}_2 = (m_1 + m_2)\mathbf{u}_c \tag{8.3.1}$$

where, as before, the velocity $\mathbf{u}_c$ of the mass centre is unchanged by the collision. It also follows that the corresponding velocities relative to the centre of mass, say $\mathbf{u}_1', \mathbf{u}_2', \mathbf{v}_1', \mathbf{v}_2'$, satisfy the relations

$$m_1\mathbf{u}_1' = -m_2\mathbf{u}_2' = \frac{m_1 m_2}{m_1 + m_2}(\mathbf{u}_1 - \mathbf{u}_2), \tag{8.3.2}$$

$$m_1\mathbf{v}_1' = -m_2\mathbf{v}_2' = \frac{m_1 m_2}{m_1 + m_2}(\mathbf{v}_1 - \mathbf{v}_2). \tag{8.3.3}$$

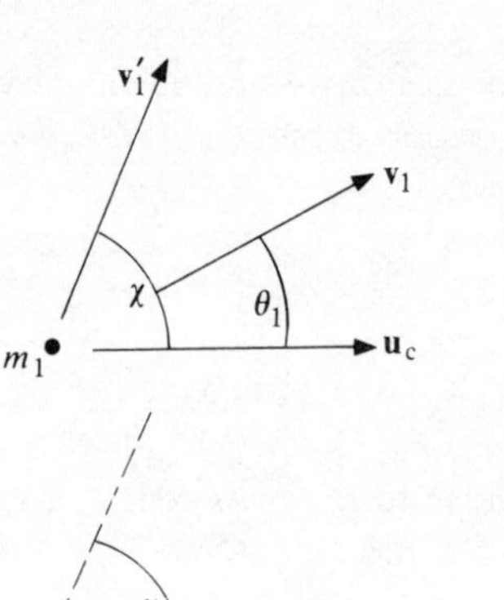

Figure 8.3.2 Illustration for Example 8.3.2: collision of two particles.

If we assume that the collision is perfectly elastic then kinetic energy is also conserved. Effectively this means that the kinetic energy of the motion relative to the mass centre is conserved, since conservation of momentum ensures that the energy associated with the motion of the mass centre is unchanged. We therefore have the further relation

$$\tfrac{1}{2}m_1u_1'^2 + \tfrac{1}{2}m_2u_2'^2 = \tfrac{1}{2}m_1v_1'^2 + \tfrac{1}{2}m_2v_2'^2. \tag{8.3.4}$$

Relations (8.3.2), (8.3.3) and (8.3.4) can only be satisfied simultaneously if $|\mathbf{u}_1 - \mathbf{u}_2| = |\mathbf{v}_1 - \mathbf{v}_2|$, so that

$$m_1u_1' = m_2u_2' = m_1v_1' = m_2v_2' = \frac{m_1m_2}{m_1 + m_2}|\mathbf{u}_1 - \mathbf{u}_2|. \tag{8.3.5}$$

Hence the momenta of two particles relative to the mass centre are unchanged in magnitude by the collision, the effect of which is to rotate both the relative momentum vectors through an angle of scatter χ, say.

Laboratory experiments are often such that particle 2 is initially at rest, and then (8.3.1), (8.3.2) and (8.3.5) give, with $\mathbf{u}_2 = \mathbf{0}$,

$$m_1u_1' = m_2u_2' = m_1v_1' = m_2v_2' = m_2u_c = \frac{m_1m_2u_1}{m_1 + m_2}. \tag{8.3.6}$$

In the original (laboratory) frame of reference we denote the angles of scatter by θ_1 and θ_2. Reference to Figure 8.3.2 and relations (8.3.6) shows that they are given in terms of χ, the angle of scatter in the centre of mass frame, by the relations

$$\tan\theta_1 = \frac{v_1'\sin\chi}{u_c + v_1'\cos\chi} = \frac{\sin\chi}{(m_1/m_2)+\cos\chi} \tag{8.3.7}$$

and

$$\theta_2 = \tfrac{1}{2}(\pi - \chi), \tag{8.3.8}$$

since $v_2' = u_c$. If $m_1 < m_2$, any value for θ_1 is possible, but if $m_1 > m_2$ there is a maximum value given by $\sin^{-1}(m_2/m_1)$.

The kinetic energy transferred to the second particle is

$$\tfrac{1}{2}m_2v_2^2 = \frac{2m_1^2m_2}{(m_1+m_2)^2}u_1^2\cos^2\theta_2.$$

We compare this with the total kinetic energy $\tfrac{1}{2}m_1u_1^2$. The ratio is

$$\frac{4m_1m_2}{(m_1+m_2)^2}\cos^2\theta_2$$

so that the maximum transfer of energy occurs for a head-on collision ($\theta_2 = 0$). Note that it is small for $m_1 \ll m_2$ and for $m_1 \gg m_2$. It can only be substantial if m_1 and m_2 are comparable. For a head-on collision of a proton and α-particle, where $m_1/m_2 = 4$, the fraction of energy transferred is 64%. For a proton–electron collision it is about 0.2% ($m_1/m_2 = 1836$).

When bodies of macroscopic size collide impulsively there is invariably a loss of kinetic energy. A head-on collision is often dealt with by using the empirical rule

$$v_2 - v_1 = e(u_1 - u_2), \tag{8.3.9}$$

where e is called the coefficient of restitution and differs for different materials. This relation used in conjunction with the conservation of momentum is sufficient to give v_1 and v_2 in terms of u_1 and u_2. By (8.3.2) the kinetic energy relative to the mass centre is

$$\tfrac{1}{2}m_1u_1'^2 + \tfrac{1}{2}m_2u_2'^2 = \frac{1}{2}\frac{m_1m_2}{m_1+m_2}(u_1-u_2)^2$$

before collision and, by analogy,

$$\frac{1}{2}\frac{m_1m_2}{m_1+m_2}(v_1-v_2)^2 = \frac{1}{2}\frac{m_1m_2e^2}{m_1+m_2}(u_1-u_2)^2$$

after collision. The loss in kinetic energy is

$$\frac{1}{2}\frac{m_1 m_2}{m_1 + m_2}(1 - e^2)(u_1 - u_2)^2 .$$

For a perfectly elastic collision with no loss of energy, $e = 1$. The case $e = 0$ corresponds to an inelastic collision, when $v_1 = v_2$ and all the kinetic energy relative to the mass centre is dissipated.

Example 8.3.3

When dissipative processes, such as friction, are present, conservation of energy can be invoked only if the energy converted into heat is taken into account. However, on a molecular level, this heat energy can be identified with the kinetic energy of the random motion of the molecules. We investigate this on the basis of an idealised model of a simple gas consisting of molecules all having the same mass m. The gas is assumed to be in equilibrium, which means that on average its energy does not change. It is also assumed to be isotropic, which means that there is no preferential direction of motion of the molecules. Let us consider the force on the walls of the container enclosing the gas. Suppose that when a molecule strikes one of the walls it is reflected elastically, so that its normal component of velocity is reversed in sign and its tangential component is unaltered. If the wall is parallel to the (y, z) plane, a molecule moving towards the wall with velocity $u\mathbf{i} + v\mathbf{j} + w\mathbf{k}$ before impact will then have a velocity $-u\mathbf{i} + v\mathbf{j} + w\mathbf{k}$ away from the wall after impact. The net change in momentum is $2mu$ normal to the wall. Let n_u be the number of molecules per unit volume having an x component of velocity u. The number of such molecules hitting the wall per unit area per unit time is then un_u. The rate of change of momentum per unit area is $2mu^2 n_u$. This is now summed over all positive u to get

$$2m \sum_{u>0} u^2 n_u = mn\overline{u^2}$$

say, where n is the number of molecules per unit volume, and $\overline{u^2}$ is the mean value of u^2 over *all* values of u. Note that the molecules for which u is positive are assumed to be moving towards the wall. Since there will be an equal number of molecules for which u is negative, we must have

$$2 \sum_{u>0} u^2 n_u = \sum_{\text{all } u} u^2 n_u = n\overline{u^2}.$$

From the assumption of isotropy we can infer that

$$\overline{u^2} = \overline{v^2} = \overline{w^2} = \tfrac{1}{3}\overline{c^2},$$

where c is the speed of a molecule. We now have the result that the rate of change of momentum per unit area of the containing wall, which is the force per unit area, or the pressure, is given by

$$p = \tfrac{1}{3}mn\overline{c^2} = \tfrac{1}{3}\rho\overline{c^2},$$

where ρ is the density of the gas.

This result should be compared with Charles's law in thermodynamics, namely

$$p = \rho RT,$$

where T is the absolute temperature and R is the gas constant. We now see that the concept of temperature is closely associated with the kinetic energy of the random motion of the molecules. In fact the kinetic energy per unit volume is

$$\tfrac{1}{2}mn\overline{c^2} = \tfrac{3}{2}\rho RT.$$

The above result might seem at first sight to depend on the idealised assumption that the molecules are all reflected elastically from the wall. Such elastic reflection does not take place in practice, indeed for some molecules there is no correlation between the velocity of approach and the velocity with which they leave the wall. Nevertheless if the isotropy of the gas is to be conserved, the average effect at the wall is the same as if such reflection had taken place. The reflecting wall is the simplest mechanism which will maintain the isotropy of the gas, but the mathematical result depends on this isotropy and not the particular means by which it is achieved.

Example 8.3.4

A particle P_1, of mass m_1, is projected from a great distance with a velocity $\mathbf{u}$ along a line whose perpendicular distance from a stationary particle P_2, of mass m_2, is d. We consider the motion of P_1 and P_2 when subjected to a mutual attracting force inversely proportional to the square of the distance (see Fig. 8.3.3).

The initial velocity of the mass centre C is $m_1\mathbf{u}/(m_1 + m_2)$ and, since there is no external force, C continues to move with this constant velocity. We therefore consider the motion of P_1 relative to C as a convenient origin. The motion of P_2 can be dealt with by a similar argument.

Let the force on P_1 towards C be

$$\frac{km_1m_2}{(P_1P_2)^2} = \frac{km_1m_2^3}{(m_1 + m_2)^2r^2},$$

where $P_1C = r$. The value of k will depend on the particular mechanism from which the force arises. A measurable effect in the laboratory is obtainable if, for example, the particles are electrically charged. Particle P_1 is therefore subject to

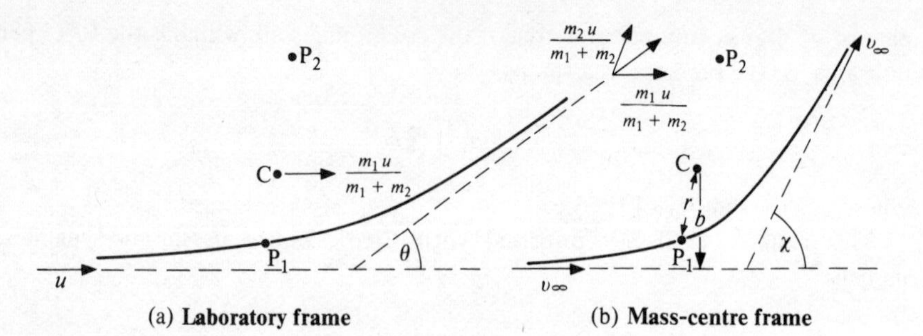

Figure 8.3.3 Illustration for Example 8.3.4: the motion of mutually attracting particles.

an inverse square law central force towards C of magnitude μ/r^2 per unit mass, where

$$\mu = km_2^3/(m_1 + m_2)^2 .$$

Relative to C, P_1 has an initial velocity

$$\mathbf{v}_\infty = m_2\mathbf{u}/(m_1 + m_2).$$

The distance from C on to the line of $\mathbf{v}_\infty$ is

$$b = m_2 d/(m_1 + m_2).$$

The relative path is a hyperbola and the details are deducible from §7.5. In particular, the angle through which the path turns is, by equation (7.5.27),

$$\chi = 2\tan^{-1}\left(\frac{\mu}{bv_\infty^2}\right) = 2\tan^{-1}\left(\frac{k(m_1 + m_2)}{du^2}\right).$$

The result is not immediately useful for comparison with observation, because in practice the deflection angle will be measured relative to axes fixed in the laboratory. To calculate this angle we note that the ultimate velocity of P_1 consists of two components, the velocity of C and the velocity relative to C. The velocity of C is

$$\mathbf{v}_c = m_1\mathbf{u}/(m_1 + m_2).$$

The magnitude of the velocity relative to C depends only on distance, see equation (7.5.24). Ultimately it is the same as the relative initial speed, namely

$$v_\infty = m_2 u/(m_1 + m_2).$$

The direction of this component makes an angle χ with the initial direction. The resultant of these two components makes an angle θ with the initial direction, where

$$\frac{m_1 u}{m_1 + m_2} \sin \theta = \frac{m_2 u}{m_1 + m_2} \sin(\chi - \theta).$$

Thus
$$m_1 \sin \theta = m_2 (\sin \chi \cos \theta - \cos \chi \sin \theta)$$

or
$$\tan \theta = \frac{\sin \chi}{(m_1/m_2) + \cos \chi},$$

in agreement with (8.3.7). If m_1/m_2 is negligibly small, we have $\theta = \chi$ and the result for a fixed centre of force is recovered. At the other extreme, when $m_1 \gg m_2$, the deflection is small and there is little effect on the motion of the approaching particle P_1. If $m_1 = m_2$, then

$$\tan \theta = \frac{\sin \chi}{1 + \cos \chi} = \tan \tfrac{1}{2}\chi$$

and so $\theta = \chi/2$.

Note that the scattering cross-section for this problem is still given by relation (7.5.29) relative to the centre of mass frame. But to obtain a result which is useful in the laboratory frame, the variation of (7.5.29) with θ and $d\theta$, rather than χ and $d\chi$, needs to be calculated using the above relation between θ and χ.

Example 8.3.5
A wedge, of mass m_2, rests with one face on a horizontal plane and with the inclined face at an angle α to the horizontal. It is hit on its inclined face by a particle, of mass m_1, moving along the plane with speed u in a direction perpendicular to the edge of the wedge. Show that the speed of the wedge immediately after impact is $m_1 u \sin^2 \alpha/(m_2 + m_1 \sin^2 \alpha)$.

The particle then slides up the face of the wedge and back to the horizontal plane. Find the final velocities of the particle and the wedge.

It is to be assumed that there are no frictional forces and that the impacts occurring are inelastic, in other words relative motion in the direction of the impact is destroyed.

(i) Since the impact between the particle and the wedge is inelastic, there will be no rebound and the particle will proceed to slide up the face of the wedge with relative initial speed u_1, say. Let the initial speed of the wedge be u_2. The impact between the particle and wedge is perpendicular to the wedge face. Hence the momentum of the particle parallel to the wedge face is unchanged. This yields the relation

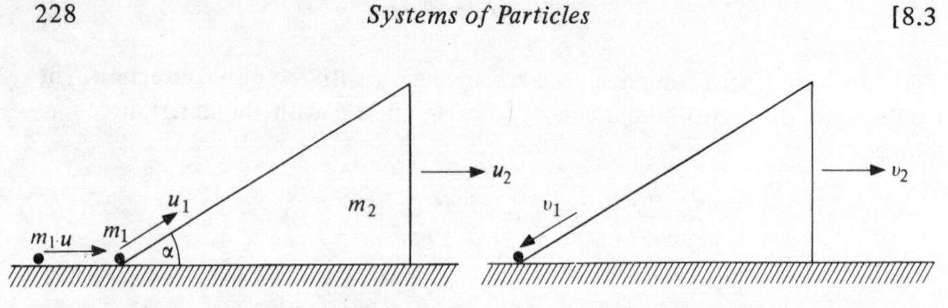

Figure 8.3.4 Illustration for Example 8.3.5: collision of particle and wedge.

$$u \cos \alpha = u_1 + u_2 \cos \alpha. \tag{8.3.10}$$

The impulsive reaction between the wedge and the plane is vertical. Hence the horizontal momentum of the system consisting of the wedge and the particle is unchanged. This yields the relation

$$m_1 u = m_1 u_1 \cos \alpha + (m_1 + m_2) u_2. \tag{8.3.11}$$

Equations (8.3.10) and (8.3.11) can be solved for u_1 and u_2. The result is

$$u_1 = \frac{m_2 u \cos \alpha}{m_1 \sin^2 \alpha + m_2}, \quad u_2 = \frac{m_1 u \sin^2 \alpha}{m_1 \sin^2 \alpha + m_2}. \tag{8.3.12}$$

(ii) In the next stage of the motion the particle slides smoothly up the face of the wedge and returns to its original position. There are two conservation equations which are sufficient to determine the final speeds v_1 of the particle relative to the wedge and v_2 of the wedge (see Fig. 8.3.4). These equations arise from the conservation of horizontal momentum, since all external forces acting on the system are vertical, and from the conservation of energy. Since there is no net work done on the system the latter equation amounts to conservation of kinetic energy, and we have

$$m_1 u = m_1 u_1 \cos \alpha + (m_1 + m_2) u_2 = -m_1 v_1 \cos \alpha + (m_1 + m_2) v_2,$$

$$\tfrac{1}{2} m_1 \{(u_2 + u_1 \cos \alpha)^2 + u_1^2 \sin^2 \alpha\} + \tfrac{1}{2} m_2 u_2^2$$
$$= \tfrac{1}{2} m_1 \{(v_2 - v_1 \cos \alpha)^2 + v_1^2 \sin^2 \alpha\} + \tfrac{1}{2} m_2 v_2^2.$$

These equations can now, in principle, be used to calculate v_1 and v_2 in terms of u_1 and u_2 and hence, by (8.3.12), in terms of u. However, it happens to be simpler to use the results of Example 3.3.1. There it was shown that the particle and the wedge have constant accelerations. Reference to equations (3.3.3) and (3.3.4) show these to be

$$a_1 = \frac{(m_1 + m_2) g \sin \alpha}{m_1 \sin^2 \alpha + m_2}$$

for the downward acceleration of the particle relative to the wedge, and

$$a_2 = \frac{m_1 g \sin\alpha\cos\alpha}{m_1\sin^2\alpha + m_2}$$

for the wedge itself.

For a particle moving with uniform acceleration we have the results of §2.3. In particular equation (2.3.8), with $u = u_1$ and $a = -a_1$, gives

$$v^2 = u_1^2 - 2a_1 x,$$

where v is the speed of the particle after travelling a distance x up the plane. In particular $x = 0$ when $v = v_1$. Hence $v_1 = u_1$ and the particle returns to the base of the wedge with the same relative speed, but with a reversal of direction. From (2.3.7) the time taken is $2u_1/a_1$, and from (2.3.6) the wedge in this time acquires a speed

$$\begin{aligned} v_2 &= u_2 + a_2 \cdot \frac{2u_1}{a_1} = u_2 + \frac{2m_1 u_1 \cos\alpha}{m_1 + m_2} \\ &= \frac{m_1 u\{(m_1 - m_2)\sin^2\alpha + 2m_2\}}{(m_1 + m_2)(m_1\sin^2\alpha + m_2)}. \end{aligned}$$

(iii) Finally the particle leaves the wedge and returns to the horizontal plane. This involves another inelastic collision which destroys the vertical component of velocity of the particle. The final speed of the particle is therefore

$$\begin{aligned} &v_2 - v_1\cos\alpha = v_2 - u_1\cos\alpha \\ &= \frac{u(m_1^2\sin^2\alpha + m_1 m_2 - m_2^2\cos^2\alpha)}{(m_1 + m_2)(m_1\sin^2\alpha + m_2)}. \end{aligned}$$

8.4 VARIABLE MASS PROBLEMS

When the equation of motion

$$\mathbf{F} = m\mathbf{a} \qquad (8.4.1)$$

was introduced for a single particle, it was assumed that the mass of the particle was constant. When the mass varies, the equation is modified, and the modification depends on the details of the process by which the variation is effected. For example, a raindrop may increase its mass by condensation as it falls through a cloud, and a rocket decrease its mass by the ejection of burnt gases. When such problems are examined from the point of view of the appropriate system, it will be seen that the mass of the system as a whole is not changing,

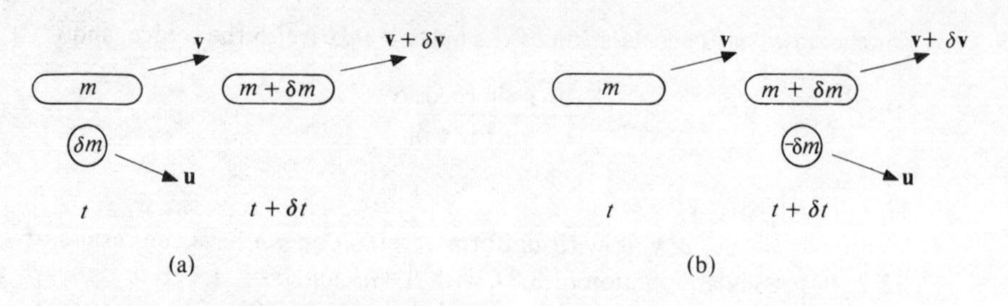

Figure 8.4.1 The changes which occur in a small interval of time when: (a) the mass increases; (b) the mass decreases.

but that individual parts of the system are changing their momentum. For example, the condensed water vapour is moving with the cloud before condensation, and with the raindrop after condensation, which implies a change in momentum. Similarly the fuel of the rocket moves with the rocket before ejection and with the velocity it acquires from combustion afterwards. Such problems can be treated as dynamical systems in which the impulse of the external forces is equated to the changing momentum of the system.

Let us consider a mass m which, during the interval of time δt, changes its velocity from $\mathbf{v}$ to $\mathbf{v} + \delta\mathbf{v}$ and combines with a mass δm. Let the velocity of δm before amalgamation with m be $\mathbf{u}$ (see Fig. 8.4.1 (a)). If $\mathbf{F}_e$ is the external force acting on the system, it follows from (8.2.4) that

$$\int_t^{t+\delta t} \mathbf{F}_e \, dt = (m + \delta m)(\mathbf{v} + \delta\mathbf{v}) - \delta m \, \mathbf{u} - m\mathbf{v} \qquad (8.4.2)$$

or, equivalently,

$$\frac{1}{\delta t}\int_t^{t+\delta t} \mathbf{F}_e \, dt = (m + \delta m)\frac{\delta\mathbf{v}}{\delta t} + (\mathbf{v} - \mathbf{u})\frac{\delta m}{\delta t}. \qquad (8.4.3)$$

The left-hand side of equation (8.4.3) represents the average value of $\mathbf{F}_e$ over the interval of time δt, and will tend to the value of $\mathbf{F}_e$ at time t as $\delta t \to 0$. Thus in the limit $\delta t \to 0$ equation (8.4.3) becomes

$$\mathbf{F}_e = m\frac{d\mathbf{v}}{dt} + (\mathbf{v} - \mathbf{u})\frac{dm}{dt}. \qquad (8.4.4)$$

Figure 8.4.1(b) represents the situation when mass is ejected, rather than acquired, as in rocket propulsion. Because δm conventionally denotes an increment in m, the ejected mass is denoted by $-\delta m$. In problems where m is actually decreasing, δm and dm/dt are negative. The same equation (8.4.2) also governs this situation and leads to the same limiting equation (8.4.4). Two particular cases are noteworthy. The first is when $\mathbf{u} = \mathbf{0}$, and then

$$\mathbf{F}_e = \frac{d}{dt}(m\mathbf{v}). \tag{8.4.5}$$

This is the equation governing the fall of a raindrop through a cloud which is at rest.

The second case is $\mathbf{u} = \mathbf{v}$, when

$$\mathbf{F}_e = m d\mathbf{v}/dt \tag{8.4.6}$$

and the equation of motion (8.4.1) for a particle of fixed mass is recovered. The motion of an object from which particles are being released would be governed by such an equation. But if the particles are ejected with a finite relative velocity, the more general equation (8.4.4) must be used.

Consider, for example, the rectilinear motion of a rocket in free flight, that is with no external forces. We assume that the exhaust velocity relative to the rocket is constant and equal to $\mathbf{c}$. The velocity of the rocket is $\mathbf{v} = v\mathbf{k}$ say, and $\mathbf{u} - \mathbf{v} = \mathbf{c} = -c\mathbf{k}$ since the gases are ejected in the direction opposite to $\mathbf{v}$. Equation (8.4.4) then becomes, with $\mathbf{F}_e = \mathbf{0}$,

$$0 = m\frac{dv}{dt} + c\frac{dm}{dt}. \tag{8.4.7}$$

This equation integrates immediately to

$$v_1 - v_0 = -c \log(m_1/m_0) = c \log R, \tag{8.4.8}$$

where R $(= m_0/m_1)$ is the ratio of the initial to the final mass. Note that the change in speed depends only on the exhaust speed c and the mass ratio R. It does not depend on the rate at which the fuel is consumed.

Example 8.4.1

In a two stage rocket, the first stage is detached after exhausting its fuel in order to minimise the mass before the second stage is fired. Such a rocket is to accelerate a 100 kg payload to a speed of 6000 m s^{-1} in free flight. The exhaust speed is 1500 m s^{-1} relative to the rocket. For structural reasons, the mass of either stage, without fuel or payload, must be 10% of the fuel it carries. We consider the best choice of masses for the two stages so that the initial mass is a minimum.

Let the masses of fuel carried in the first and second stages be x and y kg respectively as shown in Figure 8.4.2, so that the initial mass is

$$m_0 = 100 + 1{\cdot}1(x + y)$$

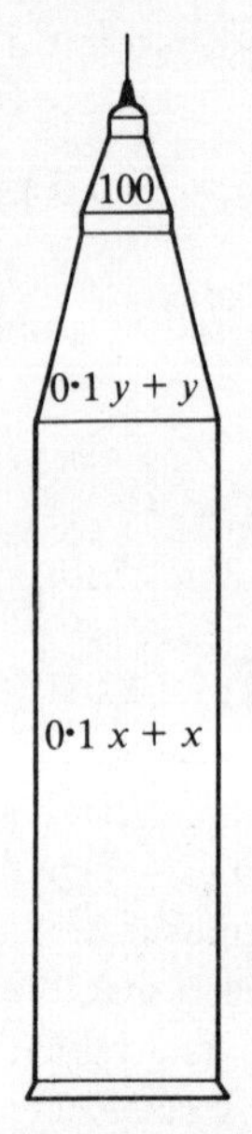

Figure 8.4.2 Illustration for Example 8.4.1: two stage rocket.

and the mass after the first stage has exhausted its fuel is

$$m_1 = 100 + 1 \cdot 1y + 0 \cdot 1\,x.$$

By equation (8.4.8) the velocity acquired at this point is

$$v_1 = 1500 \log(m_0/m_1).$$

One immediate deduction is that the desired speed of 6000 m s^{-1} cannot be achieved by a single stage rocket. For with $y = 0$, m_0/m_1 cannot be larger than 11, and 1500 log 11 is only 3597.

If we proceed to the second stage, after first jettisoning the burnt-out first stage, the initial and final masses are, respectively,

$$m_2 = 100 + 1 \cdot 1y, \quad m_3 = 100 + 0 \cdot 1y$$

and there is a further increment in speed to v_2, where

$$v_2 = v_1 + 1500 \log \frac{m_2}{m_3} = 1500 \log \frac{m_0 m_2}{m_1 m_3}.$$

If v_2 is to be 6000, we require that $(m_0 m_2)/(m_1 m_3) = \mathrm{e}^4$. Since $11m_1 = m_0 + 10m_2$, we can eliminate m_1 to get the relation

$$\frac{11 m_0 m_2}{(m_0 + 10m_2)m_3} = \mathrm{e}^4$$

and this can, in turn, be rearranged to give

$$m_0\left(\frac{11}{m_3}-\frac{e^4}{m_2}\right)=m_0\left(\frac{11}{100+0{\cdot}1y}-\frac{e^4}{100+1{\cdot}1y}\right)=10e^4.$$

If m_0 is to be a minimum, y must be chosen so that the expression in brackets is a maximum. This is so when

$$\frac{100+1{\cdot}1y}{100+0{\cdot}1y}=e^2 \quad \text{or} \quad y=\frac{1000(e^2-1)}{11-e^2}=1769.$$

From this, one can calculate $m_0 = 41\,873$ and $x = 36\,206$.

Example 8.4.2

A raindrop falls vertically from rest at time $t = 0$ through an atmosphere containing water vapour at rest. The mass of the raindrop increases by condensation, so that the mass is $m_0 + \lambda t$ at time t, where m_0 and λ are constants. The motion of the raindrop is resisted by a force equal to kv per unit mass, where v is the speed and k is constant.

Since the cloud is at rest, the equation of motion of the raindrop is

$$\frac{d}{dt}\{(m_0+\lambda t)v\}=(m_0+\lambda t)g-k(m_0+\lambda t)v$$

or

$$\frac{d}{dt}\{(m_0+\lambda t)ve^{kt}\}=(m_0+\lambda t)ge^{kt},$$

which integrates to

$$(m_0+\lambda t)ve^{kt}=k^{-1}(m_0+\lambda t)ge^{kt}-k^{-2}\lambda ge^{kt}+\text{constant}.$$

Since $v = 0$ when $t = 0$, we have

$$(m_0+\lambda t)ve^{kt}=k^{-1}(m_0+\lambda t)ge^{kt}-k^{-2}\lambda ge^{kt}-k^{-1}m_0g+k^{-2}\lambda g,$$

and so

$$v=\frac{g}{k}-\frac{g}{k^2(m_0+\lambda t)}\{\lambda-(\lambda-m_0k)e^{-kt}\}.$$

EXERCISES

8.1 A ring, of mass m_1, slides on a smooth, horizontal wire and is attached to a particle, of mass m_2, by a string of length l which passes over a small, smooth, fixed pulley. The pulley is at a depth d below the wire and in the same vertical plane. The system starts from rest from a position in which the upper part of the string makes an angle $\pi/3$ with the vertical. Prove that the greatest inclination of the string to the vertical after it is free from the pulley is

$$\cos^{-1}\left(1 - \frac{m_2 d}{(m_1 + m_2)l}\right)$$

8.2 Three equal particles A, B, C lie on a smooth, horizontal table in a straight line. They are joined together by two taut light inelastic strings AB, BC, where AB = BC = l. At time $t = 0$ the particle B is given a horizontal velocity u perpendicular to ABC.

Write down equations of conservation of energy and momentum for the subsequent motion of the system and show that, when either string has turned through an angle θ, its angular speed is given by

$$l\dot{\theta} = u(1 + 2\sin^2\theta)^{-1/2}.$$

Hence show that A and C collide after a time lI/u, where

$$I = \int_0^{\pi/2} (1 + 2\sin^2\theta)^{1/2}\, d\theta,$$

and that B has then travelled a distance $l(2 + I)/3$.

Show that

$$I = 3^{1/2} E\{(2/3)^{1/2}\},$$

where $E\{k\}$ is the complete elliptic integral of the second kind, namely

$$E\{k\} = \int_0^{\pi/2} \{1 - k^2 \sin^2\theta\}^{1/2}\, d\theta.$$

8.3 A smooth semi-circular thin wire of radius a has two rings fixed at its ends A and B. The total mass of the rings and the wire is m_t. The rings can slide freely on a fixed, smooth, horizontal rod. A bead of mass m can move freely on the wire and starts from rest at the end A of the wire. The motion takes place in a vertical plane, and θ is the angle which the radius through the bead makes with the horizontal radius through A.

State, with reasons, whether or not the following quantities are conserved:

(i) the horizontal momentum of the system;
(ii) the vertical momentum of the system;
(iii) the mechanical energy of the system;
(iv) the mechanical energy of the bead;
(v) the angular momentum of the bead about the centre of the semi-circle.

Use the conservation laws to derive the equation

$$\dot{\theta}^2 = \frac{2g(m_t + m)\sin\theta}{a(m_t + m\cos^2\theta)}.$$

Hence show that the time for one complete oscillation of the system is $I(a/g)^{1/2}$, where

$$I = \int_0^{\pi} \left(\frac{2(m_t + m\cos^2\theta)}{(m_t + m)\sin\theta}\right)^{1/2} d\theta.$$

8.4 Two particles have masses m_1, m_2 and position vectors $\mathbf{r}_1$, $\mathbf{r}_2$ respectively. They influence each other by gravitational attraction and there are no other forces.

Show that if $\mathbf{r}_c$ is the position vector of the mass centre and G is the constant of gravitation,

(i) $\dot{\mathbf{r}}_c$ is constant,

(ii) $\dfrac{d^2}{dt^2}(\mathbf{r}_1 - \mathbf{r}_2) = -\mu \dfrac{(\mathbf{r}_1 - \mathbf{r}_2)}{|\mathbf{r}_1 - \mathbf{r}_2|^3}$,

(iii) $(\mathbf{r}_1 - \mathbf{r}_2) \times (\dot{\mathbf{r}}_1 - \dot{\mathbf{r}}_2)$ is constant,

(iv) $\dfrac{d^2}{dt^2}(\mathbf{r}_1 - \mathbf{r}_c) = -\dfrac{Gm_2^3}{(m_1 + m_2)^2} \dfrac{(\mathbf{r}_1 - \mathbf{r}_c)}{|\mathbf{r}_1 - \mathbf{r}_c|^3}$,

where μ is a constant which should be determined.

Show that if the orbit of m_1 relative to the mass centre is an ellipse, the periodic time is the same as that for m_1 in its orbit relative to m_2.

8.5 A system of mutually gravitating particles moves in such a way that the position vector $\mathbf{r}_i$ of the ith particle is given by $\mathbf{r}_i = \mathbf{a}_i\lambda(t)$, where the $\mathbf{a}_i$ are constant vectors. Write down the equation of motion of the ith particle. Obtain conditions which are necessary for the given motion to be possible. Give the differential equation to be satisfied by λ when the conditions hold.

Given that all the particles recede to infinity, show that ultimately their velocities will be constant.

8.6 Two particles P_1, P_2, of masses m_1, m_2 respectively, lie on a smooth horizontal plane and are joined by a spring, of modulus λ and natural length l. Initially the particles are at a distance l apart and have velocities, of magnitude u_1, u_2 respectively, in the direction P_1P_2. The motion is wholly one-dimensional along the line joining P_1 and P_2.

Show that the maximum and minimum lengths of the spring are $l(1 \pm k)$ provided $k < 1$, where

$$k^2 = \frac{m_1 m_2 (u_1 - u_2)^2}{\lambda l (m_1 + m_2)}.$$

8.7 Two particles, each of mass m, are connected by a light elastic string of natural length l and modulus of elasticity λ. Initially the particles are at rest on a smooth horizontal table with the string just taut. One of the particles is then given a horizontal velocity of magnitude u at right angles to the string. Prove that during the subsequent motion the length x of the string satisfies the equation

$$\left(\frac{dx}{dt}\right)^2 = n^2(x-l)\left(\frac{\alpha l^2}{x^2}(x+l)-(x-l)\right),$$

where $n^2 = 2\lambda/ml$ and $\alpha = mu^2/2\lambda l$.

Show that if $\alpha \ll 1$,

$$\left(\frac{dx}{dt}\right)^2 = n^2\{l^2\alpha^2 - (l + l\alpha - x)^2\}$$

approximately. Hence show that

$$x = l + l\alpha(1 - \cos nt).$$

Show that if $\alpha \gg 1$ the length of the string varies between the limits l and $l(\alpha^{1/2} - 1)$ approximately.

8.8 A smooth block has mass m_b and its upper and lower faces are horizontal planes. It is free to move in a groove in a horizontal plane and a particle of mass m is attached to a point on its upper face by an elastic string of natural length l and modulus λ. The string is stretched parallel to the groove to a length $(\alpha + 1)$ times its natural length and the system is released from rest. The particle always remains in contact with the upper surface of the block.

Show that the motion is periodic with the periodic time

$$\left(\frac{lmm_b}{\lambda(m+m_b)}\right)^{1/2}\left(2\pi + \frac{4}{\alpha}\right).$$

8.9 Three particles, each of mass m, are free to move on a smooth horizontal plane. The mutual force of attraction between each pair of particles is of magnitude $m^2\mu$ times their distance apart. At time $t = 0$ the particles are at points P, Q, R and have velocities $\lambda\overline{QR}$, $\lambda\overline{RP}$, $\lambda\overline{PQ}$ respectively. Show that their centre of mass remains at rest, and that each particle describes an ellipse with periodic time $T = 2\pi/(3\mu m)^{1/2}$.

Show that the area of the ellipse is $2\lambda AT/3$, where A is the area of the triangle PQR.

8.10 Two gravitating particles, of masses m_1 and m_2, start from rest an infinite distance apart and are allowed to fall freely towards one another. Show that, when their distance apart is a, their relative velocity of approach is of magnitude $\{2G(m_1 + m_2)/a\}^{1/2}$, where G is the gravitational constant.

Impulses of magnitude F are applied to the particles in opposite directions perpendicular to their line of motion at the instant when their distance apart is a. Show how to determine b, their distance of closest approach during the subsequent motion, and prove that when $b \ll a$, it is given approximately by the relation

$$\frac{b}{a} = \frac{\bar{F}^2 a}{2G} \frac{m_1 + m_2}{m_1^2 m_2^2}.$$

8.11 A train consists of an engine, of mass m_e, and n trucks, each of mass m. The train is initially at rest and each coupling has a length l of slack. Show that if the engine starts to move forward and exerts a constant pull F, the speed u_j of the train just after the jth truck has been jerked (inelastically) into motion satisfies the difference equation

$$\{m_e + jm\}^2 u_j^2 = \{m_e + (j-1)m\}^2 u_{j-1}^2 + 2Fl\{m_e + (j-1)m\}.$$

Solve this difference equation by the substitution

$$\{m_e + jm\}^2 u_j^2 = Aj^2 + Bj + C,$$

where A, B, C are constants to be calculated.

Hence show that the speed just after the last truck has been jerked into motion is

$$[Fln\{2m_e + (n-1)m\}]^{1/2}/(m_e + nm).$$

8.12 A shell of mass $2m$ is fired vertically upwards with velocity v from a point on a level stretch of ground. When it reaches the top of its trajectory it is split into two equal fragments by an internal explosion which supplies kinetic energy equal to mu^2 to the system. Show that the greatest possible distance between the points where the two fragments hit the ground is $2uv/g$ if $u \leqslant v$ or $(u^2 + v^2)/g$ if $u \geqslant v$.

8.13 A shell of mass m explodes when in flight into n fragments, whose masses are $m_1, m_2, \ldots, m_n$. Show that the energy developed by the explosion is such that the kinetic energy associated with the mass centre is unchanged. Show that the kinetic energy relative to the mass centre is

$$\frac{1}{4m} \sum_i \sum_j m_i m_j v_{ij}^2,$$

where $\mathbf{v}_{ij}$ is the velocity of m_i relative to m_j ($\mathbf{v}_{ii} = \mathbf{0}$).

8.14 A shell moving through the air is broken into two portions, of masses m_1 and m_2, by an internal explosion which imparts kinetic energy E to the system. Show that the relative speed of the portions after the explosion is $\{2(m_1 + m_2)E/m_1 m_2\}^{1/2}$.

8.15 A particle of mass m is describing a circular orbit under an inverse square law central force of attraction. It is split by an inner explosion into two parts, of masses $(1 - k)m$, km. Show that if the first part is instantaneously brought to rest, the second part describes an ellipse provided $2k^2 > 1$.

8.16 An initially parallel beam of particles collides with a fixed smooth sphere of radius a, and the particles are scattered without loss of energy. Show that the effective scattering cross-section is

$$d\sigma = \tfrac{1}{2}\pi a^2 \sin \chi \, d\chi,$$

where χ is the scattering angle.

8.17 A particle, of mass m_1 and velocity $\mathbf{v}_1$, is captured by a stationary nucleus. A particle, of mass m_2, is ejected from the nucleus with velocity $\mathbf{v}_2$ perpendicular to $\mathbf{v}_1$. The rest of the nucleus has mass m_3 and recoils with velocity $\mathbf{v}_3$.

Let T_1, T_2 be the kinetic energies of m_1, m_2 respectively, and let E be the energy absorbed in the reaction. Show that

$$(m_2 + m_3)T_2 = m_3 E + (m_3 - m_1)T_1.$$

8.18 A photon, of momentum $\mathbf{k}_1$, is absorbed by an electron initially at rest, which instantly recoils and emits a second photon of momentum $\mathbf{k}_2$ in a direction making an angle θ with the direction of $\mathbf{k}_1$. The electron at rest has an energy $c^2 m_0$ and when moving with momentum $\mathbf{p}$ has an energy $c(c^2 m_0^2 + p^2)^{1/2}$. The photons have energies ck_1 and ck_2 respectively.

Prove that

$$\frac{1}{k_2} - \frac{1}{k_1} = \frac{1}{cm_0}(1 - \cos\theta).$$

8.19 The mass m of a moving particle is a function of the speed v, and the force required to accelerate the particle is equal to the rate of change of momentum. Show that the energy equation can be written

$$\frac{1}{2}mv^2 + \frac{1}{2}\int v^2 \frac{dm}{dv}\, dv + V = \text{constant},$$

where V is the potential energy.

Let

$$m = m_0\left(1 - \frac{v^2}{c^2}\right)^{-1/2}$$

Show that the appropriate energy equation is

$$m_0 c^2 \left(1 - \frac{v^2}{c^2}\right)^{-1/2} + V = \text{constant}.$$

Suppose that the particle moves along a straight line with $V = \frac{1}{2}\mu m_0 x^2$, where x is the distance of the particle from a centre of attraction O. Show that the motion is oscillatory with the periodic time

$$\frac{4}{c}\int_0^a \frac{(\alpha^2 - x^2)\, dx}{\{(a^2 - x^2)(\beta^2 - x^2)\}^{1/2}},$$

where a is the maximum value of $|x|$,

$$\alpha^2 = a^2 + 2c^2/\mu \quad \text{and} \quad \beta^2 = a^2 + 4c^2/\mu.$$

8.20 (i) Material is fed continuously from a hopper at rest on to a moving belt at a mass rate $\dot{m}$. Find the force required to keep the belt moving at a constant speed v. Show that the power required is twice the rate of change of kinetic energy. Why is the extra power required?

It may be assumed that the material sticks to the belt without sliding.

(ii) An endless belt moves on rollers at each end. The axes of the rollers are horizontal and parallel. The upper part of the belt is plane and inclined at an angle θ to the horizontal. It is fed from a hopper, at a constant rate $\dot{m}$, with material which sticks to the belt until it reaches the lower end, where it falls off. The distance travelled by the material before it falls off is l. There is a constant force R resisting the motion.

Show that the steady state speed of the roller is

$$v = \frac{\{R^2 + 4\dot{m}^2 gl \sin\theta\}^{1/2} - R}{2\dot{m}}.$$

8.21 A uniform layer of snow, whose surface is a rectangle with two sides horizontal, rests on a mountain slope of uniform inclination α, the adhesion being just sufficient to hold the snow while at rest. At a certain instant, the uppermost line of snow starts to move downwards and collects with it the snow it meets on the way. Kinetic friction between the snow and the slope is negligible.

Prove that if v is the speed when a distance x of the slope has been uncovered, then

$$\frac{\mathrm{d}}{\mathrm{d}x}(x^2 v^2) = 2gx^2 \sin\alpha.$$

Hence show that the moving snow has a constant acceleration of $\frac{1}{3}g \sin\alpha$.

8.22 A boat, of total mass m, is propelled at constant speed u through water which offers a resistance equal to k times the speed.

At a given instant the boat springs a leak and subsequently takes in water at a constant mass rate λ. The boat sinks when the mass of water in it is m. Show that the boat travels a distance

$$\frac{mu}{k+\lambda}\left(\frac{k}{\lambda} + \frac{\lambda}{k}(1 - 2^{-k/\lambda})\right)$$

after springing the leak before sinking.

8.23 A train has speed v and mass $m - \lambda t$ at time t. The decrease in mass arises because of leakage in the wagons which allows the contents to pour out. The train is subject to a driving force F and a resistance R. Show that

$$(m - \lambda t)\frac{\mathrm{d}v}{\mathrm{d}t} = F - R.$$

Given that the engine does work at the constant rate P, that there is a resistance λkv, and that $v = v_0$ when $t = 0$, show that

$$v^2 = v_0^2(1 - \lambda t/m)^{2k} + \{1 - (1 - \lambda t/m)^{2k}\}P/\lambda k.$$

8.24 A cloud of water vapour moves with constant velocity $\mathbf{u}$. A spherical drop in the cloud increases its mass, by condensation, at a rate $k_1 S$, where S is its surface area. It is subject to a constant gravitational acceleration $\mathbf{g}$ and a resistance equal to $k_2 S$ times the velocity of the cloud relative to the drop, where k_1 and k_2 are positive constants. Given that the drop is initially very small, show that its trajectory is similar to that of a particle subject to a uniform gravitational acceleration equal to $k_1 \mathbf{g}/(4k_1 + 3k_2)$.

8.25 A rocket, of initial total mass m_0, is fired from rest vertically upwards. It propels itself by ejecting mass at a constant rate k with constant speed c relative to itself. When all the fuel has burnt, the mass of the rocket is m_1. Show that if $kc/m_0 g \gg 1$, the speed of the rocket at the instant of burning out is $c \log(m_0/m_1)$ approximately.

A two-stage rocket is constructed from two rockets with the same exhaust speed c. They each have a mass ratio m_0/m_1 equal to 4, but the initial total mass of the first is twice that of the second. The second is mounted on top of the first in such a way that it begins to fire just when the first is burnt out and at this instant it is automatically released from the first. Show that the speed attained by the second rocket is the same as that of a single-stage rocket of mass ratio 8.

8.26 A rocket is fired vertically, starting from rest, and is propelled upwards by projecting propellant at a constant relative speed c downwards. The propellant burns for a time T and the ratio of the initial mass of the rocket to the mass when the propellant is all burnt is R.

Write down the equation of motion and deduce that the rocket will not rise from the ground initially unless

$$(R - 1)/R > gT/c$$

where g is the acceleration due to gravity, assumed constant.

Show that if this condition is satisfied, the maximum height attained is

$$\frac{c^2 \log^2 R}{2g} - cT\left(\frac{R \log R}{R - 1} - 1\right).$$

Deduce that the maximum height can be increased by reducing the burning time T.

8.27 A rocket leaves the earth's atmosphere with speed u at time $t = 0$. It is then at a distance a from the centre of the earth and moving radially. The gravitational acceleration varies inversely as the square of the distance and is equal to g at the distance a. The rocket ejects propellant backwards at a constant speed c relative to itself. The total mass of the rocket and remaining propellant is $m_0 e^{-kt}$ at time t, where m_0 and k are positive constants such that $kc < g$.

Show that the minimum value of u for which the rocket will escape from the earth's gravitational field is $(2a)^{1/2}\{g^{1/2} - (kc)^{1/2}\}$.

8.28 A ram-jet aeroplane, of *constant* mass m, moves horizontally with speed $v(t)$. It takes in air, which is at rest, and ejects an equivalent mass of air at the back with constant speed c relative to itself. The mass rate at which air is taken in and ejected is mav per unit time, where a is a constant. There is a resistance to the motion equal to mkv^n.

Calculate the change in momentum during a small increment in time, and hence show that

$$\frac{dv}{dt} + av(v - c) + kv^n = 0.$$

For the case $n = 2$, show that the speed v after the aeroplane has travelled a distance x is given by

$$v = w - (w - v_0)e^{-\beta x},$$

where $\beta = a + k$, $v_0(<w)$ is the speed when $x = 0$ and w is the terminal speed.

8.29 A pail, of large cross-section and mass m_p, is made to move vertically upwards with constant speed u. Water falls into it from a tank, a mass m leaving the tank per unit time. The initial speed of the water is zero and the first of the water reaches the pail when it is a distance h below the tank. Show that if, after a further time t, the pail is at a depth x below the tank, it then contains water of mass

$$m' = m\{t + (2h/g)^{1/2} - (2x/g)^{1/2}\}.$$

Determine the rate at which mass is entering the pail and hence show that the force required to maintain the speed of the pail is

$$(m_p + m')g + m\{u + (2gx)^{1/2}\}^2(2gx)^{-1/2}.$$

8.30 A uniform chain of length l and weight W is hanging vertically from its ends A, B which are close together. At a given instant the end B is released. Find the tension at A when B has fallen a distance $x(<l)$.
(Ans.: $\frac{1}{2}W(1 + 3x/l)$.)

9 Angular Vectors

9.1 EXISTENCE

The purpose of this chapter is to introduce a vector associated with rigid body displacements, and to discuss its properties and applications.

Definition

A system of vectors $\mathbf{r}_i (i = 1, 2, \ldots, n)$, which are displaced subject to the constraints

$$(\mathbf{r}_i - \mathbf{r}_j) \cdot (\mathbf{r}_i - \mathbf{r}_j) = \text{constant} \tag{9.1.1}$$

for all i, j is said to be displaced as a rigid body.

For the moment we add the further constraint

$$\mathbf{r}_i \cdot \mathbf{r}_i = \text{constant.} \tag{9.1.2}$$

We shall discuss later the effect of relaxing (9.1.2).

Conditions (9.1.1) and (9.1.2) are together equivalent to

$$\mathbf{r}_i \cdot \mathbf{r}_j = \text{constant} \tag{9.1.3}$$

for all i, j, including $i = j$. Geometrically the vectors are all of constant magnitude and the angle between any two of them is constant. For future reference we note that the vectors $\mathbf{r}_i \times \mathbf{r}_j$ can be added to the system in the sense that the augmented system will still satisfy (9.1.3). This is so because the magnitude of a typical vector $\mathbf{r}_i \times \mathbf{r}_j$ will depend on the constant magnitudes of $\mathbf{r}_i$, $\mathbf{r}_j$ and the angle between them. Furthermore the constant angles between three vectors $\mathbf{r}_i$, $\mathbf{r}_j$, $\mathbf{r}_k$ of the system are sufficient to determine the angle between $\mathbf{r}_k$ and $\mathbf{r}_i \times \mathbf{r}_j$, which is therefore constant (see Example 1.5.4).

The vectors $\mathbf{r}_i$ need not be interpreted as position vectors, but if they are it will be seen that they are displaced according to the intuitive notion of a rigid body, for which the distance between any two points remains invariant. The effect of the additional constraint (9.1.2) is to restrict the displacement to a rotation about an origin O, which is fixed.

Let each $\mathbf{r}_i$ be displaced to $\mathbf{r}_i + \delta\mathbf{r}_i$ in a way compatible with (9.1.3). Let the displacement $\delta\mathbf{r}_i$ be of the form

$$\delta\mathbf{r}_i = (\mathbf{v}_i + \boldsymbol{\epsilon}_i)\delta t, \tag{9.1.4}$$

where $\mathbf{v}_i$ is independent of δt and $|\boldsymbol{\epsilon}_i| \to 0$ as $\delta t \to 0$ for all i. Roughly speaking, if δt is a measure of the magnitude of a small displacement, then $\delta \mathbf{r}_i = \mathbf{v}_i \, \delta t$ to the first order and $\boldsymbol{\epsilon}_i \, \delta t$ is the error term. From (9.1.3) we must have

$$(\mathbf{r}_i + \delta \mathbf{r}_i) \, . \, (\mathbf{r}_j + \delta \mathbf{r}_j) = \mathbf{r}_i \, . \, \mathbf{r}_j$$

or, with the help of (9.1.4),

$$\mathbf{r}_i \, . \, (\mathbf{v}_j + \boldsymbol{\epsilon}_j) + \mathbf{r}_j \, . \, (\mathbf{v}_i + \boldsymbol{\epsilon}_i) + (\mathbf{v}_i + \boldsymbol{\epsilon}_i) \, . \, (\mathbf{v}_j + \boldsymbol{\epsilon}_j) \, \delta t = 0.$$

Since this equation must be satisfied in the limit $\delta t \to 0$, and since $\mathbf{r}_i \, . \, \mathbf{v}_j + \mathbf{r}_j \, . \, \mathbf{v}_i$ is independent of δt, it follows that

$$\mathbf{r}_i \, . \, \mathbf{v}_j + \mathbf{r}_j \, . \, \mathbf{v}_i = 0 \tag{9.1.5}$$

and in particular

$$\mathbf{r}_i \, . \, \mathbf{v}_i = 0. \tag{9.1.6}$$

Theorem

If the rigid body displacements are of the form (9.1.4), and if the $\mathbf{r}_i$ are not all parallel, there exists a unique vector $\boldsymbol{\omega}$ such that

$$\mathbf{v}_i = \boldsymbol{\omega} \times \mathbf{r}_i \tag{9.1.7}$$

for all the vectors $\mathbf{r}_i$ of the system.

If such a vector exists, it is unique. For suppose that both $\boldsymbol{\omega}$ and $\boldsymbol{\omega}'$ satisfy (9.1.7), so that

$$(\boldsymbol{\omega} - \boldsymbol{\omega}') \times \mathbf{r}_i = \mathbf{0} \tag{9.1.8}$$

for all vectors of the system. Since the vectors $\mathbf{r}_i$ are not all parallel, (9.1.8) can only be satisfied if $\boldsymbol{\omega} - \boldsymbol{\omega}'$ is a null vector, so that $\boldsymbol{\omega}' = \boldsymbol{\omega}$.

Let $\mathbf{r}_1$, $\mathbf{r}_2$, $\mathbf{r}_3$ be three linearly independent vectors of the system. Such a choice is always possible since, by assumption, we can choose $\mathbf{r}_1$ and $\mathbf{r}_2$ to be non-parallel. If an appropriate third vector does not already exist we can augment the system with the vector $\mathbf{r}_3 = \mathbf{r}_1 \times \mathbf{r}_2$ which satisfies all the required conditions. The associated vectors $\mathbf{v}_1$, $\mathbf{v}_2$ and $\mathbf{v}_3$ cannot all be parallel since, by (9.1.6), this would imply that $\mathbf{r}_1$, $\mathbf{r}_2$ and $\mathbf{r}_3$ are all perpendicular to $\mathbf{v}_1$ and therefore are not linearly independent. Without loss of generality we assume that $\mathbf{v}_1$ and $\mathbf{v}_2$ are not parallel.

If there is a vector $\boldsymbol{\omega}$ satisfying (9.1.7) for all i it must be perpendicular to all the $\mathbf{v}_i$. In particular it must be perpendicular to $\mathbf{v}_1$ and $\mathbf{v}_2$ and therefore of the form $\boldsymbol{\omega} = c\mathbf{v}_1 \times \mathbf{v}_2$ for some constant c. Now

$$\boldsymbol{\omega} \times \mathbf{r}_1 = c(\mathbf{v}_1 \times \mathbf{v}_2) \times \mathbf{r}_1 = -c(\mathbf{r}_1 \,.\, \mathbf{v}_2)\mathbf{v}_1,$$

$$\boldsymbol{\omega} \times \mathbf{r}_2 = c(\mathbf{v}_1 \times \mathbf{v}_2) \times \mathbf{r}_2 = c(\mathbf{r}_2 \,.\, \mathbf{v}_1)\mathbf{v}_2 = -c(\mathbf{r}_1 \,.\, \mathbf{v}_2)\mathbf{v}_2,$$

where relations (9.1.5) and (9.1.6) have been used. It follows that $\boldsymbol{\omega} \times \mathbf{r}_1 = \mathbf{v}_1$ and $\boldsymbol{\omega} \times \mathbf{r}_2 = \mathbf{v}_2$ provided that $c = -(\mathbf{r}_1 \,.\, \mathbf{v}_2)^{-1} = (\mathbf{r}_2 \,.\, \mathbf{v}_1)^{-1}$ or

$$\boldsymbol{\omega} = -\frac{\mathbf{v}_1 \times \mathbf{v}_2}{\mathbf{r}_1 \,.\, \mathbf{v}_2} = \frac{\mathbf{v}_1 \times \mathbf{v}_2}{\mathbf{r}_2 \cdot \mathbf{v}_1}. \tag{9.1.9}$$

We also have the relation

$$(\boldsymbol{\omega} \times \mathbf{r}_i - \mathbf{v}_i) \,.\, \mathbf{r}_j = -\mathbf{r}_i \,.\, (\boldsymbol{\omega} \times \mathbf{r}_j) + \mathbf{r}_i \,.\, \mathbf{v}_j. \tag{9.1.10}$$

Since we have shown that $\boldsymbol{\omega} \times \mathbf{r}_j = \mathbf{v}_j$ for $j = 1, 2$, it follows that (9.1.10) is zero for $j = 1$, 2 and for all i. It is also zero for $i = j$ since both $\boldsymbol{\omega} \times \mathbf{r}_i$ and $\mathbf{v}_i$ are perpendicular to $\mathbf{r}_i$. We can therefore deduce that $(\boldsymbol{\omega} \times \mathbf{r}_3 - \mathbf{v}_3) \,.\, \mathbf{r}_j = 0$ for $j =$ 1, 2, 3. Now a vector whose scalar products with three linearly independent vectors are zero must be a null vector (see Example 1.5.5). Since $\mathbf{r}_1, \mathbf{r}_2, \mathbf{r}_3$ are linearly independent we must have $\boldsymbol{\omega} \times \mathbf{r}_3 = \mathbf{v}_3$. We have now shown that $\boldsymbol{\omega} \times \mathbf{r}_j = \mathbf{v}_j$ for $j = 1, 2, 3$. It follows from (9.1.10) that $(\boldsymbol{\omega} \times \mathbf{r}_i - \mathbf{v}_i) \,.\, \mathbf{r}_j = 0$ for $j = 1, 2, 3$ and for all i. Hence $\boldsymbol{\omega} \times \mathbf{r}_i - \mathbf{v}_i = \mathbf{0}$ and the theorem is proved.

The situation when the $\mathbf{r}_i$ are all parallel is that a linearly independent reference triad of vectors, on which the above argument depends, cannot be defined within the system. However, we can augment the system with vectors which do have the desired property, for example if $\mathbf{r}_2$ and $\mathbf{r}_3$ are added such that they form a mutually orthogonal triad with a vector $\mathbf{r}_1$ of the system, then an angular vector $\boldsymbol{\omega}$ can be defined for the extended system and hence for the original system. Because of the flexibility in the choice of the added vectors there is not a unique vector $\boldsymbol{\omega}$ associated with the original system and it is clear that, for any scalar λ, $(\boldsymbol{\omega} + \lambda\mathbf{r}_1) \times \mathbf{r}_i = \mathbf{v}_i$ if $\boldsymbol{\omega} \times \mathbf{r}_i = \mathbf{v}_i$ for any vector $\mathbf{r}_i$ parallel to $\mathbf{r}_1$. The component of $\boldsymbol{\omega}$ parallel to the system is therefore arbitrary and can be chosen, for example, to be zero by appropriate choice of the displacements $\mathbf{v}_2$ and $\mathbf{v}_3$ associated with the added vectors $\mathbf{r}_2$ and $\mathbf{r}_3$.

To interpret $\boldsymbol{\omega}$ geometrically, let $\mathbf{r}_i$ refer to the position vector of a point P_i, and consider a representation of $\boldsymbol{\omega}$ passing through the origin. Then $\boldsymbol{\omega} \times \mathbf{r}_i \,\delta t$ represents a displacement perpendicular to $\boldsymbol{\omega}$ and $\mathbf{r}_i$ with magnitude equal to $\omega\,\delta t$ multiplied by the length of the perpendicular P_iN from P_i on to $\boldsymbol{\omega}$ (see Fig. 9.1.1). The rigid body displacement is thus equivalent to an angular displacement of magnitude $\omega\,\delta t$ about ON, at least to first order. For this reason an alternative presentation of a rigid body displacement about a fixed origin O is by the introduction of a vector $\delta\theta\,\mathbf{n}$ whose magnitude is the angular displacement $\delta\theta$, and whose direction is that of ON, the axis of rotation. The rigid body

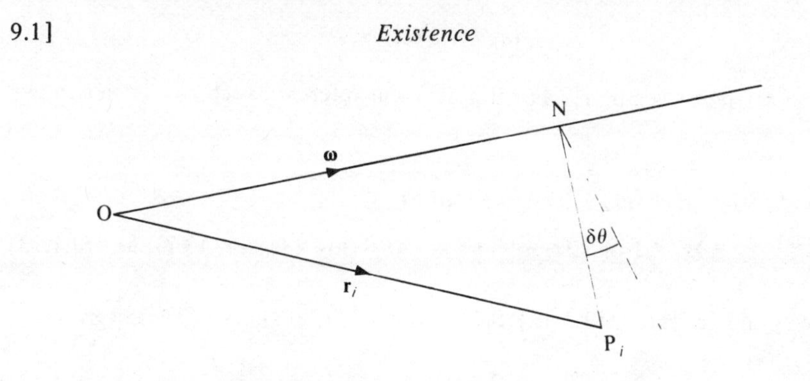

Figure 9.1.1 Geometrical interpretation of angular vector.

displacement is then $\delta\theta\mathbf{n} \times \mathbf{r}$, correct to first order. It should not, however, be inferred that there is a finite angular displacement which has the same properties. While it is possible to use an angular displacement and the direction of its axis as the magnitude and direction of a vector, the direct sum of two such vectors does not in general represent the resultant of the two angular displacements carried out successively on a rigid body (see Example 9.1.1). Small angular displacements can be combined in such a manner, and if $\delta\theta_1\mathbf{n}_1, \delta\theta_2\mathbf{n}_2$ are two small angular displacements of a rigid body, the final displacement is equivalent to $\delta\theta_1\mathbf{n}_1 + \delta\theta_2\mathbf{n}_2$ and

$$\delta\mathbf{r} = (\delta\theta_1\mathbf{n}_1 + \delta\theta_2\mathbf{n}_2) \times \mathbf{r} \tag{9.1.11}$$

for all points of the rigid body. However, the relation is correct only to first order. It is not possible to define vectors for which (9.1.11) is true exactly when $\delta\theta_1\mathbf{n}_1$ and $\delta\theta_2\mathbf{n}_2$ are small but non-zero displacements. In spite of this we shall see that the vector $\boldsymbol{\omega}$ plays a significant role, especially when we consider the continuous motion of rigid bodies.

So far the analysis applies only to a restricted rigid body displacement for which both conditions (9.1.1) and (9.1.2) are satisfied. When (9.1.2) is relaxed, we choose a reference vector, say $\mathbf{r}_1$, from the system. Then the relative vector system $\mathbf{r}_i' = \mathbf{r}_i - \mathbf{r}_1$ will satisfy the stronger conditions, including (9.1.2), and an angular velocity $\boldsymbol{\omega}_1$ can be defined such that

$$\delta\mathbf{r}_i' = (\boldsymbol{\omega}_1 \times \mathbf{r}_i' + \boldsymbol{\epsilon}_i)\,\delta t$$

and so

$$\delta\mathbf{r}_i = \delta\mathbf{r}_1 + \{\boldsymbol{\omega}_1 \times (\mathbf{r}_i - \mathbf{r}_1) + \boldsymbol{\epsilon}_i\}\,\delta t. \tag{9.1.12}$$

When $\mathbf{r}_i$ refers to the position vector of a point P_i, the more general displacement $\delta\mathbf{r}_i$ of P_i consists of the displacement $\delta\mathbf{r}_1$ of P_1, which is a translation of all the P_i as a rigid body, together with the displacement $\delta\mathbf{r}_i'$ relative to P_1, which is a rotation $\boldsymbol{\omega}_1\delta t$ to first order. In fact $\boldsymbol{\omega}_1$ is independent of the reference vector $\mathbf{r}_1$. For if a different reference vector $\mathbf{r}_2$ is chosen, we have

$$\begin{aligned}\delta \mathbf{r}_i &= \delta \mathbf{r}_2 + \{\boldsymbol{\omega}_2 \times (\mathbf{r}_i - \mathbf{r}_2) + \boldsymbol{\epsilon}_i'\}\,\delta t \\ &= \delta \mathbf{r}_1 + \{\boldsymbol{\omega}_1 \times (\mathbf{r}_2 - \mathbf{r}_1) + \boldsymbol{\epsilon}_2\}\,\delta t + \{\boldsymbol{\omega}_2 \times (\mathbf{r}_i - \mathbf{r}_2) + \boldsymbol{\epsilon}_i'\}\,\delta t \end{aligned} \tag{9.1.13}$$

since $\mathbf{r}_2$ and $\delta\mathbf{r}_2$ satisfy (9.1.12). Subtraction of (9.1.13) from (9.1.12) gives

$$\mathbf{0} = (\boldsymbol{\omega}_1 - \boldsymbol{\omega}_2) \times (\mathbf{r}_i - \mathbf{r}_2) + \boldsymbol{\epsilon}_i - \boldsymbol{\epsilon}_i' - \boldsymbol{\epsilon}_2\,. \tag{9.1.14}$$

Since (9.1.14) must hold in the limit $\delta t \to 0$, we must have

$$\mathbf{0} = (\boldsymbol{\omega}_1 - \boldsymbol{\omega}_2) \times (\mathbf{r}_i - \mathbf{r}_2) \tag{9.1.15}$$

for all vectors $\mathbf{r}_i$ of the system. Provided the vectors $\mathbf{r}_i$ are not all parallel, (9.1.13) can hold only if $\boldsymbol{\omega}_1 = \boldsymbol{\omega}_2 = \boldsymbol{\omega}$ say. Thus $\boldsymbol{\omega}$ is a property of the displacement independent of the reference vector.

Example 9.1.1

In this example we consider a vector defined by a finite angular displacement and its implications.

Let a rigid body be capable of rotation about a fixed point O. In an angular displacement θ_1 about an axis in the direction of the unit vector $\mathbf{n}_1$, let the point P of the body, with position vector $\mathbf{r}$, move to the point P_1, with position

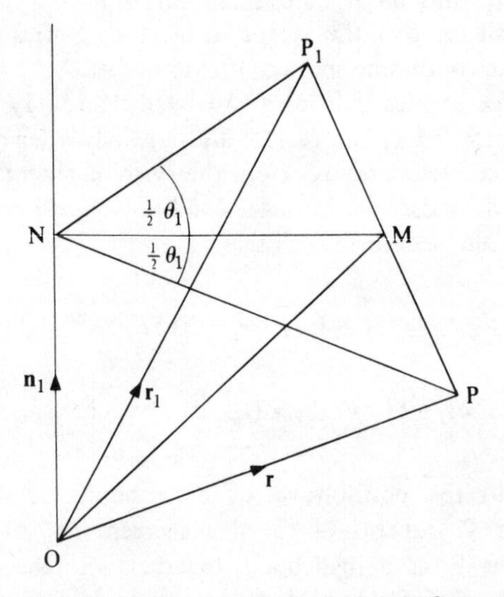

Figure 9.1.2 Illustration for Example 9.1.1: finite rotations.

vector $\mathbf{r}_1$. We can define a vector $\mathbf{\Omega}_1$ by the relation

$$\mathbf{\Omega}_1 = 2\mathbf{n}_1 \tan \tfrac{1}{2}\theta_1, \tag{9.1.16}$$

which has the property that

$$\mathbf{r}_1 - \mathbf{r} = \tfrac{1}{2}\mathbf{\Omega}_1 \times (\mathbf{r} + \mathbf{r}_1) \tag{9.1.17}$$

and serves to determine $\mathbf{r}_1$ in terms of $\mathbf{r}$ and $\mathbf{\Omega}_1$, at least implicitly. To see that relation (9.1.17) is satisfied, we first note that

$$\tfrac{1}{2}\mathbf{\Omega}_1 \times (\mathbf{r} + \mathbf{r}_1) = 2 \tan \tfrac{1}{2}\theta_1 \; \mathbf{n}_1 \times \overline{\mathrm{OM}}$$

where M is the mid point of PP_1 (see Fig. 9.1.2). Also $\mathbf{n}_1 \times \overline{\mathrm{OM}}$ has the direction of $\overline{\mathrm{PP}}_1 = \mathbf{r}_1 - \mathbf{r}$ and the magnitude of MN, where MN is the perpendicular to M from the axis. Finally

$$2\mathrm{NM} \tan \tfrac{1}{2}\theta_1 = 2\mathrm{MP}_1 = \mathrm{PP}_1 = |\mathbf{r}_1 - \mathbf{r}|$$

and so the two sides of (9.1.17) have the same magnitude and the same direction.

Note that, when θ_1 is small so that $|\mathbf{r}_1 - \mathbf{r}|$ is also small, we have

$$\mathbf{\Omega}_1 = \theta_1 \mathbf{n}_1, \quad \mathbf{r}_1 - \mathbf{r} = \mathbf{\Omega}_1 \times \mathbf{r}, \tag{9.1.18}$$

approximately. Thus $\mathbf{\Omega}_1$ is consistent with our previous description of a small angular displacement, but it is clear that relations (9.1.18) are not valid for a finite rotation.

Let $\mathbf{\Omega}_2$ be a second such displacement. If $\mathbf{\Omega}_1$ and $\mathbf{\Omega}_2$ are performed successively, so that P moves to P_1 and then to P_2, we have

$$\mathbf{r}_2 - \mathbf{r}_1 = \tfrac{1}{2}\mathbf{\Omega}_2 \times (\mathbf{r}_1 + \mathbf{r}_2). \tag{9.1.19}$$

Now there is a single vector, $\mathbf{\Omega}_{12}$ say, which represents a single angular displacement that moves P to P_2 so that

$$\mathbf{r}_2 - \mathbf{r} = \tfrac{1}{2}\mathbf{\Omega}_{12} \times (\mathbf{r} + \mathbf{r}_2). \tag{9.1.20}$$

The relationship between $\mathbf{\Omega}_1$, $\mathbf{\Omega}_2$ and $\mathbf{\Omega}_{12}$ is not the simple law of addition for vectors, that is to say

$$\mathbf{\Omega}_{12} \neq \mathbf{\Omega}_1 + \mathbf{\Omega}_2.$$

We proceed to investigate the correct relation between these vectors. It follows, from (9.1.17) and (9.1.19), that

$$\mathbf{\Omega}_1 \,.\, (\mathbf{r}_1 - \mathbf{r}) = \mathbf{\Omega}_2 \,.\, (\mathbf{r}_2 - \mathbf{r}_1) = 0 \tag{9.1.21}$$

and that

$$\begin{aligned}
4(\mathbf{r}_2 - \mathbf{r}) &= 2\mathbf{\Omega}_1 \times (\mathbf{r} + \mathbf{r}_1) + 2\mathbf{\Omega}_2 \times (\mathbf{r}_1 + \mathbf{r}_2) \\
&= 2(\mathbf{\Omega}_1 + \mathbf{\Omega}_2) \times (\mathbf{r} + \mathbf{r}_2) - 2\mathbf{\Omega}_1 \times (\mathbf{r}_2 - \mathbf{r}_1) + 2\mathbf{\Omega}_2 \times (\mathbf{r}_1 - \mathbf{r}) \\
&= 2(\mathbf{\Omega}_1 + \mathbf{\Omega}_2) \times (\mathbf{r} + \mathbf{r}_2) - \mathbf{\Omega}_1 \times \{\mathbf{\Omega}_2 \times (\mathbf{r}_1 + \mathbf{r}_2)\} \\
&\quad + \mathbf{\Omega}_2 \times \{\mathbf{\Omega}_1 \times (\mathbf{r} + \mathbf{r}_1)\} \\
&= 2(\mathbf{\Omega}_1 + \mathbf{\Omega}_2) \times (\mathbf{r} + \mathbf{r}_2) - \{\mathbf{\Omega}_1 \,.\, (\mathbf{r}_1 + \mathbf{r}_2)\}\mathbf{\Omega}_2 + \{\mathbf{\Omega}_2 \,.\, (\mathbf{r} + \mathbf{r}_1)\}\mathbf{\Omega}_1 \\
&\quad + \mathbf{\Omega}_1 \,.\, \mathbf{\Omega}_2 (\mathbf{r}_2 - \mathbf{r}).
\end{aligned} \tag{9.1.22}$$

With the help of (9.1.21), (9.1.22) can be written

$$\begin{aligned}
(4 - \mathbf{\Omega}_1 \,.\, \mathbf{\Omega}_2)(\mathbf{r}_2 - \mathbf{r}) &= 2(\mathbf{\Omega}_1 + \mathbf{\Omega}_2) \times (\mathbf{r} + \mathbf{r}_2) - \{\mathbf{\Omega}_1 \,.\, (\mathbf{r} + \mathbf{r}_2)\}\mathbf{\Omega}_2 \\
&\quad + \{\mathbf{\Omega}_2 \,.\, (\mathbf{r} + \mathbf{r}_2)\}\mathbf{\Omega}_1 \\
&= \{2(\mathbf{\Omega}_1 + \mathbf{\Omega}_2) - \mathbf{\Omega}_1 \times \mathbf{\Omega}_2\} \times (\mathbf{r} + \mathbf{r}_2).
\end{aligned} \tag{9.1.23}$$

Comparison of (9.1.20) with (9.1.23) shows that

$$\mathbf{\Omega}_{12} = \frac{\mathbf{\Omega}_1 + \mathbf{\Omega}_2 - \frac{1}{2}\mathbf{\Omega}_1 \times \mathbf{\Omega}_2}{1 - \frac{1}{4}\mathbf{\Omega}_1 \,.\, \mathbf{\Omega}_2}. \tag{9.1.24}$$

Thus $\mathbf{\Omega}_{12} \neq \mathbf{\Omega}_{21}$ unless $\mathbf{\Omega}_1 \times \mathbf{\Omega}_2 = \mathbf{0}$, so that even the order in which the displacements occur affects the result. Again, however, we see that the simple vector relation

$$\mathbf{\Omega}_{12} = \mathbf{\Omega}_1 + \mathbf{\Omega}_2$$

does hold approximately for small angular displacements, provided that the second order terms are neglected.

9.2 ANGULAR VELOCITY

Let $\mathbf{r}_i(t)$ $(i = 1, 2, \ldots, n)$ be a system of differentiable vector functions of t. Let

$$\delta\mathbf{r}_i = \mathbf{r}_i(t + \delta t) - \mathbf{r}_i(t). \tag{9.2.1}$$

Then $\mathbf{v}_i$, as defined in (9.1.4), is just the derivative of $\mathbf{r}_i(t)$ since, by definition,

$$\frac{d\mathbf{r}_i}{dt} = \lim_{\delta t \to 0} \frac{\delta \mathbf{r}_i}{\delta t} = \lim_{\delta t \to 0} (\mathbf{v}_i + \boldsymbol{\epsilon}_i) = \mathbf{v}_i. \tag{9.2.2}$$

If the system $\mathbf{r}_i$ moves as a rigid body it follows, by a similar limiting process applied to (9.1.12), that

$$\frac{d\mathbf{r}_i}{dt} = \frac{d\mathbf{r}_1}{dt} + \boldsymbol{\omega} \times (\mathbf{r}_i - \mathbf{r}_1) \tag{9.2.3}$$

for an appropriate vector $\boldsymbol{\omega}$. To give (9.2.3) a physical interpretation, let t be the time and let the $\mathbf{r}_i$ be the position vectors of points P_i moving as a rigid body. Then (9.2.3) gives the velocity of P_i as the vector sum of the velocity of a reference particle P_1 and the velocity relative to P_1. The latter is determined by a rate of rotation vector $\boldsymbol{\omega}$, now called the angular velocity. When $d\mathbf{r}_1/dt = \mathbf{0}$ and $\boldsymbol{\omega}$ is a constant vector, P_i describes a circle with $\boldsymbol{\omega}$ as axis and with constant angular speed ω. The speed of P_i is ωP_iN, where P_iN is the length of the perpendicular from P_i on to the representation of $\boldsymbol{\omega}$ passing through P_1. When $\boldsymbol{\omega}$ varies with time, it defines at time t an instantaneous axis of rotation, about which P_i instantaneously turns with speed ωP_iN. Note that relation (9.2.3) is exact, since it is obtained by a limiting process in which $\delta t \to 0$, and the term $\boldsymbol{\omega} \times (\mathbf{r}_i - \mathbf{r}_1)$ gives the relative velocity exactly. The angular velocity of a rigid body, as opposed to an angular displacement, is definable as a vector without approximation.

As before, the angular velocity $\boldsymbol{\omega}$ is independent of the particular choice of reference vector $\mathbf{r}_1$ since, with a second reference vector $\mathbf{r}_2$, we have

$$\begin{aligned}
\frac{d\mathbf{r}_i}{dt} &= \frac{d\mathbf{r}_1}{dt} + \boldsymbol{\omega} \times (\mathbf{r}_i - \mathbf{r}_1) \\
&= \frac{d\mathbf{r}_1}{dt} + \boldsymbol{\omega} \times (\mathbf{r}_2 - \mathbf{r}_1) + \boldsymbol{\omega} \times (\mathbf{r}_i - \mathbf{r}_2) \\
&= \frac{d\mathbf{r}_2}{dt} + \boldsymbol{\omega} \times (\mathbf{r}_i - \mathbf{r}_2).
\end{aligned}$$

It is sometimes possible to find a point P_1 which, though not permanently at rest, instantaneously has zero velocity. Such a point does not always exist but when it does it is called an instantaneous centre, and the velocity of every other point at the instant concerned is just the velocity relative to this point, namely $\boldsymbol{\omega} \times (\mathbf{r}_i - \mathbf{r}_1)$. Note that $\mathbf{r}_1$ need not be one of the original vectors of the system, but when added it must satisfy the rigid body relationships so that P_1 moves with the rest of the points P_i as a rigid body.

It is clear that if $d\mathbf{r}_1/dt$ is zero, then the velocities of all other points are either perpendicular to $\boldsymbol{\omega}$ or zero. A simple example is that of a body rolling without slipping along a fixed surface. The point of contact of the body with the surface is then instantaneously at rest since there is no relative motion. An instantaneous centre can always be found in two-dimensional problems, where $d\mathbf{r}_i/dt$ has no z component for all i, and $\boldsymbol{\omega} = \omega\mathbf{k}$. Strictly speaking the instantaneous centre is not unique, for if one point of the rigid body has zero velocity, it can be regarded as rotating about that point and all other points on the axis of rotation will also have zero velocity. However two-dimensional problems can usually be regarded mathematically as problems in the (x, y) plane, and the instantaneous centre is then taken to be in that plane. Given two points P_1, P_2 of the body in the (x, y) plane, with non-parallel velocities $\mathbf{v}_1$, $\mathbf{v}_2$, the instantaneous centre is at the intersection of the lines through P_1 and P_2 perpendicular to $\mathbf{v}_1$ and $\mathbf{v}_2$.

Example 9.2.1
We consider the motion of a rod AB in a vertical plane such that its ends are always in contact with the ground and with a vertical wall, as in Figure 9.2.1.

Let the instantaneous centre be I. Since the velocity of A is horizontal and that of B is vertical, it follows that I must be at the intersection of the vertical through A and the horizontal through B. Let the length of the rod be l and let its inclination to the vertical be θ. Then the velocity of A is $\mathrm{IA}\dot{\theta} = l\dot{\theta}\cos\theta$ along OA and the velocity of B is $\mathrm{IB}\dot{\theta} = l\dot{\theta}\sin\theta$ along BO. The point of the rod which has minimum speed is the foot of the perpendicular from I onto AB.

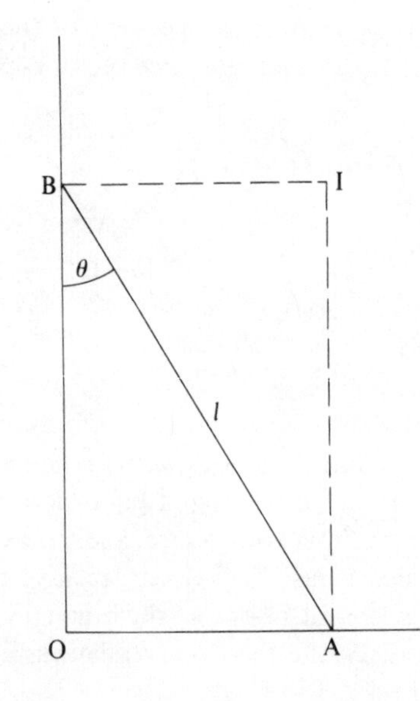

Figure 9.2.1 Illustration for Example 9.2.1: construction of the instantaneous centre I.

Example 9.2.2
A circular cylinder of radius a_1 is enclosed inside a coaxial hollow circular cylinder of radius a_2 ($>a_1$). The space between them contains ball bearings of radius $\frac{1}{2}(a_2 - a_1)$. The inner and outer cylinders are made to turn with constant angular speeds ω_1 and ω_2 respectively. Show that if there is no slipping the angular speed of a ball is $(\omega_2 a_2 - \omega_1 a_1)/(a_2 - a_1)$ and that the centre of a ball moves in a circle with angular speed $(\omega_1 a_1 + \omega_2 a_2)/(a_1 + a_2)$. What is the length of the circular arc on either cylinder with which the ball is in contact in unit time?

In Figure 9.2.2 the rotations are assumed to be clockwise. In this particular example the results hold for clockwise or anti-clockwise rotations provided that the angular speeds are allowed to take both positive values (for clockwise rotations) and negative values (for anti-clockwise rotations).

Let ω_b, ω_c be the angular speeds of the ball bearing and of the radius OC. Let v_q, v_r be the speeds of the points of the ball bearing in contact with the inner and outer cylinders at Q and R (see Fig. 9.2.2). We have

$$v_q = \omega_1 a_1, \quad v_r = \omega_2 a_2 = v_q + \omega_b \mathrm{QR}.$$

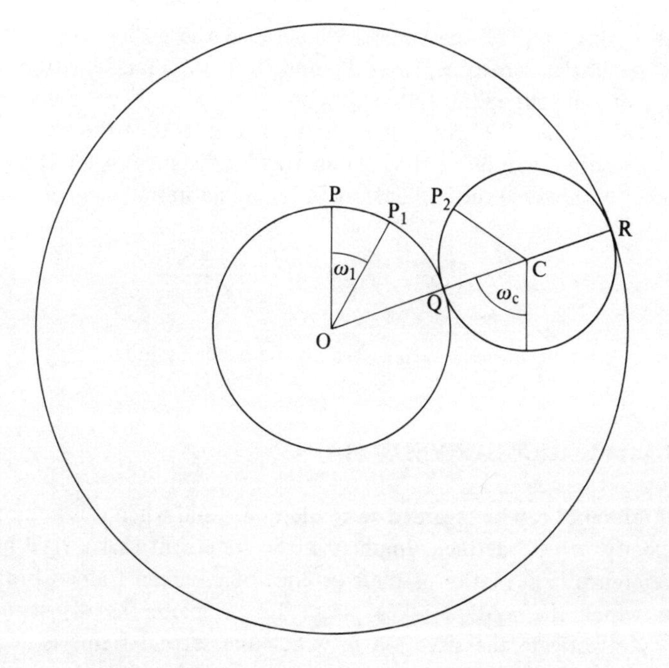

Figure 9.2.2 Illustration for Example 9.2.2: the motion of a ball bearing between coaxial cylinders.

Hence

$$\omega_2 a_2 = \omega_1 a_1 + \omega_b(a_2 - a_1), \quad \omega_b = \frac{\omega_2 a_2 - \omega_1 a_1}{a_2 - a_1}.$$

The period of rotation of a ball is

$$\frac{2\pi}{|\omega_b|} = \frac{2\pi(a_2 - a_1)}{|\omega_2 a_2 - \omega_1 a_1|}.$$

Also

$$v_c = v_q + \omega_b QC = \omega_1 a_1 + \tfrac{1}{2}(\omega_2 a_2 - \omega_1 a_1) = \tfrac{1}{2}(\omega_1 a_1 + \omega_2 a_2),$$

$$\omega_c = \frac{v_c}{OC} = \frac{\omega_1 a_1 + \omega_2 a_2}{a_1 + a_2}.$$

Hence C moves in a circle with periodic time

$$\frac{2\pi}{|\omega_c|} = \frac{2\pi(a_1 + a_2)}{|\omega_1 a_1 + \omega_2 a_2|}.$$

Let the radius OP_1 of the inner cylinder and the radius CP_2 of the ball bearing be vertical at zero time, when P_1 and P_2 both coincide with P (see Fig. 9.2.2). In unit time the radius OP_1 will move through an angular distance ω_1. The radius OQ, where OQC is the line joining the centres of the coaxial circles and the ball bearing, will move through an angular distance ω_c. The arc on the inner cylinder with which the ball bearing is in contact in unit time is

$$P_1 Q = a_1 |\omega_c - \omega_1| = \frac{a_1 a_2 |\omega_2 - \omega_1|}{a_1 + a_2}.$$

It is in contact with the same length of arc on the outer cylinder.

9.3 RELATIVE ANGULAR VELOCITY

A frame of reference can be regarded as a rigid body, and when it is in motion an angular velocity can be defined. Implicit in the statement that a rigid body or frame of reference is in motion is the existence of a second frame of reference relative to which the motion takes place. To complete the discussion it is necessary to investigate the situation in which the second frame is in motion relative to a third, giving rise to the composition of angular velocities.

Let F_1, F_2, F_3 refer to three frames of reference in relative rotatory motion about a common origin which is regarded as fixed. Let the angular velocity of F_i

relative to F_j be $\boldsymbol{\omega}_{ij}$, so that $\boldsymbol{\omega}_{ij} = -\boldsymbol{\omega}_{ji}$, and $\boldsymbol{\omega}_{ij} = \mathbf{0}$ if $i = j$. The question to be investigated is whether or not the angular velocities combine in such a way that

$$\boldsymbol{\omega}_{ik} = \boldsymbol{\omega}_{ij} + \boldsymbol{\omega}_{jk}. \tag{9.3.1}$$

This relation is, in fact, satisfied and enables us to attach a meaning to the statement that a rigid body or frame of reference can have simultaneously two or more angular velocities which are additive in the usual vector sense.

To verify (9.3.1) we consider three vector functions of time $\mathbf{r}_1, \mathbf{r}_2, \mathbf{r}_3$, which are fixed relative to F_1, F_2, F_3 respectively, and such that

$$\mathbf{r}_1(t) = \mathbf{r}_2(t) = \mathbf{r}_3(t) = \mathbf{r}$$

at time t. At time $t + \delta t$, we write

$$(\delta \mathbf{r}_i)_k = \mathbf{r}_i(t + \delta t) - \mathbf{r}_k(t + \delta t)$$

to denote the increment in $\mathbf{r}_i$ relative to F_k. In particular the increment is zero if $i = k$ since $\mathbf{r}_i$ is fixed relative to F_i. It follows that

$$\begin{aligned}(\delta \mathbf{r}_i)_k &= \mathbf{r}_i(t + \delta t) - \mathbf{r}_j(t + \delta t) + \mathbf{r}_j(t + \delta t) - \mathbf{r}_k(t + \delta t) \\ &= (\delta \mathbf{r}_i)_j + (\delta \mathbf{r}_j)_k.\end{aligned}$$

If this relation is divided by δt we have, in the limit $\delta t \to 0$,

$$\left(\frac{d\mathbf{r}_i}{dt}\right)_k = \left(\frac{d\mathbf{r}_i}{dt}\right)_j + \left(\frac{d\mathbf{r}_j}{dt}\right)_k. \tag{9.3.2}$$

Now $\mathbf{r}_i$ moves as a rigid body with F_i, and F_i has an angular velocity $\boldsymbol{\omega}_{ik}$ relative to F_k, so that

$$\left(\frac{d\mathbf{r}_i}{dt}\right)_k = \boldsymbol{\omega}_{ik} \times \mathbf{r}. \tag{9.3.3}$$

When this is substituted in (9.3.2), we get

$$\boldsymbol{\omega}_{ik} \times \mathbf{r} = \boldsymbol{\omega}_{ij} \times \mathbf{r} + \boldsymbol{\omega}_{jk} \times \mathbf{r}$$

or

$$(\boldsymbol{\omega}_{ik} - \boldsymbol{\omega}_{ij} - \boldsymbol{\omega}_{jk}) \times \mathbf{r} = \mathbf{0}.$$

This must be true for any vector $\mathbf{r}$, which is only possible if

$$\boldsymbol{\omega}_{ik} - \boldsymbol{\omega}_{ij} - \boldsymbol{\omega}_{jk} = \mathbf{0}$$

and relation (9.3.1) is verified.

9.4 ROTATING AXES

We have seen that if a frame of reference F_1 has an angular velocity $\boldsymbol{\omega}$ relative to F_2, then the derivative of a vector $\mathbf{r}$ fixed in F_1 is $\omega \times \mathbf{r}$ relative to F_2. We now relax the condition that $\mathbf{r}$ is fixed in F_1 and consider the relation between the derivatives of $\mathbf{r}$ in F_1 and F_2.

We may regard F_2 as fixed, and denote the derivative of $\mathbf{r}$ relative to F_2 as $\mathrm{d}\mathbf{r}/\mathrm{d}t$. To distinguish the derivative of $\mathbf{r}$ relative to F_1 we use the notation $\dot{\mathbf{r}}$; this will be the rate of change of $\mathbf{r}$ as measured by an observer using F_1 as his frame of reference. To be explicit, suppose that F_1 is a cartesian frame of reference with unit vectors $\mathbf{i}, \mathbf{j}, \mathbf{k}$ parallel to the axes. If

$$\mathbf{r} = x\mathbf{i} + y\mathbf{j} + z\mathbf{k}$$

we have

$$\dot{\mathbf{r}} = \frac{\mathrm{d}x}{\mathrm{d}t}\mathbf{i} + \frac{\mathrm{d}y}{\mathrm{d}t}\mathbf{j} + \frac{\mathrm{d}z}{\mathrm{d}t}\mathbf{k}, \tag{9.4.1}$$

but

$$\frac{\mathrm{d}\mathbf{r}}{\mathrm{d}t} = \frac{\mathrm{d}x}{\mathrm{d}t}\mathbf{i} + \frac{\mathrm{d}y}{\mathrm{d}t}\mathbf{j} + \frac{\mathrm{d}z}{\mathrm{d}t}\mathbf{k} + x\frac{\mathrm{d}\mathbf{i}}{\mathrm{d}t} + y\frac{\mathrm{d}\mathbf{j}}{\mathrm{d}t} + z\frac{\mathrm{d}\mathbf{k}}{\mathrm{d}t}. \tag{9.4.2}$$

Note that although the components (x, y, z) of a vector $\mathbf{r}$ depend on the axes chosen, there is no ambiguity about their derivatives provided it is understood that they are referred always to the same set of axes.

Since $\mathbf{i}, \mathbf{j}, \mathbf{k}$ are three vectors which move as a rigid body with angular velocity $\boldsymbol{\omega}$, $\mathrm{d}\mathbf{i}/\mathrm{d}t = \boldsymbol{\omega} \times \mathbf{i}$ etc., and so (9.4.2) may be written

$$\mathrm{d}\mathbf{r}/\mathrm{d}t = \dot{\mathbf{r}} + \boldsymbol{\omega} \times \mathbf{r}. \tag{9.4.3}$$

If $\mathbf{r}$ is the position vector of a particle P, then (9.4.3) gives the velocity of P relative to F_2 in terms of its velocity relative to F_1 and the angular velocity of F_1 relative to F_2. But the result is applicable to all vectors. For example

$$\mathrm{d}\boldsymbol{\omega}/\mathrm{d}t = \dot{\boldsymbol{\omega}} + \boldsymbol{\omega} \times \boldsymbol{\omega} = \dot{\boldsymbol{\omega}}$$

so that the rate of change of $\boldsymbol{\omega}$ is the same in both frames. If $\mathrm{d}\mathbf{r}/\mathrm{d}t = \mathbf{v}$, we have

$$\mathrm{d}\mathbf{v}/\mathrm{d}t = \dot{\mathbf{v}} + \boldsymbol{\omega} \times \mathbf{v}. \tag{9.4.4}$$

But, by (9.4.3),

$$\dot{\mathbf{v}} = \ddot{\mathbf{r}} + \boldsymbol{\omega} \times \dot{\mathbf{r}} + \dot{\boldsymbol{\omega}} \times \mathbf{r}, \tag{9.4.5}$$

$$\boldsymbol{\omega} \times \mathbf{v} = \boldsymbol{\omega} \times \dot{\mathbf{r}} + \boldsymbol{\omega} \times (\boldsymbol{\omega} \times \mathbf{r}), \tag{9.4.6}$$

and substitution in (9.4.4) gives

$$\mathrm{d}^2\mathbf{r}/\mathrm{d}t^2 = \ddot{\mathbf{r}} + \dot{\boldsymbol{\omega}} \times \mathbf{r} + 2\boldsymbol{\omega} \times \dot{\mathbf{r}} + \boldsymbol{\omega} \times (\boldsymbol{\omega} \times \mathbf{r}). \tag{9.4.7}$$

The first term $\ddot{\mathbf{r}}$ in (9.4.7) gives the acceleration of the particle as interpreted by an observer using the rotating frame of reference. The second term $\dot{\boldsymbol{\omega}} \times \mathbf{r}$ arises only when the angular velocity varies, but the third and fourth terms, called respectively the Coriolis acceleration and the centripetal acceleration, are present even when the angular velocity is constant.

In Chapter 3 it was stressed that Newton's laws of motion are formulated with respect to an inertial frame of reference, and such a frame has always been assumed in previous applications. This does not mean, however, that we are necessarily limited to such frames in all problems, and sometimes it is convenient to use a frame which is not inertial. In such a case one must make appropriate allowance for the acceleration in the equations of motion. If the position vector $\mathbf{r}$ of a particle of mass m is defined relative to a frame of reference whose origin O has an acceleration $\mathbf{a}_0$ relative to an inertial frame, the equation of motion for the particle is

$$\mathbf{F} = m\left(\mathbf{a}_0 + \frac{\mathrm{d}^2\mathbf{r}}{\mathrm{d}t^2}\right),$$

where $\mathbf{F}$ is the force. If the frame of reference also rotates with angular velocity $\boldsymbol{\omega}$ relative to the inertial frame then, by (9.4.7), the equation of motion will be

$$\mathbf{F} = m\{\mathbf{a}_0 + \ddot{\mathbf{r}} + \dot{\boldsymbol{\omega}} \times \mathbf{r} + 2\boldsymbol{\omega} \times \dot{\mathbf{r}} + \boldsymbol{\omega} \times (\boldsymbol{\omega} \times \mathbf{r})\}. \tag{9.4.8}$$

This equation will be interpreted in different ways by different observers. An observer using the inertial frame would interpret the force on the particle as $\mathbf{F}$, but an observer using the rotating frame, and interpreting the acceleration of the particle as $\ddot{\mathbf{r}}$, would regard the particle as subject to the apparent force

$$\mathbf{F} - m\{\mathbf{a}_0 + \dot{\boldsymbol{\omega}} \times \mathbf{r} + 2\boldsymbol{\omega} \times \dot{\mathbf{r}} + \boldsymbol{\omega} \times (\boldsymbol{\omega} \times \mathbf{r})\}.$$

The difference would arise from the apparent extra force

$$-m\{\mathbf{a}_0 + \dot{\boldsymbol{\omega}} \times \mathbf{r} + 2\boldsymbol{\omega} \times \dot{\mathbf{r}} + \boldsymbol{\omega} \times (\boldsymbol{\omega} \times \mathbf{r})\}$$

that the second observer would attribute to the particle. Whether these terms are regarded as modifying the acceleration or the force is a question of point of view. For axes fixed on the earth's surface, gravity is regarded as a force of attraction between the earth and a neighbouring mass. The earth's rotation also produces a centripetal acceleration, which is often regarded as a small modification to the earth's gravitational force. It is then called a centrifugal force.

Example 9.4.1
A smooth wire is bent into the form of a circle, of radius a, and rotates about a vertical diameter with a prescribed angular velocity $\boldsymbol{\omega}$ (see Fig. 9.4.1). A smooth bead slides freely on the wire.

Let $\mathbf{r}$ be the position vector of the bead relative to the centre of the circular wire, and let $\dot{\mathbf{r}}$ be the rate of change of $\mathbf{r}$ in the plane of the wire. The equation of motion is then

$$\ddot{\mathbf{r}} + \dot{\boldsymbol{\omega}} \times \mathbf{r} + 2\boldsymbol{\omega} \times \dot{\mathbf{r}} + \boldsymbol{\omega} \times (\boldsymbol{\omega} \times \mathbf{r}) = m^{-1}\mathbf{F} + \mathbf{g}, \tag{9.4.9}$$

where $\mathbf{F}$ is the reaction of the wire, which will be perpendicular to the wire. Of the terms on the left-hand side of the equation, $\dot{\boldsymbol{\omega}} \times \mathbf{r}$ and $2\boldsymbol{\omega} \times \dot{\mathbf{r}}$ are normal to the plane of the wire, and serve to determine the appropriate component of $\mathbf{F}$ when $\mathbf{r}$ and $\dot{\mathbf{r}}$ have been found. The component of the equation of motion perpendicular to $\mathbf{r}$ in the plane of the wire gives

$$a\ddot{\theta} - a\omega^2 \sin\theta \cos\theta = -g \sin\theta. \tag{9.4.10}$$

If we now assume that $\boldsymbol{\omega}$ is constant in magnitude as well as direction, equation (9.4.10) can be integrated further to give

$$\tfrac{1}{2}\dot{\theta}^2 - \tfrac{1}{2}\omega^2 \sin^2\theta - ga^{-1}\cos\theta = \text{constant}. \tag{9.4.11}$$

Equation (9.4.11) can be dealt with using the general theory for an energy equation with the pseudo-potential

$$U(\theta) = -\tfrac{1}{2}\omega^2 \sin^2\theta - ga^{-1}\cos\theta.$$

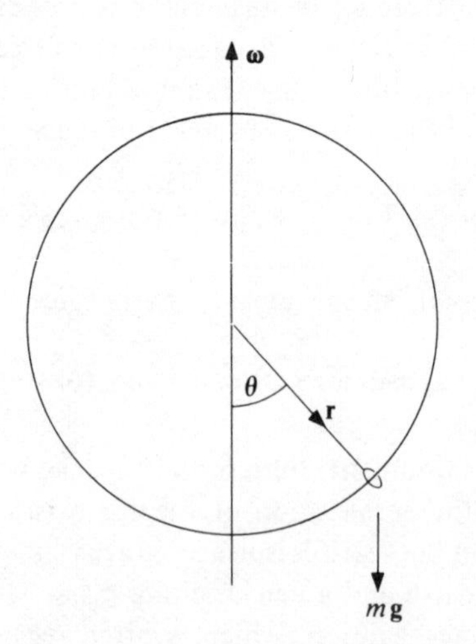

Figure 9.4.1 Illustration for Example 9.4.1.

In fact $-\frac{1}{2}ma^2\omega^2 \sin^2\theta$ can be regarded as the potential energy of a centrifugal force, which is conservative when $\boldsymbol{\omega}$ is constant.

The general theory of §6.5 shows that there are equilibrium positions when

$$U'(\theta) = -\omega^2 \sin\theta \cos\theta + ga^{-1} \sin\theta = 0,$$

that is when

$$\theta = 0,\ \pi,\ \cos^{-1}(g/a\omega^2),$$

provided that $g < a\omega^2$.

Equilibrium is unstable when

$$U''(\theta) = -\omega^2 \cos 2\theta + ga^{-1} \cos\theta < 0.$$

Thus $\theta = \pi$ is always unstable, and $\theta = 0$ is unstable if $g < a\omega^2$. A small perturbation about $\cos\theta = g/a\omega^2$ is then oscillatory with frequency

$$\omega(1 - g^2/a^2\omega^4)^{1/2}.$$

If $g > a\omega^2$, there is equilibrium only when $\theta = 0$ or π, and $\theta = 0$ is stable. The frequency of a small perturbation about $\theta = 0$ is

$$\omega(g/a\omega^2 - 1)^{1/2}.$$

If $g = a\omega^2$ it is necessary to use a more accurate approximation. Near $\theta = 0$ the first approximation to equation (9.4.11) in this case is

$$\tfrac{1}{2}\dot{\theta}^2 = (\alpha^4 - \theta^4)\omega^2/8, \tag{9.4.12}$$

where α is the value of θ when $\dot{\theta} = 0$. This equation also has oscillatory solutions, although they are not harmonic. The periodic time is

$$T = \frac{4}{\omega}\int_{-\alpha}^{\alpha} \frac{d\theta}{(\alpha^4 - \theta^4)^{1/2}} = \frac{2^{5/2}}{\alpha\omega}\int_0^{\pi/2} \frac{d\phi}{(1 - \frac{1}{2}\sin^2\phi)^{1/2}}$$

$$= \frac{2^{5/2}}{\alpha\omega} K(2^{-1/2}) = \frac{10{\cdot}49}{\alpha\omega}$$

where K is the complete elliptic integral with argument $2^{-1/2}$ (see Example 6.5.2).

Example 9.4.2 Larmor Precession

A transformation to a rotating frame of reference can sometimes simplify the equation of motion, as the following example illustrates. Suppose that an electron, which has a charge e and mass m, moves in an electric field $\mathbf{E}$ and a magnetic field **B**. Its equation of motion is

$$m\frac{d^2\mathbf{r}}{dt^2} = -e\mathbf{E} - e\frac{d\mathbf{r}}{dt} \times \mathbf{B}. \tag{9.4.13}$$

If the acceleration is expressed relative to a frame of reference rotating with angular velocity $\boldsymbol{\omega}$, the equation becomes

$$m\{\ddot{\mathbf{r}} + \dot{\boldsymbol{\omega}} \times \mathbf{r} + 2\boldsymbol{\omega} \times \dot{\mathbf{r}} + \boldsymbol{\omega} \times (\boldsymbol{\omega} \times \mathbf{r})\} = -e\mathbf{E} - e(\dot{\mathbf{r}} + \boldsymbol{\omega} \times \mathbf{r}) \times \mathbf{B}. \tag{9.4.14}$$

The terms in (9.4.14) which involve $\dot{\mathbf{r}}$ can be eliminated by the choice $\boldsymbol{\omega} = e\mathbf{B}/2m$ and then

$$\ddot{\mathbf{r}} = -\frac{e\mathbf{E}}{m} - \dot{\boldsymbol{\omega}} \times \mathbf{r} + \boldsymbol{\omega} \times (\boldsymbol{\omega} \times \mathbf{r}).$$

If the magnetic field is substantially constant and sufficiently weak for the second and third terms on the right-hand side to be neglected compared with the first, the equation approximates to

$$\ddot{\mathbf{r}} = -e\mathbf{E}/m.$$

It follows that the effect of a weak magnetic field is to cause the trajectory obtained by its neglect to precess slowly, with angular velocity $\boldsymbol{\omega}$. This is known as Larmor precession. If, for example, the electron is in orbit about a nucleus of charge e', then $\mathbf{E} = e'\mathbf{r}/r^3$ and so

$$\ddot{\mathbf{r}} = -ee'\mathbf{r}/mr^3.$$

This is an inverse square law of force and the orbit of the electron is an ellipse. The addition of a weak magnetic field causes this ellipse to precess slowly with angular velocity $\boldsymbol{\omega}$ about an axis through the nucleus.

The conditions required for the validity of the approximation follow from the equation of motion. They are

$$\omega^2 \ll ee'/mr^3, \quad |\dot{\boldsymbol{\omega}}| \ll ee'/mr^3,$$

where r is a typical length defining the orbit. For an ellipse this can be taken as the semi-major axis a, and then the conditions are

$$\omega \ll 2\pi/T, \quad |\dot{\boldsymbol{\omega}}| \ll 4\pi^2/T^2,$$

where
$$T = 2\pi a^{3/2}(ee'/m)^{-1/2}.$$

Reference to (7.5.15) shows that T is the periodic time of the electron in orbit, so that $2\pi/T$ is the mean angular speed.

9.5 AXES ON THE EARTH'S SURFACE

For many practical problems it is convenient to use a set of axes fixed on the earth's surface. The acceleration of a particle, as it appears to an observer using such a frame of reference, will differ slightly from the acceleration relative to an inertial frame because of the motion of the earth. The most significant effect arises from the rotation of the earth about its axis, which amounts to 2π radians per sidereal day. This is equivalent to an angular velocity of magnitude

$$\frac{2\pi}{86\,164} = 7{\cdot}29 \times 10^{-5} \text{ rad s}^{-1}. \tag{9.5.1}$$

To define the frame of reference, we choose an origin O fixed on, or near, the earth's surface at a distance BO from the earth's axis of rotation (see Fig. 9.5.1). Let $\overrightarrow{BO} = \mathbf{b}$ and let $\mathbf{r}$ be the position vector of a particle P relative to O. Let $\mathbf{i}, \mathbf{j}, \mathbf{k}$ be a mutually orthogonal right-handed set of unit vectors such that $\mathbf{i}$ is horizontal and points south, $\mathbf{j}$ is horizontal and points east, and $\mathbf{k}$ points vertically upwards. Here horizontal means perpendicular to the vertical, and vertical means the direction defined by a plumb line, which is a string fixed relative to the

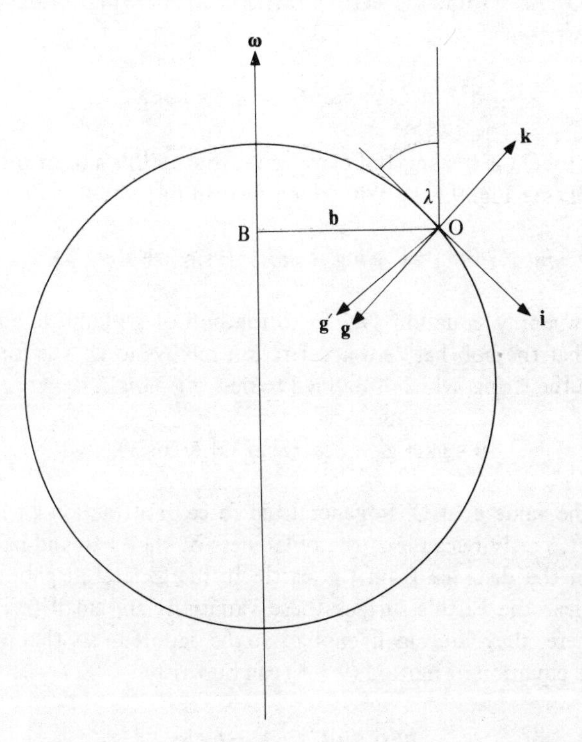

Figure 9.5.1 Illustration of the notation associated with axes on the earth's surface.

earth's surface and supporting a particle suspended from one end. Because of the rotation of the earth, the direction of the vertical differs slightly from the direction defined solely by the earth's gravitational attraction. We therefore denote the latter by $m\mathbf{g}'$, where m is the mass of the plumb bob, and distinguish it from $m\mathbf{g} = -mg\mathbf{k}$, which is equal and opposite to the force measured by the tension in the string of the plumb line.

If we assume that the point B has no acceleration relative to an inertial frame of reference, the acceleration of the particle P is

$$\frac{\mathrm{d}^2\mathbf{b}}{\mathrm{d}t^2} + \frac{\mathrm{d}^2\mathbf{r}}{\mathrm{d}t^2} \tag{9.5.2}$$

and the equation of motion is

$$m\left(\frac{\mathrm{d}^2\mathbf{b}}{\mathrm{d}t^2} + \frac{\mathrm{d}^2\mathbf{r}}{\mathrm{d}t^2}\right) = \mathbf{F} + m\mathbf{g}', \tag{9.5.3}$$

where m now denotes the mass of the particle. The force $\mathbf{F}$ is the resultant of all the forces except the earth's gravitational attraction $m\mathbf{g}'$. The acceleration $\mathrm{d}^2\mathbf{b}/\mathrm{d}t^2$ of O arises from the earth's rotation with angular velocity $\boldsymbol{\omega}$. It may therefore be written

$$\mathrm{d}^2\mathbf{b}/\mathrm{d}t^2 = -b\omega^2(\mathbf{i}\sin\lambda + \mathbf{k}\cos\lambda), \tag{9.5.4}$$

where the latitude λ is the angle of elevation of the earth's axis of rotation above the horizontal (see Fig. 9.5.1). With this substitution, equation (9.5.3) becomes

$$m(\mathrm{d}^2\mathbf{r}/\mathrm{d}t^2) = \mathbf{F} + m\mathbf{g}' + mb\omega^2(\mathbf{i}\sin\lambda + \mathbf{k}\cos\lambda). \tag{9.5.5}$$

Let us now apply equation (9.5.5) to the bob of a plumb line fixed at the point O, so that the bob has zero acceleration relative to O. The force $\mathbf{F}$ will be the tension in the string, which is defined to be $-m\mathbf{g} = mg\mathbf{k}$. Hence

$$\mathbf{0} = g\mathbf{k} + \mathbf{g}_0' + b\omega^2(\mathbf{i}\sin\lambda + \mathbf{k}\cos\lambda), \tag{9.5.6}$$

where $\mathbf{g}_0'$ is the value $\mathbf{g}'$ at O. In general the force of attraction $\mathbf{g}$ will vary from point to point, partly because of the oblateness of the earth and partly because it depends on the distance from the earth. In the general neighbourhood of a fixed point near the earth's surface these variations are small (see §2.4). For problems where they are small enough to be ignored, so that $|\mathbf{g}' - \mathbf{g}_0'|$ is negligible, the equation of motion (9.5.5) can be written

$$m(\mathrm{d}^2\mathbf{r}/\mathrm{d}t^2) = \mathbf{F} - mg\mathbf{k} \tag{9.5.7}$$

for the general motion of a particle near the earth's surface. Since it is $mg\mathbf{k}$

which is readily measurable and which has in general the more immediate practical significance, equation (9.5.7) is the form usually to be preferred.

To get some idea of the difference between $\mathbf{g}'$ and $\mathbf{g}$, we can assume the earth to be a sphere of radius r_e and that, approximately, $b = r_e \cos\lambda$. Then, from (9.5.6),

$$-\mathbf{g}_0' = g\mathbf{k} + r_e\omega^2(\mathbf{i}\sin\lambda\cos\lambda + \mathbf{k}\cos^2\lambda).$$

The maximum difference in magnitude occurs at the equator ($\lambda = 0$), where

$$\frac{g_0' - g}{g} = \frac{r_e\omega^2}{g}$$

approximately. The maximum deviation of $\mathbf{g}_0'$ from the vertical occurs when $\lambda = \pi/4$ and is $r_e\omega^2/2g$ rad approximately. With

$$r_e = 6{\cdot}37 \times 10^6 \text{ m}, \quad \omega = 7{\cdot}29 \times 10^{-5} \text{ rad s}^{-1}, \quad g = 9{\cdot}81 \text{ m s}^{-2},$$

we get

$$r_e\omega^2/g = 3{\cdot}45 \times 10^{-3}, \tag{9.5.8}$$

so that the modification is quite small.

We proceed on the basis of equation (9.5.7). Since the axes defined at O by $\mathbf{i}$, $\mathbf{j}$, $\mathbf{k}$ are rotating with angular velocity $\boldsymbol{\omega}$, we have that

$$d^2\mathbf{r}/dt^2 = \ddot{\mathbf{r}} + 2\boldsymbol{\omega}\times\dot{\mathbf{r}} + \boldsymbol{\omega}\times(\boldsymbol{\omega}\times\mathbf{r}),$$

and the equation of motion now reads

$$m\ddot{\mathbf{r}} = \mathbf{F} - mg\mathbf{k} - 2m\boldsymbol{\omega}\times\dot{\mathbf{r}} - m\boldsymbol{\omega}\times(\boldsymbol{\omega}\times\mathbf{r}). \tag{9.5.9}$$

Of these terms it is customary to neglect the centrifugal force $-m\boldsymbol{\omega}\times(\boldsymbol{\omega}\times\mathbf{r})$. The ratio of its magnitude to mg is of the order $r\omega^2/g$. Reference to equation (9.5.8) giving the value of $r_e\omega^2/g$ shows that $r\omega^2/g$ will be negligible in all problems where $r \ll r_e$. The approximate equation of motion is then

$$m\ddot{\mathbf{r}} = \mathbf{F} - mg\mathbf{k} - 2m\boldsymbol{\omega}\times\dot{\mathbf{r}}. \tag{9.5.10}$$

On the other hand the Coriolis force, $-2m\boldsymbol{\omega}\times\dot{\mathbf{r}}$, gives rise to substantial meteorological effects. When a pressure gradient is set up in the atmosphere the air will initially tend to move along the gradient. But the Coriolis force, which is always perpendicular to the velocity, tends to give the motion of the air a component perpendicular to the pressure gradient. This in turn gives the Coriolis force a component opposing the pressure gradient, and equilibrium is approached as the two forces tend to cancel each other. The air is then moving at right angles to the pressure gradient rather than along it. This is called a geostrophic wind. When a localised centre of low pressure is set up, the air will tend ultimately to circle round it, giving rise to a cyclone. Consideration of the earth's angular velocity shows that the air will circulate anti-clockwise in the northern hemisphere, clockwise in the southern hemisphere, as viewed from above. At the equator cyclones do not occur.

On a larger scale, the heating of the air at the equator causes it to rise, and to be replaced by cooler air flowing from the poles. Thus a circulation is set up in which the air flows from the poles to the equator, rises and returns in the upper atmosphere. Because of the Coriolis force, the air does not flow due south or north; it has a component of velocity towards the west, which causes it to blow from the northeast in the northern hemisphere giving rise to the northeast trade winds. Their counterpart in the southern hemisphere are the southeast trade winds. The effect is more readily appreciated from the point of view of an observer using an inertial frame of reference. Underneath the air flowing from the poles to the equator is the earth rotating from west to east, and because of this rotation the air has a component of velocity from east to west relative to the earth.

Example 9.5.1

We consider the free trajectory of a particle under gravity, whose equation of motion will be, from (9.5.9) with $\mathbf{F} = \mathbf{0}$,

$$\ddot{\mathbf{r}} = -g\mathbf{k} - 2\boldsymbol{\omega} \times \dot{\mathbf{r}} - \boldsymbol{\omega} \times (\boldsymbol{\omega} \times \mathbf{r}). \tag{9.5.11}$$

Now the effect of rotation is small, and the terms on the right-hand side of (9.5.11) are in decreasing order of significance. The crudest approximation is therefore obtained by neglecting rotation completely and gives

$$\ddot{\mathbf{r}} = -g\mathbf{k}, \quad \dot{\mathbf{r}} = \mathbf{u} - gt\mathbf{k}, \tag{9.5.12}$$

where $\mathbf{u}$ is the initial velocity. An improved approximation is possible by the substitution of (9.5.12) in the next most significant term of (9.5.11), namely $-2\boldsymbol{\omega} \times \dot{\mathbf{r}}$. This gives

$$\ddot{\mathbf{r}} = -g\mathbf{k} - 2\boldsymbol{\omega} \times (\mathbf{u} - gt\mathbf{k}). \tag{9.5.13}$$

The integral of (9.5.13) will now include a first approximation to the correction for the earth's rotation and we have

$$\dot{\mathbf{r}} = \mathbf{u} - gt\mathbf{k} - 2\boldsymbol{\omega} \times (\mathbf{u}t - \tfrac{1}{2}gt^2\mathbf{k}), \tag{9.5.14}$$

$$\mathbf{r} = \mathbf{u}t - \tfrac{1}{2}gt^2\mathbf{k} - \boldsymbol{\omega} \times (\mathbf{u}t^2 - \tfrac{1}{3}gt^3\mathbf{k}). \tag{9.5.15}$$

The term $-\boldsymbol{\omega} \times (\mathbf{u}t^2 - \tfrac{1}{3}gt^3\mathbf{k})$ represents the small correction to the position vector. For a particle which falls from rest relative to the surface of the earth,

$\mathbf{u} = \mathbf{0}$ and equation (9.5.15) gives

$$x = 0, \quad y = \tfrac{1}{3}\omega g t^3 \cos\lambda, \quad z = -\tfrac{1}{2}gt^2.$$

The value of y represents the deviation from the vertical towards the east. If the particle falls from rest at a height h above the ground then $t = (2h/g)^{1/2}$ and

$$y = \tfrac{1}{3}\omega(8h^3/g)^{1/2} \cos\lambda,$$

when $z = -h$. For $h = 100$ m this gives $y = 0{\cdot}022 \cos\lambda$ m. It is very small, and difficult to detect because of extraneous disturbances in the atmosphere.

It is instructive to consider such problems using an inertial set of axes. For simplicity we consider the case of a particle at the equator dropped from relative rest at a height h above the surface of the earth. Let r_e be the radius of the earth and let $\boldsymbol{\omega}$ be its angular velocity. We use inertial axes and specify the position of the particle by polar coordinates $(r_e + r, \theta)$ based on an origin at the centre of the earth. Initially $r = h, \dot{r} = 0, \theta = 0, \dot{\theta} = \omega$ and the equations of motion are

$$(r_e + r)^2 \dot{\theta} = \text{constant} = (r_e + h)^2 \omega, \quad \ddot{r} - (r_e + r)\dot{\theta}^2 = -g,$$

from which

$$\dot{\theta} = \left(1 + \frac{2}{r_e}(h - r)\right)\omega + \text{smaller terms}, \quad \ddot{r} = -g + \text{smaller terms}.$$

A first approximation to r is $h - \tfrac{1}{2}gt^2$ and this is sufficient to calculate the correction term in θ. We now have

$$\dot{\theta} = \left(1 + \frac{gt^2}{r_e}\right)\omega \quad \text{or} \quad \theta = \left(t + \frac{gt^3}{3r_e}\right)\omega$$

approximately. In this time the earth has rotated through an angle ωt. The relative angular displacement is therefore $\omega g t^3/3r_e$. This gives a relative linear displacement on the surface of the earth of magnitude

$$r_e(\theta - \omega t) = \tfrac{1}{3}\omega g t^3 = \tfrac{1}{3}\omega(8h^3/g)^{1/2}$$

as before.

For a projectile given a large initial speed in a nearly horizontal trajectory, as in the flight of a rifle bullet, the term $\mathbf{u}t^2$ will dominate the term $gt^3\mathbf{k}/3$, at least initially, and equation (9.5.15) approximates to

$$\mathbf{r} = \mathbf{u}t - \tfrac{1}{2}gt^2\mathbf{k} + \omega t^2 \cos\lambda\, \mathbf{i} \times \mathbf{u} - \omega t^2 \sin\lambda\, \mathbf{k} \times \mathbf{u}. \tag{9.5.16}$$

The term $\omega t^2 \cos\lambda\, \mathbf{i} \times \mathbf{u}$ represents a correction to the vertical displacement, since $\mathbf{u}$ is almost horizontal. But $\mathbf{k} \times \mathbf{u}$ represents a deflection to the left of the line of flight. It follows that the deflection is to the left if $-\sin\lambda$ is positive, that is in the southern hemisphere. In the northern hemisphere there is a deflection to the right.

Example 9.5.2 Foucault's Pendulum

For a spherical pendulum we have seen in Example 7.4.1, equation (7.4.9), that the equation of motion takes the form

$$m\frac{d^2\mathbf{l}}{dt^2} = m\mathbf{g} - \frac{T\mathbf{l}}{l}, \tag{9.5.17}$$

where $\mathbf{l}$ is the position vector of the bob relative to the point of suspension and T is the tension (see Fig. 7.4.1). We now see, however, that this equation is in a form suitable for use with an inertial set of axes. Relative to axes fixed on the earth's surface, equation (9.5.17) becomes

$$\ddot{\mathbf{l}} + 2\boldsymbol{\omega} \times \dot{\mathbf{l}} = -g\mathbf{k} - T\mathbf{l}/ml, \tag{9.5.18}$$

where the small centripetal acceleration has been neglected. The vertical component of (9.5.18) gives

$$-T\mathbf{l}\,.\,\mathbf{k}/ml = \mathbf{k}\,.\,(g\mathbf{k} + \ddot{\mathbf{l}} + 2\boldsymbol{\omega} \times \dot{\mathbf{l}})$$

$$= g - \ddot{z} + 2\boldsymbol{\omega}\,.\,(\dot{\mathbf{r}} \times \mathbf{k}), \tag{9.5.19}$$

where $-z\mathbf{k}$ is the vertical component of $\mathbf{l}$ and $\mathbf{r}$ is the horizontal component. For small perturbations about the equilibrium position, the most significant term on the right-hand side of (9.5.19) is g, the other terms being small in comparison. Also $\mathbf{l}\,.\,\mathbf{k} = -l$ approximately and so, if only the most significant terms are retained, we have, as before,

$$T/m = g.$$

The horizontal component of (9.5.18) then reads

$$\ddot{\mathbf{r}} + 2\omega\sin\lambda\,\mathbf{k} \times \dot{\mathbf{r}} = -T\mathbf{r}/ml = -g\mathbf{r}/l \tag{9.5.20}$$

since, on a linear theory, only the most significant approximation for T/m is required. For the linear theory, then, it is sufficient to regard the inertial axes as rotating relative to the earth with the effective angular velocity $-\mathbf{k}\omega\sin\lambda$. Since the bob performs elliptic harmonic motion relative to inertial axes, it follows that, relative to the earth, the ellipse rotates with an angular velocity $-\omega\sin\lambda$ about the vertical.

Alternatively one can argue directly from equation (9.5.20). In polar coordinates, the components of (9.5.20) along and perpendicular to $\mathbf{r}$ are

$$\ddot{r} - r\dot{\theta}^2 - 2\omega r\dot{\theta}\sin\lambda = -gr/l,$$

$$r\ddot{\theta} + 2\dot{r}\dot{\theta} + 2\omega\dot{r}\sin\lambda = 0. \tag{9.5.21}$$

The substitution

$$\dot{\theta} = \dot{\phi} - \omega\sin\lambda \tag{9.5.22}$$

gives, when, as usual, terms involving ω^2 are ignored,

$$\ddot{r} - r\dot{\phi}^2 = -gr/l, \quad r\ddot{\phi} + 2\dot{r}\dot{\phi} = 0. \tag{9.5.23}$$

These are just the equations which describe elliptic harmonic motion using the polar coordinates (r, ϕ). Equation (9.5.22) implies that the resulting elliptic trajectory will, in the (r, θ) plane, rotate about the vertical with angular speed $-\omega\sin\lambda$. A more direct interpretation of the result is to regard the pendulum as oscillating in space with the earth rotating underneath it with the effective angular velocity $\mathbf{k}\omega\sin\lambda$, which gives the pendulum an equal and opposite relative angular velocity.

Foucault first suggested that the effect could be used to demonstrate the rotation of the earth. Care is required in setting up such a demonstration, because ω is so small that the effect can be concealed by other small higher order corrections.

EXERCISES

9.1 A rigid body is free to rotate about a fixed point O, which may be taken as the origin of a set of cartesian axes.

Angular displacements, each of magnitude $\pi/2$, are performed successively, first about Ox and then about Oy. Show that the result is equivalent to a single angular displacement $2\pi/3$ about an axis in the direction $\mathbf{i}+\mathbf{j}-\mathbf{k}$.

9.2 A rigid body is given an angular displacement θ about an axis through the origin O in the direction of the unit vector $\mathbf{n}$. If the point with position vector $\mathbf{r}$ moves to the point with position vector $\mathbf{r}_1$, show that

$$\mathbf{r}_1 = \cos\theta\,\mathbf{r} + \sin\theta\,\mathbf{n}\times\mathbf{r} + (1-\cos\theta)(\mathbf{n}\,.\,\mathbf{r})\mathbf{n}.$$

9.3 The vectors $\mathbf{r}_1$, $\mathbf{r}_2$, $\mathbf{r}_3 = \mathbf{r}_1 \times \mathbf{r}_2$ are three vectors of a system which is displaced as a rigid body subject to the constraints $\mathbf{r}_i\,.\,\mathbf{r}_j$ = constant for all i, j. The displacement $\delta\mathbf{r}_i$ of $\mathbf{r}_i$ is of the form $\delta\mathbf{r}_i = (\mathbf{v}_i + \boldsymbol{\epsilon}_i)\,\delta t$, where $\mathbf{v}_i$ is independent of δt and $|\boldsymbol{\epsilon}_i| \to 0$ as $\delta t \to 0$.

(i) Show that

$$\mathbf{v}_3 = \mathbf{r}_1 \times \mathbf{v}_2 + \mathbf{v}_1 \times \mathbf{r}_2,$$

$$\boldsymbol{\epsilon}_3 = \mathbf{r}_1 \times \boldsymbol{\epsilon}_2 + \boldsymbol{\epsilon}_1 \times \mathbf{r}_2 + (\mathbf{v}_1 + \boldsymbol{\epsilon}_1) \times (\mathbf{v}_2 + \boldsymbol{\epsilon}_2)\,\delta t.$$

(ii) A vector $\boldsymbol{\omega}$ is defined by the relation

$$\boldsymbol{\omega} = r_3^{-2}\{\mathbf{r}_3 \times \mathbf{v}_3 + (\mathbf{v}_1 \cdot \mathbf{r}_2)\mathbf{r}_3\}.$$

Verify that $\boldsymbol{\omega} \times \mathbf{r}_i = \mathbf{v}_i$ for $i = 1, 2, 3$ and hence deduce that $\boldsymbol{\omega} \times \mathbf{r}_i = \mathbf{v}_i$ for all i.

This is an alternative existence proof for the angular vector $\boldsymbol{\omega}$ which uses a representation in terms of its components perpendicular and parallel to $\mathbf{r}_3$.

9.4 Particles P_1, P_2, P_3 are instantaneously at the points $(0, 1, 1)$, $(1, 1, 0)$, $(1, 0, 1)$ with velocities (u, v, w), (v, w, u), (w, u, v) respectively, referred to a fixed frame of reference. What condition is required for the particles to be moving as a rigid body? What is the angular velocity when this condition is satisfied?
(Ans.: $2u = v + w$, $\frac{1}{2}(w - v)(1, 1, 1)$.)

9.5 A spool of film is free to rotate about its axis, and film is fed off the spool at a constant rate. Show that the angular velocity of the spool is proportional to $(c - t)^{-1/2}$, where t is the time and c is some constant.

9.6 A sphere, of radius a, rolls without slipping on a horizontal plane. The velocity of the centre of the sphere is given to be $a\boldsymbol{\omega} \times \mathbf{k}$, where $\mathbf{k}$ is a unit vector vertically upwards. Show that the angular velocity of the sphere is $\boldsymbol{\omega} + s\mathbf{k}$, where s is its arbitrary spin about the vertical axis.

9.7 Two discs, each of radius a, are attached to an axle of length l. The discs are able to rotate independently about the axle. The system rolls without slipping on a horizontal plane and the centres of the discs have velocities $\mathbf{v}_1$, $\mathbf{v}_2$. Show that the angular velocities of the discs are $l^{-1}(v_1 - v_2)\mathbf{k} + a^{-1}\,\mathbf{k} \times \mathbf{v}_1$ and $l^{-1}\,(v_1 - v_2)\mathbf{k} + a^{-1}\,\mathbf{k} \times \mathbf{v}_2$, where $\mathbf{k}$ is a unit vector vertically upwards.

9.8 A rigid sphere, of radius a, is pressed between two rough parallel planes which rotate with angular velocities $\boldsymbol{\omega}_1$, $\boldsymbol{\omega}_2$ about fixed axes perpendicular to the planes. A parallel plane through the centre C of the sphere intersects these axes at O_1 and O_2, where $\overline{O_1C} = \mathbf{r}_1$, $\overline{O_2C} = \mathbf{r}_2$.
Show that the velocity of C is $\frac{1}{2}(\boldsymbol{\omega}_1 \times \mathbf{r}_1 + \boldsymbol{\omega}_2 \times \mathbf{r}_2)$.
Show that C moves in a circle, with centre O and with angular velocity $\frac{1}{2}(\boldsymbol{\omega}_1 + \boldsymbol{\omega}_2)$, where O is such that

$$\omega_1 \overline{OO_1} + \omega_2 \overline{OO_2} = \mathbf{0}.$$

Show that the angular velocity of the sphere is

$$\frac{1}{2a}(\omega_2 \mathbf{r}_2 - \omega_1 \mathbf{r}_1) + \lambda\boldsymbol{\omega}_1,$$

where λ is an arbitrary scalar.

9.9 Two concentric spherical shells, of radii a_1, a_2 $(>a_1)$, rotate with constant angular velocities $\boldsymbol{\omega}_1$, $\boldsymbol{\omega}_2$ respectively. A sphere of radius $(a_2 - a_1)$ rolls without slipping in the annulus between the first two spheres. Show that the centre of the third sphere describes a circle with angular velocity $(a_1\boldsymbol{\omega}_1 + a_2\boldsymbol{\omega}_2)/(a_1 + a_2)$.

9.10 A right circular cone, of semi-angle α, rolls without slipping on a horizontal plane with constant angular velocity $\boldsymbol{\omega}$. If the contact generator of the cone revolves around the vertical with angular velocity $\boldsymbol{\omega}_1$, show that $\omega \sin \alpha = \omega_1 \cos \alpha$.

Show that the time for a complete revolution of a generator of the cone is $(2\pi \sin \alpha)/\omega_1$.

9.11 Two points O, P of a rigid body have velocities $\mathbf{v}_o$, $\mathbf{v}_p$ and accelerations $\mathbf{a}_o$, $\mathbf{a}_p$ respectively. The body has angular velocity $\boldsymbol{\omega}$.

Show that

$$\mathbf{v}_p = \mathbf{v}_o + \boldsymbol{\omega} \times \mathbf{r},$$

$$\mathbf{a}_p = \mathbf{a}_o + \frac{d\boldsymbol{\omega}}{dt} \times \mathbf{r} + \boldsymbol{\omega} \times (\boldsymbol{\omega} \times \mathbf{r}),$$

where $\mathbf{r} = \overline{\mathrm{OP}}$.

Hence show that, for some scalar λ,

$$\mathbf{r} = \omega^{-2}(\mathbf{v}_p - \mathbf{v}_o) \times \boldsymbol{\omega} + \lambda\boldsymbol{\omega},$$

$$|\boldsymbol{\omega} \times \dot{\boldsymbol{\omega}}|^2 \mathbf{r} = \omega^2\{(\mathbf{a}_o - \mathbf{a}_p) \cdot \boldsymbol{\omega}\}\boldsymbol{\omega} + \{(\mathbf{a}_o - \mathbf{a}_p) \cdot \dot{\boldsymbol{\omega}}\}\dot{\boldsymbol{\omega}}$$
$$+ (\mathbf{a}_o - \mathbf{a}_p) \times \{\boldsymbol{\omega} \times \dot{\boldsymbol{\omega}})\}.$$

Deduce that there is no point for which $\mathbf{v}_p = \mathbf{0}$ unless $\mathbf{v}_o \cdot \boldsymbol{\omega} = 0$ and that in general there is one point for which $\mathbf{a}_p = \mathbf{0}$.

9.12 A wheel moves in a vertical plane in a straight line along the ground. It slips and rolls so that at time t the horizontal displacement of its centre is x and its angular displacement is θ.

Show that the point which instantaneously has zero velocity is at a distance $\dot{x}/\theta$ from the centre.

Show that the point which instantaneously has zero acceleration is at a distance $\ddot{x}/(\dot{\theta}^2 + \dot{\theta}^4)^{1/2}$ from the centre.

9.13 A disc, of radius a, rolls without slipping on a horizontal plane. The disc is inclined at a constant angle α to the horizontal and its centre C describes a horizontal circle, of radius c, with constant speed v.

(i) Find the angular velocity of the disc, and show that it is horizontal if

$$\cos \alpha = \frac{(c^2 + 4a^2)^{1/2} - c}{2a}.$$

(ii) Find the acceleration of the highest point of the disc, and show that it is vertical if $\cos \alpha = a/c$.

9.14 When the position of a particle is specified by spherical polar coordinates (r, θ, ϕ) at time t, the components of the acceleration in the directions of r, θ, ϕ increasing are respectively

$$\ddot{r} - r\dot{\theta}^2 - r \sin^2 \theta\, \dot{\phi}^2,$$

$$\frac{1}{r}\frac{d}{dt}(r^2\dot{\theta}) - r\sin\theta\cos\theta\,\dot{\phi}^2,$$

$$\frac{1}{r\sin\theta}\frac{d}{dt}(r^2\sin^2\theta\,\dot{\phi}).$$

Derive these results from the theory of rotating axes.

9.15 Let

$$d\mathbf{r}/dt = \boldsymbol{\omega}\times\mathbf{r}, \quad \boldsymbol{\omega}\cdot\mathbf{r} = 0,$$

where $\boldsymbol{\omega}$ is a vector constant in magnitude and direction. Show that

$$d^2\mathbf{r}/dt^2 = -\omega^2\mathbf{r}$$

and hence that the general solution is of the form

$$\mathbf{r} = \mathbf{a}\cos\omega t + \omega^{-1}\boldsymbol{\omega}\times\mathbf{a}\sin\omega t.$$

Deduce from this result that a particle whose position vector is $\mathbf{r}$ describes a circle of radius a with angular speed ω.

9.16 A beam of particles, each of mass m and charge e, is emitted from a point source. The particles are subject to a uniform magnetic field $\mathbf{B}$ and they have initial velocities whose components parallel to $\mathbf{B}$ are all of magnitude v. The equation of motion for a particle is

$$m\frac{d^2\mathbf{r}}{dt^2} = e\frac{d\mathbf{r}}{dt}\times\mathbf{B}.$$

Show that the effect of the field is to focus the beam at a point whose distance is $2\pi mv/eB$ from the source.

This result is used in electromagnetic focusing.

9.17 A bead slides on a smooth circular wire which rotates in its own horizontal plane with constant angular speed ω about a vertical axis passing through a point A of the wire. Let O be the centre of the circular wire and let θ be the angular displacement of the bead from OA. Show that

$$\dot{\theta}^2 = 4\omega^2(c^2 - \cos^2\tfrac{1}{2}\theta),$$

where c is a constant. Deduce that the position of stable equilibrium is $\theta = \pi$ and that small oscillations about this equilibrium position have frequency ω.

9.18 A smooth tube of narrow bore is bent into a circle of radius a. The circle is fixed in a vertical plane which is constrained to rotate with constant angular speed ω about a fixed vertical axis not in the plane. A particle P, of mass m, is free to move in the tube and is initially at the highest point of the tube with zero relative velocity.

Show that when the radius vector to P from the centre of the circle makes an angle θ with the upward vertical,

$$\dot{\theta}^2 = \omega^2\{(\alpha + 1)^2 - (\alpha + \cos\theta)^2\},$$

where $\alpha = g/a\omega^2$.

9.19 A particle is constrained to move in a straight line due north at latitude λ with constant speed u. Calculate the force on the particle.

(i) Show that if a train of mass m travels due north at 100 km h^{-1}, there is a small horizontal force on the eastern rail of about $2 \times 10^{-4}\ mg \sin\lambda$. ($\omega = 7{\cdot}29 \times 10^{-5}$ rad s^{-1}, $g = 9{\cdot}81$ m s^{-2}).

(ii) Show that if the point of suspension of a plumb line is constrained to move due north with constant speed u, the line will be inclined at an angle $\tan^{-1}(2u\omega \sin\lambda/g)$ west of the downward vertical.

(iii) The free surface of a liquid moving in the same way sets itself perpendicular to the direction of the plumb line, so that there is no apparent tangential force on the liquid. Show that if a tidal current runs due north in a channel of breadth b, the height on the east coast exceeds that on the west coast by an amount $2bu\omega \sin\lambda/g$.

For the Irish Channel show that this is about 1·6 m. ($b = 90$ km, $u = 1{\cdot}5$ m s^{-1}, $\sin\lambda = 0{\cdot}8$).

9.20 A particle is projected with speed u on a smooth horizontal plane at latitude λ. Show that, approximately, the particle will describe an arc of a circle of radius $u/(2\omega \sin\lambda)$, where ω is the angular speed of the earth.

9.21 Relative to axes rotating with angular velocity $\boldsymbol{\omega}$, the equation of motion of a particle subject to a constant gravitational force can be written in the form

$$\ddot{\mathbf{r}} = -g\mathbf{k} - 2\boldsymbol{\omega} \times \dot{\mathbf{r}} - \boldsymbol{\omega} \times (\boldsymbol{\omega} \times \mathbf{r}).$$

On the assumption that the effect of rotation is small, show that the third approximation to the trajectory of a particle falling from rest at $\mathbf{r} = \mathbf{0}$ is

$$\mathbf{r} = -\tfrac{1}{2}gt^2\mathbf{k} + \tfrac{1}{3}gt^3\,\boldsymbol{\omega} \times \mathbf{k} - \tfrac{1}{8}gt^4\,\boldsymbol{\omega} \times (\boldsymbol{\omega} \times \mathbf{k}).$$

In the case of the earth the final term is negligible. Indeed the refinement is not strictly justified. It is comparable in magnitude with errors already incurred in the neglect of local variations in g.

9.22 A projectile is fired due north, from a point on the earth's surface whose latitude is λ, at an angle α to the horizontal, where $\alpha > 0, \lambda < \pi/2$. Show that the point where it strikes the earth will be east of the vertical plane through the point of projection if $\tan\alpha < 3\tan\lambda$.

9.23 A particle moves freely near the surface of the earth. Derive an approximate equation of motion for the particle in the form

$$\ddot{\mathbf{r}} + 2\boldsymbol{\omega} \times \dot{\mathbf{r}} = \mathbf{g}.$$

A particle is projected vertically upwards with speed u. Show that when it returns to the ground there is a deviation to the west equal to $4\omega u^3 \cos\lambda/3g^2$.

9.24 Let O be a point fixed on the earth's surface. Let $\mathbf{a}$ be the position vector of O relative to the centre of the earth. Let the force on a particle of mass m, arising from the attraction of the earth, be $m\mathbf{g}'(\mathbf{r})$ when the position vector of the particle relative to O is $\mathbf{r}$.

Show that the equation of motion of the particle is

$$\ddot{\mathbf{r}} + 2\boldsymbol{\omega} \times \dot{\mathbf{r}} + \boldsymbol{\omega} \times (\boldsymbol{\omega} \times \mathbf{r}) + \boldsymbol{\omega} \times (\boldsymbol{\omega} \times \mathbf{a}) = \mathbf{g}'(\mathbf{r}).$$

Show that when $r \ll a$ this can be approximated to

$$\ddot{\mathbf{r}} + 2\boldsymbol{\omega} \times \dot{\mathbf{r}} + \boldsymbol{\omega} \times (\boldsymbol{\omega} \times \mathbf{r}) = -g\mathbf{k} + g\{\mathbf{r} - 3(\mathbf{k} \, . \, \mathbf{r})\mathbf{k}\}/a$$

where $-g\mathbf{k} = \mathbf{g}'(0) - \boldsymbol{\omega} \times (\boldsymbol{\omega} \times \mathbf{a})$.

Show that for a projectile the term $g\{\mathbf{r} - 3(\mathbf{k} \, . \, \mathbf{r})\mathbf{k}\}/a$ will be more significant than the term $\boldsymbol{\omega} \times (\boldsymbol{\omega} \times \mathbf{r})$.

Solve the equation

$$\ddot{\mathbf{r}} + g\mathbf{k} = -2\boldsymbol{\omega} \times \dot{\mathbf{r}} + g\{\mathbf{r} - 3(\mathbf{k} \, . \, \mathbf{r})\mathbf{k}\}/a$$

on the basis that the terms on the right-hand side are small and of the same order of magnitude.

9.25 A satellite travels round the earth in a circular orbit of radius a with angular velocity $\boldsymbol{\omega}$. A set of axes is defined with the origin at the satellite, the x-axis along the radius vector from the centre of the earth to the satellite and the y-axis in the direction of motion. Show that a particle moving freely in the vicinity of the satellite has the approximate equation of motion

$$\ddot{\mathbf{r}} + 2\boldsymbol{\omega} \times \dot{\mathbf{r}} - 3\omega^2 x\mathbf{i} = \mathbf{0}.$$

Assume that $r \ll a$ and that the mutual attraction between the particle and the satellite is negligible.

10 Moments

10.1 FIRST MOMENT

This chapter is devoted to the introduction of three different types of moment and their properties. The definitions are independent of the postulates of dynamics; rather they are extensions of the vector algebra introduced in Chapter 1. However, the discussion is a necessary preliminary to the next chapter which deals with the dynamics of rigid bodies.

For a set of scalar quantities m_i $(i = 1, 2, \ldots, n)$ associated with a discrete set of points P_i, the first moment relative to some origin O is defined as $\Sigma_i m_i \overline{OP}_i$. We have already met a special case of this notion when the mass centre of a point mass distribution was introduced in §8.1 as the point C such that

$$\left(\sum_i m_i\right)\overline{OC} = \sum_i m_i \overline{OP}_i, \tag{10.1.1}$$

or, equivalently,

$$\sum_i m_i \overline{CP}_i = \mathbf{0}. \tag{10.1.2}$$

Another special case is the geometrical centre or centroid, for which each point is given equal weighting. Then we have

$$n\overline{OC} = \sum_i \overline{OP}_i.$$

For economy of notation we shall continue to use m as an arbitrary weighting factor, and refer to C as the weighted centroid. When a mass distribution with its associated mass centre is being discussed, this will be made clear in the context. Note that the position of C is well defined by (10.1.1) provided $\Sigma_i m_i \neq 0$.

The following general results are of practical use in the calculation of the position of the weighted centroid.

Let a distribution m_i at P_i $(i = 1, 2, \ldots, n)$ have the weighted centroid C_1. Let a second distribution m_j at P_j $(j = n + 1, n + 2, \ldots, p)$ have the weighted centroid C_2. Then the two distributions taken together have a weighted centroid which coincides with that of $\Sigma_i m_i$ at C_1 and $\Sigma_j m_j$ at C_2. For we have that

$$\overline{OC}_1 = \frac{\Sigma_i m_i \overline{OP}_i}{\Sigma_i m_i}, \quad \overline{OC}_2 = \frac{\Sigma_j m_j \overline{OP}_j}{\Sigma_j m_j},$$

$$\overline{OC} = \frac{\Sigma_i m_i \overline{OP}_i + \Sigma_j m_j \overline{OP}_j}{\Sigma_i m_i + \Sigma_j m_j} = \frac{(\Sigma_i m_i)\overline{OC}_1 + (\Sigma_j m_j)\overline{OC}_2}{(\Sigma_i m_i) + (\Sigma_j m_j)}$$

from which the result follows. The result clearly generalises to a distribution involving several subdivisions. As a simple example, the distribution m_1, m_2, m_3 at the points P_1, P_2, P_3 is such that m_2 at P_2 and m_3 at P_3 have a weighted centroid C_1 on P_2P_3, where $m_2 P_2C_1 = m_3 C_1P_3$. Then m_1 at P_1 and $m_2 + m_3$ at C_1 have a weighted centroid C on P_1C_1, where $m_1 P_1C = (m_2 + m_3)CC_1$. The weighted centroid of the original system coincides with C.

An immediate corollary of the above result is that if a distribution can be divided in such a way that the weighted centroids of the separate subgroups are at a given point, or on a given line, or on a given plane, then the weighted centroid of the whole distribution coincides with the given point, or lies on the given line, or on the given plane. Thus, for example, a distribution of pairs of points, each pair having the same weighting m and equal and opposite position vectors relative to some point C, has its weighted centroid at C. Such a distribution is said to have central symmetry. Again two equally weighted points which lie on either side of a given line (or plane) and are equidistant from the line (or plane) have a weighted centroid on the line (or plane). Any distribution of such pairs of points will also have a weighted centroid on the line (or plane). Note that it is not necessary for the pairs of points to be mirror images.

Equation (10.1.1) is readily generalised to include a continuous distribution. This extension will be required when dealing with the dynamics of rigid bodies. Let each point P of a given region have associated with it a density $\rho(\mathbf{r})$, which is a function of the position vector $\mathbf{r}$ of P (see Fig. 10.1.1). If an elementary volume $\delta\tau$ containing P has a weighting δm, the density at P may be regarded as the limit

$$\rho = \lim_{\delta\tau \to 0} \frac{\delta m}{\delta\tau} = \frac{dm}{d\tau},$$

on the assumption that such a limit exists. In particular, if δm is the mass of the element $\delta\tau$, then ρ is just the mass density. It follows that the mass of the given region is

$$\int \rho(\mathbf{r})\, d\tau,$$

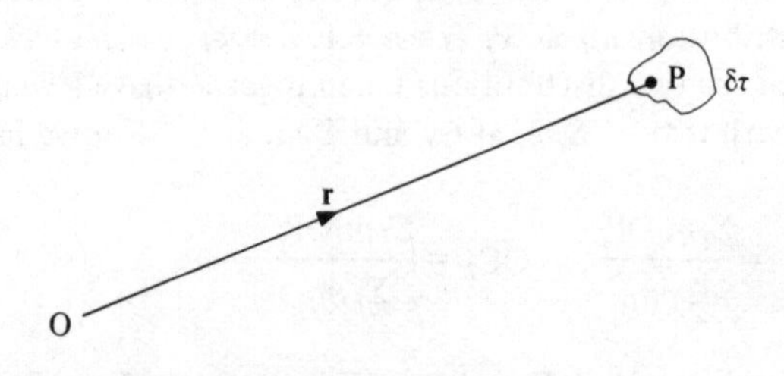

Figure 10.1.1 The position vector of a point P contained in an elementary volume $\delta\tau$ of a continuous distribution.

where the integration is taken over the volume of the region concerned. The first moment is defined as

$$\int \mathbf{r}\rho(\mathbf{r})\, \mathrm{d}\tau$$

and the weighted centroid is the point C whose position vector is given by

$$\overline{\mathrm{OC}} = \frac{\int \mathbf{r}\rho(\mathbf{r})\, \mathrm{d}\tau}{\int \rho(\mathbf{r})\, \mathrm{d}\tau}. \tag{10.1.3}$$

An important special case is the uniform distribution, for which ρ = constant. The centroid is then given by

$$\overline{\mathrm{OC}} = \frac{\int \mathbf{r}\, \mathrm{d}\tau}{\int \mathrm{d}\tau}. \tag{10.1.4}$$

With an appropriate interpretation of $\mathrm{d}\tau$, the results can be used to calculate the first moment and the position of C for a continuous distribution in one, two, or three dimensions (line, surface, or volume distributions).

Example 10.1.1
Let the line density along the x-axis be proportional to x^n. It follows, from (10.1.3), with $\mathrm{d}\tau = \mathrm{d}x$, that the weighted centroid of a line distribution OA, of length a, along the x-axis is a point C_1 such that

$$\mathrm{OC}_1 = \frac{\int_0^a x^{n+1}\, \mathrm{d}x}{\int_0^a x^n\, \mathrm{d}x} = \frac{n+1}{n+2}\, a. \tag{10.1.5}$$

The calculation is similar for a surface density distribution $\rho = \rho_0 x^n$ over the rectangular surface with adjacent sides OA, of length a, and OB, of length b (see Fig. 10.1.2). Relation (10.1.5) then gives the x coordinate of the weighted

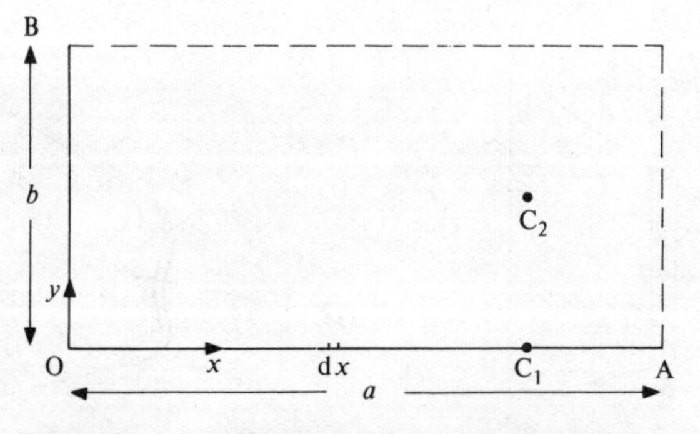

Figure 10.1.2 Illustration for Example 10.1.1.

centroid C_2. The formal calculation, with $d\tau = dx\,dy$ in (10.1.3), is

$$\overline{OC_2} = \frac{\int_0^a \int_0^b \mathbf{r}\rho_0 x^n\,dx\,dy}{\int_0^a \int_0^b \rho_0 x^n\,dx\,dy}$$

$$= \frac{(\mathbf{i}\int_0^a x^{n+1}\,dx \int_0^b dy + \mathbf{j}\int_0^a x^n\,dx \int_0^b y\,dy)}{\int_0^a x^n\,dx \int_0^b dy}$$

$$= \frac{n+1}{n+2}a\mathbf{i} + \tfrac{1}{2}b\mathbf{j}. \qquad (10.1.6)$$

Example 10.1.2
Let us consider the centroid of a uniform density distribution over the following regions (see Fig. 10.1.3):

(i) the two straight lines OA, OB;
(ii) the area of the triangle OAB;
(iii) the surface of a cone with vertex O and plane base;
(iv) the volume of a cone with vertex O and plane base.

In (iii) and (iv) the cross-section of the cone can be arbitrary.

(i) For uniform line distributions along OA, OB we consider an elementary strip, parallel to AB, which cuts off elements δa from OA and δb from OB. Since, by similarity, $\delta a/\text{OA} = \delta b/\text{OB}$ it follows that the weightings of δa and δb are in the ratio OA/OB. Hence the weighted centroid of δa and δb is a point C_1, say, which divides the strip in the same ratio. By similarity, the weighted centroid of each such elementary strip will lie on the line OC_2, where C_2 lies on AB and $BC_2/C_2A = \text{OA}/\text{OB}$. It follows that the centroid C of the original distribution lies on OC_2. For a given width, the weighting of each elementary strip is independent of its position and so the determination of C is equivalent to the determination of the centroid of a uniform line density distribution along OC_2. This is a particular case of Example 10.1.1 with $n = 0$, and (10.1.5) gives $OC = \frac{1}{2}OC_2$.

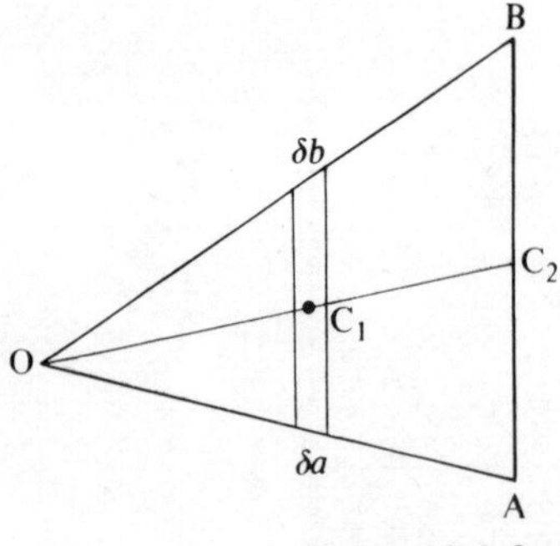

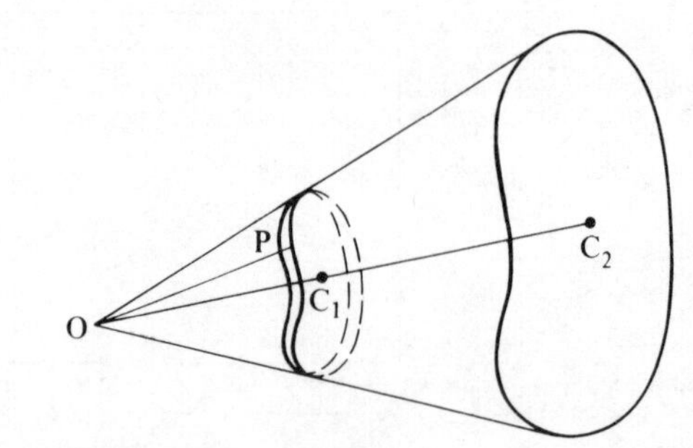

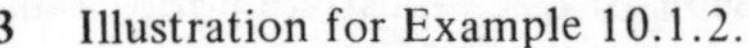
Figure 10.1.3 Illustration for Example 10.1.2.

(ii) When the strip refers to an element of area, the centroid C_1 of the strip clearly lies on the median of the triangle OAB, and C_2 is then the mid point of AB. It follows that the centroid of the triangle lies on OC_2. To find the position of C on OC_2 we note that, as the position of C_1 varies along OC_2, the weighting of an elementary strip varies in proportion to $x = OC_1$ since, by similarity, the length of the strip parallel to AB varies in proportion to x. We therefore consider a line density distribution along OC_2 proportional to x. With $n = 1$ in (10.1.5) this gives $OC = \frac{2}{3}OC_2$.

(iii) For a distribution over the surface of a cone, the elementary slice refers to a closed contour, whose thickness at a point P on the circumference is proportional to OP, the distance from the vertex O. The weighted centroid C_1 of this slice is that of a line distribution around the circumference, with a density proportional to the distance from O of the points on the circumference. By similarity, C_1 will lie on OC_2, where C_2 is the weighted centroid of the circumference of the base, weighted according to the distance from O. This part of the calculation will depend on the details of the base profile.

When C_2 is known, the position of C on OC_2 is obtained from a line distribution along OC_2 with density varying in proportion to $x = OC_1$ since, as x varies, the circumference of the ring, and hence its weighting, varies in proportion to x. This gives $OC = \frac{2}{3}OC_2$.

(iv) For a solid cone, C_2 is clearly the centroid of a uniform density distribution over the surface of the base of the cone. To calculate the position of C on OC_2, a line distribution along OC_2 with weighting proportional to the cross-sectional area of an elementary slice is required. Since the variation of this area is proportional to x^2, we have, from (10.1.5) with $n = 2$, $OC = \frac{3}{4}OC_2$.

10.2 SECOND MOMENT

The second moment about a given axis, of a set of scalar quantities m_i associated with a set of points P_i, is defined as

$$I = \sum_i m_i p_i^2, \tag{10.2.1}$$

where p_i is the perpendicular distance of P_i from the axis. It is often referred to as the moment of inertia, particularly when the m_i refer to point masses.

Let P_i have position vector $\mathbf{r}_i$ relative to an origin O on the given axis, let $\mathbf{n}$ be a unit vector along the axis and let θ be the angle between $\mathbf{r}_i$ and $\mathbf{n}$, as in Figure 10.2.1. We have

$$\begin{aligned} p_i^2 &= r_i^2 \sin^2\theta = (\mathbf{n} \times \mathbf{r}_i)^2 \\ &= r_i^2 - r_i^2\cos^2\theta = r_i^2 - (\mathbf{n} \cdot \mathbf{r}_i)^2, \end{aligned}$$

and so

$$I = \sum_i m_i p_i^2 = \sum_i m_i(\mathbf{n} \times \mathbf{r}_i)^2 = \sum_i m_i\{r_i^2 - (\mathbf{n} \cdot \mathbf{r}_i)^2\}. \tag{10.2.2}$$

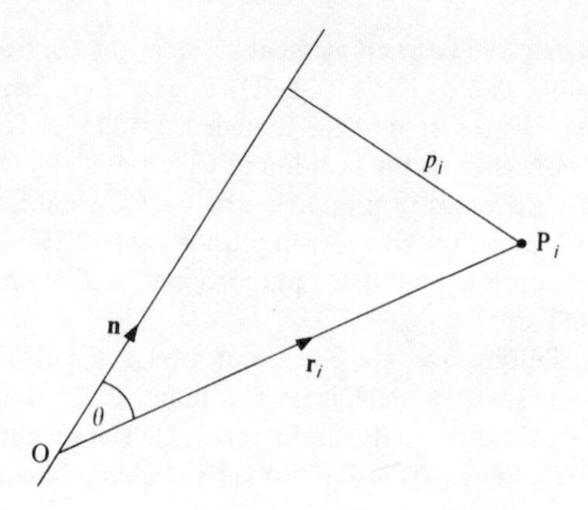

Figure 10.2.1 The notation used in the discussion of second moments.

The analysis may be carried still further with the choice of a set of cartesian axes based on O such that $\mathbf{r}_i$ has components (x_i, y_i, z_i) and $\mathbf{n}$ has components (n_x, n_y, n_z), with $n_x^2 + n_y^2 + n_z^2 = 1$. Then we have

$$I = \sum_i m_i \{(n_x^2 + n_y^2 + n_z^2)(x_i^2 + y_i^2 + z_i^2) - (n_x x_i + n_y y_i + n_z z_i)^2\}$$

$$= I_{xx}n_x^2 + I_{yy}n_y^2 + I_{zz}n_z^2 + 2I_{yz}n_y n_z + 2I_{zx}n_z n_x + 2I_{xy}n_x n_y \qquad (10.2.3)$$

where, with some judicious rearrangement, it may be shown that

$$\left.\begin{aligned} I_{xx} &= \sum_i m_i(y_i^2 + z_i^2), \quad I_{yy} = \sum_i m_i(z_i^2 + x_i^2), \quad I_{zz} = \sum_i m_i(x_i^2 + y_i^2), \\ I_{yz} &= -\sum_i m_i y_i z_i, \quad I_{zx} = -\sum_i m_i z_i x_i, \quad I_{xy} = -\sum_i m_i x_i y_i. \end{aligned}\right\} \qquad (10.2.4)$$

The coefficients I_{xx}, I_{yy}, I_{zz} are just the moments of inertia about the respective coordinate axes Ox, Oy, Oz. The coefficients I_{yz}, I_{zx}, I_{xy} are called products of inertia. It follows that, with a knowledge of these six coefficients, the moment of inertia about any line through O, with direction cosines (n_x, n_y, n_z), may be calculated using (10.2.3).

Although we shall not prove it here, it is in fact true that there always exists a set of axes based on O for which $I_{yz} = I_{zx} = I_{xy} = 0$. They are called principal axes and the associated moments I_{xx}, I_{yy}, I_{zz} are called principal moments of inertia. Relation (10.2.3) for the moment of inertia about a line through O with direction cosines (n_x, n_y, n_z) simplifies to

$$I = I_{xx}n_x^2 + I_{yy}n_y^2 + I_{zz}n_z^2 \qquad (10.2.5)$$

when principal axes are used. The simplicity resulting from their use makes such axes a desirable choice in practical applications.

It is sometimes possible to recognise the principal axes. For example suppose that the distribution has symmetry about two perpendicular planes, so that to each point there corresponds an equally weighted image point relative to either plane. The principal axes at any point on the line of intersection of the planes are the line of intersection itself and the lines in the planes that are perpendicular to the line of intersection.

If two principal moments of inertia I_{xx} and I_{yy} are equal, the body is said to have kinetic symmetry about the third axis. It may be clear from geometrical symmetry when this is the case. For example, at a point on the axis of a body of revolution, or of a pyramid with square base, the moments of inertia are the same about all axes perpendicular to the axis of symmetry.

It follows from (10.2.5) that, if $I_{xx} = I_{yy}$, then $I = I_{xx}$ for all lines through O in the (x, y) plane, for then $n_z = 0$ and $n_x^2 + n_y^2 = 1$.

If all three principal moments of inertia are equal, then

$$I = I_{xx}(n_x^2 + n_y^2 + n_z^2) = I_{xx},$$

the moment of inertia is the same for all axes through O without restriction and the products of inertia will vanish for any set of axes. This is referred to as total kinetic symmetry. It is well to remember that for this to be so it is not sufficient that $I_{xx} = I_{yy} = I_{zz}$. It is also necessary that the axes are principal ones so that the products of inertia vanish.

It is often convenient to introduce a length k, called the radius of gyration, defined by

$$I = \sum_i m_i p_i^2 = \left(\sum_i m_i\right) k^2. \tag{10.2.6}$$

It is an immediate consequence of the definition that several distributions, having second moments $I_1 = (\Sigma m)_1 k_1^2$, $I_2 = (\Sigma m)_2 k_2^2$ ... about a given axis, have together a second moment given by

$$I = I_1 + I_2 + \ldots = (\Sigma m)_1 k_1^2 + (\Sigma m)_2 k_2^2 + \ldots$$

when considered as a single distribution. A useful special case is that in which each subdivision has the same radius of gyration k. Then

$$I = \{(\Sigma m)_1 + (\Sigma m)_2 + \ldots\} k^2 \tag{10.2.7}$$

and so k is also the radius of gyration of the complete distribution.

Theorem of Parallel Axes

Let k be the radius of gyration of a system about a given axis. Let k_c be the

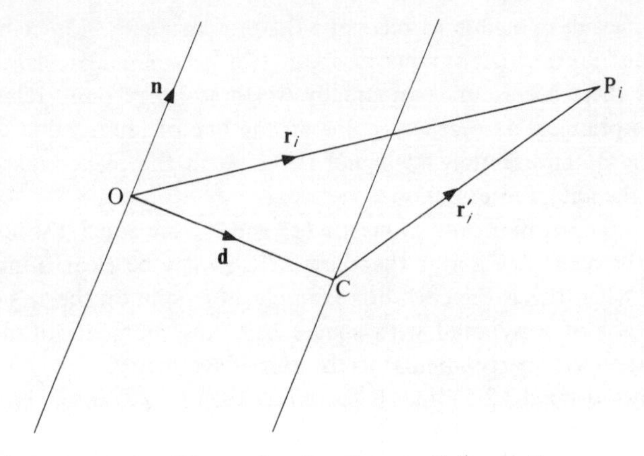

Figure 10.2.2 Illustration for the parallel axes theorem.

radius of gyration about a parallel axis through the weighted centroid C of the system. Then

$$k^2 = d^2 + k_c^2, \tag{10.2.8}$$

where d is the perpendicular distance between the axes.

In order to prove this simple and useful result, it is convenient to choose an origin O on the given axis such that $\overline{OC}$ is along the common perpendicular to the two axes and hence $\overline{OC} = \mathbf{d}$. Let $\mathbf{n}$ be a unit vector parallel to the axis, let $\mathbf{r}_i$ be the position vector relative to O of a typical point P_i with weighting m_i and let $\mathbf{r}'_i$ be the position vector of P_i relative to C (see Fig. 10.2.2). The radius of gyration about the given axis is given by

$$\begin{aligned}\left(\sum_i m_i\right) k^2 &= \sum_i m_i(\mathbf{n} \times \mathbf{r}_i)^2 = \sum_i m_i(\mathbf{n} \times \mathbf{d} + \mathbf{n} \times \mathbf{r}'_i)^2 \\ &= \sum_i m_i\{d^2 + (\mathbf{n} \times \mathbf{r}'_i)^2 + 2(\mathbf{n} \times \mathbf{d}) \cdot (\mathbf{n} \times \mathbf{r}'_i)\} \\ &= \left(\sum_i m_i\right)(d^2 + k_c^2) + 2(\mathbf{n} \times \mathbf{d}) \cdot \left(\mathbf{n} \times \sum_i m_i\mathbf{r}'_i\right)\end{aligned}$$

Since $\Sigma_i m_i \mathbf{r}'_i = \mathbf{0}$, by (10.1.2), the last term is zero and so, finally, we obtain the result $k^2 = d^2 + k_c^2$.

Theorem of Perpendicular Axes

This theorem is applicable to *plane* distributions only. Let the plane be defined by the axes Ox, Oy of a cartesian system Oxyz. Let P_i be a typical point with weighting m_i and coordinates $(x_i, y_i, 0)$ (see Fig. 10.2.3). Let k_x, k_y, k_z be the

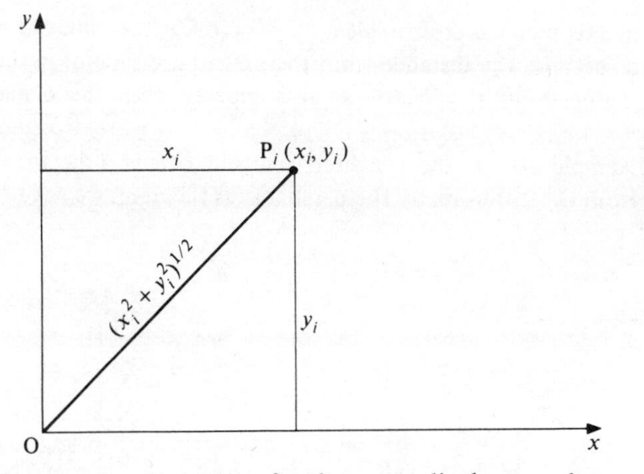

Figure 10.2.3 Illustration for the perpendicular axes theorem.

radii of gyration about the axes Ox, Oy, Oz respectively. It follows at once that

$$\left(\sum_i m_i\right) k_x^2 = \sum_i m_i y_i^2, \quad \left(\sum_i m_i\right) k_y^2 = \sum_i m_i x_i^2, \quad \left(\sum_i m_i\right) k_z^2 = \sum_i m_i(x_i^2 + y_i^2)$$

and hence that

$$k_z^2 = k_x^2 + k_y^2. \tag{10.2.9}$$

Finally we note that the results of this section generalise immediately to continuous distributions of density and all the theorems so far discussed are applicable. The generalisation of (10.2.6) is simply

$$I = \int p_i^2 \rho \, \mathrm{d}\tau = k^2 \int \rho \, \mathrm{d}\tau, \tag{10.2.10}$$

where p_i is the distance from the given axis, and ρ is the density at the field point associated with the continuous distribution.

Example 10.2.1
As in Example 10.1.1 and Figure 10.1.2, let the line density along Ox be proportional to x^n. The radius of gyration $k_{\mathrm{o}y}$ about an axis Oy perpendicular to Ox is given by

$$k_{\mathrm{o}y}^2 = \frac{\int_0^a x^{n+2} \, \mathrm{d}x}{\int_0^a x^n \, \mathrm{d}x} = \frac{n+1}{n+3} a^2. \tag{10.2.11}$$

We can also infer the following related results.

(i) About an axis inclined at an angle α ($< \pi/2$) to Ox, the radius of gyration is $k_{oy} \sin \alpha$, because the distance from the axis of every point in the distribution is reduced by the factor $\sin \alpha$ compared with the distance when $\alpha = \pi/2$.

(ii) From Example 10.1.1, the weighted centroid C is at a distance $(n + 1)a/(n + 2)$ from O. Therefore, by the parallel axes theorem, we have

$$k_{oy}^2 = k_{cy}^2 + \left(\frac{n+1}{n+2}\right)^2 a^2,$$

and so

$$k_{cy}^2 = \left\{\frac{n+1}{n+3} - \left(\frac{n+1}{n+2}\right)^2\right\} a^2 = \frac{(n+1)a^2}{(n+2)^2(n+3)}, \qquad (10.2.12)$$

where k_{cy} is the radius of gyration about an axis through C parallel to Oy. In particular, for a uniform density distribution ($n = 0$), we have

$$k_{oy}^2 = a^2/3, \quad k_{cy}^2 = a^2/12. \qquad (10.2.13)$$

(iii) The calculation is essentially unchanged for a surface density $\rho = \rho_0 x^n$ over the rectangular surface with sides OA $= a$, OB $= b$ (Fig. 10.1.2). The formal calculation is

$$\begin{aligned} k_{oy}^2 &= \frac{\int_0^a \int_0^b \rho x^2 \, dx \, dy}{\int_0^a \int_0^b \rho \, dx \, dy} \\ &= \frac{\int_0^a x^{n+2} \, dx \int_0^b dy}{\int_0^a x^n \, dx \int_0^b dy} \\ &= \frac{n+1}{n+3} a^2 \end{aligned}$$

as before. The result might also be inferred from a subdivision of the rectangle into elementary strips parallel to OA, each of which has the radius of gyration k_{oy} given in (10.2.11), which is therefore the radius of gyration of the complete distribution. The product of inertia I_{xy} at O is given by

$$I_{xy} = -\int_0^a \int_0^b \rho xy \, dx \, dy = -\int_0^a \int_0^b \rho_0 x^{n+1} y \, dx \, dy = -\rho_0 \frac{a^{n+2} b^2}{2(n+2)}.$$

(iv) For a uniform density distribution over the rectangle, the radius of gyration about the centroid C, where $\overline{\text{OC}} = \frac{1}{2}(\overline{\text{OA}} + \overline{\text{OB}})$, is given by $k_{cy}^2 = a^2/12$ for an axis parallel to Oy. By analogy, we have $k_{cx}^2 = b^2/12$ about an axis through C parallel to Ox. By the perpendicular axes theorem, the radius of gyration about an axis through C perpendicular to the plane Oxy is therefore given by

$$k_{cz}^2 = (a^2 + b^2)/12. \tag{10.2.14}$$

The principal axes at C are parallel to Ox, Oy and their mutual perpendicular Oz, since the products of inertia vanish by symmetry. The radius of gyration k about any other line through C with direction cosines (n_x, n_y, n_z) is given, with the help of (10.2.5), by

$$k^2 = \frac{b^2}{12} n_x^2 + \frac{a^2}{12} n_y^2 + \frac{a^2 + b^2}{12} n_z^2. \tag{10.2.15}$$

Note that, for a uniform square of side a, the radius of gyration about any line through C in the plane of the square (so that $n_z = 0$) is such that

$$k^2 = \frac{a^2}{12} n_x^2 + \frac{a^2}{12} n_y^2 = \frac{a^2}{12}, \tag{10.2.16}$$

since $n_x^2 + n_y^2 = 1$. Thus the radius of gyration of a uniform square is the same about any coplanar line through C.

(v) Now consider a uniform density distribution through the volume of a rectangular parallelepiped with sides parallel to Ox, Oy, Oz. The radius of gyration about an axis parallel to Oz through the centroid C is still given by $k_{cz}^2 = (a^2 + b^2)/12$. In particular, for a uniform cube of mass m and side a, the principal moments of inertia at C are all equal to $ma^2/6$. From (10.2.5), the moment of inertia about any line through C with direction cosines (n_x, n_y, n_z) is

$$\frac{ma^2}{6}(n_x^2 + n_y^2 + n_z^2) = \frac{ma^2}{6}, \tag{10.2.17}$$

since $\mathbf{n}$ is a unit vector. A uniform cube therefore has the same moment of inertia about any line through its centroid.

Example 10.2.2

(i) The radius of gyration of a uniform line density around the circumference of a circle, of radius a, is equal to a about an axis Oz through the centre O and perpendicular to its plane. This follows immediately from the fact that every element of the distribution has this radius of gyration.

About perpendicular axes Ox, Oy in the plane of the circle, the radii of gyration are equal by symmetry. From the perpendicular axes theorem, they must therefore be equal to $a/2^{1/2}$. It is clear from symmetry that this is the radius of gyration about any axis through O in the plane of the circle.

(ii) For a uniform surface density over a circular disc, of radius a, the radius of

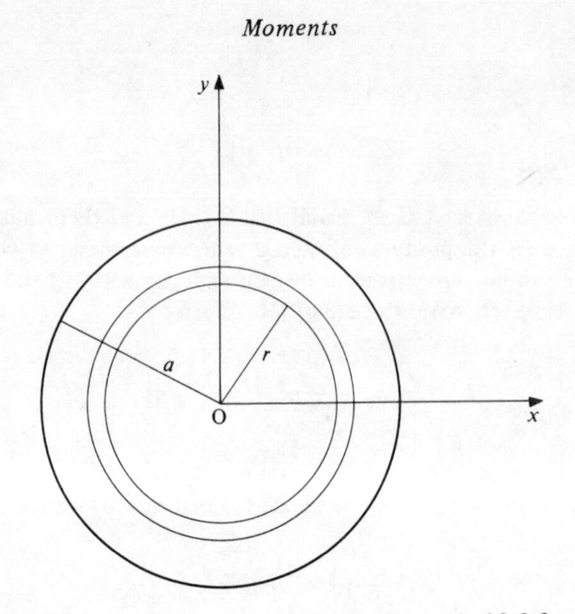

Figure 10.2.4 Illustration for Example 10.2.2.

gyration about Oz can be calculated from a decomposition into elementary rings, of radius r and thickness δr, each of which has a radius of gyration r and area $2\pi r\,\delta r$ (see Fig. 10.2.4). Their moment of inertia is $2\pi\rho r^3\,\delta r$. By integration we have

$$mk_z^2 = \int_0^a 2\pi\rho r^3\,\mathrm{d}r = \tfrac{1}{2}\pi a^4\rho,$$

where
$$m = \int_0^a 2\pi\rho r\,\mathrm{d}r = \pi a^2\rho.$$

Hence
$$k_z^2 = \tfrac{1}{2}a^2.$$

By symmetry and the perpendicular axes theorem, it follows that the radius of gyration about an axis through O in the plane of the disc is $a/2$.

(iii) For a uniform surface density over the surface of a spherical shell of radius a, we consider the radii of gyration about three mutually perpendicular axes Ox, Oy, Oz through the centre O. From symmetry they are all equal. Let (x, y, z) be coordinates of a typical point on the surface of the shell, subject to the condition $x^2 + y^2 + z^2 = a^2$. We have

$$mk_x^2 = \int \rho(y^2 + z^2)\,\mathrm{d}\tau, \quad mk_y^2 = \int \rho(z^2 + x^2)\,\mathrm{d}\tau, \quad mk_z^2 = \int \rho(x^2 + y^2)\,\mathrm{d}\tau$$

and so

$$m(k_x^2 + k_y^2 + k_z^2) = \int 2\rho(x^2 + y^2 + z^2)\,\mathrm{d}\tau = \int 2\rho a^2\,\mathrm{d}\tau = 2ma^2.$$

Hence, by symmetry,

$$k_x^2 = k_y^2 = k_z^2 = \tfrac{2}{3}a^2 \tag{10.2.18}$$

and $k^2 = \frac{2}{3}a^2$ for any axis through O.

(iv) For a uniform volume density throughout the volume of a sphere of radius a, we can calculate the radius of gyration about any axis through O from that of elementary spherical shells of radius r. The volume of the elementary shell is $4\pi r^2\ \delta r$, its moment of inertia is

$$\tfrac{2}{3}r^2 \,.\, 4\pi\rho r^2\ \delta r = \tfrac{8}{3}\pi\rho r^4\ \delta r$$

and so, by integration, we have

$$mk^2 = \int_0^a \tfrac{8}{3}\pi\rho r^4\ \mathrm{d}r = \tfrac{8}{15}\pi a^5 \rho,$$

where $$m = \int_0^a 4\pi\rho r^2\ \mathrm{d}r = \tfrac{4}{3}\pi a^3 \rho.$$

Hence $$k^2 = \tfrac{2}{5}a^2. \tag{10.2.19}$$

(v) The radii of gyration calculated in (i) – (iv) are also appropriate for, respectively: (i) a semicircle; (ii) a semicircular disc; (iii) a hemispherical shell; (iv) a solid hemisphere. This follows from the fact that the two halves which constitute a complete circle or sphere have, by symmetry, the same radius of gyration. It must therefore be the same radius of gyration as that of the complete configuration. Note that the semicircular distributions have the same radius of gyration about any coplanar axis through O. The hemispherical configurations have the same radius of gyration about any axis through O.

10.3 VECTOR MOMENT

The moment of the ordered pair of vectors **r**, **F** is defined to be **r** x **F**. The moment is simply a function of the two vectors defining it, but in applications it is useful to regard **r** as the position vector of some point P relative to some origin O, and to regard the line of action of **F** as passing through P. Vectors with different lines of action are then to be distinguished and they are called line vectors to indicate the restriction. The advantage of this interpretation is that the line of action of **F** is sufficient to define the moment about a particular origin, since **r** x **F** is the same for the position vector of any point on the line of action of **F**. For if **r**′ is the position vector of P′, as in Figure 10.3.1, we have

$$\mathbf{r}' \times \mathbf{F} = (\mathbf{r} + \overline{\mathrm{PP}'}) \times \mathbf{F} = \mathbf{r} \times \mathbf{F},$$

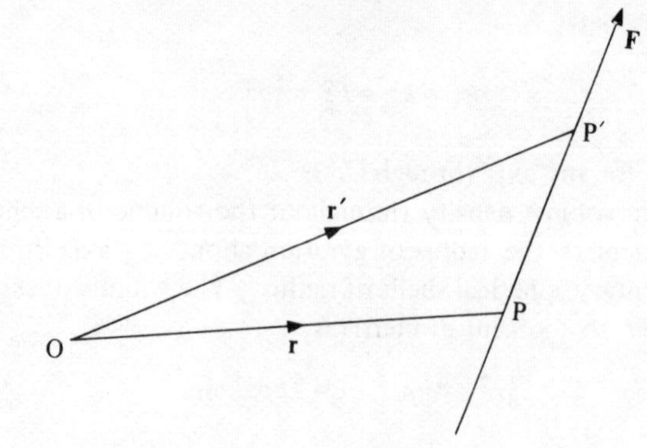

Figure 10.3.1 The notation used in the discussion of vector moments.

since $\overline{PP'}$ and $\mathbf{F}$ are parallel vectors with zero vector product.

The combined moment of several line vectors $\mathbf{F}_1, \mathbf{F}_2, \ldots$ all of whose lines of action pass through the point P is

$$\mathbf{r} \times \mathbf{F}_1 + \mathbf{r} \times \mathbf{F}_2 + \cdots = \mathbf{r} \times (\mathbf{F}_1 + \mathbf{F}_2 + \cdots),$$

which is the moment of a line vector passing through P equal to their resultant. For a general system of vectors the notion of a resultant is still applicable, in the usual sense of a free vector. But for a system of line vectors it is useful to introduce the following definition of equivalence.

Definition

Two systems of line vectors are said to be equivalent if: (i) they have the same resultant; (ii) they have the same moment.

It is sufficient that they have the same moment about one point; it can then be shown that they have the same moment about all other points. For let $\mathbf{F}_i$, $\mathbf{F}'_j$ be typical members of the two systems, and let P_i, P'_j be points on their respective lines of action. If we are given that

$$\sum_i \mathbf{F}_i = \sum_j \mathbf{F}'_j, \quad \sum_i \overline{OP_i} \times \mathbf{F}_i = \sum_j \overline{OP'_j} \times \mathbf{F}'_j$$

then, for any other origin O′, we have that

$$\begin{aligned}\sum_i \overline{O'P_i} \times \mathbf{F}_i &= \sum_i (\overline{O'O} + \overline{OP_i}) \times \mathbf{F}_i \\ &= \overline{O'O} \times \sum_i \mathbf{F}_i + \sum_i \overline{OP_i} \times \mathbf{F}_i \\ &= \overline{O'O} \times \sum_j \mathbf{F}'_j + \sum_j \overline{OP'_j} \times \mathbf{F}'_j\end{aligned}$$

$$= \sum_j (\overline{\mathrm{O'O}} + \overline{\mathrm{OP}'_j}) \times \mathbf{F}'_j$$

$$= \sum_j \overline{\mathrm{O'P}'_j} \times \mathbf{F}'_j.$$

Any system of concurrent vectors, with resultant **R**, is equivalent to a single line vector **R** passing through the common point. However, we shall see that for a general system of vectors it is not always possible to obtain a single equivalent vector.

10.4 PARALLEL VECTORS

Let a typical vector of a system of parallel vectors be $F_i\,\mathbf{n}$, where $\mathbf{n}$ is a fixed unit vector, and let P_i be a point on its line of action. The moment of the system about an origin O is

$$\sum_i \overline{\mathrm{OP}_i} \times \mathbf{n}F_i = \left(\sum_i F_i\overline{\mathrm{OP}_i}\right) \times \mathbf{n}.$$

Now $$\sum_i F_i\overline{\mathrm{OP}_i} = R\,\overline{\mathrm{OC}}, \quad R = \sum_i F_i,$$

and this defines a point C, the weighted centroid of the F_i associated with the points P_i. A necessary condition here is that the resultant R must not be zero, and provided this is satisfied the system is equivalent to a single vector $R\mathbf{n}$ whose line of action passes through C. Note that the position of C is independent of the direction of $\mathbf{n}$.

A particular application of this result is to the gravitational forces acting on a system of particles (or, in the limit, to the continuous distribution of parallel forces acting on the elements of a continuous distribution of mass). Here $F_i = m_i g$, $R = \Sigma_i m_i g$ is the total weight, $\mathbf{n}$ is in the direction of the downward vertical and C is the mass centre. Thus a system of gravitational forces is equivalent to a single force, namely the total weight acting through the mass centre. In this context C is called the centre of gravity.

Note that if the mass system moves as a rigid body, the position of each element is fixed relative to a frame of reference moving with it. Hence so is the position of the mass centre, which is therefore fixed in the body. But whereas the result is precise for the mass centre, it is only approximate for the centre of gravity because the gravitational attraction of the earth is towards its centre, and the forces on masses in different positions are not strictly parallel. If this is taken into account, the concept of a centre of gravity ceases to be appropriate. However, the error incurred in the neglect of this refinement is small for situations involving length scales which are small compared with the radius of the earth.

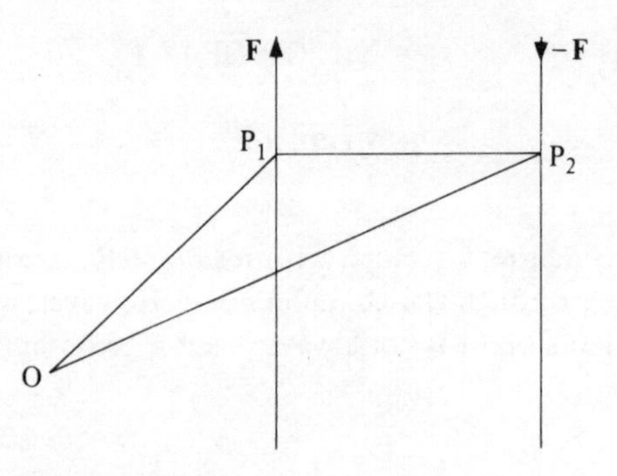

Figure 10.4.1 Illustration of a couple.

We return to the general system of parallel vectors, and to the special case $R = 0$ for which there is no single equivalent vector. The most that is possible by way of reduction is to separate the vectors into two groups, those for which $F_i > 0$, and those for which $F_i < 0$. Since $\Sigma_i F_i \neq 0$ for each group, there is a single equivalent vector for each group separately. These two vectors will be equal in magnitude, opposite in direction and each will have a definite line of action. Apart from the trivial case in which the two lines of action coincide and the system is equivalent to a null vector, the equivalent system consists of two equal and opposite parallel vectors a definite distance apart. Such an entity is called a couple and no further reduction is possible. We list below three important properties of couples.

(i) A couple has the same moment about any point. Let P_1P_2 be the common perpendicular to the lines of action of equal and opposite line vectors **F** and −**F** (see Fig. 10.4.1). The moment about any point O is

$$\overline{OP_1} \times \mathbf{F} + \overline{OP_2} \times (-\mathbf{F}) = (\overline{OP_1} - \overline{OP_2}) \times \mathbf{F}$$

$$= \overline{P_2P_1} \times \mathbf{F}.$$

This is a vector perpendicular to the plane of the couple and of magnitude $|\overline{P_2P_1}|\,|\mathbf{F}|$, which is independent of O.

(ii) Two couples are equivalent if they have the same moment, since the resultants are zero in both cases and so equal.

(iii) Two couples are together equivalent to a third couple. Again the resultants are zero in both systems, and we have merely to find a couple whose moment is the sum of the separate moments. Note that the moment is by definition a vector, and it is the vector sum of the constituent moments that is required. The result can clearly be generalised to include an arbitrary number of couples.

It is clear that a couple is defined by its moment, say **M**, and in future it will be referred to as a couple **M**.

10.5 REDUCTION OF A SYSTEM OF LINE VECTORS

An arbitrary system of line vectors is not in general equivalent to a single line vector, as we have seen in the particular case of a couple. The simplest equivalent system is a combination of a line vector and a couple. To see this let $\mathbf{F}_i$, localised at $\mathbf{r}_i$, be a typical line vector of the system with moment $\mathbf{r}_i \times \mathbf{F}_i$ about some chosen origin O. This line vector is equivalent to a line vector $\mathbf{F}_i$ at O together with a couple $\mathbf{M}_i = \mathbf{r}_i \times \mathbf{F}_i$, since the resultants and their moments about O are the same. The original system is thus equivalent to a system of line vectors $\mathbf{F}_i$, localised at O, and couples $\mathbf{M}_i$. The concurrent line vectors are themselves equivalent to a single line vector $\mathbf{R} = \Sigma_i \mathbf{F}_i$ through O and the couples are equivalent to a single couple $\mathbf{M} = \Sigma_i \mathbf{M}_i$.

The couple $\mathbf{M}$ depends on the choice of origin but the resultant $\mathbf{R}$ does not. The system $(\mathbf{R}, \mathbf{M})$ at O is equivalent to $(\mathbf{R}, \mathbf{M} + \overline{\mathrm{O'O}} \times \mathbf{R})$ at O′ (see Fig. 10.5.1). Only in the special case $\mathbf{R} = \mathbf{0}$ is the system equivalent to a couple independent of the origin.

If $\mathbf{R} \neq \mathbf{0}$ let us suppose that $\mathbf{M}$ has a component $p\mathbf{R}$ parallel to $\mathbf{R}$ and a component $\mathbf{d} \times \mathbf{R}$ perpendicular to $\mathbf{R}$. A change of origin to O′ produces the couple

$$p\mathbf{R} + \mathbf{d} \times \mathbf{R} + \overline{\mathrm{O'O}} \times \mathbf{R}.$$

A choice of O′ such that $\overline{\mathrm{OO'}} = \mathbf{d}$ reduces to zero the component of $\mathbf{M}$ perpendicular to $\mathbf{R}$. No choice of origin is possible which will eliminate the component of $\mathbf{M}$ parallel to $\mathbf{R}$. The system $(\mathbf{R}, p\mathbf{R})$ is called a wrench, and p is called the pitch of the wrench.

Coplanar systems of line vectors are always equivalent to a couple if $\mathbf{R} = \mathbf{0}$ or to a single line vector appropriately chosen if $\mathbf{R} \neq \mathbf{0}$. For $\mathbf{M}$ must be perpendicular to the common plane, and hence to the resultant $\mathbf{R}$, so that p is always zero.

Example 10.5.1
We have already seen that, in general, a system of line vectors is not equivalent

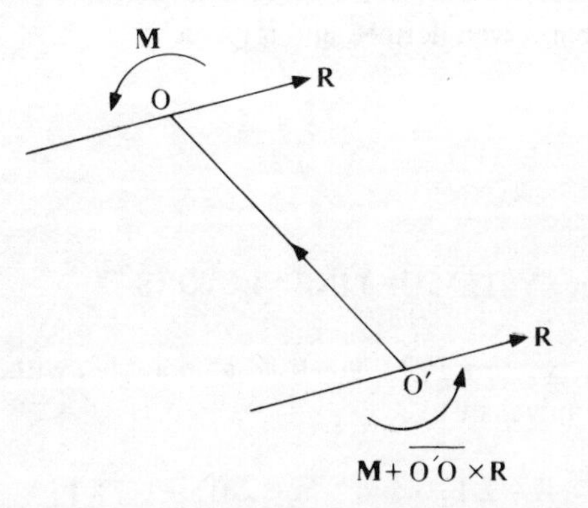

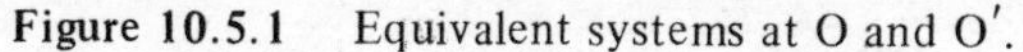
Figure 10.5.1 Equivalent systems at O and O′.

to a single line vector. One might, however, consider the possibility that such a system is equivalent to two line vectors, one of which acts along a prescribed line.

Let O be a point on the prescribed line, and let $\mathbf{n}$ be a unit vector defining its direction. We assume that the given system is equivalent to the force and couple combination $(\mathbf{R}, \mathbf{M})$ at O, and look for an equivalent system consisting of $\mathbf{F}_1 = F_1\mathbf{n}$ at O and $\mathbf{F}_2 = \mathbf{R} - F_1\mathbf{n}$ along a suitable line. Let the position vector of a typical point on the line of $\mathbf{F}_2$ be $\mathbf{r}$ relative to O.

The couple $\mathbf{M}$ can arise only from $\mathbf{F}_2$, since $\mathbf{F}_1$ has no moment about O. Hence we have

$$\mathbf{r} \times \mathbf{F}_2 = \mathbf{r} \times (\mathbf{R} - F_1\mathbf{n}) = \mathbf{M}.$$

This condition requires that $\mathbf{F}_2$ be perpendicular to $\mathbf{M}$, so that

$$\mathbf{M} \cdot \mathbf{F}_2 = \mathbf{M} \cdot (\mathbf{R} - F_1\mathbf{n}) = 0.$$

In order to satisfy this condition F_1, which is not yet determined, is chosen so that

$$F_1 = \frac{\mathbf{M} \cdot \mathbf{R}}{\mathbf{M} \cdot \mathbf{n}}.$$

Note that this can only be done provided $\mathbf{M} \cdot \mathbf{n} \neq 0$ and so the prescribed line cannot be chosen in a completely arbitrary fashion. It must not be perpendicular to $\mathbf{M}$.

With $\mathbf{F}_1$ and hence $\mathbf{F}_2$ determined, the relation $\mathbf{r} \times \mathbf{F}_2 = \mathbf{M}$ is, in effect, an equation giving the line of action of $\mathbf{F}_2$. To derive an explicit expression for $\mathbf{r}$, we write it in the form

$$\mathbf{r} = \mathbf{a} + \lambda\mathbf{F}_2,$$

where λ is an arbitrary scalar, and choose $\mathbf{a}$ so that $\mathbf{a}$, $\mathbf{F}_2$ and $\mathbf{M}$ are mutually perpendicular. Then $\mathbf{a}$ is well defined and is given by

$$\mathbf{a} = \frac{\mathbf{F}_2 \times \mathbf{M}}{F_2^2}.$$

10.6 WORK OF A SYSTEM OF LINE VECTORS

Let a system of line vectors $\mathbf{F}_i$ be localised at points P_i with position vectors $\mathbf{r}_i$. Let the system be equivalent to

$$\mathbf{R} = \sum_i \mathbf{F}_i \quad \text{and} \quad \mathbf{M} = \sum_i (\mathbf{r}_i - \mathbf{r}_1) \times \mathbf{F}_i$$

at P_1. Let the $\mathbf{r}_i$ suffer a small rigid body displacement $\delta\mathbf{r}_i$ defined by a translation $\delta\mathbf{r}_1$ of P_1 and an angular displacement $\delta\theta\mathbf{n}$ about P_1. We then have

$$\delta\mathbf{r}_i = \delta\mathbf{r}_1 + \delta\theta\mathbf{n} \times (\mathbf{r}_i - \mathbf{r}_1), \tag{10.6.1}$$

$$\sum_i \mathbf{F}_i \cdot \delta\mathbf{r}_i = \sum_i \mathbf{F}_i \cdot \delta\mathbf{r}_1 + \sum_i \mathbf{F}_i \cdot \{\delta\theta\mathbf{n} \times (\mathbf{r}_i - \mathbf{r}_1)\}$$

$$= \mathbf{R} \cdot \delta\mathbf{r}_1 + \mathbf{M} \cdot \mathbf{n}\delta\theta. \tag{10.6.2}$$

This expression is the work of the system correct to the first order, which is sufficient for our purposes. Note that only **R**, **M** and the small rigid body displacement are required to evaluate (10.6.2). Since **R**, **M** are the same for all equivalent systems, it follows that the work done in any given rigid body displacement is the same, to first order, for all equivalent systems.

We note in particular that the work done in any small rigid body displacement is zero, to first order, for a null system in which $\mathbf{R} = \mathbf{M} = \mathbf{0}$. Conversely (10.6.2) can only be zero for all small rigid body displacements if $\mathbf{R} = \mathbf{M} = \mathbf{0}$ and the system is equivalent to a null vector. These results form the basis of the discussion of virtual work in Chapter 12.

If the $\mathbf{r}_i$ are differentiable functions of a parameter t and, as in §9.1,

$$\delta\mathbf{r}_i = (\mathbf{v}_i + \epsilon_i)\,\delta t = \{\mathbf{v}_1 + \omega \times (\mathbf{r}_i - \mathbf{r}_1) + \epsilon_i\}\,\delta t,$$

where $|\epsilon_i| \to 0$ as $\delta t \to 0$, we can divide (10.6.2) by δt and take the limit as $\delta t \to 0$ to get, exactly,

$$\sum_i \mathbf{F}_i \cdot \mathbf{v}_i = \mathbf{R} \cdot \mathbf{v}_1 + \mathbf{M} \cdot \boldsymbol{\omega}. \tag{10.6.3}$$

The work of the system over a finite interval of t is then

$$\int \sum_i \mathbf{F}_i \cdot \mathrm{d}\mathbf{r}_i = \int \sum_i \mathbf{F}_i \cdot \mathbf{v}_i \,\mathrm{d}t = \int (\mathbf{R} \cdot \mathbf{v}_1 + \mathbf{M} \cdot \boldsymbol{\omega})\,\mathrm{d}t. \tag{10.6.4}$$

EXERCISES

10.1 The lines QS, RT intersect at O, and P is a point such that $\overline{PT} = \overline{RO}$. Show that the centroids of uniform density distributions over the areas QRST and PQS coincide.

10.2 Establish the following rule, known as Routh's rule.

For the bodies listed below, the moment of inertia about an axis of symmetry of a uniform density distribution with total mass m is

$$\frac{m}{n}(\text{sum of squares of the perpendicular semi-axes}),$$

where

$$n = \begin{cases} 3 & \text{for a rectangle or a rectangular parallelepiped,} \\ 4 & \text{for an ellipse,} \\ 5 & \text{for an ellipsoid.} \end{cases}$$

10.3 An isosceles triangle ABC is such that AB = AC, BC = a and the perpendicular height from A to BC is h. Show that a uniform density distribution over the area ABC has a radius of gyration equal to $(h^2/2 + a^2/24)^{1/2}$ about an axis through A perpendicular to the plane of ABC. What is the interpretation of this result as: (i) $h \to 0$; (ii) $a \to 0$?

Deduce that the radius of gyration of an n sided regular polygon, about an axis through its centroid perpendicular to its plane, is given by

$$k^2 = \tfrac{1}{2}h^2\left(1 + \tfrac{1}{3}\tan^2\frac{\pi}{n}\right),$$

where h is the radius of the inscribed circle.

Verify the correctness of this result for the special cases $n = 4, n \to \infty$.

10.4 A wheel consists of a rim, which is a ring of mass m_r and radius a_r, a hub, which is a disc of mass m_h and radius a_h, and n spokes, which are rods each of mass m_s and length $(a_r - a_h)$.

What is the moment of inertia of the wheel about its axle?

(Ans.: $m_r a_r^2 + \frac{1}{2}m_h a_h^2 + \frac{1}{3}nm_s(a_r^2 + a_h^2 - a_r a_h)$.)

10.5 Show that the moment of inertia of a uniform density distribution over the surface of a parallelogram is:

(i) $mh^2/12$ about an axis through the centroid parallel to one pair of sides, where m is the total mass and h is the perpendicular distance between the pair of sides;

(ii) $m(a^2 + b^2)/3$ about an axis through C perpendicular to the plane of the distribution, where $2a$, $2b$ are the lengths of the adjacent sides.

10.6 The two sets of axes Oxy, O$x'y'$ have a common origin O and lie in the plane of a two-dimensional distribution of matter. The angle between Ox' and Ox is α. Show that

$$x' = x\cos\alpha + y\sin\alpha, \quad y' = y\cos\alpha - x\sin\alpha.$$

Show that

$$I_{x'y'} = I_{xy}\cos 2\alpha - \tfrac{1}{2}(I_{xx} - I_{yy})\sin 2\alpha.$$

Deduce that the two values of α for which $I_{x'y'} = 0$ differ by $\pi/2$ and hence that the principal axes are perpendicular.

10.7 A plane distribution of matter has the same amount of inertia about two lines which lie in the plane and meet at a point O. Show that the distribution either has kinetic symmetry about an axis through O perpendicular to the plane, or that the two principal axes in the plane are equally inclined to the two given lines. Show that if the distribution has the same moment of inertia about three lines through O in the plane then it must have kinetic symmetry.

10.8 Let a typical point P of a system have coordinates (x, y, z) and weighting m. Show that

$$\Sigma myz = (\Sigma m)y_c z_c + \Sigma my'z',$$

where y_c, z_c refer to the coordinates of the weighted centroid C and $y = y_c + y'$, $z = z_c + z'$.

Suppose that one of the principal axes at O, say Ox, passes through C. Verify that the principal axes at C are parallel to those at O. If the principal moments of inertia at O are I_{xx}, I_{yy}, I_{zz}, show that, at C, they are I_{xx}, $I_{yy} - (\Sigma m)c^2$, $I_{zz} - (\Sigma m)c^2$, where $c = OC$.

Show that the principal moments of inertia at one of the corners of a uniform cube of mass m and side $2a$ are $\frac{2}{3}ma^2$, $\frac{11}{3}ma^2$, $\frac{11}{3}ma^2$.

10.9 Find the centroid of a uniform density distribution over: (i) a semicircular wire; (ii) a semicircular disc; (iii) a hemispherical shell; (iv) a solid hemisphere.

Calculate the principal radii of gyration of the above distributions about the geometrical centre O and the centroid C.

(Ans.: The centroid C lies on the radius of symmetry. As a fraction of the radius, OC is: (i) $2/\pi$; (ii) $4/3\pi$; (iii) $\frac{1}{2}$; (iv) $\frac{3}{8}$.

As a fraction of the radius the principal radii of gyration at O are:

(i) $(\frac{1}{2})^{1/2}$, $(\frac{1}{2})^{1/2}$, 1; (ii) $\frac{1}{2}$, $\frac{1}{2}$, $(\frac{1}{2})^{1/2}$; (iii) $(\frac{2}{3})^{1/2}$, $(\frac{2}{3})^{1/2}$, $(\frac{2}{3})^{1/2}$;
(iv) $(\frac{2}{5})^{1/2}$, $(\frac{2}{5})^{1/2}$, $(\frac{2}{5})^{1/2}$

At C the corresponding results are:

(i) $1/2^{1/2}$, $((\pi^2 - 8)/2\pi^2)^{1/2}$, $((\pi^2 - 4)/\pi^2)^{1/2}$;
(ii) $1/2$, $((9\pi^2 - 64)/36\pi^2)^{1/2}$, $((9\pi^2 - 32)/18\pi^2)^{1/2}$;
(iii) $(2/3)^{1/2}$, $(5/12)^{1/2}$, $(5/12)^{1/2}$;
(iv) $(2/5)^{1/2}$, $(83/320)^{1/2}$, $(83/320)^{1/2}$.)

10.10 A circle in the (x, y) plane has radius a and centre at the origin of coordinates. A second circle has radius $a/2$ and centre at $(a/2, 0)$.

Find the centroid of the area between the two circles.

Find the principal radii of gyration at the origin of coordinates and at the centroid.

$$\left(\text{Ans.: } \left(-\frac{a}{6}, 0\right), \left(\left(\frac{5}{16}\right)^{1/2} a, \left(\frac{11}{48}\right)^{1/2} a, \left(\frac{13}{24}\right)^{1/2} a\right),\right.$$

$$\left.\left(\left(\frac{5}{16}\right)^{1/2} a, \left(\frac{29}{144}\right)^{1/2} a, \left(\frac{37}{72}\right)^{1/2} a\right)\right)$$

10.11 The mutually perpendicular edges of a uniform rectangular block are of lengths $2a$, $2b$, $2c$, respectively. If these edges are used as coordinate axes Ox, Oy, Oz, respectively, show that

$$I_{xx} = \tfrac{4}{3}m(b^2 + c^2), \quad I_{yz} = -mbc.$$

10.12 Three uniform line distributions OX, OY, OZ, each of length l, lie along the coordinate axes. Show that any axis through O is a principal axis with associated moment of inertia $2ml^2/9$, where m is the total mass. Show that the principal moments of inertia at the centroid C are

$$\tfrac{2}{9}ml^2, \quad \tfrac{5}{36}ml^2, \quad \tfrac{5}{36}ml^2.$$

Show that relative to axes parallel to the original axes, but with the origin at C, the associated moments of inertia are all equal to $ml^2/6$ and the products of inertia are all equal to $ml^2/36$.

10.13 Three line vectors have lines of action passing through P_1, P_2, P_3 respectively. They are represented in magnitude and direction by $\overline{OQ_1}$, $\overline{OQ_2}$, $\overline{OQ_3}$, where Q_1, Q_2, Q_3 are the mid points of P_2P_3, P_3P_1, P_1P_2 respectively. Show that they are equivalent to the system represented by $\overline{OP_1}$, $\overline{OP_2}$, $\overline{OP_3}$ passing through Q_1, Q_2, Q_3.

10.14 The system of line vectors

$$\mathbf{i} \times (\mathbf{F} \times \mathbf{i}), \quad \mathbf{j} \times (\mathbf{F} \times \mathbf{j}), \quad \mathbf{k} \times (\mathbf{F} \times \mathbf{k})$$

pass through the points with position vectors

$$\mathbf{i} \times (\mathbf{r} \times \mathbf{i}), \quad \mathbf{j} \times (\mathbf{r} \times \mathbf{j}), \quad \mathbf{k} \times (\mathbf{r} \times \mathbf{k}),$$

respectively. Prove that they are equivalent to a single line vector $2\mathbf{F}$ passing through the point $\frac{1}{2}\mathbf{r}$.

10.15 A system of three wrenches is equivalent to a null vector. Show that their axes are all parallel to one plane and that there is a common transversal perpendicular to all the axes.

10.16 A system of line vectors has a resultant $\mathbf{R}$ and a moment $\mathbf{M}$ about an origin O. Given that the system is equivalent to a wrench $(\mathbf{R}, p\mathbf{R})$, show that $p = \mathbf{R} \cdot \mathbf{M}/R^2$ and that the axis of the wrench is given by

$$\mathbf{r} = \frac{\mathbf{R} \times \mathbf{M}}{R^2} + \lambda \mathbf{R}$$

for arbitrary λ.

10.17 Show that a system of line vectors is equivalent to a null vector if their moments about two points are both zero and their resultant has no component in the direction of the line joining these two points.

10.18 Investigate the possibility of equivalence between a general system of line vectors, and a combination of two line vectors acting at two prescribed points.

10.19 A typical mass m of a system has a position vector $\mathbf{r}$ relative to the mass centre C. A distant point P has position vector $\mathbf{r}_\mathrm{p}$ and $r_\mathrm{p} \gg r$ for all $\mathbf{r}$ of the system.

Show that the gravitational potential at P arising from the mass system is

$$V_\mathrm{p} = -\Sigma \frac{Gm}{|\mathbf{r}_\mathrm{p} - \mathbf{r}|} = -\frac{G}{r_\mathrm{p}} \Sigma m \left(1 + \frac{\mathbf{r}_\mathrm{p} \cdot \mathbf{r}}{r_\mathrm{p}^2} + \frac{3(\mathbf{r}_\mathrm{p} \cdot \mathbf{r})^2 - r_\mathrm{p}^2 r^2}{2r_\mathrm{p}^4} + \cdots \right)$$

$$= -G\left(\frac{\Sigma m}{r_\mathrm{p}} + \frac{1}{2r_\mathrm{p}^3}(I_{xx} + I_{yy} + I_{zz} - 3I) + \cdots \right),$$

where $I_{xx} + I_{yy} + I_{zz} = \Sigma m r^2$ and I is the moment of inertia of the mass system about CP.

Let there be a particle m_p at P. Show that the moment about C of the forces on the system arising from the attraction of m_p is approximately

$$\mathbf{M} = \frac{3Gm_\mathrm{p}}{r_\mathrm{p}^5} \Sigma\, m(\mathbf{r} \times \mathbf{r}_\mathrm{p})(\mathbf{r} \cdot \mathbf{r}_\mathrm{p}).$$

Show that, relative to the principal axes at C of the system, this expression can be written

$$\mathbf{M} = -\frac{3Gm_\mathrm{p}}{r_\mathrm{p}^5}\{(I_{yy} - I_{zz})y_\mathrm{p}z_\mathrm{p}\mathbf{i} + (I_{zz} - I_{xx})z_\mathrm{p}x_\mathrm{p}\mathbf{j} + (I_{xx} - I_{yy})x_\mathrm{p}y_\mathrm{p}\mathbf{k}\}.$$

10.20 A system of n particles, of mass m_i and position vector $\mathbf{r}_i$ ($i = 1, 2, \ldots, n$), are in motion under the action of the mutual forces of gravitation which act between each pair of particles. The centre of mass of the system is at rest at the origin O. Show that

$$m_i\ddot{\mathbf{r}}_i = \sum_{\substack{j=1 \\ j \neq i}}^{n} \frac{Gm_i m_j(\mathbf{r}_j - \mathbf{r}_i)}{|\mathbf{r}_j - \mathbf{r}_i|^3}.$$

Show that the potential energy of the system is

$$V = -\frac{1}{2}\sum_{i=1}^{n}\sum_{\substack{j=1 \\ i \neq j}}^{n} \frac{Gm_i m_j}{|\mathbf{r}_j - \mathbf{r}_i|}.$$

Let the three moments of inertia referred to coordinates at O be I_{xx}, I_{yy}, I_{zz}. Derive the Lagrange–Jacobi identity, namely

$$\frac{\mathrm{d}^2}{\mathrm{d}t^2}(I_{xx} + I_{yy} + I_{zz}) = 4(T + E),$$

where T is the kinetic energy and $E = T + V$ is the total energy of the system.

Note that $I_{xx} + I_{yy} + I_{zz}$, and hence the identity itself, does not depend on the particular choice of coordinate system at O.

11 Rigid Bodies

11.1 EQUATIONS OF MOTION

We start from the fundamental equation $\mathbf{F} = m\mathbf{a}$ for the motion of a single particle. It follows immediately that

$$\Sigma\mathbf{F} = \Sigma m\mathbf{a}, \quad \Sigma\mathbf{r} \times \mathbf{F} = \Sigma\mathbf{r} \times m\mathbf{a}, \tag{11.1.1}$$

where the summation is over a system of particles. For economy of notation the suffix i labelling a particular particle is omitted in (11.1.1), so that $\mathbf{F}$ represents the resultant of the external and internal forces on a typical particle of mass m, position vector $\mathbf{r}$ and acceleration $\mathbf{a}$. When the internal forces form a null system (that is, a system with zero resultant and zero moment), as is the case when they occur in equal and opposite pairs in the same line, they do not contribute to the left-hand sides of either of equations (11.1.1) and the summation is then confined to the external forces only. The same is true of a rigid body if it is regarded as a system of particles obeying the same rules. There are, however, logical difficulties associated with this assumption, one of which is that there are limiting processes involved in the mathematics since rigid bodies are modelled as continuous distributions of matter. Another difficulty is that a model in which the internal forces occur in equal and opposite pairs acting along the same line is not realistic, since there are electromagnetic forces involved which do not have this property. Indeed the internal forces in real solids are not adequately described by classical mechanics. However, we can proceed with the dynamics of rigid bodies if we assume that equations (11.1.1) are valid, with the summation confined to the external forces only. Since this assumption is justified under conditions which are weaker than those invoked for a system of particles, it is better to accept the equations themselves as basic postulates and to regard their derivation from the motion of a system of particles as a motivation rather than a proof. This point of view makes the minimum necessary assumptions, and no discussion of the internal forces of cohesion is required. The ultimate justification, as always, depends on the correspondence between theory and observation.

We use the notation

$$\Sigma\mathbf{F} = \mathbf{R}, \quad \Sigma\mathbf{r} \times \mathbf{F} = \mathbf{M}, \tag{11.1.2}$$

where $\mathbf{F}$ now refers to a typical external force. Note that the resultant $\mathbf{R}$ depends only on the external force system but that, in general, the couple $\mathbf{M}$ depends also on the reference point O relative to which $\mathbf{r}$ is measured. When this dependence on O needs to be stressed the couple will be written $\mathbf{M}_o$, but when no confusion is likely to arise the suffix will be omitted.

We shall henceforth assume that the equations

$$\Sigma \mathbf{F} = \mathbf{R} = \Sigma m\mathbf{a}, \quad \Sigma \mathbf{r} \times \mathbf{F} = \mathbf{M} = \Sigma \mathbf{r} \times m\mathbf{a} \tag{11.1.3}$$

are valid for any system of particles or rigid bodies or combinations of particles and rigid bodies. These equations imply that the system of external forces is equivalent, in the mathematical sense of §10.3, to the force and couple combination $(\mathbf{R}, \mathbf{M})$ which is in turn equivalent to the system of mass accelerations $m\mathbf{a}$. They do not contain all the information about the dynamics of a general system, but for the particular case of a rigid body they are just sufficient to determine the motion completely when combined with appropriate initial conditions. It is for this reason that they are now to be discussed.

When we calculate the moment in (11.1.3) it is not essential that the position vector $\mathbf{r}$ be referred to a fixed origin, and for the motion of rigid bodies it is often convenient to take an origin moving with the body. However, care is then required in the calculation of the acceleration $\mathbf{a}$, because $d^2\mathbf{r}/dt^2$ is merely the relative acceleration. In Figure 11.1.1 let O have acceleration $\mathbf{a}_o$ and let a typical particle m have position vector $\mathbf{r}$ relative to O. The acceleration of m is then

$$\mathbf{a} = \mathbf{a}_o + \frac{d^2\mathbf{r}}{dt^2}$$

and so

$$\Sigma \mathbf{r} \times m\mathbf{a} = (\Sigma m\mathbf{r}) \times \mathbf{a}_o + \Sigma \mathbf{r} \times m \frac{d^2\mathbf{r}}{dt^2},$$

$$= m_c \mathbf{r}_c \times \mathbf{a}_o + \frac{d}{dt} \Sigma \mathbf{r} \times m \frac{d\mathbf{r}}{dt}, \tag{11.1.4}$$

where the suffix c in general refers to the mass centre C and, in particular, $m_c = \Sigma m$ is the total mass. Note that the second term on the right-hand side depends only on the motion relative to O. It is the moment of the relative mass acceleration and can be written in the alternative form

$$\Sigma \mathbf{r} \times m \frac{d^2\mathbf{r}}{dt^2} = \frac{d}{dt} \Sigma \mathbf{r} \times m \frac{d\mathbf{r}}{dt}$$

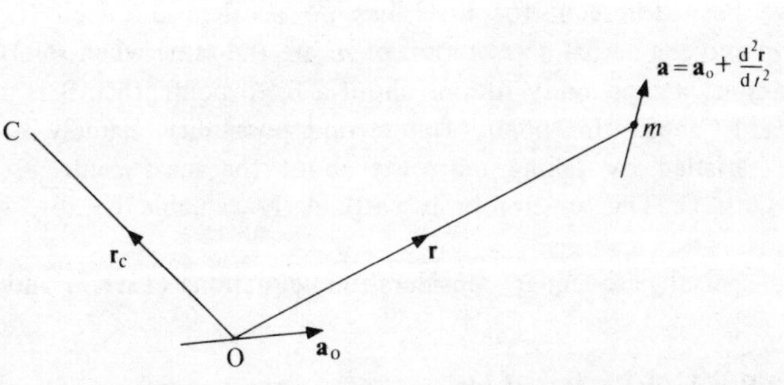

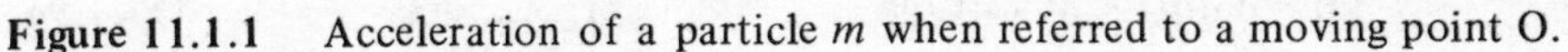
Figure 11.1.1 Acceleration of a particle m when referred to a moving point O.

because $d\mathbf{r}/dt \times m\, d\mathbf{r}/dt = \mathbf{0}$. Now $\Sigma(\mathbf{r} \times m\, d\mathbf{r}/dt)$ is just the moment of the relative momentum about O, more usually called the relative angular momentum. It follows that the moment of the relative mass acceleration is equal to the rate of change of relative angular momentum.

We use the notation

$$\mathbf{p} = \Sigma m\mathbf{v} = m_c\mathbf{v}_c \tag{11.1.5}$$

for the linear momentum of the system, and

$$\mathbf{L} = \Sigma\mathbf{r} \times m\frac{d\mathbf{r}}{dt} \tag{11.1.6}$$

for the relative angular momentum. Note that $\mathbf{L}$, like $\mathbf{M}$, depends on the reference point O and will be written $\mathbf{L}_o$ when this dependence needs to be stressed.

We can summarise the foregoing notation and results in the equations

$$\Sigma\mathbf{F} = \mathbf{R} = \Sigma m\mathbf{a} = m_c\mathbf{a}_c = d\mathbf{p}/dt, \tag{11.1.7}$$

$$\Sigma\mathbf{r} \times \mathbf{F} = \mathbf{M} = \Sigma\mathbf{r} \times m\mathbf{a} = m_c\mathbf{r}_c \times \mathbf{a}_o + d\mathbf{L}/dt. \tag{11.1.8}$$

The dynamics of rigid bodies is founded on equations (11.1.7) and (11.1.8). In applications it is necessary to choose a reference point O about which moments are taken, and a wise choice can simplify the mathematics considerably. It is particularly recommended that moments are calculated about a point for which the term $m_c\mathbf{r}_c \times \mathbf{a}_o$ in (11.1.8) vanishes. We discuss this case separately.

(i) The special case $\mathbf{r}_c \times \mathbf{a}_o = \mathbf{0}$.

When this condition is satisfied, only the relative motion need be considered in the moment equation (11.1.8), since $\mathbf{L}$ refers to the relative angular momentum as given by (11.1.6). We shall see later that there are a few applications, such as the rolling of a circular body, where it will be useful to use a reference point O for which $\mathbf{r}_c \times \mathbf{a}_o = \mathbf{0}$ because $\mathbf{r}_c$ is parallel to $\mathbf{a}_o$. But the two most useful conditions in practice are $\mathbf{a}_o = \mathbf{0}$ and $\mathbf{r}_c = \mathbf{0}$. That $\mathbf{a}_o = \mathbf{0}$ is a possibility might have been foreseen, for if O has no acceleration itself the relative acceleration and the actual acceleration of m are the same when referred to O. If, for example, a rigid body rotates about a fixed point, then it is natural to take moments about this point. The second possibility, namely $\mathbf{r}_c = \mathbf{0}$, can always be satisfied by taking moments about the mass centre, so that O coincides with C. This possibility is particularly valuable because it always exists.

For the special case under consideration, equations (11.1.7) and (11.1.8) reduce to

$$\mathbf{R} = d\mathbf{p}/dt, \quad \mathbf{M} = d\mathbf{L}/dt, \qquad \textit{if} \qquad \mathbf{r}_c \times \mathbf{a}_o = \mathbf{0}. \tag{11.1.9}$$

(ii) The general case $\mathbf{r}_c \times \mathbf{a}_o \neq \mathbf{0}$.

In most applications it is safer and advisable to take moments about a point O for which $\mathbf{r}_c \times \mathbf{a}_o = \mathbf{0}$, otherwise this term makes it awkward to handle (11.1.8) since $\mathbf{a}_o$ is not known *a priori*. If there is a compelling reason to take moments about a general point for which $\mathbf{r}_c \times \mathbf{a}_o \neq \mathbf{0}$, it is better to proceed as follows.

We first note, from case (i) with moments taken about the mass centre C, that

$$\mathbf{R} = \mathrm{d}\mathbf{p}/\mathrm{d}t, \quad \mathbf{M}_c = \mathrm{d}\mathbf{L}_c/\mathrm{d}t, \tag{11.1.10}$$

where $\mathbf{M}_c$ and $\mathbf{L}_c$ now refer specifically to the moment of the external forces and the relative angular momentum about C. Next we appeal to §10.5 to infer that the system of external forces, which is equivalent to the vector couple combination $(\mathbf{R}, \mathbf{M}_c)$ at C, is therefore equivalent to $(\mathbf{R}, \mathbf{M} = \mathbf{M}_c + \mathbf{r}_c \times \mathbf{R})$ at O (compare Fig. 11.1.2 with Fig. 10.5.1). The corresponding results for the mass acceleration system are that it is equivalent to $(\mathrm{d}\mathbf{p}/\mathrm{d}t,\ \mathrm{d}\mathbf{L}_c/\mathrm{d}t)$ at C and $(\mathrm{d}\mathbf{p}/\mathrm{d}t, \mathrm{d}\mathbf{L}_c/\mathrm{d}t + \mathbf{r}_c \times \mathrm{d}\mathbf{p}/\mathrm{d}t)$ at O. It follows immediately that, without restriction,

$$\mathbf{R} = \frac{\mathrm{d}\mathbf{p}}{\mathrm{d}t}, \quad \mathbf{M} = \frac{\mathrm{d}\mathbf{L}_c}{\mathrm{d}t} + \mathbf{r}_c \times \frac{\mathrm{d}\mathbf{p}}{\mathrm{d}t}, \tag{11.1.11}$$

where $\mathbf{M}$ is now the moment of the external forces about O. The rate of change of angular momentum, however, is calculated relative to C with a correction equal to the moment about O of $\mathrm{d}\mathbf{p}/\mathrm{d}t$ localised at C.

It is also possible to deduce the moment equation given in (11.1.11) from first principles, and it is perhaps instructive to see what lies behind the analysis leading to this equation. The basic relation is $\mathbf{M} = \Sigma \mathbf{r} \times m\mathbf{a}$ and we rewrite this relation in the form

$$\mathbf{M} = \Sigma \mathbf{r} \times m\mathbf{a} = \Sigma\,\{(\mathbf{r} - \mathbf{r}_c) \times m(\mathbf{a} - \mathbf{a}_c)\} + \Sigma \mathbf{r}_c \times m\mathbf{a} + \Sigma\,\{(\mathbf{r} - \mathbf{r}_c) \times m\mathbf{a}_c\}.$$

On the right-hand side, the first term is

$$\Sigma\left((\mathbf{r} - \mathbf{r}_c) \times m\,\frac{\mathrm{d}^2}{\mathrm{d}t^2}(\mathbf{r} - \mathbf{r}_c)\right) = \frac{\mathrm{d}}{\mathrm{d}t}\Sigma\left((\mathbf{r} - \mathbf{r}_c) \times m\,\frac{\mathrm{d}}{\mathrm{d}t}(\mathbf{r} - \mathbf{r}_c)\right) = \frac{\mathrm{d}\mathbf{L}_c}{\mathrm{d}t}.$$

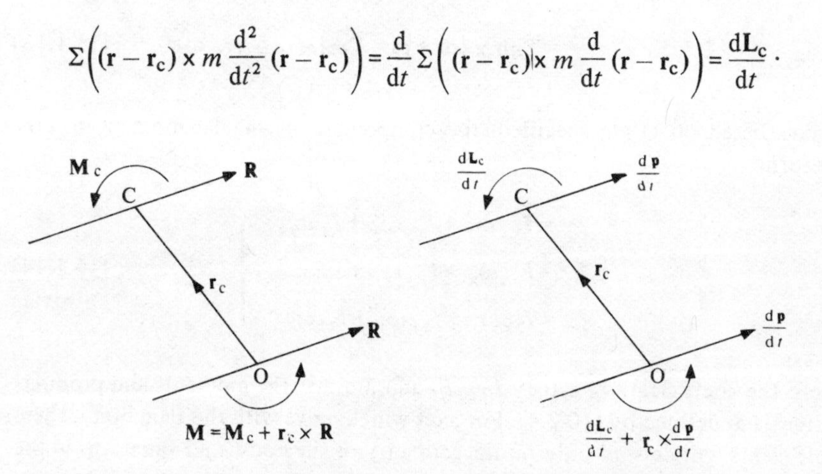

Figure 11.1.2 Equivalent systems at C and O.

The second term $\Sigma\mathbf{r}_c \times m\mathbf{a}$ is, by (11.1.7), $\mathbf{r}_c \times \mathrm{d}\mathbf{p}/\mathrm{d}t$ and the third term $\Sigma\{(\mathbf{r}-\mathbf{r}_c)\times m\mathbf{a}_c\}$ is zero because $\Sigma m\mathbf{r} = (\Sigma m)\mathbf{r}_c$. The final result agrees with the second of equations (11.1.11).

One useful application of equations (11.1.11) is that for which the system divides naturally into several component subsystems, for example two or more rigid bodies moving under constraints. In these circumstances the notion of a single mass centre for the whole system is not so convenient, and it can be more useful to preserve the identity of each component in the calculation of the mass accelerations and their moments. For this purpose we observe that, although (11.1.11) was derived for the system as a single entity, the arguments are still valid if the system is split into several components. Equations (11.1.11) are then replaced by

$$\left.\begin{aligned} \mathbf{R} &= (\Sigma m\mathbf{a})_1 + (\Sigma m\mathbf{a})_2 + \cdots = \left(\frac{\mathrm{d}\mathbf{p}}{\mathrm{d}t}\right)_1 + \left(\frac{\mathrm{d}\mathbf{p}}{\mathrm{d}t}\right)_2 + \cdots, \\ \mathbf{M} &= (\Sigma\mathbf{r}\times m\mathbf{a})_1 + (\Sigma\mathbf{r}\times m\mathbf{a})_2 + \cdots = \left(\frac{\mathrm{d}\mathbf{L}_c}{\mathrm{d}t} + \mathbf{r}_c \times \frac{\mathrm{d}\mathbf{p}}{\mathrm{d}t}\right)_1 \\ &\qquad + \left(\frac{\mathrm{d}\mathbf{L}_c}{\mathrm{d}t} + \mathbf{r}_c \times \frac{\mathrm{d}\mathbf{p}}{\mathrm{d}t}\right)_2 + \cdots, \end{aligned}\right\} \qquad (11.1.12)$$

where the notation implies that each term is calculated from the linear momentum and the relative angular momentum of each subsystem separately. Note that the calculation for a particular subsystem is referred to its own mass centre.

The results so far are applicable to any system of particles. We now specialise to the case of a rigid body and consider the point O to be moving with the body. Then the velocity relative to O of any particle of the body is just $\boldsymbol{\omega}\times\mathbf{r}$, where $\boldsymbol{\omega}$ is its angular velocity. The relative angular momentum about O is

$$\mathbf{L} = \Sigma\mathbf{r}\times m\frac{\mathrm{d}\mathbf{r}}{\mathrm{d}t} = \Sigma m\mathbf{r}\times(\boldsymbol{\omega}\times\mathbf{r}) = \Sigma m\{r^2\boldsymbol{\omega} - (\mathbf{r}\cdot\boldsymbol{\omega})\mathbf{r}\}. \qquad (11.1.13)$$

If axes based on O are specified, the components of angular momentum take the form

$$\left.\begin{aligned} L_x &= I_{xx}\omega_x + I_{xy}\omega_y + I_{xz}\omega_z, \\ L_y &= I_{yx}\omega_x + I_{yy}\omega_y + I_{yz}\omega_z, \\ L_z &= I_{zx}\omega_x + I_{zy}\omega_y + I_{zz}\omega_z, \end{aligned}\right\} \qquad (11.1.14)$$

where the coefficients I_{ij} satisfy $I_{ij} = I_{ji}$ and are just the moments and products of inertia as defined by (10.2.4). For axes which move with the rigid body, these coefficients will depend only on the geometry of the body, and variations in the angular momentum will arise only through variations in

Equations (11.1.14) simplify when the axes are chosen as principal axes, for then the products of inertia all vanish. To denote that principal axes are being used, we introduce the notation

$$I_{xx} = I_1, \quad I_{yy} = I_2, \quad I_{zz} = I_3, \tag{11.1.15}$$

$$\boldsymbol{\omega} = \omega_1 \mathbf{i} + \omega_2 \mathbf{j} + \omega_3 \mathbf{k}, \tag{11.1.16}$$

and then

$$\mathbf{L} = I_1 \omega_1 \mathbf{i} + I_2 \omega_2 \mathbf{j} + I_3 \omega_3 \mathbf{k}. \tag{11.1.17}$$

The components of the angular momentum are now proportional to the corresponding components of the angular velocity. Note, however, that $\mathbf{L}$ and $\boldsymbol{\omega}$ are not in general parallel vectors. For this to be so the body must rotate about a principal axis.

The simplest application of the foregoing theory is to a rigid body which is at rest. Then the linear and angular momenta are both zero and so, by (11.1.7) and (11.1.8), the external forces form a null system. Conversely, if the external forces form a null system, it follows from (11.1.7) and (11.1.8) that the linear and angular momenta are constant. Thus the velocity of the mass centre is constant and, by (11.1.14) or more directly (11.1.17), the angular velocity is constant. In particular if they are zero at one particular instant, they remain zero and the body is permanently at rest.

The dynamical applications will be confined initially to two-dimensional motion, in which all particles move parallel to the (x, y) plane and $\boldsymbol{\omega} = \omega\mathbf{k}$, say. Usually only the component of relative angular momentum perpendicular to the plane of the motion is required, namely $L_z = I_{zz}\omega$, where $I_{zz} = \Sigma m(x^2 + y^2)$ is the moment of inertia about the axis through O perpendicular to the plane of the motion. If M is the moment of the external forces about this axis we have

$$\Sigma\mathbf{F} = \frac{\mathrm{d}}{\mathrm{d}t} m_c \mathbf{v}_c, \quad M = \frac{\mathrm{d}}{\mathrm{d}t} I_{zz}\omega \tag{11.1.18}$$

provided that O is a point such that $\mathbf{r}_c \times \mathbf{a}_o = \mathbf{0}$. If it is not we replace (11.1.18) by the appropriate components of (11.1.11).

The more difficult three-dimensional problems are left until the end of the chapter and are discussed in §§11.5, 11.6.

Example 11.1.1

A uniform cube, of mass m_1 and side d, stands on a rough plane. A uniform circular cylinder, of mass m_2, diameter d and length d, rests with its curved surface in contact with the plane and with one of the vertical sides of the cube. The plane is gradually tilted about an axis parallel to the line of contact of the cube and cylinder. The coefficient of friction for every contact is μ.

Show that if $\mu < 1$, equilibrium is broken by the cube slipping down the plane and the cylinder rolling down the plane. The angle of inclination α of the

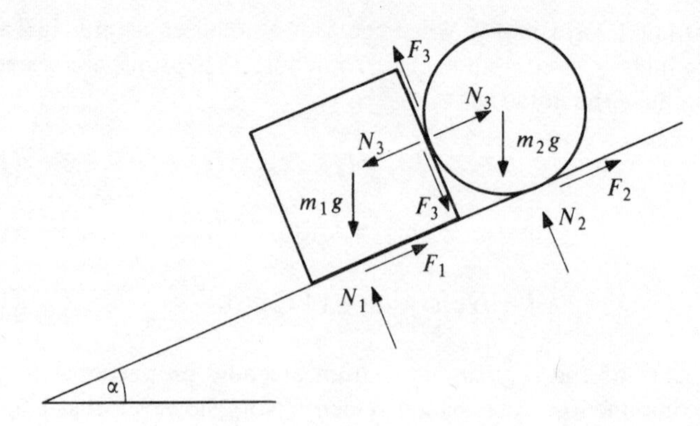

Figure 11.1.3 Illustration for Example 11.1.1.

plane to the horizontal is then given by

$$\tan \alpha = \frac{\mu m_1}{m_1 + (1 - \mu)m_2}.$$

The system of forces is shown in Figure 11.1.3. Note that the frictional forces F_1 and F_2 must act up the plane to prevent sliding. The tangential component F_3 of the contact force between the cube and the cylinder must then be in the direction shown in order that the force system should have zero moment, for equilibrium, about the axis of the cylinder. In fact since all the forces on the cylinder act through its mass centre except F_2 and F_3, we must have

$$F_2 = F_3 = F \text{ (say)}.$$

Resolving along and perpendicular to the plane for the cube and the cylinder, we get

$$m_1 g \sin \alpha + N_3 - F_1 = 0, \quad m_1 g \cos \alpha - N_1 + F_3 = 0,$$

$$m_2 g \sin \alpha - N_3 - F_2 = 0, \quad m_2 g \cos \alpha - N_2 - F_3 = 0.$$

Accordingly we have

$$N_2 = m_2 g \cos \alpha - F, \quad N_3 = m_2 g \sin \alpha - F,$$

and so $N_2 > N_3$ if $\tan \alpha < 1$. Since $F_2 = F_3$ this means that $F_2/N_2 < F_3/N_3$ if $\tan \alpha < 1$. Since μ is the same at all contacts, it follows that slipping will occur between the cylinder and the plane before it occurs between the cylinder and the cube if $\tan \alpha < 1$.

We now consider how equilibrium is broken when the cube is about to move. Let us assume that $\tan \alpha < 1$ and that the cube slips. Then

$$F_1 = \mu N_1, \quad F = \mu N_3 = \frac{\mu}{1+\mu} m_2 g \sin \alpha,$$

$$m_1 g \sin \alpha + N_3 = F_1 = \mu N_1 = \mu(m_1 g \cos \alpha + F).$$

Elimination of N_3 and F gives

$$m_1 g \sin \alpha + \frac{1}{1+\mu} m_2 g \sin \alpha = \mu m_1 g \cos \alpha + \frac{\mu^2}{1+\mu} m_2 g \sin \alpha,$$

or

$$\tan \alpha = \frac{\mu m_1}{m_1 + (1-\mu) m_2}$$

and $\tan \alpha < 1$ if $\mu < 1$, which is consistent with our original assumption.

It remains to consider the condition for which equilibrium is broken by the cube toppling about its lowest edge. When this is about to happen N_1 will have no moment about this edge, and so the moment equation about it gives

$$m_1 g(\cos \alpha - \sin \alpha) + 2F_3 - N_3 = 0,$$

or

$$m_1 g(\cos \alpha - \sin \alpha) - \frac{1-2\mu}{1+\mu} m_2 g \sin \alpha = 0,$$

so that

$$\tan \alpha = \frac{(1+\mu) m_1}{(1+\mu) m_1 + (1-2\mu) m_2}.$$

Hence equilibrium will be broken by slipping between the cube and the plane if

$$\frac{\mu m_1}{m_1 + (1-\mu) m_2} < \frac{(1+\mu) m_1}{(1+\mu) m_1 + (1-2\mu) m_2}$$

or

$$\frac{\mu(1-2\mu)}{1-\mu^2} < 1 + \frac{m_1}{m_2}.$$

For $0 \leqslant \mu \leqslant 1$, the maximum value of the left-hand side of this inequality is attained when $\mu = 2 - 3^{1/2}$ and is $(2 - 3^{1/2})^2/(3^{1/2} - 1)^2 < 1$. It follows that the above inequality is always satisfied for $0 \leqslant \mu \leqslant 1$ and so the cube will slip before it topples.

Example 11.1.2

A uniform rod AB, of mass m and length $2l$, is suspended by two strings OA, OB of equal length attached to a fixed point O. The rod is at rest in a horizontal

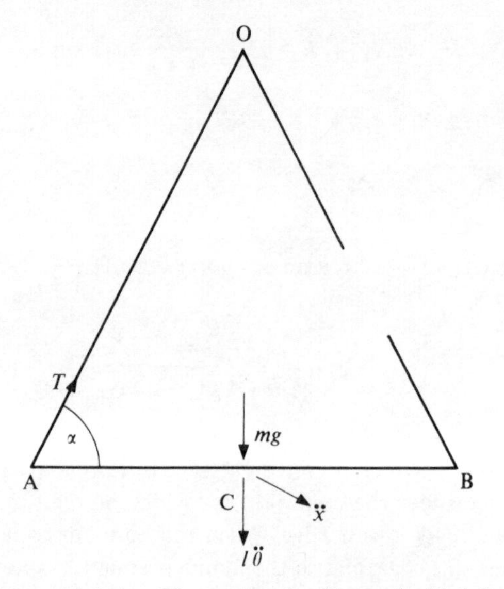

Figure 11.1.4 Illustration for Example 11.1.2.

position, and each string makes an angle α with the horizontal (see Fig. 11.1.4).

If the string OB is cut, show that the tension in OA is instantaneously reduced in the ratio $2\sin^2\alpha/(1 + 3\sin^2\alpha)$.

When OB is cut, the acceleration of the mass centre C of the rod is the acceleration of A plus the acceleration relative to A. The end A moves in a circle with the string as radius but, since the initial angular velocity is zero, it will have no radial acceleration initially. Let the transverse acceleration of A be $\ddot{x}$. Let the initial angular acceleration of the rod be $\ddot{\theta}$. Since its initial angular velocity is zero, the acceleration of the mass centre has components $\ddot{x}$ perpendicular to OA and $l\ddot{\theta}$ perpendicular to the rod.

The equations of motion are therefore

$$mg\cos\alpha = m(\ddot{x} + l\ddot{\theta}\cos\alpha),$$

$$mg\sin\alpha - T = ml\ddot{\theta}\sin\alpha,$$

$$lT\sin\alpha = \tfrac{1}{3}ml^2\ddot{\theta}.$$

The last equation is obtained by taking moments about C and using the result that the moment of inertia of the rod about a perpendicular axis through C is $\frac{1}{3}ml^2$. Elimination of $\ddot{\theta}$ from the last two equations gives

$$T = \frac{mg\sin\alpha}{1 + 3\sin^2\alpha}.$$

This compares with the value $mg/2 \sin \alpha$ when the rod is at rest supported by both strings. There is a reduction in the ratio $2 \sin^2 \alpha/(1 + 3 \sin^2 \alpha)$.

Example 11.1.3

A light, inextensible string passes over a pulley of mass m, radius a and radius of gyration k about its axis. Particles, of masses m_1 and m_2 ($m_2 > m_1$), are suspended from the ends of the string. The pulley is sufficiently rough to prevent the string from sliding and motion is resisted by a frictional couple M at the axis.

Find the acceleration of the masses when the system is released from rest.

Let x be the displacement of either particle. The equations of motion are

$$m_2 g - T_2 = m_2 \ddot{x}, \quad T_1 - m_1 g = m_1 \ddot{x}.$$

In addition, taking moments about the axis of the pulley, we have

$$(T_2 - T_1)a - M = mk^2 \ddot{x}/a,$$

since the angular acceleration of the pulley is $\ddot{x}/a$. Elimination of T_1 and T_2 from these three equations gives

$$\left(m_1 + m_2 + m\frac{k^2}{a^2}\right)\ddot{x} = (m_2 - m_1)g - \frac{M}{a}.$$

Since $m_2 > m_1$, $\ddot{x}$ cannot be negative. It follows that for motion to be possible we must have

$$M < (m_2 - m_1)ga$$

otherwise the system will be in equilibrium with $M = (m_2 - m_1)ga$.

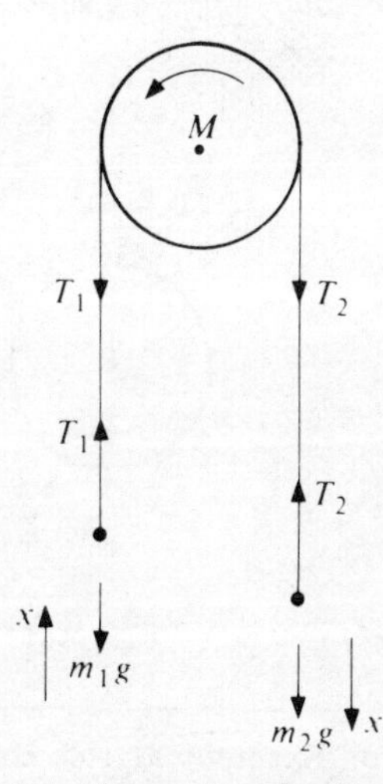

Figure 11.1.5 Illustration for Example 11.1.3.

Example 11.1.4

We consider the problem of a body of circular cross-section rolling and slipping down an inclined plane. We shall assume that the body is symmetrical about a circular cross-section, and that the mass centre coincides with the geometrical centre of this circular cross-section. This enables us to treat the problem as a two-dimensional one. The uniform sphere and the uniform cylinder are special cases.

Let a be the radius of the body in the plane of symmetry, x the displacement of the mass centre C, θ the angular displacement of the body, F the frictional force and N the normal reaction (see Fig. 11.1.6). The equations of motion are

$$mg \cos \alpha - N = 0, \quad mg \sin \alpha - F = m\ddot{x}, \quad Fa = mk^2 \ddot{\theta},$$

where k is the radius of gyration of the body about an axis through C.

In the above three equations there are four unknowns, namely F, N, x, θ. Hence a further relation is required before the equations can be solved.

We consider first the possibility of the body rolling without slipping, so that its point of contact with the plane is instantaneously at rest. The kinematical condition for this is

$$\dot{x} - a\dot{\theta} = 0 \quad \text{or} \quad \ddot{x} - a\ddot{\theta} = 0$$

which gives

$$mg \sin \alpha - F = F a^2 / k^2,$$

or

$$F = \frac{k^2}{a^2 + k^2} mg \sin \alpha,$$

and so

$$\ddot{x} = a\ddot{\theta} = \frac{a^2}{a^2 + k^2} g \sin \alpha.$$

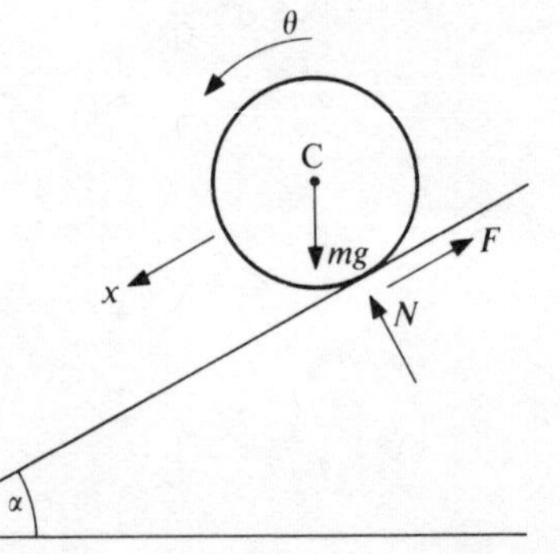

Figure 11.1.6 Illustration for Example 11.1.4: circular body rolling and slipping on an inclined plane.

Note that, by taking moments about the instantaneous centre, we can obtain directly the relation

$$mga \sin \alpha = m(k^2 + a^2)\ddot{\theta}.$$

The right-hand side here is the rate of change of relative angular momentum because the acceleration of the instantaneous centre is towards C ($\mathbf{a}_0$ parallel to $\mathbf{r}_C$ in the notation of equations (11.1.9)). To see this we can calculate the acceleration of the instantaneous centre as the resultant of the acceleration of C, which is $\ddot{x}$ down the plane, and the acceleration relative to C, which has the components $a\ddot{\theta}$ up the plane and $a\dot{\theta}^2$ towards C. The resultant has components $\ddot{x} - a\ddot{\theta} = 0$ down the plane and $a\dot{\theta}^2$ towards C.*

If μ is the coefficient of friction we must have $F \leqslant \mu N$. If there is no slipping this gives $k^2 \tan \alpha \leqslant \mu(a^2 + k^2)$. This is a necessary condition to be satisfied if there is to be no slipping, but if the motion is to start as a rolling motion it is also necessary that the initial velocity v_0 of the point of contact between the body and the plane be zero. The condition for this is $v_0 = \dot{x}_0 - a\dot{\theta}_0 = 0$, where $\dot{x}_0$ and $\dot{\theta}_0$ are the initial values of $\dot{x}$ and $\dot{\theta}$. In this case, if the condition $k^2 \tan \alpha \leqslant \mu(a^2 + k^2)$ is also satisfied, the body will continue to roll. If, however, v_0 is positive, the body will start by slipping down the plane. The frictional force F will be limiting and will act up the plane. Conversely if v_0 is negative F will be limiting and will act down the plane. To take the former case, $v_0 > 0$, the kinematical condition of no slip is replaced by $F = \mu N = \mu mg \cos \alpha$ as the fourth relation. This gives

$$\ddot{x} = g(\sin \alpha - \mu \cos \alpha), \quad \ddot{\theta} = \mu ga \cos \alpha / k^2$$

and so, by integration,

$$\dot{x} - a\dot{\theta} = v_0 + g\left\{\sin \alpha - \left(1 + \frac{a^2}{k^2}\right)\mu \cos \alpha\right\}t.$$

Unless $k^2 \tan \alpha \leqslant \mu(a^2 + k^2)$, $\dot{x} - a\dot{\theta}$ will be positive for all positive t, which verifies our previous conclusion that the inequality is a necessary condition for rolling. If $k^2 \tan \alpha = \mu(a^2 + k^2)$, rolling is possible only if $v_0 = 0$, but if $k^2 \tan \alpha < \mu(a^2 + k^2)$ there is a value of $t \geqslant 0$ for which $\dot{x} - a\dot{\theta}$ is zero. The motion for later times is one of rolling.

If $v_0 < 0$, then $F = -\mu mg \cos \alpha$ and

$$\dot{x} - a\dot{\theta} = v_0 + g\left\{\sin \alpha + \left(1 + \frac{a^2}{k^2}\right)\mu \cos \alpha\right\}t.$$

* Note that when taking moments about the instantaneous centre, the unrestricted relation (11.1.11) will in general be required. The restricted relation (11.1.9) is valid here only because of the special properties of the kinematics associated with a rolling body of circular cross-section.

The condition $\dot{x} - a\dot{\theta} = 0$ is certainly attained at some positive value of t. Whether or not the body will subsequently roll depends on whether or not the inequality $k^2 \tan \alpha \leqslant \mu(a^2 + k^2)$ is satisfied.

Example 11.1.5

In this example we investigate the dynamics of a car moving in a straight path without slipping. The mass of the car body, excluding the wheels and axles, is m and its mass centre C is at height h above the ground and a horizontal distance d from the rear axle. The distance apart of the front and rear axles is l, and the radius of the car wheels is a. The mass centre of the front wheels and axle is C_1, the mass is m_1 and the radius of gyration of the system about the axle is k_1. The corresponding quantities for the rear wheels and axle are C_2, m_2, k_2, respectively. It is assumed that the car has a vertical plane of symmetry defined by the plane containing C, C_1 and C_2.

We consider first the forces involved and we distinguish between those which are external to the car as a whole and those which are external to the wheels and axles. For the car as a whole there are the gravitational forces mg, m_1g and m_2g acting vertically through C, C_1 and C_2 in the plane of symmetry. There are also the reactions at the points of contact of the wheels with the ground. For either pair of wheels these reactions can be combined into a single equivalent resultant in the plane of symmetry. Let the resultants have components in the horizontal and vertical directions equal to F_1, N_1 at the front and F_2, N_2 at the rear, as shown in Figure 11.1.7. It is assumed that the car is being driven forward by the rear wheels, which means that the frictional force F_2 must be in the direction of motion to prevent the slip that would otherwise occur if the rear wheels rotated

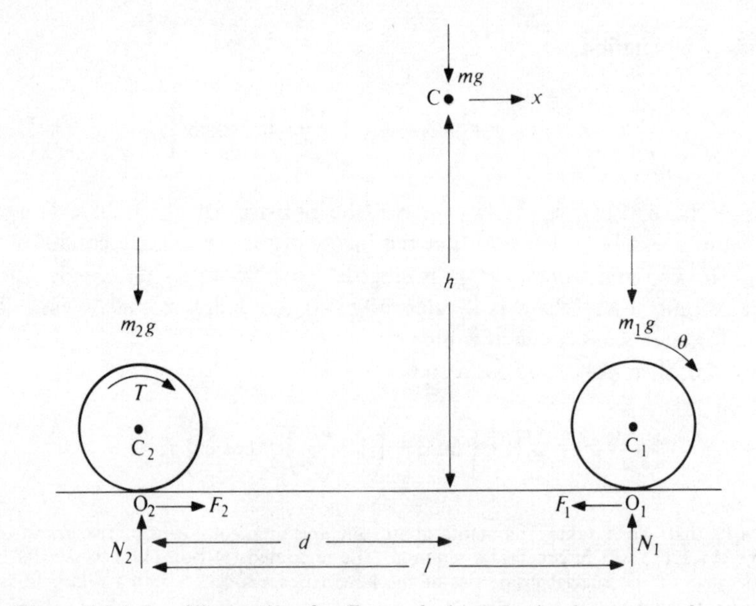

Figure 11.1.7 Illustration for Example 11.1.5: the dynamics of a car.

with the car at rest. In fact F_2 is the force which moves the car forward. Conversely the front wheels are pulled along by the car and so F_1 opposes the motion of the car to prevent slip.

Internal to the car as a whole, but external to the rear wheels and axle, is the torque or couple assumed to be applied to the rear wheels by the engine. This is denoted by T. There are also reactions where the car body is mounted on the axles. These are assumed to have no moment about C_1 or C_2.

Let the horizontal displacement of C be x, and let the angular displacement of the wheels be θ. The kinematical condition of no slip gives $x - a\theta = 0$. The equations of motion for the car as a whole are

$$F_2 - F_1 = (m + m_1 + m_2)\ddot{x}, \qquad (11.1.19)$$

$$N_1 + N_2 - (m + m_1 + m_2)g = 0. \qquad (11.1.20)$$

The moment equation for the whole system requires some care. It is possible to define a mass centre for the system as a whole, but this would not be a convenient procedure here since we are dealing with three systems. These are: (i) the front wheels and axle; (ii) the rear wheels and axle; and (iii) the car body; they do not move as a single rigid body because the wheels can rotate relative to the rest of the car. This is one of the exceptional situations where it is preferable to choose a convenient reference point and apply the unrestricted result in its more general form (11.1.12). We choose to take moments about the point in the plane of symmetry which is equivalent in the two-dimensional problem to the point of contact of the rear wheels with the ground, denoted by O_2 in Figure 11.1.7. The advantage of this is that F_1, F_2, N_2 have no moments about O_2 and the resulting relation can be used to determine N_1. The car body has no angular momentum about its mass centre C, so that $d\mathbf{L}_c/dt$ is zero and the only contribution to the moment is through the term $\mathbf{r}_c \times d\mathbf{p}/dt$. Since the linear momentum of the body is $m\dot{x}$ horizontally, and since the height of C above O_2 is h, this contribution is $hm\ddot{x}$ (the vector character of the moment is suppressed since the direction of all moments is perpendicular to the plane of the problem). For the front wheels and axle, the relative angular momentum about C_1 is $mk_1^2\dot{\theta}$, so that $d\mathbf{L}_c/dt$ for this component is of magnitude $mk_1^2\ddot{\theta}$. The linear momentum is $m_1\dot{x}$ horizontally and contributes a term $am_1\ddot{x}$ to the moment about O_2. Similarly the contribution from the rear wheels and axle is $m_2k_2^2\ddot{\theta} + am_2\ddot{x}$. The resulting equation is

$$\begin{aligned} mgd + (m_1g - N_1)l &= mh\ddot{x} + m_1k_1^2\ddot{\theta} + m_1a\ddot{x} + m_2k_2^2\ddot{\theta} + m_2a\ddot{x} \\ &= \{mh + m_1(a + k_1^2/a) + m_2(a + k_2^2/a)\}\ddot{x}. \qquad (11.1.21) \end{aligned}$$

For the wheels and axle systems, moments about C_1 and C_2 for each system separately give

$$F_1a = m_1k_1^2\ddot{\theta}, \quad T - F_2a = m_2k_2^2\ddot{\theta}. \qquad (11.1.22)$$

Note that, by assumption, the reactions between the car body and the axles do not contribute.

We now have, from (11.1.19) and (11.1.22),

$$a(F_2 - F_1) = a(m + m_1 + m_2)\ddot{x} = T - (m_1 k_1^2 + m_2 k_2^2)\ddot{\theta}$$

or

$$T = (m + m_1 + m_2 + m_1 k_1^2/a^2 + m_2 k_2^2/a^2)a\ddot{x}.$$

This equation serves to give the linear acceleration directly in terms of the torque of the engine. When $\ddot{x}$ is known we can obtain N_1 from (11.1.21), which gives

$$lN_1 = mgd + m_1 gl - \{mh + m_1(a + k_1^2/a) + m_2(a + k_2^2/a)\}\ddot{x},$$

and then, with the help of (11.1.20), we have

$$lN_2 = mg(l - d) + m_2 gl + \{mh + m_1(a + k_1^2/a) + m_2(a + k_2^2/a)\}\ddot{x},$$

which is equivalent to the moment equation about the point O_1 of Figure 11.1.7.

If the torque is applied to the front wheels, the only modification is that

$$T + F_1 a = m_1 k_1^2 \ddot{\theta}, \quad -F_2 a = m_2 k_2^2 \ddot{\theta}.$$

Here the signs of F_1 and F_2 will change but the expression for $F_2 - F_1$ is unchanged. Since $\ddot{x}$ depends only on $F_2 - F_1$, it follows that the expressions for N_1 and N_2 are unchanged. Now slipping begins at the front wheels when $|F_1| = \mu N_1$ and at the rear wheels when $|F_2| = \mu N_2$, and so it depends on the magnitudes of the normal reactions.

There are two factors affecting the magnitudes of N_1 and N_2. The first is the position of the mass centre C. In a front-engined car C will tend to be nearer the front, and so $d > l - d$. This factor will tend to increase N_1 relative to N_2 and improve the adhesion of the front wheels. On the other hand acceleration has the opposite effect, since it increases N_2 and decreases N_1. When the acceleration is substantial it will tend to be the dominant effect so that the rear wheels are able to sustain a larger frictional force than the front wheels before slipping. This is a reason in favour of using the rear wheels to drive the car. Conversely the act of braking when $\ddot{x} < 0$ tends to increase N_1 relative to N_2 and more reliable braking is obtained through the front wheels.

11.2 IMPULSE AND MOMENTUM

Equations (11.1.3), with **F** interpreted as a typical external force, can be integrated over an interval of time to give

$$\Sigma \int \mathbf{F}\, dt = \Sigma \int m\mathbf{a}\, dt, \quad \Sigma \int \mathbf{r} \times \mathbf{F}\, dt = \Sigma \int \mathbf{r} \times m\mathbf{a}\, dt. \tag{11.2.1}$$

We shall discuss equations (11.2.1) only in the limit of idealised impulses when the time interval tends to zero but

$$\bar{\mathbf{F}} = \int \mathbf{F}\, dt$$

remains finite.* We then have an instantaneous change in the velocities of the particles, but no instantaneous change in position since the velocities remain finite while the time interval tends to zero. In this case the relation between moments in (11.2.1) can be written

$$\Sigma \mathbf{r} \times \int \mathbf{F}\, dt = \Sigma \mathbf{r} \times \int m\mathbf{a}\, dt$$

and equations (11.2.1) become, in terms of $\bar{\mathbf{F}}$,

$$\left.\begin{aligned} \Sigma \bar{\mathbf{F}} &= \Sigma \int \frac{d}{dt}(m\mathbf{v})\, dt = [\Sigma m\mathbf{v}] = [m_c \mathbf{v}_c], \\ \Sigma \mathbf{r} \times \bar{\mathbf{F}} &= \Sigma \mathbf{r} \times \int \frac{d}{dt}(m\mathbf{v})\, dt = \Sigma \mathbf{r} \times [m\mathbf{v}] = [\Sigma \mathbf{r} \times m\mathbf{v}], \end{aligned}\right\} \tag{11.2.2}$$

where the square brackets denote the instantaneous change of the quantity enclosed.

In the calculation of the angular momentum $\Sigma \mathbf{r} \times m\mathbf{v}$ about a point O, we can write

$$\mathbf{v} = \mathbf{v}_o + d\mathbf{r}/dt,$$

and the analogue of (11.1.4), with $\mathbf{a}$ replaced by $\mathbf{v}$, is

$$\Sigma \mathbf{r} \times m\mathbf{v} = m_c \mathbf{r}_c \times \mathbf{v}_o + \Sigma \mathbf{r} \times m\, d\mathbf{r}/dt. \tag{11.2.3}$$

It follows that the total angular momentum about O is equal to the relative angular momentum provided that $\mathbf{r}_c \times \mathbf{v}_o = \mathbf{0}$. The practically useful cases are when moments are taken about C ($\mathbf{r}_c = \mathbf{0}$) and when moments are taken about a fixed point ($\mathbf{v}_o = \mathbf{0}$). For such points equations (11.2.2) give

$$\Sigma \bar{\mathbf{F}} = \bar{\mathbf{R}} = [\mathbf{p}], \quad \Sigma \mathbf{r} \times \bar{\mathbf{F}} = \bar{\mathbf{M}} = [\mathbf{L}] \quad \textit{if} \quad \mathbf{r}_c \times \mathbf{v}_o = \mathbf{0}, \tag{11.2.4}$$

where $\bar{\mathbf{F}}$ is a typical external impulse and $\mathbf{L}$ is the relative angular momentum

* Remember that any force which does not become correspondingly large as the time interval becomes small will not contribute to the system of impulses.

about O, as defined in (11.1.6). Equations (11.2.4) imply that the system of external impulses is equivalent to the vector couple combination [**p**], [**L**] at O.

There is also the analogue of (11.1.11), namely

$$\Sigma\bar{\mathbf{F}} = \bar{\mathbf{R}} = [\mathbf{p}], \quad \Sigma\mathbf{r} \times \bar{\mathbf{F}} = \bar{\mathbf{M}} = [\mathbf{L}_c + \mathbf{r}_c \times \mathbf{p}] \tag{11.2.5}$$

without restriction on the origin. These equations follow directly by integration of (11.1.11) if account is taken of the fact that **r** does not change during the interval of integration. They are also a consequence of the equivalence of the systems ($\mathbf{p}$, $\mathbf{L}_c$) at C and ($\mathbf{p}$, $\mathbf{L}_c + \mathbf{r}_c \times \mathbf{p}$) at O.

Example 11.2.1

Suppose a rigid body suffers an impulsive blow $\bar{F}$ in a plane of symmetry, so that the motion is two dimensional. We show that although the body is instantaneously set in motion, there can be points for which the induced velocity is zero.

Let $\bar{F}$ act along a line whose perpendicular distance from the mass centre C is x. Let the instantaneous angular speed of the body be ω and let the instantaneous speed of C be v_c (see Fig. 11.2.1). We have

$$\bar{F} = mv_c, \quad \bar{F}x = mk^2\omega.$$

If P is a point on the perpendicular, through C, to the line of $\bar{F}$ and if PC $= y$, the instantaneous speed of P is $v_c - y\omega$. This is zero if $y = v_c/\omega = k^2/x$. For example a uniform rod, of length l, will begin to turn about one end if given an impulse perpendicular to its length at a distance $\frac{2}{3}l$ from that end. A uniform sphere of radius a resting on a horizontal plane will begin to roll without slipping if it receives an impulse at a height $7a/5$ above the plane. This is the correct height for the cushion of a billiards table.

Example 11.2.2

Two uniform rods AO, OB, each of mass m, length $2l$ and radius of gyration

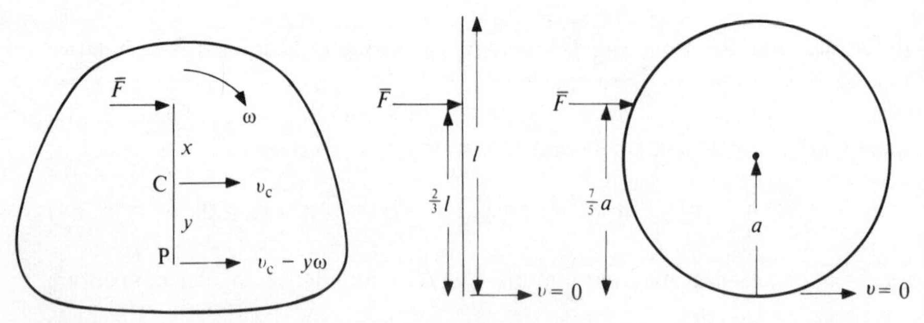

Figure 11.2.1 Illustration for Example 11.2.1: two-dimensional motion produced by an impulse.

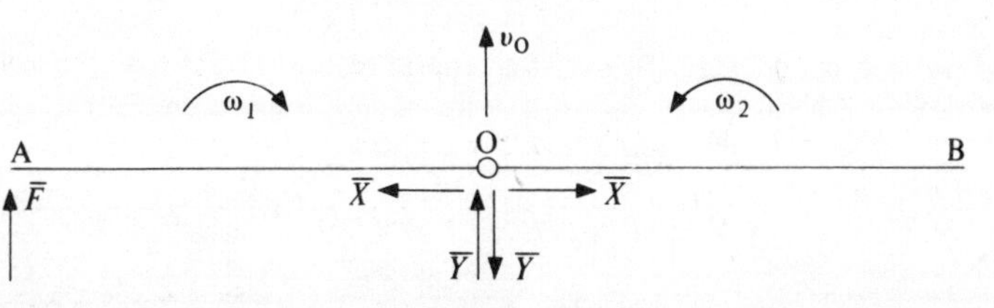

Figure 11.2.2 Illustration for Example 11.2.2: impulsive motion of two jointed rods.

$l/3^{1/2}$ about an axis through its mass centre, are smoothly jointed together at O and are at rest in a straight line. An impulse $\bar{F}$ is applied at A perpendicular to the line of the rods. We wish to determine the instantaneous motion of the two rods.

The problem is two dimensional in the plane defined by the initial line of the rods and the line of $\bar{F}$. There will be an internal impulsive reaction at the hinge producing equal and opposite impulses on the rods, say with components $\bar{X}$ along the rods and $\bar{Y}$ perpendicular to the rods in the directions shown in Figure 11.2.2.

Since the impulses on the two rods parallel to AB, namely $\pm\bar{X}$, are equal and opposite, any velocity components acquired by the mass centres of the rods parallel to AB must be equal and opposite. This is not consistent with the constraint imposed by the hinge unless $\bar{X} = 0$ and the rods move perpendicular to AB initially. Let the velocity of O be v_o perpendicular to AB and let the angular velocities of the rods be ω_1, ω_2 perpendicular to the plane of the problem, as shown in Figure 11.2.2. These quantities are sufficient to determine the motion of the rods.

The impulse equations can be written down separately for the mass centres of the rods. We have

$$\bar{F} + \bar{Y} = m(v_o + l\omega_1), \quad -\bar{Y} = m(v_o + l\omega_2),$$

and, by taking moments about the mass centres of the rods, we get

$$(\bar{F} - \bar{Y})l = \tfrac{1}{3}ml^2\omega_1, \quad \bar{Y}l = \tfrac{1}{3}ml^2\omega_2.$$

We then have four equations to determine v_o, ω_1, ω_2 and $\bar{Y}$. Alternatively, if the information required is confined to the motion of the rods, it is possible to write down three equations which do not contain $\bar{Y}$. For example, the impulse equation for the whole system is

$$\bar{F} = m(v_o + l\omega_1) + m(v_o + l\omega_2).$$

Two further equations not containing $\bar{Y}$ are obtainable by taking moments about the hinge for each rod separately. Since O is then neither the mass centre nor a fixed point it is necessary to use the general result (11.2.5), which gives

$$2l\bar{F} = ml(v_0 + l\omega_1) + \tfrac{1}{3}ml^2\omega_1,$$

$$0 = ml(v_0 + l\omega_2) + \tfrac{1}{3}ml^2\omega_2.$$

Either set of equations can be solved to give $v_0 = -\bar{F}/m$, $\omega_1 = 9\bar{F}/(4ml)$, $\omega_2 = 3\bar{F}/(4ml)$.

Example 11.2.3
A uniform inelastic sphere of radius a rolls without slipping along a horizontal plane with speed u. It strikes a kerb, of height $a/5$, which is at right angles to its path. Show that the conditions for the sphere to turn without slipping until it has surmounted the kerb are

$$18u^2 > 7ga, \quad \mu(98ga - 90u^2) > 21ga,$$

where μ is the coefficient of friction.

As the sphere strikes the kerb, there is an impulsive change from rolling with speed u to rotation about the point of contact with angular speed ω, say. Since the only impulsive force arises at the kerb, there will be no change in angular momentum about this point. Since this is not a fixed point before impact, we must again use (11.2.5) to get

$$mu(a - h) + mk^2u/a = m(a^2 + k^2)\omega,$$

where $h = a/5$, $k^2 = 2a^2/5$, and so $\omega = 6u/7a$.

Let the impulsive reaction on the sphere at the kerb have components $\bar{N}$ radially and $\bar{F}$ tangentially, as shown in Figure 11.2.3. Then we have

$$\bar{F} = m(a\omega - u\sin\theta), \quad \bar{N} = mu\cos\theta,$$

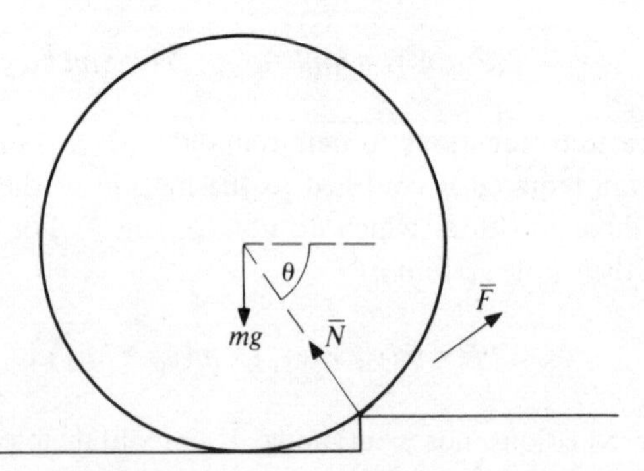

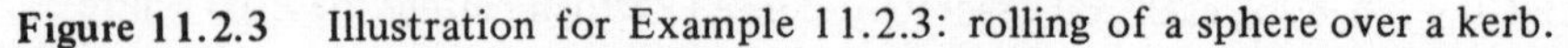

Figure 11.2.3 Illustration for Example 11.2.3: rolling of a sphere over a kerb.

where $\sin\theta = 4/5$, $\cos\theta = 3/5$. If the sphere is not to slip initially, it is necessary that $\bar{F} < \mu\bar{N}$, or

$$\frac{6u}{7} - \frac{4u}{5} < \frac{3\mu u}{5},$$

which gives $\mu > 2/21$.

In the subsequent motion, we continue to use θ for the angle made by the radius vector to the kerb with the horizontal, and we denote by F, N the tangential and radial components of the reaction at the kerb. The equations of motion are

$$F - mg\cos\theta = ma\ddot{\theta},$$

$$N - mg\sin\theta = -ma\dot{\theta}^2,$$

$$mga\cos\theta = -\tfrac{7}{5}ma^2\ddot{\theta} = -\tfrac{7}{5}ma^2\dot{\theta}\,\frac{d\dot{\theta}}{d\theta}.$$

The final equation integrates to

$$mga\left(\sin\theta - \frac{4}{5}\right) = -\tfrac{7}{10}ma^2\left(\dot{\theta}^2 - \frac{36u^2}{49a^2}\right).$$

In order that the sphere surmounts the kerb, $\dot{\theta}^2$ must remain positive until $\theta = \pi/2$, so that

$$36u^2/49a^2 > 2g/7a, \quad \text{or} \quad 18u^2 > 7ga.$$

In order that there should be no slipping during this motion, it is necessary that $F < \mu N$, and so

$$m(g\cos\theta + a\ddot{\theta}) < \mu m(g\sin\theta - a\dot{\theta}^2).$$

The most critical case occurs initially, since $\cos\theta$, $\dot{\theta}$, $\ddot{\theta}$ all decrease, and $\sin\theta$ increases. Substitution of the values

$$\sin\theta = \tfrac{4}{5}, \quad \cos\theta = \tfrac{3}{5}, \quad \dot{\theta} = 6u/7a, \quad \ddot{\theta} = -3g/7a$$

gives

$$\tfrac{3}{5}g - \tfrac{3}{7}g < \mu\left(\frac{4}{5}g - \frac{36u^2}{49a}\right),$$

or

$$21 < \mu\left(98 - \frac{90u^2}{ga}\right).$$

The inequality required for no slip during impact is clearly covered by this.

11.3 WORK AND ENERGY

It has already been noted that the system of vectors consisting of the external forces $\mathbf{F}$ and the system of vectors consisting of the mass accelerations $m\mathbf{a}$ are equivalent in the sense of the definition introduced in §10.3. Hence the work done by these two systems in a rigid body displacement is the same (see §10.6). The work done by the mass acceleration system is the increase in kinetic energy since

$$\Sigma \int m\mathbf{a} \,.\, \mathrm{d}\mathbf{r} = \Sigma \int m \frac{\mathrm{d}\mathbf{v}}{\mathrm{d}t} \,.\, \mathbf{v}\, \mathrm{d}t = [\Sigma \tfrac{1}{2} m v^2].$$

Thus we arrive again at the energy relation, namely that the work done by the external forces is equal to the increase in kinetic energy. Note, however, the slight change in the argument compared with that used in §8.2. Here there is no mention of the internal forces. The equivalence of the two systems of vectors is itself sufficient to ensure that they do the same work in any rigid body displacement.

The result can be generalised to a system consisting of several particles and rigid bodies by summation over the constituent parts. However, the calculation of the work done by the forces must then include the contribution, if any, from the forces of constraint. Such a contribution might arise, for example, from the stretching of a string, or the sliding of two rough bodies in contact. On the other hand there are a number of situations for which the constraints are workless. For example the net work done by the reactions $\pm\mathbf{X}$ between the two surfaces in contact is

$$\int \mathbf{X} \,.\, \mathrm{d}\mathbf{r},$$

where $\mathrm{d}\mathbf{r}$ refers to the relative displacement. If the contact is smooth $\mathbf{X}$ has no component in the direction of relative displacement, so that the work done is zero. If the contact is rough, but there is no sliding, the relative velocity is zero and so

$$\int \mathbf{X} \,.\, \mathrm{d}\mathbf{r} = \int \mathbf{X} \,.\, \mathbf{v}\, \mathrm{d}t = 0.$$

Another example arises from the tensions at the two ends of a light inextensible string, which are equal in magnitude. The displacement of the two ends is composed of two components, along and perpendicular to the string. The component perpendicular to the string is also perpendicular to the tension, and therefore makes no contribution to the work integral. The component parallel to the string will make equal and opposite contributions to the work integral from the two ends and so these contributions cancel. This will be the case if the string is free or if it passes over a smooth peg but not if the peg is rough (see Example 5.3.1).

Workless constraints do not contribute to the work equation

$$\Sigma \int \mathbf{F} \,.\, \mathrm{d}\mathbf{r} = [\Sigma \tfrac{1}{2} m v^2].$$

This is particularly useful in applications since the forces of constraint are usually not known *a priori*. Those forces which do contribute to the work equation are called applied forces, in order to distinguish them from the forces involved in the workless constraints. When the applied forces are conservative, the work can be evaluated from the loss in potential energy, as described in Chapter 5, and the work equation is simply the energy equation

kinetic energy + potential energy = constant.

To calculate the kinetic energy of a rigid body, let O be a point moving with the body and let $\boldsymbol{\omega}$ be the angular velocity. A point of the body with position vector $\mathbf{r}$ relative to O has velocity $\boldsymbol{\omega} \times \mathbf{r}$ relative to O. The kinetic energy of the motion relative to O is therefore $\Sigma \frac{1}{2} m(\boldsymbol{\omega} \times \mathbf{r})^2$. With the help of (11.1.13), this is seen to be equal to $\frac{1}{2}\mathbf{L} \,.\, \boldsymbol{\omega}$, since

$$\tfrac{1}{2}\mathbf{L} \,.\, \boldsymbol{\omega} = \tfrac{1}{2}\boldsymbol{\omega} \,.\, \Sigma m \mathbf{r} \times (\boldsymbol{\omega} \times \mathbf{r}) = \Sigma \tfrac{1}{2} m(\boldsymbol{\omega} \times \mathbf{r})^2 \tag{11.3.1}$$

with the help of the relation $\boldsymbol{\omega} \,.\, (\mathbf{r} \times \mathbf{v}) = \mathbf{v} \,.\, (\boldsymbol{\omega} \times \mathbf{r})$. The most general expression for the kinetic energy of the motion relative to O is therefore, from (11.1.14),

$$\Sigma \tfrac{1}{2} m(\boldsymbol{\omega} \times \mathbf{r})^2 = \tfrac{1}{2}\mathbf{L} \,.\, \boldsymbol{\omega} = \tfrac{1}{2}(I_{xx}\omega_x^2 + I_{yy}\omega_y^2 + I_{zz}\omega_z^2 + 2I_{yz}\omega_y\omega_z + 2I_{zx}\omega_z\omega_x + 2I_{xy}\omega_x\omega_y). \tag{11.3.2}$$

It simplifies to

$$\tfrac{1}{2}\mathbf{L} \,.\, \boldsymbol{\omega} = \tfrac{1}{2} I \omega^2, \tag{11.3.3}$$

where I is the moment of inertia about the axis defined by $\boldsymbol{\omega}$. This is readily demonstrated by taking instantaneous axes such that $\boldsymbol{\omega} = (\omega, 0, 0)$. It also simplifies to

$$\tfrac{1}{2}\mathbf{L} \,.\, \boldsymbol{\omega} = \tfrac{1}{2}(I_1 \omega_1^2 + I_2 \omega_2^2 + I_3 \omega_3^2) \tag{11.3.4}$$

when the axes are principal axes, so that the products of inertia vanish.

In applications, the above results are most useful when O is a fixed point, so that the relative kinetic energy is the total kinetic energy, or when O is the mass centre. In the latter case the total kinetic energy T can be written, with the help of (8.2.8), (11.1.5) and (11.3.3), as

$$T = \tfrac{1}{2} m_c v_c^2 + \tfrac{1}{2} I_c \omega^2 = \tfrac{1}{2}\mathbf{p} \,.\, \mathbf{v}_c + \tfrac{1}{2}\mathbf{L}_c \,.\, \boldsymbol{\omega}, \tag{11.3.5}$$

where $\mathbf{L}_c$ is the relative angular momentum about the mass centre C and I_c is the moment of inertia about an axis through C parallel to $\boldsymbol{\omega}$.

Example 11.3.1

A uniform circular disc turns about a horizontal axis through its centre O, the motion being resisted by a frictional couple whose magnitude is constant and equal to M when the disc rotates. A particle is attached to a point P on the rim of the disc. The system is released from rest with OP horizontal and swings through an angle $(\frac{1}{2}\pi + \theta)$ before coming to rest (see Fig. 11.3.1).

Show that: (i) the disc will not move from rest again unless $(\frac{1}{2}\pi + \theta)\tan\theta > 1$; (ii) if this condition is satisfied, OP will not reach the vertical again unless

$$\cos\theta < (\pi + 2\theta)/(\pi + 4\theta).$$

The loss in potential energy of the particle is $mga\cos\theta$. The work done by the frictional couple M is $-M(\frac{1}{2}\pi + \theta)$. Hence

$$mga\cos\theta = M(\tfrac{1}{2}\pi + \theta).$$

The particle will not move unless the moment of the force mg can overcome the couple M, or $mga\sin\theta > M$. Therefore we must have

$$(\tfrac{1}{2}\pi + \theta)\tan\theta > 1.$$

If OP is to reach the vertical, the loss in potential energy must not be less than the work done by the frictional couple, since the kinetic energy cannot be negative. This means that

$$mga(1 - \cos\theta) \geqslant M\theta$$

or

$$\cos\theta \leqslant \frac{\pi + 2\theta}{\pi + 4\theta}.$$

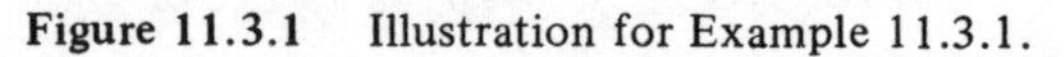

Figure 11.3.1 Illustration for Example 11.3.1.

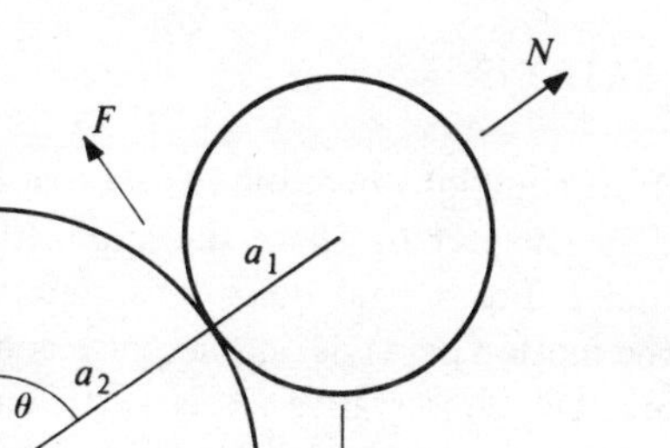

Figure 11.3.2 Illustration for Example 11.3.2: rolling of a sphere on a fixed sphere.

Example 11.3.2

A uniform solid sphere is slightly displaced from its position of unstable equilibrium at the topmost point of a fixed sphere and it then rolls on this sphere (see Fig. 11.3.2). Show that, irrespective of the magnitude of the coefficient of friction, slipping must occur before the angle θ which the join of the centres of the spheres makes with the vertical attains the value $\cos^{-1}(10/17)$.

Let the radii of the spheres be a_1, a_2. The centre of the rolling sphere describes a circle of radius $(a_1 + a_2)$. Its acceleration therefore has radial and transverse components $-(a_1 + a_2)\dot{\theta}^2$ and $(a_1 + a_2)\ddot{\theta}$ respectively. The equations of motion are

$$mg \cos \theta - N = m(a_1 + a_2)\dot{\theta}^2,$$

$$mg \sin \theta - F = m(a_1 + a_2)\ddot{\theta}.$$

The velocity of the centre of the rolling sphere is $(a_1 + a_2)\dot{\theta}$. It is also equal to $a_1\omega$, where ω is the angular speed of the sphere, since the sphere is instantaneously rotating about the common point of contact. Hence $\omega = (a_1 + a_2)\dot{\theta}/a_1$ and the energy equation gives

$$\tfrac{1}{2}m(a_1 + a_2)^2\dot{\theta}^2 + \tfrac{1}{2}(\tfrac{2}{5}ma_1^2)\omega^2 - mg(a_1 + a_2)(1 - \cos\theta) = 0,$$

$$(a_1 + a_2)\dot{\theta}^2 = \frac{10}{7}g(1 - \cos\theta),$$

$$N = mg\left(\cos\theta - \frac{10}{7}(1 - \cos\theta)\right) = \frac{mg}{7}(17\cos\theta - 10).$$

Accordingly N is zero when $\cos\theta = 10/17$. Slipping must occur before this.

11.4 OSCILLATIONS

The mathematical equations involved in the theory of oscillations have already been discussed in Chapter 6. There the applications were wholly to the oscillations of particles. For a rigid body it is simply a matter of deriving similar equations. If the motion depends on one space coordinate, and if the only forces which do work are conservative, it is sufficient to write down the energy equation, as the following examples illustrate.

Example 11.4.1

A rigid body, allowed to oscillate about a horizontal axis through O, is called a compound pendulum (see Fig. 11.4.1). Moments about the fixed axis give

$$mgl \sin \theta = -m(k^2 + l^2)\, d^2\theta/dt^2 ,$$

where θ is the angular displacement, k is the radius of gyration about the mass centre C, and $OC = l$. The integral of the above equation gives the energy equation, namely

$$\tfrac{1}{2}m(k^2 + l^2)(d\theta/dt)^2 - mgl \cos \theta = \text{constant}.$$

The equations are similar to those for the simple pendulum (see Example 6.5.1) and can be integrated in similar fashion. For small oscillations we have

$$\frac{d^2\theta}{dt^2} = -n^2\theta, \quad n^2 = \frac{gl}{k^2 + l^2},$$

so that the oscillations are simple harmonic with period $2\pi/n$.

If the pendulum is subject to a resisting couple M, the equation of motion is

$$(k^2 + l^2)\frac{d^2\theta}{dt^2} + gl \sin \theta + \frac{M}{m} = 0.$$

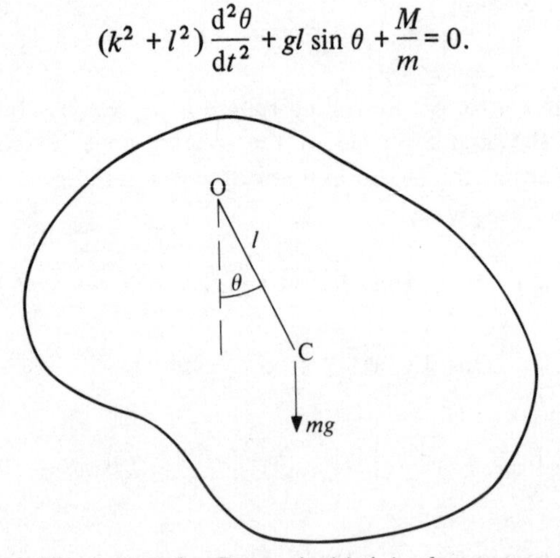

Figure 11.4.1 Illustration for Example 11.4.1: the compound pendulum.

When M arises from solid friction, it is constant in magnitude (but its sign changes with that of $d\theta/dt$). The analysis for small oscillations corresponds to that of Example 6.1.3. For viscous damping M is proportional to $d\theta/dt$. The analysis for small oscillations corresponds to that of §6.2. For quadratic damping M is proportional to $d\theta/dt \mid d\theta/dt \mid$. The equation in this case is essentially non-linear even when the amplitude is small. It is discussed in Chapter 13.

Example 11.4.2

We consider the equilibrium of a cylinder resting on a fixed cylinder. The generators of the two cylinders are horizontal, their perimeters are assumed to have continuous curvatures, otherwise their cross-sections are arbitrary. The contact is assumed to be sufficiently rough to prevent sliding.

We can treat this as a two-dimensional problem in the vertical plane through the mass centre C of the movable cylinder. Let P denote the point in this plane on the generator of contact, and let the common normal at P make an angle ψ with the upward vertical. The mass centre C of the first cylinder must be vertically above P, say at a height h, for otherwise the moment of the gravitational force through C about P will be non-zero and prevent equilibrium.

Let us consider a small displacement from such a position of equilibrium, in which the point of contact moves to Q. Let P_1, P_2 be the points on the two cylinders which started together at P in equilibrium. Let the normals at P_1 and Q meet in O_1, and let the normals at P_2 and Q meet in O_2 as shown in Figure 11.4.2. We use the notation

$$P_1\hat{O}_1Q = \theta_1, \quad P_2\hat{O}_2Q = \theta_2, \quad \phi = \theta_1 + \theta_2,$$

where ϕ is the angle through which the cylinder turns.

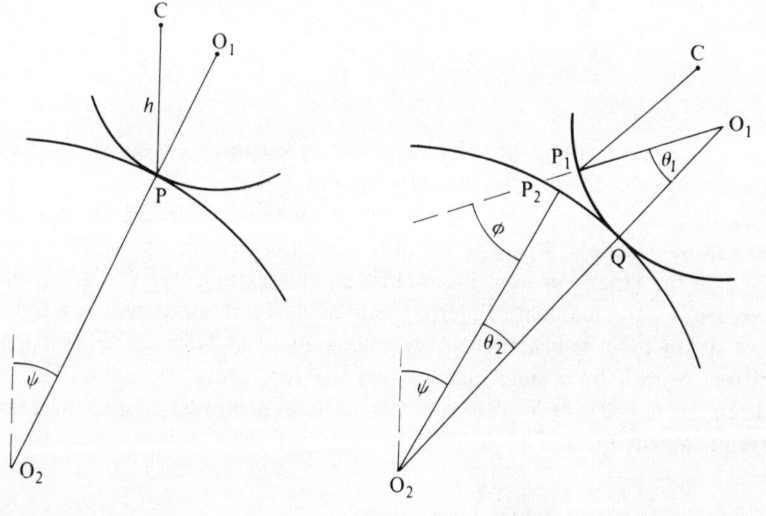

Figure 11.4.2 Illustration for Example 11.4.2: equilibrium of one cylinder resting on another.

In order to discuss the stability of equilibrium we require an estimate of the change in potential energy arising from the displacement. Since this is a second order quantity in general, it is necessary to calculate the change in height of the mass centre C to this order of accuracy. Initially the height of C above P_2 is h. In the displaced position C is at a height $h \cos \phi$ above P_1 since the body rotates through an angle ϕ, and P_1 is at a height $P_1P_2 \cos \psi$ above P_2 since P_1 is displaced along the normal at P_2 in a small displacement. Hence we must now calculate P_1P_2 which is already a small quantity of the second order. We first note that, correct to the first order, the straight line segment P_1Q is parallel to the tangent at the mid point of the arc P_1Q. Hence the angle between P_1Q and the common tangent at Q is $\frac{1}{2}\theta_1$. Similarly the angle between the straight segment P_2Q and the tangent at Q is $\frac{1}{2}\theta_2$. It follows that $P_1\hat{Q}P_2 = \frac{1}{2}(\theta_1 + \theta_2) = \frac{1}{2}\phi$ with an error of second order. With the same order of accuracy we also have $P_1Q = P_2Q = \delta s$, where δs is the length of arc, since to first order this is equal to the length of arc of either cylinder which is in contact during the displacement. We can therefore infer that $P_1P_2 = \frac{1}{2}\phi\delta s$ which verifies that it is indeed of second order. Finally we derive a more useful expression for δs from the first order relations $P_1Q = r_1\theta_1$, $P_2Q = r_2\theta_2$, where r_1 and r_2 are the radii of curvature at P_1 and P_2 of the moving and the fixed cylinder respectively. These relations give $\delta s(1/r_1 + 1/r_2) = \theta_1 + \theta_2 = \phi$, or $\delta s = r_1r_2\phi/(r_1 + r_2)$, and so the increment in potential energy arising from the displacement is

$$mg(P_1P_2 \cos \psi + h \cos \phi - h) = \frac{1}{2}mg\left(\frac{r_1r_2 \cos \psi}{r_1 + r_2} - h\right)\phi^2$$

correct to second order. This expression is positive, and so the equilibrium position is stable, provided that

$$\frac{\cos \psi}{h} > \frac{1}{r_1} + \frac{1}{r_2}.$$

(In the critical case of equality the increment in potential energy is of a higher order and a more accurate calculation is required.)

If equilibrium is stable the cylinder will oscillate when disturbed. The oscillations can be described with the aid of the energy equation, for which we need to calculate the kinetic energy. From (11.3.5) this is $\frac{1}{2}mv_c^2 + \frac{1}{2}I_c(\mathrm{d}\phi/\mathrm{d}t)^2$, where $I_c = mk^2$ (say) is the moment of inertia about an axis through C, and $v_c = \mathrm{QC}\,\mathrm{d}\phi/\mathrm{d}t$ since the cylinder is instantaneously rolling about Q. Now $\mathrm{d}\phi/\mathrm{d}t$ is small and QC differs from h by a small quantity of the first order. It follows that the kinetic energy is $\frac{1}{2}m(k^2 + h^2)(\mathrm{d}\phi/\mathrm{d}t)^2$ correct to second order, and so the energy equation takes the form

$$\left(\frac{\mathrm{d}\phi}{\mathrm{d}t}\right)^2 + n^2\phi^2 = \text{constant},$$

where
$$n = \frac{r_1 r_2 \cos\psi - h(r_1 + r_2)}{(r_1 + r_2)(k^2 + h^2)} g.$$

This is the standard form of the energy equation for simple harmonic motion.

If one of the cylinders is locally concave, the formulae are still applicable with a change of sign for the appropriate radius. If one of the cylinders degenerates into a plane, so that the appropriate radius tends to infinity, we have

$$n^2 = \frac{r\cos\psi - h}{k^2 + h^2} g,$$

where r is the radius of curvature of the non-degenerate cylinder. The stability condition is then $r\cos\psi - h > 0$.

As a specific example consider a cylinder, whose cross-section is a semi-circle, resting in equilibrium with its curved surface on a horizontal plane. Here we have

$$\psi = 0, \quad h = r\left(1 - \frac{4}{3\pi}\right), \quad k^2 = \frac{r^2}{2} - \frac{16r^2}{9\pi^2},$$

$$n^2 = \frac{8}{9\pi - 16}\frac{g}{r} = \frac{0{\cdot}6518g}{r}.$$

If the cross-section of the cylinder is an ellipse, with major and minor semi-axes a and b $(<a)$, it will rest in stable equilibrium on a horizontal plane with the major axis horizontal. In this case we have

$$\psi = 0, \quad h = b, \quad k^2 = \tfrac{1}{4}(a^2 + b^2), \quad r = a^2/b,$$

$$n^2 = \frac{(a^2/b) - b}{\tfrac{1}{4}(a^2 + b^2) + b^2} g = \frac{a^2 - b^2}{a^2 + 5b^2}\frac{4g}{b}.$$

Note that when the cylinder rests with its minor axis horizontal we have $h = a$, $r = b^2/a$. Hence $r - h < 0$ and equilibrium is consequently unstable.

11.5 THREE-DIMENSIONAL PROBLEMS

The applications discussed so far have been confined to two-dimensional problems, although the theory is quite general. We now discuss some examples which involve three-dimensional motion. To deal with the motion of the mass centre is relatively straightforward, since it is formally the same as that of the motion of a particle subject to the same external forces. To deal with the motion relative to the mass centre can be substantially more complicated than it is for two-dimensional motion, since the moment equation now takes on its full vector form. It is helpful to consider the formulation of this equation in its various stages.

(i) Choose an appropriate origin. In most examples this will be either a fixed point of the body or the mass centre.
(ii) Calculate the moments and products of inertia at the origin. The analysis is considerably simplified by choosing principal axes. In the following examples this is done by recognising the appropriate axes through appeal to symmetry properties of the body.
(iii) Formulate the angular momentum vector $\mathbf{L}$ using (11.1.14) or preferably (11.1.17).
(iv) Calculate $d\mathbf{L}/dt$. Remember that if the axes rotate, as they will if they are fixed in the body, the derivative will contain a contribution from the rotation of the axes, as discussed in §9.4. In general principal axes are fixed in the body, and so have the angular velocity of the body. It will be noticed, however, that in several of the examples the axes and the body have different angular velocities. This arises when advantage is taken of the symmetry properties of the body. For example the principal axes at the centre of a uniform sphere need not rotate with the sphere. We can even choose them to be fixed in direction since all axes are principal axes at the centre of the sphere.
(v) Formulate the moment equation using (11.1.9), namely,

$$\mathbf{M} = \Sigma \mathbf{r} \times \mathbf{F} = d\mathbf{L}/dt \quad \textit{if } \mathbf{r}_c \times \mathbf{a}_o \doteq \mathbf{0},$$

or, more generally, using (11.1.11)

$$\mathbf{M} = \Sigma \mathbf{r} \times \mathbf{F} = \frac{d\mathbf{L}_c}{dt} + \mathbf{r}_c \times \frac{d\mathbf{p}}{dt}$$

without restriction on the origin O.

For impulsive problems equate the moment of the external impulses to the change in momentum, as in (11.2.4), namely

$$\bar{\mathbf{M}} = \Sigma \mathbf{r} \times \bar{\mathbf{F}} = [\mathbf{L}] \quad \textit{if } \mathbf{r}_c \times \mathbf{v}_o = \mathbf{0},$$

or, more generally, using (11.2.5)

$$\bar{\mathbf{M}} = \Sigma \mathbf{r} \times \bar{\mathbf{F}} = [\mathbf{L}_c + \mathbf{r}_c \times \mathbf{p}]$$

without restriction on the origin O.

Example 11.5.1

A body rotates with an angular velocity $\boldsymbol{\omega}$ about a fixed point O. The forces on the body have moment $\mathbf{M}$ about O and the angular momentum of the body is $\mathbf{L}$.

Show that if $\mathbf{M} \cdot \boldsymbol{\omega} = 0$, the kinetic energy is constant and that if in addition $\boldsymbol{\omega}$ is fixed in direction, then $\boldsymbol{\omega}$ is constant.

Show that if $\mathbf{L} \cdot \mathbf{M} = 0$ then $\mathbf{L}$ is constant in magnitude.

The principal axes at O are fixed in the body and therefore rotate with angular velocity $\boldsymbol{\omega}$. Let the principal moments of inertia at O be I_1, I_2, I_3, so that $\mathbf{L} = I_1\omega_1\mathbf{i} + I_2\omega_2\mathbf{j} + I_3\omega_3\mathbf{k}$.

When the equation of motion is formulated, account must be taken of the rotation of the axes so that, in the notation of §9.4, we have

$$\mathrm{d}\mathbf{L}/\mathrm{d}t = \dot{\mathbf{L}} + \boldsymbol{\omega} \times \mathbf{L} = \mathbf{M}$$

with components

$$I_1\dot{\omega}_1 - (I_2 - I_3)\omega_2\omega_3 = M_1,$$
$$I_2\dot{\omega}_2 - (I_3 - I_1)\omega_3\omega_1 = M_2,$$
$$I_3\dot{\omega}_3 - (I_1 - I_2)\omega_1\omega_2 = M_3.$$

These are called Euler's equations. From them we see that

$$\mathbf{M} \cdot \boldsymbol{\omega} = I_1\omega_1\dot{\omega}_1 + I_2\omega_2\dot{\omega}_2 + I_3\omega_3\dot{\omega}_3 = \mathrm{d}T/\mathrm{d}t,$$

where $T = \frac{1}{2}(I_1\omega_1^2 + I_2\omega_2^2 + I_3\omega_3^2)$ is the kinetic energy. It follows that the kinetic energy is constant if $\mathbf{M} \cdot \boldsymbol{\omega} = 0$. If $\boldsymbol{\omega}$ is also fixed in direction, say $\boldsymbol{\omega} = (l_1, l_2, l_3)\omega$, where (l_1, l_2, l_3) are fixed direction cosines, then $T = \frac{1}{2}(I_1 l_1^2 + I_2 l_2^2 + I_3 l_3^2)\omega^2$ is constant and so $\boldsymbol{\omega}$ is also constant in magnitude. Note that the couple required to maintain the fixed direction of spin is

$$\mathbf{M} = -(I_2 - I_3)\omega_2\omega_3\mathbf{i} - (I_3 - I_1)\omega_3\omega_1\mathbf{j} - (I_1 - I_2)\omega_1\omega_2\mathbf{k}.$$

From Euler's equations, or directly from the vector equation of motion, we have

$$\mathbf{L} \cdot \mathbf{M} = \mathbf{L} \cdot \frac{\mathrm{d}\mathbf{L}}{\mathrm{d}t} = \frac{1}{2}\frac{\mathrm{d}}{\mathrm{d}t}L^2$$

and if $\mathbf{L} \cdot \mathbf{M} = 0$ we have L^2 = constant so that $\mathbf{L}$ is constant in magnitude.

Example 11.5.2

A uniform rectangular plate, with sides of length $2a$, $2b$, spins with constant angular speed ω about a diagonal. In order to maintain this motion a couple has to be exerted by the bearings, which we proceed to calculate.

The principal axes at the mass centre C of the plate are parallel to the sides and perpendicular to the plane of the rectangle, since the products of inertia for these axes clearly vanish, by symmetry. The principal moments of inertia at C

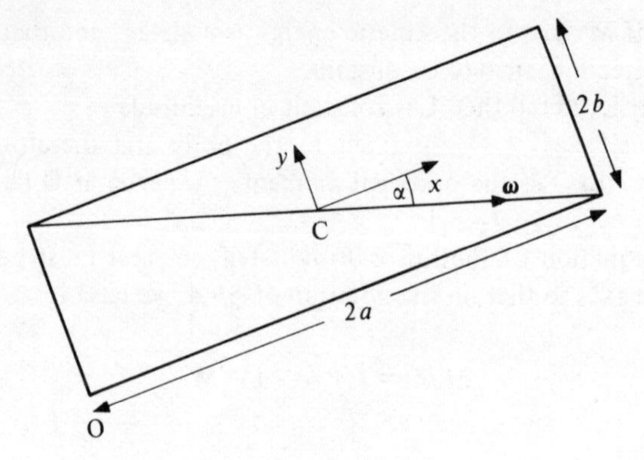

Figure 11.5.1 Illustration for Example 11.5.2: rotating rectangular plate.

are

$$I_1 = \tfrac{1}{3}mb^2, \quad I_2 = \tfrac{1}{3}ma^2, \quad I_3 = \tfrac{1}{3}m(a^2 + b^2).$$

The angular velocity is

$$\boldsymbol{\omega} = \omega(\mathbf{i}\cos\alpha - \mathbf{j}\sin\alpha),$$

where $\tan\alpha = b/a$.

The angular momentum vector at C is

$$\begin{aligned}\mathbf{L}_c &= I_1\omega_1\mathbf{i} + I_2\omega_2\mathbf{j} + I_3\omega_3\mathbf{k} \\ &= \tfrac{1}{3}m\omega(\mathbf{i}b^2\cos\alpha - \mathbf{j}a^2\sin\alpha) \\ &= \frac{m\omega ab}{3(a^2+b^2)^{1/2}}(b\mathbf{i} - a\mathbf{j}).\end{aligned} \tag{11.5.1}$$

The couple exerted on the plate is $d\mathbf{L}_c/dt$, where

$$\begin{aligned}\frac{d\mathbf{L}_c}{dt} &= \dot{\mathbf{L}}_c + \boldsymbol{\omega} \times \mathbf{L}_c = \boldsymbol{\omega} \times \mathbf{L}_c \\ &= -\tfrac{1}{3}m\omega^2(a^2 - b^2)\mathbf{k}\sin\alpha\cos\alpha \\ &= -\frac{m\omega^2 ab(a^2 - b^2)\mathbf{k}}{3(a^2+b^2)}.\end{aligned}$$

This is the couple required to be exerted by the bearings.

Let us now consider the effect of releasing the plate from its bearings and instantaneously fixing the corner O (see Fig. 11.5.1). To find the new angular velocity, say $\boldsymbol{\omega}$, we use the fact that the angular momentum about O must remain unchanged, because the impulsive reaction required to fix O will have no moment about O. To calculate the angular momentum both before and after the corner is fixed, it is instructive to use the unrestricted result (11.2.5) and evaluate $\mathbf{L}_c + \mathbf{r}_c \times \mathbf{p}$. Before the corner is fixed $\mathbf{p} = \mathbf{0}$, since the mass centre has no motion, and so the angular momentum about O is just $\mathbf{L}_c$. This gives, from (11.5.1),

$$\mathbf{L}_o = \mathbf{L}_c + \mathbf{r}_c \times \mathbf{p} = \mathbf{L}_c = \frac{m\omega ab}{3(a^2 + b^2)^{1/2}} (b\mathbf{i} - a\mathbf{j}). \tag{11.5.2}$$

Let $\boldsymbol{\omega}'$ be the angular velocity immediately after O is fixed. The new angular momentum relative to C is

$$\mathbf{L}_c' = \tfrac{1}{3}mb^2\omega_1'\mathbf{i} + \tfrac{1}{3}ma^2\omega_2'\mathbf{j} + \tfrac{1}{3}m(a^2 + b^2)\omega_3'\mathbf{k}.$$

Relative to axes based on O, we have $\mathbf{r}_c = a\mathbf{i} + b\mathbf{j}$, and $\mathbf{p}' = m\mathbf{v}_c' = m\boldsymbol{\omega}' \times \mathbf{r}_c$ since the plate is now rotating about O with angular velocity $\boldsymbol{\omega}'$. Accordingly

$$\begin{aligned}\mathbf{r}_c \times \mathbf{p}' &= m\mathbf{r}_c \times (\boldsymbol{\omega}' \times \mathbf{r}_c)\\ &= m(b^2\omega_1' - ab\omega_2')\mathbf{i} - m(ab\omega_1' - a^2\omega_2')\mathbf{j} + m(a^2 + b^2)\omega_3'\mathbf{k}\end{aligned}$$

and so*

$$\begin{aligned}\mathbf{L}_o' = \mathbf{L}_c' + \mathbf{r}_c \times \mathbf{p}' &= (\tfrac{4}{3}mb^2\omega_1' - mab\omega_2')\mathbf{i} + (-mab\omega_1' + \tfrac{4}{3}ma^2\omega_2')\mathbf{j}\\ &\quad + \tfrac{4}{3}m(a^2 + b^2)\omega_3'\mathbf{k}.\end{aligned} \tag{11.5.3}$$

Since the expressions calculated in (11.5.2) and (11.5.3) are equal, it follows that $\omega_3' = 0$ and

$$\frac{4}{3}b^2\omega_1' - ab\omega_2' = \frac{ab^2\omega}{3(a^2 + b^2)^{1/2}},$$

$$-ab\omega_1' + \frac{4}{3}a^2\omega_2' = -\frac{a^2b\omega}{3(a^2 + b^2)^{1/2}}.$$

When solved for ω_1', ω_2', these equations give

$$\boldsymbol{\omega}' = \frac{\omega}{7(a^2 + b^2)^{1/2}} (a\mathbf{i} - b\mathbf{j}) = \frac{1}{7}\boldsymbol{\omega},$$

*The result can also be deduced from (11.1.14). With axes based on O parallel to the original axes, we have

$$I_{xx} = \tfrac{4}{3}mb^2, \quad I_{yy} = \tfrac{4}{3}ma^2, \quad I_{zz} = \tfrac{4}{3}m(a^2 + b^2),$$
$$I_{yz} = 0, \quad I_{zx} = 0, \quad I_{xy} = -mab,$$

from which (11.5.3) follows.

so that $\boldsymbol{\omega}'$ is parallel to $\boldsymbol{\omega}$, and is reduced in magnitude by a factor 1/7. Immediately before O is fixed the kinetic energy is $\frac{1}{2}\mathbf{L}_c \cdot \boldsymbol{\omega}$. Immediately after O is fixed the kinetic energy is $\frac{1}{2}\mathbf{L}'_o \cdot \boldsymbol{\omega}'$. Since $\mathbf{L}_c = \mathbf{L}'_o$ and $\boldsymbol{\omega}' = \boldsymbol{\omega}/7$, there is a reduction in kinetic energy by a factor 1/7.

Example 11.5.3

A uniform circular disc is suspended from a point O on its circumference. It is given an impulsive blow, perpendicular to its plane, at a point P of its circumference (see Fig. 11.5.2). We wish to find the instantaneous axis of rotation.

The principal axes at O are horizontal and vertical in the plane of the disc and perpendicular to the plane of the disc. The principal moments of inertia are

$$I_1 = \tfrac{5}{4}ma^2, \quad I_2 = \tfrac{1}{4}ma^2, \quad I_3 = \tfrac{3}{2}ma^2,$$

where a is the radius of the disc. The angular momentum is

$$\mathbf{L}_o = I_1\omega_1\mathbf{i} + I_2\omega_2\mathbf{j} + I_3\omega_3\mathbf{k} = ma^2(\tfrac{5}{4}\omega_1\mathbf{i} + \tfrac{1}{4}\omega_2\mathbf{j} + \tfrac{3}{2}\omega_3\mathbf{k}).$$

The impulse is $\bar{F}\mathbf{k}$, say, and is applied at the point

$$\mathbf{r} = \mathbf{i}a\sin\alpha - \mathbf{j}a(1 - \cos\alpha),$$

where α is the inclination of the radius vector CP to the vertical. The moment of

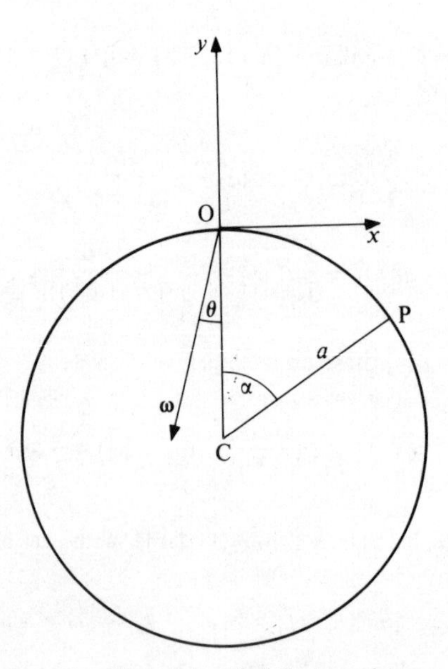

Figure 11.5.2 Illustration for Example 11.5.3: impulsive motion of suspended disc.

the impulse about O is

$$\mathbf{r} \times \bar{F}\mathbf{k} = -\mathbf{i}a\bar{F}(1 - \cos\alpha) - \mathbf{j}a\bar{F}\sin\alpha,$$

and the moment equation $\mathbf{r} \times \bar{F}\mathbf{k} = \mathbf{L}_\mathrm{O}$ gives

$$-\mathbf{i}a\bar{F}(1 - \cos\alpha) - \mathbf{j}a\bar{F}\sin\alpha = ma^2(\tfrac{5}{4}\omega_1\mathbf{i} + \tfrac{1}{4}\omega_2\mathbf{j} + \tfrac{3}{2}\omega_3\mathbf{k}).$$

It follows that $$\boldsymbol{\omega} = -\frac{4\bar{F}}{ma}\left(\frac{1}{5}\mathbf{i}(1 - \cos\alpha) + \mathbf{j}\sin\alpha\right).$$

The instantaneous axis of rotation is therefore in the plane of the disc. Let it make an angle θ with OC so that

$$\boldsymbol{\omega} = -\omega(\mathbf{i}\sin\theta + \mathbf{j}\cos\theta).$$

Comparison of the two expressions for $\boldsymbol{\omega}$ shows that

$$\tan\theta = \frac{1 - \cos\alpha}{5\sin\alpha} = \frac{1}{5}\tan\frac{1}{2}\alpha.$$

It is clear that θ remains small over a substantial range of α. In fact $0 \leqslant \theta < \pi/15$ for $0 \leqslant \alpha \leqslant \pi/2$.

To find the impulsive reaction at O, say $\bar{\mathbf{X}}$, we require the velocity of the mass centre. This is

$$\mathbf{v}_c = \boldsymbol{\omega} \times (-a\mathbf{j}) = \frac{4\bar{F}}{5m}(1 - \cos\alpha)\mathbf{k}.$$

Then from the equation for the motion of the mass centre, namely

$$\bar{F}\mathbf{k} + \bar{\mathbf{X}} = m\mathbf{v}_c,$$

we have $$\bar{\mathbf{X}} = \{\tfrac{4}{5}\bar{F}(1 - \cos\alpha) - \bar{F}\}\mathbf{k} = -\tfrac{1}{5}\bar{F}(1 + 4\cos\alpha)\mathbf{k}.$$

Note that $\bar{\mathbf{X}}$ is zero when $\cos\alpha = -\frac{1}{4}$ so that for this value of α the disc would still rotate about the same axis if free from the constraint at O.

Example 11.5.4

We consider the motion of a spherically symmetric sphere rolling and slipping on a horizontal plane. Let $m, a, \mathbf{r}$ denote respectively the mass, radius and position vector of the mass centre of the sphere. Let $-F\mathbf{n}$ denote the frictional force at the point of contact of the sphere and plane, where $\mathbf{n}$ is a horizontal unit vector. When the sphere slips we have $F = \mu mg$ and the velocity of slip will be in the direction of $\mathbf{n}$ and so of the form $v\mathbf{n}$ (see Fig. 11.5.3).

The calculation of the angular momentum is straightforward for the sphere

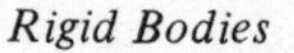

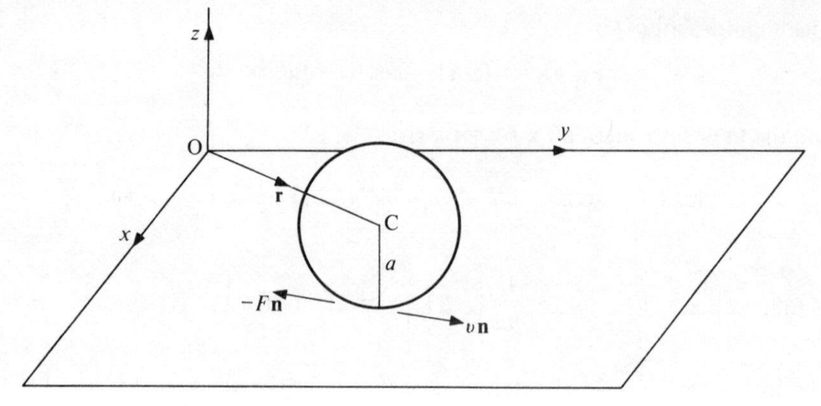

Figure 11.5.3 Illustration for Example 11.5.4: a sphere rolling and slipping on a horizontal plane.

because all sets of orthogonal axes through the mass centre are principal axes. If the angular velocity of the sphere is $\boldsymbol{\omega}$, the angular momentum about the mass centre is

$$\mathbf{L}_c = I_1 \omega_1 \mathbf{i} + I_2 \omega_2 \mathbf{j} + I_3 \omega_3 \mathbf{k} = I \boldsymbol{\omega}$$

since $I_1 = I_2 = I_3 = I$ say, at the mass centre. Note that the symmetry properties of the sphere allow a choice of principal axes which are fixed in direction and not rotating with the sphere.

The equations arising from the kinematical condition at the point of slip, the motion of the mass centre and the motion relative to the mass centre are

$$\frac{\mathrm{d}\mathbf{r}}{\mathrm{d}t} + \boldsymbol{\omega} \times (-a\mathbf{k}) = v\mathbf{n},$$

$$-\mu m g \mathbf{n} = m\, \mathrm{d}^2\mathbf{r}/\mathrm{d}t^2,$$

$$a\mathbf{k} \times \mu m g \mathbf{n} = I\, \mathrm{d}\boldsymbol{\omega}/\mathrm{d}t.$$

Note that the gravitational force is cancelled by the normal reaction, and neither of these forces is included in the equations. Differentiation of the first equation and elimination of $\mathrm{d}^2\mathbf{r}/\mathrm{d}t^2$ and $\mathrm{d}\boldsymbol{\omega}/\mathrm{d}t$ with the help of the second and third equations gives

$$\frac{\mathrm{d}v}{\mathrm{d}t}\mathbf{n} + v\frac{\mathrm{d}\mathbf{n}}{\mathrm{d}t} = -\mu g \mathbf{n} - \frac{ma^2}{I}\mu g(\mathbf{k} \times \mathbf{n}) \times \mathbf{k}$$

$$= -\mu g \mathbf{n} - \frac{ma^2}{I}\mu g \mathbf{n},$$

where the fact that $\mathbf{k}$ and $\mathbf{n}$ are perpendicular has been used. Since $\mathbf{n}$ is a unit vector, $\mathrm{d}\mathbf{n}/\mathrm{d}t$ is also perpendicular to $\mathbf{n}$ and so the above equation implies that $\mathrm{d}\mathbf{n}/\mathrm{d}t = \mathbf{0}$ or $\mathbf{n}$ is a constant vector. The equation then reduces to

$$\frac{\mathrm{d}v}{\mathrm{d}t} = -\mu g\left(1 + \frac{ma^2}{I}\right),$$

from which it follows that v decreases uniformly to zero. Note that $\mathrm{d}^2\mathbf{r}/\mathrm{d}t^2 = -\mu g\mathbf{n}$ is a constant vector, so that

$$\mathbf{r} = \mathbf{r}_0 + \mathbf{u}t - \tfrac{1}{2}\mu g\mathbf{n}t^2,$$

where $\mathbf{r}_0$, $\mathbf{u}$ are the values of $\mathbf{r}$, $\mathrm{d}\mathbf{r}/\mathrm{d}t$ at $t = 0$. The above expression for $\mathbf{r}$ represents a parabolic path for the centre of the sphere.

The equation

$$I(\mathrm{d}\boldsymbol{\omega}/\mathrm{d}t) = a\mathbf{k} \times \mu mg\mathbf{n}$$

shows that $\mathbf{k} \cdot \boldsymbol{\omega}$ and $\mathbf{n} \cdot \boldsymbol{\omega}$ are constant and that the horizontal component of $\boldsymbol{\omega}$ perpendicular to $\mathbf{n}$ is proportional to the time.

After slipping ceases, the equations become

$$\frac{\mathrm{d}\mathbf{r}}{\mathrm{d}t} - a\boldsymbol{\omega} \times \mathbf{k} = \mathbf{0},$$

$$-F\mathbf{n} = m\,\mathrm{d}^2\mathbf{r}/\mathrm{d}t^2,$$

$$I\frac{\mathrm{d}\boldsymbol{\omega}}{\mathrm{d}t} = a\mathbf{k} \times F\mathbf{n} = -a\mathbf{k} \times m\,\mathrm{d}^2\mathbf{r}/\mathrm{d}t^2$$

$$= -ma^2\mathbf{k} \times \left(\frac{\mathrm{d}\boldsymbol{\omega}}{\mathrm{d}t} \times \mathbf{k}\right)$$

$$= -ma^2\left\{\frac{\mathrm{d}\boldsymbol{\omega}}{\mathrm{d}t} - \left(\mathbf{k} \cdot \frac{\mathrm{d}\boldsymbol{\omega}}{\mathrm{d}t}\right)\mathbf{k}\right\},$$

so that
$$(I + ma^2)\left(\frac{\mathrm{d}\omega_1}{\mathrm{d}t}\mathbf{i} + \frac{\mathrm{d}\omega_2}{\mathrm{d}t}\mathbf{j}\right) + I\frac{\mathrm{d}\omega_3}{\mathrm{d}t}\mathbf{k} = \mathbf{0}.$$

Therefore $\mathrm{d}\boldsymbol{\omega}/\mathrm{d}t = \mathbf{0}$ and $\boldsymbol{\omega}$ is a constant vector. Hence $\mathrm{d}\mathbf{r}/\mathrm{d}t = a\boldsymbol{\omega} \times \mathbf{k}$ is also a constant vector and the mass centre moves with constant velocity.

Example 11.5.5

Two uniform rods OA_1, OA_2, each of mass m and length $2a$, are freely jointed together at O. We shall derive the equations of motion when the system falls freely under gravity.

Let C_1, C_2 be the mass centres of OA_1, OA_2 respectively and let $\mathbf{a}_1 = \overline{OC}_1$,

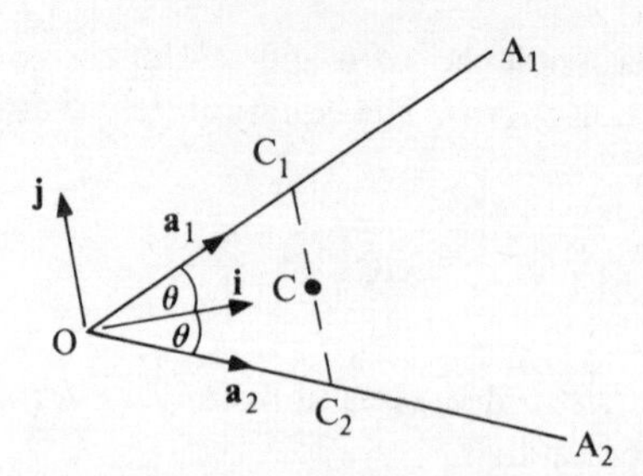

Figure 11.5.4 Illustration for Example 11.5.5: two jointed rods falling under gravity.

$\mathbf{a}_2 = \overline{OC_2}$, so that $a_1 = a_2 = a$ (see Fig. 11.5.4). The acceleration of C_1 relative to the mass centre C of the whole system is $\frac{1}{2}d^2(\mathbf{a}_1 - \mathbf{a}_2)/dt^2$. Since the only external force is gravitational, the acceleration of C is $\mathbf{g}$, so that the acceleration of C_1 is $\mathbf{g} + \frac{1}{2}d^2(\mathbf{a}_1 - \mathbf{a}_2)/dt^2$. Let $\mathbf{R}$, $-\mathbf{R}$ be the reactions at the hinge on the rods OA_1, OA_2 respectively. The motion of C_1 is then governed by the equation

$$\mathbf{R} + m\mathbf{g} = m\left(\mathbf{g} + \frac{1}{2}\frac{d^2}{dt^2}(\mathbf{a}_1 - \mathbf{a}_2)\right) \quad \text{or} \quad \mathbf{R} = \frac{1}{2}m\frac{d^2}{dt^2}(\mathbf{a}_1 - \mathbf{a}_2).$$

The moment equation for OA_1 about C_1 gives $-\mathbf{a}_1 \times \mathbf{R} = d\mathbf{L}/dt$ where $\mathbf{L}$ depends on the angular velocity $\boldsymbol{\omega}_1$ say of OA_1. Now the angular velocity for a rod is not uniquely defined and we can in fact specify that $\boldsymbol{\omega}_1 \cdot \mathbf{a}_1 = 0$. Since $\boldsymbol{\omega}_1 \times \mathbf{a}_1 = d\mathbf{a}_1/dt$, we then have

$$\mathbf{a}_1 \times \frac{d\mathbf{a}_1}{dt} = \mathbf{a}_1 \times (\boldsymbol{\omega}_1 \times \mathbf{a}_1) = a^2\boldsymbol{\omega}_1$$

and so

$$\mathbf{L} = I\boldsymbol{\omega}_1 = \tfrac{1}{3}m\mathbf{a}_1 \times d\mathbf{a}_1/dt,$$

where $I = ma^2/3$ is the moment of inertia of OA_1 about a perpendicular axis through C_1. The moment equation now becomes

$$-\mathbf{a}_1 \times \mathbf{R} = -\frac{1}{2}m\mathbf{a}_1 \times \frac{d^2}{dt^2}(\mathbf{a}_1 - \mathbf{a}_2) = \frac{1}{3}m\frac{d}{dt}\left(\mathbf{a}_1 \times \frac{d\mathbf{a}_1}{dt}\right) = \frac{1}{3}m\mathbf{a}_1 \times \frac{d^2\mathbf{a}_1}{dt^2}.$$

The corresponding equation for OA_2 is

$$-\mathbf{a}_2 \times (-\mathbf{R}) = \frac{1}{2}m\mathbf{a}_2 \times \frac{d^2}{dt^2}(\mathbf{a}_1 - \mathbf{a}_2) = \frac{1}{3}m\frac{d}{dt}\left(\mathbf{a}_2 \times \frac{d\mathbf{a}_2}{dt}\right) = \frac{1}{3}m\mathbf{a}_2 \times \frac{d^2\mathbf{a}_2}{dt^2}.$$

We therefore have a pair of equations for $\mathbf{a}_1$ and $\mathbf{a}_2$, namely

$$\mathbf{a}_1 \times \frac{d^2}{dt^2}\left(\frac{5}{3}\mathbf{a}_1 - \mathbf{a}_2\right) = \mathbf{0}, \quad \mathbf{a}_2 \times \frac{d^2}{dt^2}\left(\frac{5}{3}\mathbf{a}_2 - \mathbf{a}_1\right) = \mathbf{0}.$$

To take the analysis further we define unit vectors **i** along OC and **j** perpendicular to OC in the plane of A_1OA_2. Together with $\mathbf{k} = \mathbf{i} \times \mathbf{j}$ these unit vectors determine a moving frame of reference in which $a^{-1}\mathbf{a}_1 = \mathbf{i}\cos\theta + \mathbf{j}\sin\theta$ and $a^{-1}\mathbf{a}_2 = \mathbf{i}\cos\theta - \mathbf{j}\sin\theta$, where $2\theta = A_1\hat{O}A_2$. Addition and subtraction of the two equations for $\mathbf{a}_1$ and $\mathbf{a}_2$ then give

$$\mathbf{i}\cos\theta \times \frac{d^2}{dt^2}(\mathbf{i}\cos\theta) + 4\mathbf{j}\sin\theta \times \frac{d^2}{dt^2}(\mathbf{j}\sin\theta) = \mathbf{0},$$

$$4\mathbf{i}\cos\theta \times \frac{d^2}{dt^2}(\mathbf{j}\sin\theta) + \mathbf{j}\sin\theta \times \frac{d^2}{dt^2}(\mathbf{i}\cos\theta) = \mathbf{0}.$$

The first of these relations integrates to

$$\mathbf{i}\cos\theta \times \frac{d}{dt}(\mathbf{i}\cos\theta) + 4\mathbf{j}\sin\theta \times \frac{d}{dt}(\mathbf{j}\sin\theta) = \mathbf{c},$$

where **c** is a constant vector. This relation in turn can be written

$$\mathbf{i} \times \frac{d\mathbf{i}}{dt}\cos^2\theta + 4\mathbf{j} \times \frac{d\mathbf{j}}{dt}\sin^2\theta = \mathbf{c}.$$

If the reference system has angular velocity $\boldsymbol{\omega}$ with components $(\omega_x, \omega_y, \omega_z)$, we have $d\mathbf{i}/dt = \boldsymbol{\omega} \times \mathbf{i}$ and $d\mathbf{j}/dt = \boldsymbol{\omega} \times \mathbf{j}$, and so the last equation implies that the quantities $\omega_x \sin^2\theta$, $\omega_y \cos^2\theta$, $\omega_z(\cos^2\theta + 4\sin^2\theta)$ are all constants of the motion which are therefore determined by the initial conditions. Further details of the motion are determined, at least in principle, by the remaining equation.

11.6 GYROSCOPIC PROBLEMS

The application of rigid body dynamics to gyroscopic problems is one of the most striking successes of the theory, and warrants a separate account. These problems are concerned with the motion of a rigid body in which the main constituent of the motion is a basic spin about a principal axis. This underlying spin causes curious effects which are not intuitively obvious.

Let O be a point on the axis of spin, and let **k** be a unit vector along this axis. Let $\boldsymbol{\omega}$ be the angular velocity of the body, We have

$$d\mathbf{k}/dt = \boldsymbol{\omega} \times \mathbf{k}, \tag{11.6.1}$$

$$\mathbf{k} \times d\mathbf{k}/dt = \mathbf{k} \times (\boldsymbol{\omega} \times \mathbf{k}) = \boldsymbol{\omega} - (\boldsymbol{\omega} . \mathbf{k})\mathbf{k}, \tag{11.6.2}$$

and so we can write
$$\boldsymbol{\omega} = \mathbf{k} \times \frac{d\mathbf{k}}{dt} + s\mathbf{k}, \tag{11.6.3}$$

where $s = \boldsymbol{\omega} . \mathbf{k}$ is defined as the spin. Equation (11.6.3) simply displays the components of $\boldsymbol{\omega}$ perpendicular and parallel to $\mathbf{k}$.

We shall consider only bodies having kinetic symmetry about the principal axis in the direction of $\mathbf{k}$. Let $I_1 = I$, $I_2 = I$, I_3 be the principal moments of inertia at O. The relative angular momentum is then

$$\mathbf{L} = I\mathbf{k} \times \frac{\mathrm{d}\mathbf{k}}{\mathrm{d}t} + I_3 s\mathbf{k}. \tag{11.6.4}$$

Let $\mathbf{M}$ be the moment of the external forces about O, so that

$$\mathbf{M} = \frac{\mathrm{d}\mathbf{L}}{\mathrm{d}t} = I\mathbf{k} \times \frac{\mathrm{d}^2\mathbf{k}}{\mathrm{d}t^2} + I_3 s \frac{\mathrm{d}\mathbf{k}}{\mathrm{d}t} + I_3 \frac{\mathrm{d}s}{\mathrm{d}t}\mathbf{k} \tag{11.6.5}$$

provided that O is either a fixed point or the mass centre. Equation (11.6.5) is the general equation describing the motion of a spinning symmetrical body about a fixed point. For more general motion it can be used to describe the motion relative to the mass centre and it is then augmented by an equation for the motion of the mass centre.

Note that the first two terms on the right-hand side of (11.6.5) have no component parallel to $\mathbf{k}$. In gyroscopic problems this is often true of $\mathbf{M}$ as well, in which case (11.6.5) implies that s is constant and simplifies to

$$\mathbf{M} = I\mathbf{k} \times \frac{\mathrm{d}^2\mathbf{k}}{\mathrm{d}t^2} + I_3 s \frac{\mathrm{d}\mathbf{k}}{\mathrm{d}t}. \tag{11.6.6}$$

Example 11.6.1. The spinning top

The most familiar gyroscopic problem is that of a rigid body free to rotate about a fixed point O on an axis of symmetry OC. The spin s about OC is taken to be constant and the equation of motion is consequently given by (11.6.6).

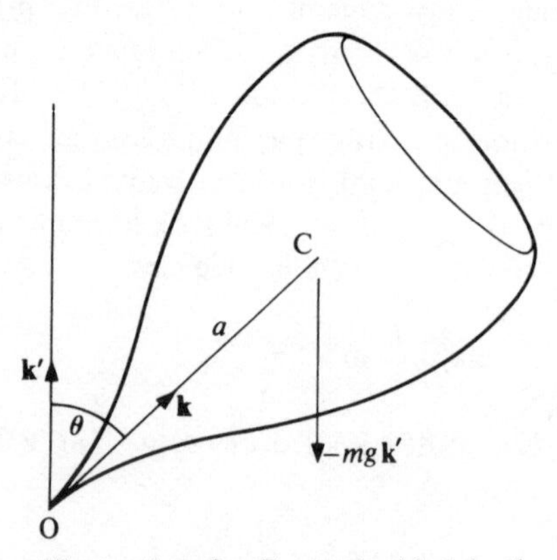

Figure 11.6.1 Illustration for Example 11.6.1: the spinning top.

Let $\mathbf{k}$ be a unit vector along OC and let $\mathbf{k}'$ be a unit vector in a fixed direction. We consider first the conditions for which a solution is possible in which the angle θ between $\mathbf{k}$ and $\mathbf{k}'$ is constant and $\mathbf{k}$ rotates steadily about $\mathbf{k}'$. This is called steady precession and the constant angular speed of precession p is defined so that $p\mathbf{k}'$ is the angular velocity with which $\mathbf{k}$ rotates about $\mathbf{k}'$. The point on OC with position vector $\mathbf{k}$ then has a velocity

$$\mathrm{d}\mathbf{k}/\mathrm{d}t = p\mathbf{k}' \times \mathbf{k} \tag{11.6.7}$$

so that

$$\mathbf{k} \times \mathrm{d}^2\mathbf{k}/\mathrm{d}t^2 = \mathbf{k} \times (p\mathbf{k}' \times \mathrm{d}\mathbf{k}/\mathrm{d}t) = -p(\mathbf{k} \cdot \mathbf{k}')\, \mathrm{d}\mathbf{k}/\mathrm{d}t. \tag{11.6.8}$$

Equation (11.6.6) then becomes

$$\mathbf{M} = \{I_3 s - Ip(\mathbf{k} \cdot \mathbf{k}')\}\, \mathrm{d}\mathbf{k}/\mathrm{d}t = \{I_3 sp - Ip^2 \cos\theta\}\mathbf{k}' \times \mathbf{k}. \tag{11.6.9}$$

It follows that a couple of constant magnitude $(I_3 sp - Ip^2 \cos\theta)\sin\theta$ is required to maintain the steady precession. Its direction is normal to the plane of $\mathbf{k}$ and $\mathbf{k}'$. It is the direction of this couple which is surprising when first encountered. Suppose the couple is produced by a force $F\mathbf{k}'$, say, acting at a point $a\mathbf{k}$ on the axis of symmetry so that $\mathbf{M} = a\mathbf{k} \times (F\mathbf{k}')$. The body then moves perpendicular to the applied force rather than yielding to it, as might have been expected. A familiar example of this effect is the fact that the rider of a bicycle must lean over on one side in order to turn a corner. The axle of the front wheel is then the axis of spin, and the gravitational couple acts about a fore and aft horizontal axis. The effect of the couple is to give the wheel an additional angular velocity of precession which has a vertical component and turns the front wheel in the required direction (the rear wheel is of course not free to rotate about a vertical axis relative to the frame).

The results are immediately applicable to the spinning top, as illustrated in Figure 11.6.1, which has its mass centre C at a distance a along the axis of symmetry from the fixed point O. The fixed direction $\mathbf{k}'$ is then vertical, the gravitational couple at O is $a\mathbf{k} \times (-mg\mathbf{k}')$, and a steady precession about the vertical is possible if

$$mga(\mathbf{k}' \times \mathbf{k}) = (I_3 sp - Ip^2 \cos\theta)(\mathbf{k}' \times \mathbf{k}). \tag{11.6.10}$$

Apart from the degenerate case, when the axis of spin is vertical, the condition is

$$Ip^2 \cos\theta - I_3 sp + mga = 0. \tag{11.6.11}$$

This is a quadratic in p^2 and so, for given values of s and θ, two values of p are possible which are real if

$$I_3^2 s^2 > 4\, mgaI \cos\theta. \tag{11.6.12}$$

When s is large, as is usually the case in practice, p approximates either to the small value $mga/I_3 s$ (slow precession) or to the large value $I_3 s/I \cos\theta$. Note that when $\cos\theta < 0$, so that the mass centre lies below O, solutions are always possible.

Example 11.6.2. The nearly vertical top
The equation of motion of the spinning top, namely

$$-mga\mathbf{k}\times\mathbf{k}' = I\mathbf{k}\times\frac{d^2\mathbf{k}}{dt^2} + I_3 s\frac{d\mathbf{k}}{dt} \tag{11.6.13}$$

has the solution $\mathbf{k}=\mathbf{k}'$. This solution refers to the case of a top spinning with constant spin s about the vertical. To discuss the stability of this solution, we assume that in a small perturbation of the above solution

$$\mathbf{k}=\mathbf{k}'+\boldsymbol{\rho},$$

where $\boldsymbol{\rho}$ is a vector of small magnitude. Since $\mathbf{k}'.\mathbf{k}=\cos\theta = 1+0(\theta^2)$, it follows that $\mathbf{k}'.\boldsymbol{\rho}=0(\theta^2)$ and that hence $\boldsymbol{\rho}$ is a small horizontal vector if second order quantities are neglected. Equation (11.6.13), when linearised, simplifies to

$$-mga\boldsymbol{\rho}\times\mathbf{k}' = I\mathbf{k}'\times\frac{d^2\boldsymbol{\rho}}{dt^2} + I_3 s\frac{d\boldsymbol{\rho}}{dt}. \tag{11.6.14}$$

This equation is best discussed with reference to axes rotating with constant angular velocity $\omega_a\mathbf{k}'$, where ω_a is a constant to be determined. Relative to such a frame we have

$$d\boldsymbol{\rho}/dt = \dot{\boldsymbol{\rho}} + \omega_a\mathbf{k}'\times\boldsymbol{\rho}, \tag{11.6.15}$$

$$d^2\boldsymbol{\rho}/dt^2 = \ddot{\boldsymbol{\rho}} + 2\omega_a\mathbf{k}'\times\dot{\boldsymbol{\rho}} - \omega_a^2\boldsymbol{\rho}, \tag{11.6.16}$$

since $\mathbf{k}'.\boldsymbol{\rho}=0$. The notation now being used is that introduced for rotating axes in §9.4. Substitution of (11.6.15), (11.6.16) in (11.6.14) gives

$$-mga\boldsymbol{\rho}\times\mathbf{k}' = I\mathbf{k}'\times(\ddot{\boldsymbol{\rho}} - \omega_a^2\boldsymbol{\rho}) - 2I\omega_a\dot{\boldsymbol{\rho}} + I_3 s(\dot{\boldsymbol{\rho}} + \omega_a\mathbf{k}'\times\dot{\boldsymbol{\rho}}). \tag{11.6.17}$$

Since ω_a is at our disposal, we choose it so that the coefficient of $\dot{\boldsymbol{\rho}}$ vanishes. Therefore $2I\omega_a = I_3 s$ and (11.6.17) then simplifies to

$$I\mathbf{k}'\times\left\{\ddot{\boldsymbol{\rho}} + \left(\frac{I_3^2 s^2 - 4Imga}{4I^2}\right)\boldsymbol{\rho}\right\} = \mathbf{0}. \tag{11.6.18}$$

The nature of the solution of (11.6.18) depends on the sign of the coefficient of $\boldsymbol{\rho}$ within the parentheses. If the coefficient is positive, say n^2, then $\boldsymbol{\rho}$ satisfies the equation

$$\ddot{\boldsymbol{\rho}} + n^2 \boldsymbol{\rho} = \mathbf{0}$$

which represents elliptic harmonic motion relative to the frame rotating with angular speed $I_3 s/2I$. It follows that any small initial disturbance will remain small. On the other hand if the coefficient is negative, say $-\mu^2$, the solution for $\boldsymbol{\rho}$ has the form

$$\boldsymbol{\rho} = \boldsymbol{\rho}_1 \mathrm{e}^{-\mu t} + \boldsymbol{\rho}_2 \mathrm{e}^{\mu t},$$

where $\boldsymbol{\rho}_1$ and $\boldsymbol{\rho}_2$ are fixed vectors in the rotating frame. Since $\boldsymbol{\rho}_2$ will not be zero for an arbitrary initial disturbance, this situation represents instability. In the critical case, for which the coefficient of $\boldsymbol{\rho}$ in (11.6.18) is zero, we have $\ddot{\boldsymbol{\rho}} = \mathbf{0}$. Hence $\boldsymbol{\rho}$ varies linearly with t and this case is also unstable. It follows that the condition for instability is

$$I_3^2 s^2 \leqslant 4Imga. \qquad (11.6.19)$$

Example 11.6.3

We now consider the motion of a uniform sphere rolling without slipping on the surface of a second fixed rough sphere. The fixed sphere has centre O, radius b. The moving sphere has centre C, radius a and mass m. The reaction on it at the point of contact with the fixed sphere is **R**. Unit vectors **k**′ and **k** are defined in the vertical direction and along the line of centres OC respectively (see Fig. 11.6.2).

In this example **k** is not fixed in the moving sphere, so that the previous analysis is not immediately applicable. However, it is clear that the position

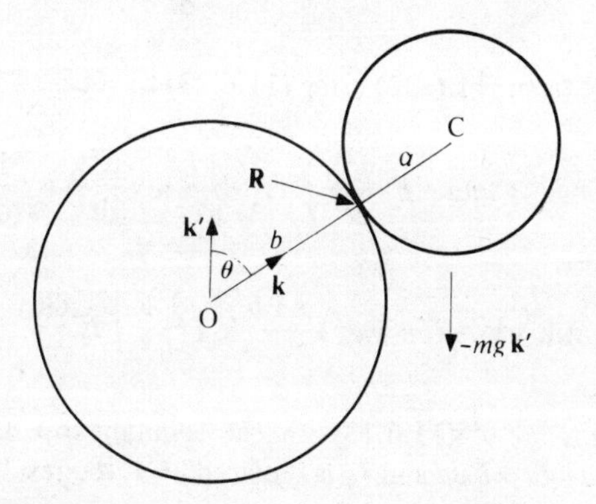

Figure 11.6.2 Illustration for Example 11.6.3: a sphere rolling and slipping on a fixed sphere.

vector of C relative to O is $(a+b)\mathbf{k}$ so that the velocity of C is $(a+b)\,\mathrm{d}\mathbf{k}/\mathrm{d}t$. Also if $\boldsymbol{\omega}$ is the angular velocity of the moving sphere, the velocity of C can be expressed alternatively as $\boldsymbol{\omega} \times a\mathbf{k}$ since the sphere is instantaneously rotating about the point of contact with the fixed sphere. It follows that

$$\frac{a+b}{a}\frac{\mathrm{d}\mathbf{k}}{\mathrm{d}t} = \boldsymbol{\omega} \times \mathbf{k}. \tag{11.6.20}$$

This equation is the same as (11.6.1) apart from the factor $(a+b)/a$ and the analysis now follows similar lines. Analogous to (11.6.2) we have

$$\frac{a+b}{a}\mathbf{k} \times \frac{\mathrm{d}\mathbf{k}}{\mathrm{d}t} = \boldsymbol{\omega} - (\boldsymbol{\omega}\,.\,\mathbf{k})\mathbf{k}$$

and so

$$\boldsymbol{\omega} = \frac{a+b}{a}\mathbf{k} \times \frac{\mathrm{d}\mathbf{k}}{\mathrm{d}t} + s\mathbf{k}, \tag{11.6.21}$$

where $s = \boldsymbol{\omega}\,.\,\mathbf{k}$ is the spin of the sphere about $\mathbf{k}$.

For the motion relative to C we have the equation

$$-a\mathbf{k} \times \mathbf{R} = \frac{\mathrm{d}\mathbf{L}_c}{\mathrm{d}t} = I\frac{\mathrm{d}\boldsymbol{\omega}}{\mathrm{d}t} = I\left(\frac{a+b}{a}\mathbf{k} \times \frac{\mathrm{d}^2\mathbf{k}}{\mathrm{d}t^2} + s\frac{\mathrm{d}\mathbf{k}}{\mathrm{d}t} + \frac{\mathrm{d}s}{\mathrm{d}t}\mathbf{k}\right), \tag{11.6.22}$$

where I is the moment of inertia of the moving sphere about a diameter. The component of the above equation parallel to $\mathbf{k}$ gives $\mathrm{d}s/\mathrm{d}t = 0$ and so s is constant.

Equation (11.6.22) must now be augmented by the equation for the motion of the mass centre, which is

$$\mathbf{R} - mg\mathbf{k}' = m(a+b)\frac{\mathrm{d}^2\mathbf{k}}{\mathrm{d}t^2}. \tag{11.6.23}$$

We can eliminate $\mathbf{R}$ from (11.6.22) using (11.6.23) to get

$$-a\mathbf{k} \times \left(mg\mathbf{k}' + m(a+b)\frac{\mathrm{d}^2\mathbf{k}}{\mathrm{d}t^2}\right) = I\left(\frac{a+b}{a}\mathbf{k} \times \frac{\mathrm{d}^2\mathbf{k}}{\mathrm{d}t^2} + s\frac{\mathrm{d}\mathbf{k}}{\mathrm{d}t}\right),$$

or

$$-mga\mathbf{k} \times \mathbf{k}' = (I + ma^2)\frac{a+b}{a}\mathbf{k} \times \frac{\mathrm{d}^2\mathbf{k}}{\mathrm{d}t^2} + Is\frac{\mathrm{d}\mathbf{k}}{\mathrm{d}t}. \tag{11.6.24}$$

This equation is similar to (11.6.13) for the spinning top. If in (11.6.13) I is replaced by $(I + ma^2)(a+b)/a$ and I_3 is replaced by I, the result is just (11.6.24). Hence we can discuss (11.6.24) on the basis of the analysis of (11.6.13) and the above substitutions.

There are, for example, solutions for which the angle θ between $\mathbf{k}$ and $\mathbf{k}'$ is constant and $\mathbf{k}$ rotates steadily about $\mathbf{k}'$ with constant angular velocity $p\mathbf{k}'$. From (11.6.11) and the appropriate substitutions the condition for this is,

$$(I + ma^2)\frac{a+b}{a}p^2 \cos\theta - Isp + mga = 0. \qquad (11.6.25)$$

There is a solution of (11.6.24), for which $\mathbf{k} = \mathbf{k}'$, which represents a spin about the highest point of the fixed sphere. From (11.6.19) this is unstable if

$$I^2 s^2 \leqslant 4(I + ma^2)(a+b)mg. \qquad (11.6.26)$$

Example 11.6.4.   The Gyrocompass

In Figure 11.6.3(a), O represents a point on the earth's surface at latitude λ, and the line SN represents the direction south to north at O, which therefore makes an angle λ with the earth's axis of rotation. Let the unit vectors $\mathbf{i}, \mathbf{j}, \mathbf{k}$ at O be such that $\mathbf{i}$ is vertical, and $\mathbf{j}$ and $\mathbf{k}$ are horizontal with $\mathbf{k}$ making an angle θ with SN (see Fig. 11.6.3(b)). If ω_e is the angular speed of the earth about its axis, the angular velocity of the earth has components $\omega_e \cos\lambda$ parallel to SN and $\omega_e \sin\lambda$ parallel to $\mathbf{i}$, and we have

$$\boldsymbol{\omega}_e = \omega_e(\mathbf{i}\sin\lambda - \mathbf{j}\cos\lambda\sin\theta + \mathbf{k}\cos\lambda\cos\theta). \qquad (11.6.27)$$

The angular velocity $\boldsymbol{\omega}_a$ of the axes differs from $\boldsymbol{\omega}_e$ by a term $-\dot{\theta}\mathbf{i}$ and so

$$\boldsymbol{\omega}_a = \mathbf{i}(\omega_e \sin\lambda - \dot{\theta}) - \mathbf{j}\omega_e\cos\lambda\sin\theta + \mathbf{k}\omega_e\cos\lambda\cos\theta. \quad (11.6.28)$$

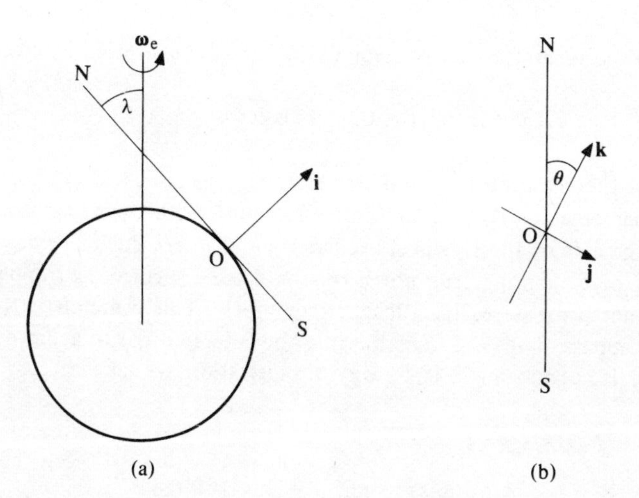

Figure 11.6.3   Illustration for Example 11.6.4: notation used in the discussion of the gyrocompass.

We now consider the motion of an axially symmetric flywheel, situated at O, whose axis is constrained by an appropriate couple to remain in a horizontal plane. If $\mathbf{k}$ is chosen to lie along the axis of the flywheel, the latter's angular velocity $\boldsymbol{\omega}$ differs from $\boldsymbol{\omega}_a$ by a spin about $\mathbf{k}$ and so

$$\boldsymbol{\omega} = \mathbf{i}(\omega_e \sin\lambda - \dot{\theta}) - \mathbf{j}\omega_e \cos\lambda \sin\theta + \mathbf{k}s, \tag{11.6.29}$$

say. The angular momentum will then be

$$\mathbf{L} = \mathbf{i}I(\omega_e \sin\lambda - \dot{\theta}) - \mathbf{j}I\omega_e \cos\lambda \sin\theta + \mathbf{k}I_3 s, \tag{11.6.30}$$

where I, I, I_3 are the principal moments of inertia at the mass centre of the flywheel. To maintain the axis in a horizontal plane, it is necessary to apply an appropriate couple of the form $M\mathbf{j}$ through the bearings in which the axis rotates. Now

$$\begin{aligned} M\mathbf{j} &= \mathrm{d}\mathbf{L}/\mathrm{d}t = \dot{\mathbf{L}} + \boldsymbol{\omega}_a \times \mathbf{L} \\ &= -\mathbf{i}I\ddot{\theta} - \mathbf{j}I\omega_e\dot{\theta}\cos\lambda\cos\theta + \mathbf{k}I_3\dot{s} \\ &\quad + \mathbf{i}\omega_e \cos\lambda \sin\theta(I\omega_e\cos\lambda\cos\theta - I_3 s) \\ &\quad + \mathbf{j}(\omega_e \sin\lambda - \dot{\theta})(I\omega_e \cos\lambda\cos\theta - I_3 s), \end{aligned} \tag{11.6.31}$$

and so there are two relations implied by the condition that the $\mathbf{i}$ and $\mathbf{k}$ components of $\mathrm{d}\mathbf{L}/\mathrm{d}t$ are zero. The second of these implies that the spin s is constant, and the first gives the equation

$$-I\ddot{\theta} + \omega_e \cos\lambda \sin\theta(I\omega_e \cos\lambda\cos\theta - \mathrm{I}_3 s) = 0. \tag{11.6.32}$$

In applications of interest, $s \gg \omega_e$ and (11.6.32) can be approximated to

$$\ddot{\theta} + n^2 \sin\theta = 0, \quad n^2 = I_3 s\omega_e \cos\lambda/I. \tag{11.6.33}$$

This is just the equation for the simple pendulum as discussed in Example 6.5.1. In particular, small oscillations are simple harmonic with a periodic time $2\pi/n$.

Since the axis of the flywheel oscillates about $\theta = 0$, such a device offers the possibility of estimating true north on the earth's surface, as opposed to the magnetic north registered by a magnetic compass. This principle is the basis of the gyrocompass. Since ω_e is small, s must be large in order to achieve a periodic time which is not too small. If, by way of illustration, we take

$$I_3 \cos\lambda/I = 1$$

$$s = 6000 \text{ rev min}^{-1} = 2\pi \times 100 \text{ rad s}^{-1},$$

$$\omega_e = 1 \text{ rev(sidereal day)}^{-1} = 2\pi/86\,164 \text{ rad s}^{-1},$$

the periodic time is

$$2\pi\left(\frac{I}{I_3 s\omega_e \cos\lambda}\right)^{1/2} = (861\cdot64)^{1/2} = 29\cdot4 \text{ s.}$$

Example 11.6.5

In this example we consider the motion of a uniform disc, of mass m and radius a, which is rolling on a rough, horizontal plane. Let $\mathbf{k}'$, $\mathbf{k}$, $\mathbf{j}$, $\mathbf{i}$, be unit vectors such that $\mathbf{k}'$ is vertically upwards, $\mathbf{k}$ is along the axis of the disc, $\mathbf{j}$ is parallel to $\mathbf{k} \times \mathbf{k}'$ and $\mathbf{i} = \mathbf{j} \times \mathbf{k}$ (see Fig. 11.6.4). Let θ be the inclination of $\mathbf{k}$ to the vertical and let ϕ be the angular displacement of the disc about the vertical. The angular velocity $\boldsymbol{\omega}_a$ of the axes is

$$\boldsymbol{\omega}_a = \mathbf{i}\dot{\phi}\sin\theta - \mathbf{j}\dot{\theta} + \mathbf{k}\dot{\phi}\cos\theta. \tag{11.6.34}$$

The angular velocity $\boldsymbol{\omega}$ of the disc is

$$\boldsymbol{\omega} = \mathbf{i}\dot{\phi}\sin\theta - \mathbf{j}\dot{\theta} + \mathbf{k}s. \tag{11.6.35}$$

The angular momentum relative to the mass centre C is

$$\mathbf{L} = \mathbf{i}I\dot{\phi}\sin\theta - \mathbf{j}I\dot{\theta} + \mathbf{k}I_3 s, \tag{11.6.36}$$

where I, I, I_3* are the principal moments of inertia at C. The equations for the motion of the mass centre, the motion relative to the mass centre and the

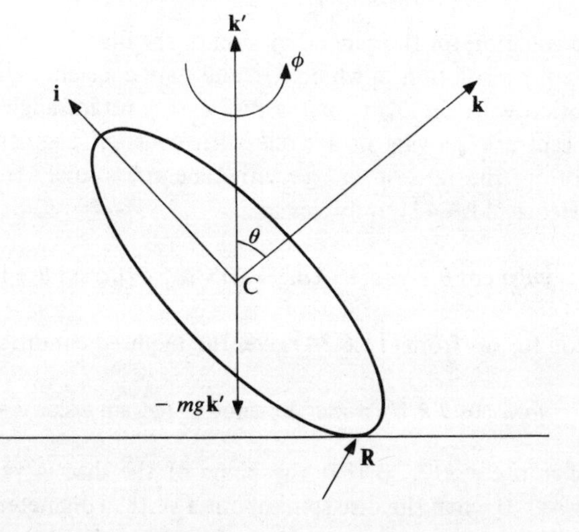

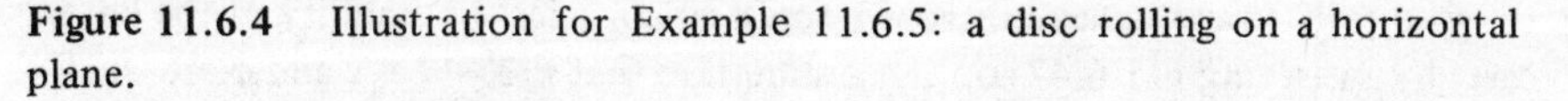

Figure 11.6.4 Illustration for Example 11.6.5: a disc rolling on a horizontal plane.

* In this particular example $I_3 = 2I$, but the mathematics is applicable to more general bodies which have the same perimeter as the disc and for which the disc is a plane of symmetry.

condition of no slip give respectively

$$\mathbf{R} - mg\mathbf{k}' = m(\mathrm{d}\mathbf{v}_c/\mathrm{d}t), \tag{11.6.37}$$

$$-a\mathbf{i} \times \mathbf{R} = \mathrm{d}\mathbf{L}/\mathrm{d}t, \tag{11.6.38}$$

$$\mathbf{v}_c - \boldsymbol{\omega} \times a\mathbf{i} = \mathbf{0}, \tag{11.6.39}$$

where $\mathbf{R}$ is the reaction at the point of contact of the disc and plane and $\mathbf{v}_c$ is the velocity of the mass centre.

Elimination of $\mathbf{R}$ from (11.6.38) using (11.6.37) gives

$$-a\mathbf{i} \times \left(mg\mathbf{k}' + m\frac{\mathrm{d}\mathbf{v}_c}{\mathrm{d}t} \right) = \frac{\mathrm{d}\mathbf{L}}{\mathrm{d}t}. \tag{11.6.40}$$

Elimination of $\mathbf{v}_c$ from (11.6.40) using (11.6.39) gives

$$-a\mathbf{i} \times \left(mg\mathbf{k}' + m\frac{\mathrm{d}}{\mathrm{d}t}(\boldsymbol{\omega} \times a\mathbf{i}) \right) = \frac{\mathrm{d}\mathbf{L}}{\mathrm{d}t}. \tag{11.6.41}$$

Substitution for $\boldsymbol{\omega}$ and $\mathbf{L}$ from (11.6.35) and (11.6.36) into (11.6.41) gives

$$\mathbf{j}mga \cos\theta - ma^2\mathbf{i} \times \frac{\mathrm{d}}{\mathrm{d}t}(\mathbf{j}s + \mathbf{k}\dot{\theta}) = \frac{\mathrm{d}}{\mathrm{d}t}(\mathbf{i}I\dot{\phi}\sin\theta - \mathbf{j}I\dot{\theta} + \mathbf{k}I_3 s). \tag{11.6.42}$$

This is the equation for the general motion of the disc. We consider a special case by looking for a solution in which θ, $\dot{\phi}$ and s are constant. This case represents a rolling motion with the plane of the disc at a constant angle to the vertical and the mass centre C moving in a circle with constant angular velocity. For this steady motion, the rate of change with time arises solely from the rotation of the axes. Hence (11.6.42) simplifies to

$$\mathbf{j}mga\cos\theta - ma^2\mathbf{i} \times (\boldsymbol{\omega}_a \times \mathbf{j}s) = \boldsymbol{\omega}_a \times (\mathbf{i}I\dot{\phi}\sin\theta + \mathbf{k}I_3 s). \tag{11.6.43}$$

Substitution for $\boldsymbol{\omega}_a$ from (11.6.34) gives the required condition

$$mga\cos\theta + (I_3 + ma^2)s\dot{\phi}\sin\theta - I\dot{\phi}^2\sin\theta\cos\theta = 0. \tag{11.6.44}$$

In particular if $\theta = \pi/2$, so that the plane of the disc is vertical, (11.6.44) is satisfied by $s = 0$ when the disc spins about a vertical diameter, or by $\dot{\phi} = 0$ when the disc rolls straight ahead. It is of interest to consider the stability of the latter case, by satisfying (11.6.42) on the assumption that $\pi/2 - \theta = \chi$ and $\dot{\phi}$ are small. Correct to first order, we have

$$\boldsymbol{\omega}_a = \mathbf{i}\dot{\phi} + \mathbf{j}\dot{\chi} \tag{11.6.45}$$

which is also small. Equation (11.6.42) then approximates to

$$jmga\chi - ma^2(\mathbf{j}\ddot{\chi} + \mathbf{k}\dot{s}) - ma^2\mathbf{i} \times (\boldsymbol{\omega}_a \times \mathbf{j}s) = \mathbf{i}I\ddot{\phi} + \mathbf{j}I\ddot{\chi} + \mathbf{k}I_3\dot{s} + \boldsymbol{\omega}_a \times \mathbf{k}I_3 s. \tag{11.6.46}$$

The three components of (11.6.46) give, respectively,

$$I\ddot{\phi} + I_3 s\dot{\chi} = 0, \tag{11.6.47}$$

$$mga\chi - ma^2\ddot{\chi} + ma^2 s\dot{\phi} = I\ddot{\chi} - I_3 s\dot{\phi}, \tag{11.6.48}$$

$$-ma^2\dot{s} = I_3\dot{s}. \tag{11.6.49}$$

Equation (11.6.49) shows that s is constant to first order. Equation (11.6.47) then integrates to

$$I\dot{\phi} + I_3 s\chi = 0$$

if it is assumed that $\chi = \dot{\phi} = 0$ initially. Elimination of $\dot{\phi}$ from (11.6.48) gives, finally,

$$I(I + ma^2)\ddot{\chi} + \{I_3(I_3 + ma^2)s^2 - Imga\}\chi = 0.$$

It follows that rolling is unstable if

$$Imga \geqslant I_3(I_3 + ma^2)s^2.$$

EXERCISES

11.1 A rigid body has angular velocity $\boldsymbol{\omega}$ and a point O of the rigid body has velocity $\mathbf{v}_O$. Show that the kinetic energy $T(\mathbf{v}_O, \boldsymbol{\omega})$ is such that

$$\mathbf{v}_O \cdot \mathbf{p} + \boldsymbol{\omega} \cdot \mathbf{L}_O = 2T(\mathbf{v}_O, \boldsymbol{\omega}),$$

$$\mathbf{v}_O \cdot \mathbf{p} - \boldsymbol{\omega} \cdot \mathbf{L}_O = 2T(\mathbf{v}_O, 0) - 2T(0, \boldsymbol{\omega}),$$

where $\mathbf{p}, \mathbf{L}_O$ are respectively the linear momentum and the angular momentum about O.

11.2 A uniform rod, of length $2l$, rests partly inside and partly outside a smooth, fixed, hemispherical bowl of radius a whose rim is horizontal. Given that the inclination of the rod to the horizontal is α, show that $l \cos \alpha = 2a \cos 2\alpha$.

11.3 A rough rod passes under a peg A and over a peg B and rests in equilibrium at an angle α to the horizontal. Given that the mass centre of the rod is at a distance a from A and b from B, and that the friction is limiting, show that $\tan \alpha = (a + b)/(a - b)$.

11.4 A rough uniform circular cylinder, of mass m, rests in a rough V-shaped groove whose sides are inclined at angles α with the horizontal. A gradually increasing horizontal force R is applied to the highest point of the cylinder at right angles to its axis and in the vertical plane containing the mass centre. Given that the coefficient of friction is $\tan \lambda$, show that equilibrium is broken by slipping when $\alpha > 2\lambda$ and

$$R = \frac{mg \sin 2\lambda}{2(\cos \alpha + \sin^2 \lambda)}.$$

11.5 Two equal rough circular cylinders rest in contact with each other on a rough inclined plane, and their axes are horizontal. The inclination of the plane to the horizontal is α. The coefficient of friction at all contacts is $\tan \lambda$.

The cylinders are prevented from moving down the plane by a force applied to the lower cylinder along the line of centres, and this force is gradually increased. Show that equilibrium is finally broken by the lower cylinder slipping up the plane if $\alpha > \frac{1}{4}\pi - \lambda$, and by slipping between the two cylinders if $\alpha < \frac{1}{4}\pi - \lambda$.

11.6 A uniform circular disc, of mass m, is suspended in a vertical plane by two vertical strings attached at points A, B on the circumference. The points A, B are on the same level and the arc AB subtends an angle 2α at the centre of the disc.

One of the strings is cut; what is the initial tension in the other? (Ans.: $mg/(1 + 2 \sin^2 \alpha)$.)

11.7 Two equal uniform rods AO, OB, are smoothly hinged together at O and the end A is smoothly hinged to a fixed point. The rods are held in a vertical plane with A and B on the same level and the angle AOB a right angle.

Show that when the rods are released the initial angular accelerations are in the ratio 3 : 4.

Note that there are two possible initial configurations, with O either above or below the level of AB. The answer is the same for each case.

11.8 A uniform rod is supported against a smooth fixed sphere by a horizontal string fastened to its upper end and to the highest point of the sphere. Show that if the string is cut, the reaction between the sphere and the rod is changed instantaneously by a factor $\cos^2 \alpha/(1 + 3 \sin^4 \alpha)$, where α is the inclination of the rod to the horizontal.

11.9 A flywheel rotates with an average angular speed ω. It is acted upon by a forcing torque $T \sin^2 \omega t$ and there is a constant torque $\frac{1}{2}T$ opposing the motion.

What is the least moment of inertia required to make the difference between the greatest and least angular speeds less than $\omega/100$? (Ans.: $50T/\omega^2$.)

11.10 A rough uniform rod, of mass m and length $2l$, is freely pivoted at one end and is initially supported at the other end in a horizontal position. A particle, of mass αm, is placed on the rod at the mid point. The coefficient of friction between the particle and the rod is μ. Show that if the support is removed:

(i) the reaction at the hinge is instantaneously reduced by a factor $2/(4 + 3\alpha)$;

(ii) the particle begins to slide when the rod is inclined to the horizontal at an angle θ given by $\mu = (10 + 9\alpha)\tan\theta$.

11.11 Two rods, each of mass m and length l, are joined together to form a single rod of length $2l$. The combination rotates on a smooth horizontal plane about one end which is fixed. The join is such that it cannot sustain a tension greater than T. Show that the angular speed must be less than $(2T/3ml)^{1/2}$.

Describe the subsequent motion of the two rods if the linkage should snap.

11.12 A straight beam, of mass m_b, rests on two uniform circular cylinders, each of mass m_c and radius a. The cylinders are at rest on a horizontal plane with their axes parallel, and the length of the beam is perpendicular to the axes. The mass centres of the beam and cylinders lie in one vertical plane. The beam is pulled with a horizontal force F parallel to its length acting through its mass centre. Show that if there is no slipping, the acceleration of the beam is $F(m_b + \frac{3}{4}m_c)^{-1}$.

11.13 An axially symmetric yo-yo consists of two equal discs joined by a short axle, of radius a. The radius of gyration of the yo-yo about its axis is k. A light string is partly wound round the axle and held vertically by its free end.

Show that if the free end of the string is stationary and the yo-yo is ascending with angular speed ω, it will climb a length l of the string provided that $\omega^2 > 2gl/(k^2 + a^2)$. Show that if this condition is not satisfied it will still climb a length l of the string if the free end is given a constant downward acceleration not less than

$$g - \frac{(k^2 + a^2)\omega^2}{2l}.$$

11.14 A pair of wheels joined by an axle stands on a rough slope, of inclination α, with the axle horizontal. The radius of the axle is a and the radius of each wheel is b. The system is axially symmetric, its mass is m and the radius of gyration of the system about its axis is k. A light string is wound round the axle and emerges from the underside. It then passes up the plane parallel to the line of greatest slope, over a smooth peg and hangs vertically with a particle, also of mass m, suspended from its end.

The system is released from rest. Show that if the wheels roll without slipping, they roll down or up the plane according as $a \gtrless b(1 - \sin\alpha)$.

If slipping should take place, show that it will be such that the points of contact move up the plane.

11.15 A non-uniform sphere, of radius a, has its mass centre C at a distance c from its geometrical centre O. It is placed on a rough horizontal plane so that OC is horizontal. On the assumption that the sphere rolls in the ensuing motion, write down the equations of motion in terms of the angle θ which OC makes with the horizontal.

Show that it will roll initially if $\mu > ac/(k^2 + a^2)$, where μ is the coefficient of friction and k is the radius of gyration about the horizontal axis through C perpendicular to OC.

If $\mu = ac/(k^2 + a^2)$, what is the condition for the sphere to roll initially?
(Ans.: $3k^2 > a^2$.)

11.16 Two uniform cog wheels have masses m_1, m_2, radii a_1, a_2 and moments of inertia I_1, I_2 about their axes.

Show that if they rotate with angular speeds ω_1, ω_2 on the same axle and are suddenly locked together, the final angular speed is $(I_1\omega_1 + I_2\omega_2)/(I_1 + I_2)$.

Show that if they rotate on parallel axles, and are suddenly meshed at their circumference, the final angular speeds are v/a_1, v/a_2,

where $$v = (I_1\omega_1 a_1 a_2^2 - I_2\omega_2 a_2 a_1^2)/(I_1 a_2^2 + I_2 a_1^2).$$

11.17 A uniform circular hoop, of radius a, is struck a blow at a point of its circumference. If it is perfectly free, find the initial axis of rotation: (i) when the blow is tangential in the plane of the hoop; (ii) when it is at right angles to the plane of the hoop.
(Ans.: A distance: (i) a; (ii) $a/2$ from the centre.)

11.18 A train begins to move with a carriage door open at right angles to the length of the train. What is the angular speed of the door as it closes if:

(i) The train moves with a constant acceleration a?
(ii) The train jerks into motion with speed v, which is maintained?

(Ans.: $\{2la/(l^2 + k^2)\}^{1/2}$, $lv/(l^2 + k^2)$, where l is the horizontal distance of the mass centre from the hinge of the door and k is the radius of gyration about a vertical axis through the mass centre.)

11.19 A uniform bar, of length $2l$, rests symmetrically on two pegs at a distance $2d$ apart in a horizontal line. One end of the bar is raised and then released. Show that if there is no slip, and no recoil when the bar strikes the peg, the bar is reduced to rest on impact if $3d^2 > l^2$. Show that if $3d^2 < l^2$, the angular velocity is instantaneously reduced at each impact in the ratio $(l^2 - 3d^2)/(l^2 + 3d^2)$.

11.20 Two uniform rods AB, BC, each of mass m and length $2l$, are hinged together at B and are free to rotate about A, which is fixed. At a certain instant they are collinear and rotating as if rigid with angular speed ω. An impulse is applied that brings both rods to rest. Find the magnitude of the impulse and the point at which it acts.
(Ans.: $11ml\omega/3$, $26l/11$ from A.)

11.21 Four equal uniform rods, of mass m, are freely jointed together to form a square and rest on a horizontal plane. A horizontal impulse $\bar{\mathrm{F}}$ is applied at one corner. What is the initial velocity of the opposite corner?
(Ans.: $-\bar{\mathrm{F}}/8m$.)

11.22 A uniform circular cylinder, of radius a, rolls with speed u along the ground, up a slope inclined at an angle α to the horizontal, back down the slope and finally along the ground.

The problem is two-dimensional, all surfaces are inelastic and rough enough to prevent sliding.

To what height does the mass centre of the cylinder rise? What is the final speed along the ground?

$$\left(\text{Ans.: } a + \frac{(1 + 2\cos\alpha)^2 u^2}{6g}, \frac{(1 + 2\cos\alpha)^2 u}{9}.\right)$$

11.23 A rod, of mass m and length l, rests on a horizontal plane and is freely hinged at one end. A particle, of mass αm, rests on the plane and touches the rod. An impulse is applied to the free end of the rod which jerks the particle into motion. Show that the initial velocity of the particle will be a maximum if it lies at a distance $k/\alpha^{1/2}$ from the hinge, where k is the radius of gyration of the rod about the hinge. Show that the total energy is then shared equally between the rod and the particle.

11.24 Two uniform rods, each of mass m and length $2l$, are hinged together and lie in a straight line. An impulse $\bar{\mathbf{F}}$ is applied to the mid point of one of the rods perpendicular to its length. What is the kinetic energy imparted to the rods?
(Ans.: $7\bar{F}^2/16m$.)

11.25 A spherically symmetric golf ball of radius a rolls without slipping directly towards the centre of a hole of diameter $d(>2a)$ with speed $v > (ga)^{1/2}$. If the ball strikes the opposite edge of the hole and no slipping takes place, show that it will overshoot if

$$\left\{\left(\frac{k^2}{a} + a\cos\alpha\right)v - \frac{ga\sin\alpha(d - a\sin\alpha)}{v}\right\}^2 > 2ga(k^2 + a^2)(1 - \cos\alpha),$$

where
$$\cos\alpha = 1 - \frac{g(d - a\sin\alpha)^2}{2av^2}$$

and k is the radius of gyration about an axis through the mass centre.

11.26 A uniform rectangular lamina is held with one edge on a smooth horizontal plane and the opposite edge against a smooth vertical wall. It is released from rest inclined at an angle α to the horizontal. Show that its angular speed when it reaches the horizontal is $\{g\sin\alpha(9 - \sin^2\alpha)/6l\}^{1/2}$, where $2l$ is the length of the inclined edge.

11.27 A uniform rod rests vertically with one end on a rough horizontal floor and is then slightly disturbed. Show that it must slip before $\theta = \cos^{-1}\frac{1}{3}$, where θ is the inclination of the rod to the upward vertical.

Show that it slips backwards or forwards according as slipping occurs before or after $\theta = \cos^{-1}\frac{2}{3}$.

11.28 A uniform sphere (or cylinder), of mass m and radius a, rolls and slides along a rough horizontal plane. The motion is two-dimensional and initially the velocity of the mass centre is of magnitude u. At the same time the sphere has an angular speed ω tending to make it move in the opposite direction.

Show that ultimately the loss in kinetic energy is

$$\frac{mk^2(u + a\omega)^2}{2(a^2 + k^2)},$$

where k is the radius of gyration about an axis through the mass centre.

Show that the mass centre of the sphere (or cylinder) will return to its initial position if $u < k^2\omega/a$.

11.29 A rigid body has radius of gyration k about an axis through the mass centre C.

(i) Show that the frequency of small oscillations about a parallel axis is a maximum if the axis is at a distance k from C.
(ii) Show that the frequency of oscillation is the same for all parallel axes at a distance c or k^2/c from C.
(iii) The points O, C, O′ lie on a straight line in that order and such that $OC = c$, $CO' = k^2/c$. Show that if a particle is placed on the line, the frequency of oscillation of the complete system about an axis through O is increased if it lies between O and O′, and decreased if it lies outside this interval.

11.30 A uniform rod, of length $2b$, is suspended by two vertical light strings, each of length l, which are fixed to the ends of the rods and to two fixed points a distance $2b$ apart.

Show that the energy equations for motions in which the strings remain taut and: (i) each string moves through an angular distance θ with the rod moving in a vertical plane; (ii) the strings move through equal and opposite angular distances θ with the mass centre of the rod moving vertically, are

(i) $\frac{1}{2}l\dot{\theta}^2 + g(1 - \cos\theta) = \text{constant}$,

(ii) $$\frac{1}{2}l\dot{\theta}^2\left(\frac{\frac{1}{3}b^2(1 + 2\sin^2\theta) - \frac{1}{4}l^2\sin^4\theta}{b^2 - \frac{1}{4}l^2\sin^2\theta}\right) + g(1 - \cos\theta) = \text{constant}.$$

Deduce that the periods of small oscillations are: (i) $2\pi(l/g)^{1/2}$; (ii) $2\pi(l/3g)^{1/2}$.

11.31 A uniform rectangular lamina, of mass m and with vertices A, B, C, D, lies in a vertical plane and is free to swing about the point A which is fixed. A particle of mass $\frac{1}{2}m$ is fixed to the lamina at B. The system is released from rest with AB horizontal and is next instantaneously at rest when AB is vertical. Show that BC = 2AB.

Show that the frequency of small oscillations about the equilibrium position of the system is

$$\left(\frac{13g}{6.2^{1/2}\,\text{AB}}\right)^{1/2}.$$

Show that if the motion is retarded by a constant frictional couple M, AB swings through an angle θ, where

$$\sin\theta + \cos\theta - \alpha\theta = 1$$

and $\alpha = M/(mg\text{AB})$. Show that if α is small $\theta = \frac{1}{2}\pi(1 - \alpha)$, correct to first order.

11.32 A uniform solid hemisphere, of radius a, stands with its curved surface in contact with a horizontal plane.

Show that the period of small oscillations about its equilibrium position is $2\pi/n$, where

$$n^2 = \frac{120g}{119a}$$

if the contact is perfectly smooth, and

$$n^2 = \frac{15g}{26a}$$

if the contact is perfectly rough.

In both cases show that if the hemisphere starts from rest with its plane surface vertical, the angular speed is $2^{1/2}n$ when the plane surface becomes horizontal.

11.33 A circular cylinder has mass m and radius a. The mass centre of the cylinder lies on its axis, about which the radius of gyration is k. The cylinder rolls inside a fixed circular cylindrical shell of radius $a + b$.

Show that the condition for rolling without slipping is that the coefficient of friction μ satisfies the inequality

$$\mu \geqslant \frac{k^2}{k^2 + a^2} \tan \alpha,$$

where α is the maximum angle which the plane joining the axes of the cylinders makes with the vertical.

Show that the period of small oscillations about the equilibrium position is then $2\pi/n$, where

$$n^2 = \frac{ga^2}{b(k^2 + a^2)}.$$

11.34 A circular cylinder, of radius a, is fixed with its axis horizontal. A uniform plank, of thickness $2h$, rests horizontally across the top of the cylinder with its mass centre C vertically above the axis of the cylinder and its length perpendicular to this axis. The radius of gyration of the plank about an axis through C parallel to the axis of the cylinder is k.

Suppose that the plank is disturbed from its position of equilibrium. Show that, when it is inclined at an angle θ to the horizontal, the energy equation is

$$\tfrac{1}{2}m(k^2 + h^2 + a^2\theta^2)\dot{\theta}^2 + g\{a\theta \sin\theta - (a + h)(1 - \cos\theta)\} = \text{constant}.$$

Show that the horizontal position is unstable for $a \leqslant h$.

Show that if $a > h$, the period of small oscillations is $2\pi\{(k^2 + h^2)/(a - h)g\}^{1/2}$.

11.35 A uniform right circular cylinder, of mass m, has height $2h$ and radius a.

Let O be a point on the circumference of the cylinder at one end. Let axes Ox, Oy, Oz be defined such that Ox is along a diameter and Oy is parallel to the axis of the cylinder.

Show that

$$I_{xx} = m\left(\frac{1}{4}a^2 + \frac{4}{3}h^2\right), \quad I_{yy} = \frac{3}{2}ma^2, \quad I_{zz} = m\left(\frac{5}{4}a^2 + \frac{4}{3}h^2\right),$$

$$I_{yz} = 0, \qquad I_{zx} = 0, \qquad I_{xy} = -mah.$$

The cylinder rotates with angular velocity $\boldsymbol{\omega} = \omega(\mathbf{i}\cos\alpha + \mathbf{j}\sin\alpha)$ about a line passing through O. Show that the angular momentum vector makes an angle β with the x-axis, where

$$\tan\beta = \frac{\frac{3}{2}a^2\tan\alpha - ah}{\frac{1}{4}a^2 + \frac{4}{3}h^2 - ah\tan\alpha}.$$

11.36 An axially symmetric disc, of mass m, radius a and radius of gyration k about its axis, rolls along the ground at a constant inclination α, and its centre C describes a horizontal circle, of radius c, with speed v.

Show that the kinetic energy is

$$\frac{1}{2}mv^2\left(1 + \frac{k^2}{2c^2}\sin^2\alpha + \frac{k^2}{a^2}\right).$$

11.37 A uniform rectangular block has adjacent edges of length $2a$, $2b$, $2c$. It rotates with angular speed ω about an axis parallel to a principal diagonal passing through a vertex not lying on the diagonal.

Show that the kinetic energy is

$$\frac{m(b^2c^2 + 7c^2a^2 + 7a^2b^2)\omega^2}{3(a^2 + b^2 + c^2)}.$$

11.38 Two wheels, each of mass m_w and radius a, are joined by an axle of mass m_a and length l. Each wheel has moments of inertia equal to I_a about its axis and I_d about a diameter. The axle has moment of inertia I at its mass centre. The system is axially symmetric.

The combination rolls on a horizontal plane and the angular speeds of the wheels are ω, ω'.

Show that the kinetic energy is

$$\tfrac{1}{4}(m_wa^2 + \tfrac{1}{2}m_aa^2 + I_a)(\omega + \omega')^2 + \tfrac{1}{4}(m_wa^2 + I_a + 4I_d + 2I)(\omega - \omega')^2.$$

11.39 A uniform circular cone, of semi-angle β, vertex O and axis OC, rolls inside a conical shell, of semi-angle $(\alpha + \beta)$, vertex O and axis OZ. The shell rotates with angular velocity $\boldsymbol{\omega}_s$ about OZ. The axis OC of the cone rotates with angular velocity $\boldsymbol{\omega}_c$ about OZ.

Show that the angular velocity of the cone is

$$\boldsymbol{\omega}_s + (\omega_s - \omega_c)\frac{\sin\alpha}{\sin\beta}\mathbf{k},$$

where $\mathbf{k}$ is the unit vector along the common generator OA.

Let the principal moments of inertia of the cone at O be I, I, I_3. Show that its kinetic energy is a minimum as a function of ω_s when

$$\omega_c = (1 + \cos\alpha\tan\beta)\omega_s.$$

Show that the kinetic energy is a minimum as a function of ω_c when

$$\omega_s = \frac{(I\sin^2\beta + I_3\cos^2\beta)\sin\alpha}{I_3\cos\beta\sin(\alpha+\beta)}\omega_c.$$

11.40 A uniform pole is supported at its lower end, which is carried round in a horizontal circle of radius a with constant angular speed ω. Prove that it can maintain a constant inclination α to the vertical provided that

$$gc\tan\alpha = \omega^2(ac - k_0^2\sin\alpha),$$

where k_0 is its radius of gyration about its supported end, and c is the distance of its mass centre from this end.

11.41 A uniform sphere rolls without slipping on an inclined plane. Show that the path of the centre of the sphere is a parabola.

11.42 A spherically symmetric sphere rolls without slipping on a horizontal plane, and the plane rotates about a fixed vertical axis with constant angular speed ω_0. Show that the centre of the sphere describes a circle with constant angular speed $\omega_0/(1 + ma^2/I)$, where m is the mass, a is the radius and I is the moment of inertia about an axis through the centre of the sphere.

11.43 Two men support a uniform pole, of mass m and length $2l$, in a horizontal position. They wish to change ends without changing their positions by throwing the pole into the air and catching it. Show that if the pole is to remain horizontal throughout its flight, and if the magnitude of the impulses applied by each man is to be a minimum, the required impulses have components $\frac{1}{2}m(\pi gl/6)^{1/2}$ vertically and $\frac{1}{2}m(\pi gl/6)^{1/2}$ horizontally.

11.44 A uniform square plate, of mass m and side $2a$, is suspended from one corner O. Let axes Oxyz be defined such that Ox is horizontal in the plane of the plate and Oy is vertically downwards. Show that these are principal axes.

Let the plate be given an impulsive blow $\bar{\mathbf{F}}$ at the corner with position vector $2^{1/2}a(\mathbf{i} + \mathbf{j})$. Show that the angular velocity imparted to the plate is

$$\frac{2^{1/2}}{ma}(\tfrac{3}{7}\bar{F}_3\mathbf{i} - 3\bar{F}_3\mathbf{j} - \tfrac{3}{8}(\bar{F}_1 - \bar{F}_2)\mathbf{k}).$$

Show that the impulsive reaction at O is

$$-\tfrac{1}{4}(\bar{F}_1 + 3\bar{F}_2)\mathbf{i} - \bar{F}_2\mathbf{j} - \tfrac{1}{7}\bar{F}_3\mathbf{k}.$$

11.45 A uniform rectangular lamina has sides of lengths $2a$, $2b$. It is given an impulsive blow perpendicular to its plane at one of its corners. Show that the lamina instantaneously rotates about a line parallel to a diagonal and at a distance $ab/3(a^2 + b^2)^{1/2}$ from it.

11.46 A uniform cube is spinning freely about a diagonal. Show that if one edge is suddenly fixed, the kinetic energy is reduced by a factor 1/12.

11.47 The centre of a spherically symmetric sphere, of mass m and moment of inertia I about a diameter, coincides with the origin. The sphere is free to turn about the point $c\mathbf{k}$. It is set in motion by an impulse $\bar{F}\mathbf{i}$ applied at the point $\mathbf{r} = x_0\mathbf{i} + y_0\mathbf{j} + z_0\mathbf{k}$.

Show that the impulsive reaction on the sphere at the fixed point is $\bar{X}\mathbf{i}$, where

$$\bar{X}(I + mc^2) + \bar{F}(I + mcz_0) = 0.$$

11.48 A body turns freely about its fixed mass centre C under no forces other than the reaction at C. The principal moments of inertia at C are I_1, I_2, I_3, where $I_1 > I_2 > I_3$. The body is set in motion so that the angular momentum $\mathbf{L}$ and the kinetic energy T satisfy the relation $L^2 = 2I_2T$. Obtain a solution of Euler's equations for the angular velocity and hence show that the angular velocity tends to $(0, L/I_2, 0)$ as $t \to \infty$.

11.49 A body turns freely about its fixed mass centre C under no forces other than the reaction at C. The principal moments of inertia at C are I_1, I_2, I_3 and the angular velocity has components $(\omega_1, \omega_2, \omega_3)$ relative to the principal axes.

Show that if $\omega_1 \ll \omega_3$, $\omega_2 \ll \omega_3$, then ω_3 is approximately constant and that if the small variation in ω_3 is ignored both ω_1 and ω_2 satisfy the equation $\ddot{x} + \nu x = 0$, where $\nu = (I_3 - I_1)(I_3 - I_2)\omega_3^2/I_1I_2$.

Hence show that a steady rotation about the third principal axis is unstable unless $(I_3 - I_1)(I_3 - I_2) > 0$.

11.50 A body turns freely about its fixed mass centre C under no forces other than the reaction at C. The principal moments of inertia at C are I_1, I_2, I_3 and the angular velocity has components $(\omega_1, \omega_2, \omega_3)$ relative to the principal axes.

Show that if $(I_2 - I_1)(I_3 - I_1 - I_2) = 0$ and $I_2 \geqslant I_1$, then

$$\omega_1 = q\cos\{(I_3 - I_2)\lambda/I_1\}, \quad \omega_2 = q\sin\{(I_3 - I_2)\lambda/I_1\}, \quad \omega_3 = \dot{\lambda},$$

where q is a constant and $2I_3\ddot{\lambda} + (I_2 - I_1)q^2\sin\{2(I_3 - I_2)\lambda/I_1\} = 0$.

(i) For a body with $I_1 = I_2 = I$ show that the magnitude of the angular velocity is constant and that the direction precesses about the axis of symmetry with a periodic time $2\pi I/(I_3 - I)s$, where s is the constant spin about the axis of symmetry.

Now let the body be subject to a couple $N\mathbf{k} \times \boldsymbol{\omega}$. Show that the above results are modified to the extent that I_3 is replaced by $I_3 + N/s$.

(ii) A uniform rectangular lamina, which is free to rotate about its centre, is set spinning about a diagonal with angular speed ω_0. Show that in the subsequent motion the lamina first spins about the other diagonal after a time

$$2\omega_0^{-1}(\sec 2\alpha)^{1/2}K(\sin \alpha)$$

where $$K(k) = \int_0^{\pi/2} (1 - k^2 \sin^2 \theta)^{-1/2} \, d\theta$$

and 2α is the acute angle between the diagonals.

11.51 A body of revolution turns freely about its fixed mass centre C, and is subject to a retarding couple equal to k times the angular velocity. The principal moments of inertia at C are I, I, I_3 and the initial spin about the axis of symmetry is s. Write down Euler's equations and show that, after a time t, the angular velocity makes an angle ϕ with the axis of symmetry given by

$$\tan \phi = \exp\{-(I_3 - I)kt/II_3\}\tan \phi_0,$$

where ϕ_0 is the initial value of ϕ. Deduce that if $I_3 > I$ the direction of the angular velocity tends to coincide with the direction of the axis of symmetry. Show that during the motion a plane through these two directions turns through a total angle $I_3(I_3 - I)s/kI$ relative to axes fixed in the body.

11.52 Two gyroscopes are fixed to a light axle, which is their common axis of symmetry. The gyroscopes can turn freely and independently about the axle and the whole system can turn freely about a fixed point O of the axle.

Show that the gyroscopes each have constant spin about the axle. Show that the motion of the axle would be the same if the whole system moved as a rigid body with spin $(I_3 s + I_3' s')/(I_3 + I_3')$ about the axle. Here s, s' refer to the constant spins of the gyroscopes and I_3, I_3' refer to their moments of inertia about the axle.

11.53 A top, of mass m, spins about a fixed point O on its axis of symmetry.

Unit vectors $\mathbf{k}', \mathbf{k}, \mathbf{j}, \mathbf{i}$ are defined so that $\mathbf{k}'$ is vertically upwards, $\mathbf{k}$ is parallel to the axis of the top, $\mathbf{j}$ is parallel to $\mathbf{k} \times \mathbf{k}'$ and $\mathbf{i} = \mathbf{j} \times \mathbf{k}$. The angle between $\mathbf{k}$ and $\mathbf{k}'$ is θ, and ϕ is the angular displacement of the vertical plane of $\mathbf{k}$ and $\mathbf{k}'$ relative to a fixed vertical plane. The principal moments of inertia at O are I, I, I_3.

Show that the angular velocity of the top and the angular momentum relative to the fixed point are, respectively,

$$\boldsymbol{\omega} = \mathbf{i}\dot{\phi} \sin \theta - \mathbf{j}\dot{\theta} + \mathbf{k}s,$$

$$\mathbf{L} = I(\mathbf{i}\dot{\phi} \sin \theta - \mathbf{j}\dot{\theta}) + I_3 \mathbf{k}s,$$

where s is the constant spin. From the conservation of angular momentum about the vertical and the conservation of energy, deduce the equations

$$I\dot{\phi}\sin^2\theta + I_3 s\cos\theta = H,$$

$$\tfrac{1}{2}I(\dot{\phi}^2\sin^2\theta + \dot{\theta}^2) + \tfrac{1}{2}I_3 s^2 + mga\cos\theta = E,$$

where H and E are constants, and a is the distance of the mass centre from the fixed point.

(i) A spinning top is released from an initial position in which its axis is horizontal and at rest. Show that the axis will rotate about O until it makes an angle β with the downward vertical such that

$$\cos\beta = (b^2 + 1)^{1/2} - b, \quad \text{where } b = I_3^2 s^2/(4Imga).$$

(ii) Given that the axis of the top passes through the vertical, and that $\dot{\theta} = \sigma$ when $\theta = 0$, show that

$$\dot{y}^2 = \frac{2mga}{I}y(\delta^2 + 2ky - y^2),$$

where

$$y = 1 - \cos\theta, \quad \delta^2 = \frac{\sigma^2 I}{mga}, \quad k = 1 - \frac{I^2\sigma^2 + I_3^2 s^2}{4mgaI}$$

Hence show that when δ is small, the axis of the top remains near the vertical provided that k is small or negative.

11.54 The top of Exercise 11.53 is subjected to a couple $M\mathbf{k}'$. The couple is adjusted to ensure that $\dot{\phi} = p$ is constant. Show that

$$(2Ip\cos\theta - I_3 s)\dot{\theta} = M\sin\theta,$$

$$-I\ddot{\theta} + Ip^2\sin\theta\cos\theta - I_3 ps\sin\theta = -mga\sin\theta,$$

$$I_3\dot{s} = M\cos\theta,$$

and that, if $\theta = \pi/2$, $\dot{\theta} = 0$, $s = s_0$ at $t = 0$,

$$I\ddot{\theta} + I_3 ps_0 + Ip^2(\theta - \pi/2) = mga\sin\theta.$$

Deduce that θ increases or decreases initially according as mga is greater or less than $I_3 ps_0$ and that if $I_3 ps_0 = mga(1+\epsilon)$, where $\epsilon \ll 1$, then approximately θ oscillates between $\pi/2$ and $\pi/2 - 2mga\epsilon/Ip^2$ with a period $2\pi/p$.

11.55 One end of a uniform rod, of mass m and length l, is constrained to slide along a smooth vertical axis. The other end slides on a smooth horizontal floor. Let θ be the angle between the rod and the downward vertical and let ϕ be the angular displacement of the vertical plane through the rod. Let $\theta = \pi/4$, $\dot{\theta} = 0$, $\dot{\phi} = \omega_0$ when $t = 0$. Show that

$$\dot{\theta}^2 = \frac{\omega_0^2}{4}(2 - \operatorname{cosec}^2\theta) + \frac{3g}{l}(2^{-1/2} - \cos\theta)$$

provided that the rod remains in contact with the floor. Show that the minimum value of ω_0 below which the rod will maintain contact with the floor during the subsequent motion is given by $\omega_0^2 l/g = 3(2 - 2^{1/2})$.

11.56 A spherically symmetric sphere, of mass m, centre C, radius a and moment of inertia I about a diameter, rolls without slipping inside a hollow fixed sphere, of centre O and radius b. Show that a steady precession is possible in which OC is inclined at a constant angle α to the downward vertical and rotates about the vertical with constant angular speed p provided that

$$-(I + ma^2)\frac{b - a}{a}p^2\cos\alpha + Isp + mga = 0,$$

where s is the constant spin of the sphere about OC.

Show that the solution is consistent with the no slip condition provided that

$$\sin\alpha\,|\,g - (b - a)p^2\cos\alpha\,| < \mu\{g\cos\alpha + (b - a)p^2\sin^2\alpha\},$$

where μ is the coefficient of friction.

11.57 The mass centre of an axially symmetric flywheel is fixed at a point O of the earth's surface. The axis of the flywheel is constrained to remain in the meridian plane at O (that is the vertical plane containing the earth's axis of rotation). Initially the flywheel has a spin s about its own axis, which is parallel to the earth's axis of rotation. Show that if the axis is slightly disturbed, it will oscillate with period

$$2\pi/(I_3\omega_e s/I - \omega_e^2)^{1/2},$$

where ω_e is the angular speed of the earth and I, I, I_3 are the principal moments of inertia at O of the flywheel.

11.58 An axially symmetric circular disc, of radius a, is in motion with its edge touching a smooth horizontal table. The axis of the disc is inclined at an angle θ to the vertical, the spin about the axis is s.

Show that in steady motion the two rates of precession p are given by the roots of

$$mga\cos\theta = Ip^2\sin\theta\cos\theta - I_3 sp\sin\theta,$$

where I, I, I_3 are the principal moments of inertia of the disc at its centre.

Show that the particular solution $\theta = \pi/2$, $s = 0$, when the disc rotates about the vertical, is unstable if $ap^2 \leqslant 4g$.

11.59 An axially symmetric circular cone, of mass m and semi-angle α, rolls on a horizontal plane. The angular speed of the cone's axis about the vertical is p. Show that the vertex of the cone tends to rise from the plane if

$$p^2(I \sin^2 \alpha + I_3 \cos^2 \alpha) > mgd \sin \alpha,$$

where d is the distance of the mass centre from the base of the cone, and I, I, I_3 are the principal moments of inertia at the mass centre.

11.60 A spherically symmetric sphere, of mass m, radius a and moment of inertia I about a diameter, rolls without slipping on the inner surface of a fixed hollow circular cylinder of radius b.

Let θ be the angular displacement of the plane through the axis of the cylinder and the centre of the sphere, and let x be the displacement of the centre of the sphere parallel to the axis.

(i) Assume that the axis of the cylinder is vertical. Show that $\dot{\theta}$ is constant and that x satisfies the equation $\ddot{x} + n^2\dot{\theta}^2(x + x_0) = 0$, where x_0 is a constant and $n^2 = I/(I + ma^2)$.

(ii) Now consider the problem when the axis of the cylinder is horizontal. Show that

$$\frac{d^2\dot{x}}{d\theta^2} + n^2\dot{x} = 0, \quad \ddot{\theta} + \frac{ma^2 g \sin \theta}{(I + ma^2)(b - a)} = 0.$$

12 Virtual Work and Lagrange's Equations

12.1 VIRTUAL WORK

In equations (11.1.3), namely

$$\Sigma \mathbf{F} = \Sigma m\mathbf{a}, \quad \Sigma \mathbf{r} \times \mathbf{F} = \Sigma \mathbf{r} \times m\mathbf{a},$$

we introduced the basic relations governing the dynamics of a system of rigid bodies. These relations imply that the system of external forces is equivalent to the system of mass accelerations. Now in §10.6 we derived the result that the work of equivalent systems of vectors is the same, to first order, for an arbitrary small rigid body displacement. In particular the work of a null system is zero, to first order, for an arbitrary small rigid body displacement. In this chapter we show how this result can be made the basis of a general theory for describing the mechanics of systems of rigid bodies.

For a single rigid body it follows at once that the work of the system of external forces is the same, to first order, as that of the system of mass accelerations for any small displacement of the body. However, a system of particles and rigid bodies is not in general restricted to rigid body displacements, so that the result is not immediately applicable to such a system. We can, of course, equate the work of the forces and the work of the mass accelerations for each component of the system which is displaced as a rigid body, and these relations can be summed over the whole system. There may then be contributions from forces of constraint, such as the tension in a stretched elastic string or the frictional force where one body slides on another. These constraints are external forces as far as a component of the system is concerned, but may be internal or external forces for the system as a whole. Provided the contribution from such forces is included, the result can be generalised to include systems of rigid bodies moving under constraints. Furthermore, there are several forces for which the net work of the constraints is zero, such as the tensions at the ends of an inelastic string or the contact force where one body rolls on another (see §11.3). If we refer to those forces, both internal and external, which do not arise from workless constraints as applied forces, we see that for a system of particles and rigid bodies, subject to applied forces and workless constraints, the work of the system of applied forces is the same, to first order, as that of the system of mass accelerations in an arbitrary small displacement of the system compatible with the constraints. Indeed it is possible to include a displacement of the system with a constraint relaxed, provided account is taken of the fact that the force of constraint may do work in such a displacement and thereby become an applied force, even though it is workless in the constrained condition. It is important to appreciate that the result is to be taken as it stands, and is irrespective of what may happen if the system undergoes a real physical displacement appropriate to the dynamics of the situation. In application it is important that the displacement of the system can be arbitrary, subject to compatibility with the workless constraints. For this reason the displacement is referred to as a virtual displacement and the work of the applied forces is called the virtual work.

We begin with a discussion of a system in equilibrium, for which the work of the system of mass accelerations is clearly zero, and we have the following result.

The Principle of Virtual Work

A system subject to applied forces and workless constraints is in equilibrium if the virtual work of the applied forces is zero for an arbitrary small displacement of the system compatible with the constraints.

The principle is also a sufficient condition for equilibrium, in other words if the virtual work of the applied forces is zero for any small displacement of the system compatible with the constraints, then the system is in equilibrium. For if the system is not in equilibrium it will begin to move. If it starts from rest each particle of the rigid body will move initially in the direction of its acceleration $\mathbf{a}$ and the distance it moves will be approximately $\frac{1}{2}\mathbf{a}t^2$, if t is small. Accordingly a possible small displacement of the system, which is compatible with the constraints, is $\delta\mathbf{r} = \epsilon\mathbf{a}$ for each particle, where ϵ is small and positive. For such a virtual displacement, the virtual work of the mass accelerations is

$$\Sigma m\mathbf{a} \,.\, \delta\mathbf{r} = \Sigma m\mathbf{a} \,.\, (\epsilon\mathbf{a}) = \epsilon\Sigma ma^2 > 0.$$

But this is equal to the virtual work of the applied forces, which contradicts the hypothesis that the virtual work of the applied forces is zero. Hence the system is in equilibrium.

It may be that the applied forces are all conservative. In such a case the work done is the loss in potential energy and the requirement that the system should do no work in all small rigid body displacements implies that the potential energy should have a stationary value with respect to such displacements.

To investigate the nature of the equilibrium, let us suppose that the potential energy $V = V_0$ in equilibrium, and that the system starts from rest in a neighbouring configuration for which $V = V_1$. If the system is not in equilibrium it will begin to move and so acquire kinetic energy. Since the sum of the kinetic and potential energies is constant, V will decrease so that $V \leqslant V_1$. If V_0 is a minimum, then $V \geqslant V_0$ in the neighbourhood of equilibrium and so V is confined within the limits $V_0 \leqslant V \leqslant V_1$. If $V_1 - V_0$ is small it follows that V never departs appreciably from its equilibrium value and the system will remain close to its equilibrium configuration.

Conversely if V_0 is a maximum, so that $V_1 < V_0$, the effect of the ensuing motion after a small displacement is to decrease V further from its equilibrium value, and hence the equilibrium is unstable.

The following examples illustrate how the principle of virtual work can be used to obtain information about systems in equilibrium. One of its merits is that the unknown forces of constraint do not enter into the work equation unless the constraint is relaxed and the associated force becomes an applied force. It is therefore an advantage to consider virtual displacements compatible with the constraints, unless the problem requires a force of constraint to be calculated.

Example 12.1.1

Three uniform rods AB, BC, CD, each of mass m and length $2l$, are smoothly hinged at B and C. They are suspended from A and a horizontal force F is applied at D (see Fig. 12.1.1). It is required to find the equilibrium configuration.

Let the inclinations of the rods to the downward vertical be denoted by θ_1, θ_2, θ_3. We consider the work done by the forces in a small virtual displacement in which θ_1, θ_2, θ_3 change by small increments. The mass centre of AB is at a depth $l\cos\theta_1$ below A and so there is a contribution to the work done by gravity of amount $mg\delta(l\cos\theta_1) = -mgl\delta\theta_1 \sin\theta_1$ arising from the displacement of AB. We take the sum of the contributions from the displacement of each rod together with the work of F to get

$$\begin{aligned}\delta W &= mg\,\delta(l\cos\theta_1) + mg\delta(2l\cos\theta_1 + l\cos\theta_2) + mg\delta(2l\cos\theta_1 + 2l\cos\theta_2 \\ &\quad + l\cos\theta_3) + F\delta(2l\sin\theta_1 + 2l\sin\theta_2 + 2l\sin\theta_3) \\ &= (2F\cos\theta_1 - 5mg\sin\theta_1)l\delta\theta_1 + (2F\cos\theta_2 - 3mg\sin\theta_2)l\delta\theta_2 \\ &\quad + (2F\cos\theta_3 - mg\sin\theta_3)l\delta\theta_3.\end{aligned}$$

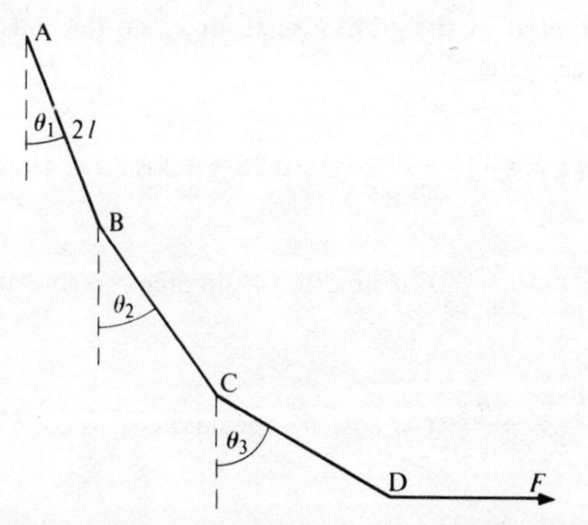

Figure 12.1.1 Illustration for Example 12.1.1: equilibrium of three jointed rods.

For equilibrium this must vanish for all displacements, which is only possible if the coefficients of $\delta\theta_1$, $\delta\theta_2$, and $\delta\theta_3$ are separately zero. Hence we have

$$\tan\theta_1 = \frac{2F}{5mg}, \quad \tan\theta_2 = \frac{2F}{3mg}, \quad \tan\theta_3 = \frac{2F}{mg}.$$

Note that because the result must be true for all displacements, we get from the one calculation the inclination of all three rods. Furthermore, since the displacements chosen are compatible with the constraints at A, B and C, the net work of the reactions at A, B and C is zero, and there is no contribution to the work equation.

Example 12.1.2
Four uniform rods AB, BC, CD, DA each of mass m and length $2l$, are smoothly hinged together at their ends to form a rhombus. The frame is maintained in a rigid configuration by a light rod BD. The rods AB and AD rest on two smooth pegs at the same level and at a distance $2d$ apart as in Figure 12.1.2(a).

We derive the condition for the stability of equilibrium of the symmetrical position and we calculate the tension in the tie rod.

Let x be the height of A above the level of the pegs, let AB be inclined at an angle $\alpha + \theta$ to the downward vertical, where $B\hat{A}D = 2\alpha$. We have

$$x\tan(\alpha+\theta) + x\tan(\alpha-\theta) = 2d$$

and hence

$$x = \frac{2d}{\tan(\alpha+\theta)+\tan(\alpha-\theta)} = \frac{d(\cos 2\theta + \cos 2\alpha)}{\sin 2\alpha}.$$

The centre of the rhombus is at a depth $2l\cos\alpha\cos\theta - x$ below the level of the pegs. The potential of the gravitational forces on the rods, measured from the level of the pegs, is therefore

$$V = 4mg(x - 2l\cos\alpha\cos\theta) = \frac{4mg}{\sin 2\alpha}(d\cos 2\theta + d\cos 2\alpha - 4l\sin\alpha\cos^2\alpha\cos\theta),$$

and so
$$\frac{dV}{d\theta} = -\frac{4mg}{\sin 2\alpha}(2d\sin 2\theta - 4l\sin\alpha\cos^2\alpha\sin\theta),$$

$$\frac{d^2V}{d\theta^2} = -\frac{4mg}{\sin 2\alpha}(4d\cos 2\theta - 4l\sin\alpha\cos^2\alpha\cos\theta).$$

It follows that $dV/d\theta = 0$ when $\theta = 0$ so that this is an equilibrium position. The equilibrium is stable if $d < l\sin\alpha\cos^2\alpha$ since this is the condition for $d^2V/d\theta^2$ to

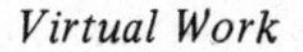

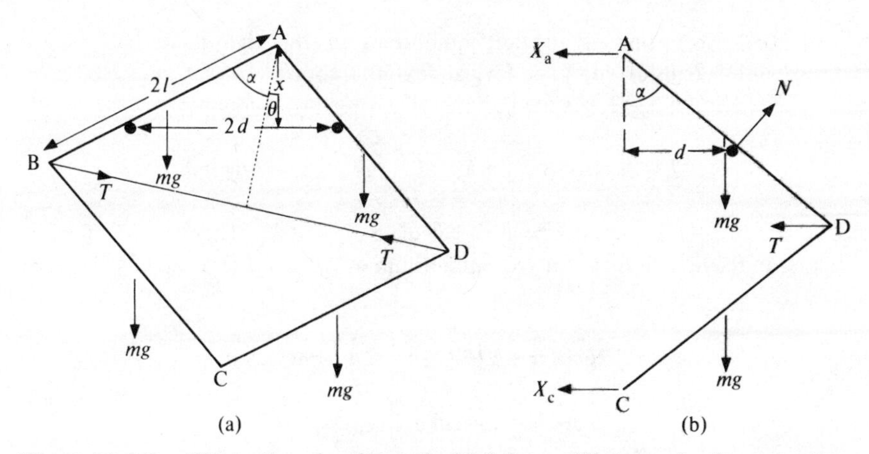

Figure 12.1.2 Illustration for Example 12.1.2: equilibrium of a frame resting on two pegs.

be positive. Equilibrium is also stable when $d = l \sin\alpha \cos^2\alpha$; in this case it is necessary to show that $d^4V/d\theta^4$ is positive at $\theta = 0$, which is now the condition for V to have a minimum value there. For $d > l \sin\alpha \cos^2\alpha$, $\theta = 0$ is an unstable position of equilibrium, but there is stable equilibrium for a value of θ given by $d \cos\theta = l \sin\alpha \cos^2\alpha$.

To calculate the tension T in the tie rod, we relax the constraint that BD is fixed in length in order to get a contribution from T to the virtual work. We consider a virtual displacement from the equilibrium position $\theta = 0$ in which α increases by $\delta\alpha$. When $\theta = 0$ we have

$$V(\alpha) = 4mg(d \cot\alpha - 2l \cos\alpha)$$

and the net work done when α changes by $\delta\alpha$ is

$$\begin{aligned}\delta W &= -\delta V - T\delta(4l \sin\alpha)\\ &= \{4mg(d \operatorname{cosec}^2\alpha - 2l \sin\alpha) - 4lT \cos\alpha\}\,\delta\alpha.\end{aligned}$$

For equilibrium this must be zero, and so

$$T = \frac{mg}{\sin^2\alpha \cos\alpha}\left(\frac{d}{l} - 2\sin^3\alpha\right).$$

Let us compare the above derivation with one working directly with the equations for equilibrium. In this approach we can take advantage of the symmetry and confine the discussion to one half of the framework. Because of the symmetry the forces on either side of AC are mirror images; in particular, the reactions X_a and X_c at A and C are horizontal, as shown in Figure

12.1.2(b). Note that when the equilibrium of the right-hand half of the framework is considered, X_a and X_c are external forces.

The relation $\Sigma \mathbf{F} = \mathbf{0}$, has the horizontal and vertical components

$$T - N\cos\alpha + X_a + X_c = 0, \quad N\sin\alpha - 2mg = 0.$$

In order to find T, we now need to calculate X_a and X_c. These are obtained by taking moments about D for the equilibrium of AD and CD separately. We have

$$X_a 2l\cos\alpha = N\left(2l - \frac{d}{\sin\alpha}\right) - mgl\sin\alpha,$$

$$X_c 2l\cos\alpha = mgl\sin\alpha.$$

These equations give

$$T = N\cos\alpha - \frac{N}{\cos\alpha}\left(1 - \frac{d}{2l\sin\alpha}\right),$$

$$= \frac{mg}{\sin^2\alpha\cos\alpha}\left(\frac{d}{l} - 2\sin^3\alpha\right).$$

Comparison of the two calculations shows how the virtual work approach avoids the calculation of unknown reactions. If it is necessary to calculate these reactions, the virtual work approach can also be used to do this. It merely requires the relaxing of a constraint so that the desired reaction contributes to the work equation. If this can be done in a way which involves only one unknown reaction, the equation gives directly an expression for the reaction. For example if it is required to calculate X_c, the constraint at C is relaxed and the virtual work equation is derived for a displacement in which the rod CD rotates through a small angle $\delta\alpha$ about D. We then have

$$mgl\,\delta\alpha\sin\alpha - X_c 2l\,\delta\alpha\cos\alpha = 0$$

for equilibrium. This gives $X_c = \frac{1}{2}mg\tan\alpha$ which is the same as the result obtained by taking moments about D for the equilibrium of CD.

12.2 GENERALISED COORDINATES

Before we proceed to a discussion of dynamical problems, it is necessary to introduce the concept of a generalised coordinate system. It is clear from earlier examples that the choice of coordinate system in any given problem is at our disposal, and that a judicious choice can simplify the mathematics considerably. For example the position of a particle moving in two dimensions can be defined by cartesian coordinates (x, y) or polar coordinates (r, θ). The latter description has obvious advantages in central force problems and especially when the particle moves in a circle, since its distance from the centre is fixed and its position can be defined by a single angular coordinate θ. If we insist on using cartesian coordinates (x, y), then we must also impose the constraint $x^2 + y^2 =$ constant for circular motion. The common feature of all descriptions, whatever coordinates are used, is that the number of coordinates less the number of constraints is a constant, called the degrees of freedom. A particle moving on a circular path has one degree of freedom. A particle free to move in three dimensions has three degrees of freedom. A system of two particles moving independently has six degrees of freedom, but if they move as a rigid body this is reduced to five by the constraint that their distance apart is invariant. The position in space of such a system can be described by three coordinates defining the position of one particle, and two angular coordinates defining the position of the second particle relative to the first. We then have five coordinates and no constraints, giving the same number of degrees of freedom as three coordinates for each particle together with one constraint. A rigid body proper actually has six degrees of freedom, because in addition to the five for the orientation of two of its points, the body can have a further angular displacement about the line joining them.

These ideas can be extended to more complex systems of rigid bodies, either moving independently or subject to certain constraints. Any set of quantities which is used to specify the configuration of a system is called a set of generalised coordinates and a knowledge of the generalised coordinates implies a knowledge of the position of every particle of the system. When the generalised coordinates are subject to constraints, it is not possible to vary all of them independently without violating the constraints. But if the constraints are of the form

$$f(q_1, q_2, \ldots, q_n, t) = 0, \tag{12.2.1}$$

where $q_1, q_2, \ldots, q_n$ are the generalised coordinates and t is the time, such a relation can be used to eliminate one of the coordinates. If there are m independent constraints, all of the form (12.2.1), it is possible in principle to eliminate m of the coordinates in this way. Any configuration of the system can then be specified in terms of the $n - m$ remaining coordinates, the number of coordinates now being equal to the degrees of freedom. There being no further coordinate constraints, these $n - m$ coordinates can be changed independently. When the configuration of a system can be defined by a set of generalised coordinates which can be varied arbitrarily and independently without violating any constraints, the system is called holonomic. This possibility does not always exist, because relation (12.2.1) does not represent the only type of constraint that can be encountered. For example, the condition of no slip when one body rolls on another involves the derivatives of the coordinates, and such constraints are of the form

$$f(q_1, q_2, \ldots, q_n, \dot{q}_1, \dot{q}_2, \ldots, \dot{q}_n, t) = 0. \tag{12.2.2}$$

Unless the expression can be integrated to the form of (12.2.1), the possibility of eliminating one of the coordinates is not available and the use of more coordinates than there are degrees of freedom is unavoidable. Systems involving such constraints are called non-holonomic.

Example 12.2.1

Figure 12.2.1 represents a system consisting of two circular discs, each of radius a, free to rotate at the opposite ends of a slender axle of length $2l$. The system is free to roll on a horizontal plane with the axle horizontal and the discs vertical.

To specify the configuration of the system we first specify the position of the mid point C of the axle by cartesian coordinates (x, y) relative to horizontal axes Ox, Oy. Relative to C the position of the axle is specified by the angular displacement θ of the axle from the fixed direction Oy. Finally the displacements of the discs relative to the axle can be specified by the angles ϕ and ψ through which they rotate about the axle.

The five quantities x, y, θ, ϕ, ψ represent one possible choice of generalised coordinates which will specify the configuration of the above system. If the discs were allowed to slip on the plane, all these coordinates could be given independent increments without violating any constraints, and so there would be five degrees of freedom. If, however, the motion is confined to that which does not involve slipping, certain constraints must be imposed to ensure that the velocity at each point of contact is zero. One condition is that the velocity of C along the line of the axle should be zero, so that there will be no slipping in the direction of the axle. This gives

$$\dot{x} \sin \theta + \dot{y} \cos \theta = 0. \tag{12.2.3}$$

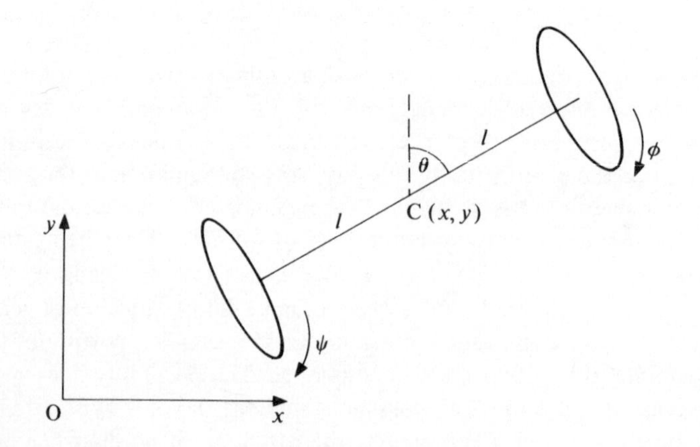

Figure 12.2.1 Illustration for Example 12.2.1: generalised coordinates for wheels and axle.

In addition the velocity of each disc perpendicular to the axle should be zero at the point of contact with the plane and this gives the two further conditions

$$\dot{x} \cos \theta - \dot{y} \sin \theta + l\dot{\theta} - a\dot{\phi} = 0, \tag{12.2.4}$$

$$\dot{x} \cos \theta - \dot{y} \sin \theta - l\dot{\theta} - a\dot{\psi} = 0. \tag{12.2.5}$$

These last two conditions are better discussed in the alternative forms

$$2(\dot{x} \cos \theta - \dot{y} \sin \theta) - a(\dot{\phi} + \dot{\psi}) = 0, \tag{12.2.6}$$

$$2l\dot{\theta} - a(\dot{\phi} - \dot{\psi}) = 0, \tag{12.2.7}$$

obtained by addition and subtraction of (12.2.4) and (12.2.5). There are in all three constraints and five coordinates and so two degrees of freedom.

Equation (12.2.7) can be integrated to give

$$2l\theta - a(\phi - \psi) = 0 \tag{12.2.8}$$

subject to suitable initial conditions. We now have the choice of eliminating one of the variables, say θ, from the system of generalised coordinates since it is seen to be specified in terms of $\phi - \psi$. No such integration is possible with the remaining pair of constraints, or any combination of them, and we cannot reduce the coordinates to less than four with two non-integrable constraints.

If, however, we consider a motion with one degree of freedom by specifying a further constraint that $\theta = 0$ and the discs roll in a straight line, the constraints simplify to

$$\dot{\theta} = 0, \quad \dot{y} = 0, \quad \dot{x} - a\dot{\phi} = 0, \quad \dot{x} - a\dot{\psi} = 0.$$

These are all integrable and give

$$\theta = 0, \quad y = 0, \quad x - a\phi = 0, \quad x - a\psi = 0$$

subject to suitable initial conditions. In this case one coordinate, say x, is sufficient to specify the configuration and this is classed as holonomic since there are no constraints on x.

12.3 LAGRANGE'S EQUATIONS

The point has been made that there is considerable merit in a careful choice of generalised coordinates, so that the configuration of a mechanical system is specified in a way that avoids unnecessary complication. But if we wish to discuss the dynamics of the system by direct application of Newton's laws of motion, it is necessary to calculate the accelerations in terms of the chosen coordinates. This is not always a straightforward matter. Even for the simple example of a particle moving in two dimensions, the components of acceleration, when using polar coordinates (r, θ), are $(\ddot{r} - r\dot{\theta}^2, r\ddot{\theta} + 2\dot{r}\dot{\theta})$. The result is not intuitively obvious and has to be calculated from first principles. For more complicated systems such calculations can be quite involved. Furthermore equations of motion have been formulated for particles or rigid bodies only. For more complicated systems it is necessary to derive the equations of motion for each separate component, and this means that unknown reactions of constraint must be included and eventually either evaluated or eliminated after the equations have been formulated.

A general procedure for obtaining the equations of motion for any system, preferably without the inclusion of unknown reactions, is much to be desired. Such is the achievement of the analysis first formulated by Lagrange. It requires a knowledge of the kinetic energy of the system, a much simpler calculation than that for the vector accelerations, and a knowledge of the work of the forces involved. Thus any force which does no work is automatically eliminated from the equations. When this information is known, there is a routine procedure for the derivation of the equations of motion. The achievement is a considerable one, even when confined in application to problems in dynamics. But the insight gained from the analysis has had a far-reaching influence in the development of both classical analytical mechanics and modern theoretical physics, and there is a considerable body of literature dealing with these more general aspects.

The present discussion will be confined to a derivation of Lagrange's equations of motion. As a preliminary, we derive some mathematical relations which will be useful later in the discussion. If the configuration of a system is defined by n generalised coordinates, it follows that the position vector $\mathbf{r}$ of any particle of the system is a function of the generalised coordinates. It may also depend explicitly on the time if there are moving constraints or a moving coordinate system. For example a natural coordinate to use for specifying the configuration of a simple pendulum is its angle of inclination θ to the downward vertical. If the point of support of the pendulum moves with a displacement $a\mathbf{i} \sin pt + b\mathbf{j} \sin pt$ at time t, the cartesian coordinates of the bob of the pendulum can be expressed as

$$x(\theta, t) = a \sin pt + l \cos \theta,$$

$$y(\theta, t) = b \sin pt - l \sin \theta,$$

and the position of the bob depends explicitly on the time as well as on the generalised coordinate θ.

We write

$$\mathbf{r} = \mathbf{r}(q_1, q_2, \ldots, q_n, t), \tag{12.3.1}$$

$$\frac{\mathrm{d}\mathbf{r}}{\mathrm{d}t} = \frac{\partial \mathbf{r}}{\partial q_1}\dot{q}_1 + \frac{\partial \mathbf{r}}{\partial q_2}\dot{q}_2 + \ldots + \frac{\partial \mathbf{r}}{\partial q_n}\dot{q}_n + \frac{\partial \mathbf{r}}{\partial t}. \tag{12.3.2}$$

Thus $\mathbf{r}$, and its partial derivatives $\partial\mathbf{r}/\partial q_i$ and $\partial\mathbf{r}/\partial t$, are functions of the coordinates and the time. The total derivative $\mathrm{d}\mathbf{r}/\mathrm{d}t$, which takes into account the variations of the coordinates q_i with t as well as the explicit variation of $\mathbf{r}$ with t, is a linear function of the generalised velocities $\dot{q}_i$ ($i = 1, 2, \ldots, n$), with coefficients which are functions of the q_i and t. If it suits our purpose to do so, we can look upon $\mathrm{d}\mathbf{r}/\mathrm{d}t$ as a function of the $2n + 1$ variables $q_i, \dot{q}_i$ ($i = 1, 2, \ldots, n$) and t, and operate on it as such. It should be stressed that this procedure is solely mathematical manipulation of the functions involved, and is quite independent of any mechanical considerations. On this basis, it follows from (12.3.2) that

$$\frac{\partial}{\partial \dot{q}_i}\frac{\mathrm{d}\mathbf{r}}{\mathrm{d}t} = \frac{\partial \mathbf{r}}{\partial q_i}. \tag{12.3.3}$$

We also have the relations

$$\frac{\partial}{\partial q_i}\frac{\mathrm{d}\mathbf{r}}{\mathrm{d}t} = \frac{\partial^2 \mathbf{r}}{\partial q_i \partial q_1}\dot{q}_1 + \frac{\partial^2 \mathbf{r}}{\partial q_i \partial q_2}\dot{q}_2 + \ldots + \frac{\partial^2 \mathbf{r}}{\partial q_i \partial q_n}\dot{q}_n + \frac{\partial^2 \mathbf{r}}{\partial q_i \partial t}, \tag{12.3.4}$$

$$\frac{\mathrm{d}}{\mathrm{d}t}\frac{\partial \mathbf{r}}{\partial q_i} = \frac{\partial^2 \mathbf{r}}{\partial q_1 \partial q_i}\dot{q}_1 + \frac{\partial^2 \mathbf{r}}{\partial q_2 \partial q_i}\dot{q}_2 + \ldots + \frac{\partial^2 \mathbf{r}}{\partial q_n \partial q_i}\dot{q}_n + \frac{\partial^2 \mathbf{r}}{\partial t \partial q_i}. \tag{12.3.5}$$

On the assumption that $\mathbf{r}$ has continuous second order partial derivatives, so that the order of differentiation is not significant, relations (12.3.4) and (12.3.5) show that

$$\frac{\partial}{\partial q_i}\frac{\mathrm{d}\mathbf{r}}{\mathrm{d}t} = \frac{\mathrm{d}}{\mathrm{d}t}\frac{\partial \mathbf{r}}{\partial q_i}. \tag{12.3.6}$$

We proceed now to consider the mechanics. To derive Lagrange's equations we use the fact that the forces acting on a mechanical system are equivalent to the mass accelerations of all the particles of the system. It follows that the work of the applied forces in a small displacement is equal to the work of the mass accelerations of the particles, as discussed in §12.1.

For the moment we consider only holonomic systems where the number of generalised coordinates is equal to the number of degrees of freedom. For such a system each coordinate can be given an arbitrary and independent increment without violating any constraint. We are therefore at liberty to discuss a small virtual displacement for which just one of the generalised coordinates q_i becomes $q_i + \delta q_i$. The work of the applied forces in such a displacement is

$$\Sigma\mathbf{F} \,.\, \delta\mathbf{r} = \left(\Sigma\mathbf{F} \,.\, \frac{\partial \mathbf{r}}{\partial q_i}\right)\delta q_i = Q_i \delta q_i, \tag{12.3.7}$$

say, where the summation is over all the forces which do work in the displacement. The quantity

$$Q_i = \Sigma \mathbf{F} \cdot \frac{\partial \mathbf{r}}{\partial q_i} \tag{12.3.8}$$

is called the generalised force* corresponding to the generalised coordinate q_i. The work of the system of mass accelerations for such a displacement is

$$\Sigma m \frac{\mathrm{d}^2\mathbf{r}}{\mathrm{d}t^2} \cdot \delta\mathbf{r} = \left(\Sigma m \frac{\mathrm{d}^2\mathbf{r}}{\mathrm{d}t^2} \cdot \frac{\partial \mathbf{r}}{\partial q_i} \right) \delta q_i, \tag{12.3.9}$$

where the summation is over all particles of the system. It follows that

$$Q_i = \Sigma m \frac{\mathrm{d}^2\mathbf{r}}{\mathrm{d}t^2} \cdot \frac{\partial \mathbf{r}}{\partial q_i}. \tag{12.3.10}$$

We transform the right-hand side of this equation by means of the following relation:

$$\Sigma m \frac{\mathrm{d}^2\mathbf{r}}{\mathrm{d}t^2} \cdot \frac{\partial \mathbf{r}}{\partial q_i} = \frac{\mathrm{d}}{\mathrm{d}t} \Sigma m \frac{\mathrm{d}\mathbf{r}}{\mathrm{d}t} \cdot \frac{\partial \mathbf{r}}{\partial q_i} - \Sigma m \frac{\mathrm{d}\mathbf{r}}{\mathrm{d}t} \cdot \frac{\mathrm{d}}{\mathrm{d}t} \frac{\partial \mathbf{r}}{\partial q_i}. \tag{12.3.11}$$

Now we have, with the help of (12.3.3),

$$\Sigma m \frac{\mathrm{d}\mathbf{r}}{\mathrm{d}t} \cdot \frac{\partial \mathbf{r}}{\partial q_i} = \Sigma m \frac{\mathrm{d}\mathbf{r}}{\mathrm{d}t} \cdot \frac{\partial}{\partial \dot{q}_i} \frac{\mathrm{d}\mathbf{r}}{\mathrm{d}t} = \frac{\partial}{\partial \dot{q}_i} \Sigma \tfrac{1}{2} m \frac{\mathrm{d}\mathbf{r}}{\mathrm{d}t} \cdot \frac{\mathrm{d}\mathbf{r}}{\mathrm{d}t} = \frac{\partial T}{\partial \dot{q}_i}, \tag{12.3.12}$$

where T is just the kinetic energy of the system. Again, with the help of (12.3.6), we have

$$\Sigma m \frac{\mathrm{d}\mathbf{r}}{\mathrm{d}t} \cdot \frac{\mathrm{d}}{\mathrm{d}t} \frac{\partial \mathbf{r}}{\partial q_i} = \Sigma m \frac{\mathrm{d}\mathbf{r}}{\mathrm{d}t} \cdot \frac{\partial}{\partial q_i} \frac{\mathrm{d}\mathbf{r}}{\mathrm{d}t} = \frac{\partial}{\partial q_i} \Sigma \tfrac{1}{2} m \frac{\mathrm{d}\mathbf{r}}{\mathrm{d}t} \cdot \frac{\mathrm{d}\mathbf{r}}{\mathrm{d}t} = \frac{\partial T}{\partial q_i}, \tag{12.3.13}$$

so that finally,

$$Q_i = \frac{\mathrm{d}}{\mathrm{d}t} \frac{\partial T}{\partial \dot{q}_i} - \frac{\partial T}{\partial q_i}. \tag{12.3.14}$$

Equation (12.3.14) is Lagrange's equation corresponding to the generalised coordinate q_i. There is one such equation corresponding to each coordinate so that the equations are equal in number to the degrees of freedom.

As anticipated earlier, Lagrange's equations in the form (12.3.14) require a knowledge of the kinetic energy as a function of q_i, $\dot{q}_i$ $(i = 1, 2, \ldots, n)$ and

* In spite of the name, the dimensions of a generalised force will depend on its derivation and need not be those of a conventional force. Similarly a generalised coordinate does not always have the dimension of length.

possibly t. In addition, the system of generalised forces Q_i $(i = 1, 2, \ldots, n)$ is required. These are best obtained in practice by calculating the work of the forces in a virtual displacement δq_i, which is then identified with $Q_i \delta q_i$. If, in particular, all the forces involved are conservative, a potential function V exists which is a function of the generalised coordinates. The work of the forces for a virtual displacement δq_i is then

$$-\delta V = -\frac{\partial V}{\partial q_i}\,\delta q_i,$$

and so we have

$$Q_i = -\partial V/\partial q_i, \tag{12.3.15}$$

hence

$$-\frac{\partial V}{\partial q_i} = \frac{\mathrm{d}}{\mathrm{d}t}\frac{\partial T}{\partial \dot{q}_i} - \frac{\partial T}{\partial q_i}. \tag{12.3.16}$$

Lagrange's equations in this form are particularly effective because all the equations are deducible from a knowledge of T and V.

The above discussion is valid only for holonomic systems where all the coordinates can be given independent displacements. This is necessary in order to justify a relation such as (12.3.7), which is obtained on the basis of a virtual displacement in which one coordinate varies independently of the rest. A non-holonomic system of the type discussed in Example 12.2.1 can be brought within the framework of the discussion by allowing all the coordinates involved to be capable of independent displacements in spite of the constraints, which will now be violated. In other words the constraints are not to be imposed during the exercise of deriving Lagrange's equations. We therefore formulate the equations of motion for a more general situation, but once they have been formulated we can reimpose the constraints and integrate the equations of motion subject to these added conditions. When this procedure is followed it should be borne in mind that the forces brought into play by a constraint may not be workless if the constraint is relaxed, even if this is so in a constrained displacement. This would be the situation in Example 12.2.1, where the system can be made holonomic by relaxing the no slip constraints. Then the contact forces between the discs and the plane will contribute to the work even though they are workless when the discs are rolling. It is further to be emphasised that the work is calculated on the basis of the actual forces involved in the physical situation when they are given a virtual displacement. If the conditions of the problem require no slipping, it is the forces appropriate to such a situation that are involved, even though the virtual displacement may violate the no slip condition.

Example 12.3.1

The position of a particle in three dimensions can be specified by the length r of the position vector, the inclination θ of the position vector to a fixed axis (the polar axis) and the angular displacement ϕ of the plane containing the position vector and the polar axis relative to a fixed plane. The quantities (r,θ,ϕ) are called spherical polar coordinates. The components of the velocity in the directions of increasing r, θ and ϕ are $\dot{r}$, $r\dot{\theta}$, $r\dot{\phi}\sin\theta$ respectively.

Let the force on the particle have components F_r, F_θ, F_ϕ in the direction of increasing r, θ, ϕ respectively. The work of the force arising from virtual displacements δr, $\delta\theta$, $\delta\phi$ in the coordinates is $F_r\delta r$, $F_\theta(r\delta\theta)$, $F_\phi(r\delta\phi\sin\theta)$. These expressions are to be identified with $Q_r\delta r$, $Q_\theta\delta\theta$, $Q_\phi\delta\phi$. Hence the generalised forces are

$$Q_r = F_r, \quad Q_\theta = rF_\theta, \quad Q_\phi = rF_\phi \sin\theta.$$

The kinetic energy is

$$T = \tfrac{1}{2}m(\dot{r}^2 + r^2\dot{\theta}^2 + r^2\dot{\phi}^2\sin^2\theta).$$

Lagrange's equations are therefore

$$F_r = \frac{\mathrm{d}}{\mathrm{d}t}\frac{\partial T}{\partial \dot{r}} - \frac{\partial T}{\partial r} = m\left(\ddot{r} - r\dot{\theta}^2 - r\dot{\phi}^2\sin^2\theta\right),$$

$$rF_\theta = \frac{\mathrm{d}}{\mathrm{d}t}\frac{\partial T}{\partial \dot{\theta}} - \frac{\partial T}{\partial \theta} = m\left(\frac{\mathrm{d}}{\mathrm{d}t}(r^2\dot{\theta}) - r^2\dot{\phi}^2\sin\theta\cos\theta\right),$$

$$rF_\phi\sin\theta = \frac{\mathrm{d}}{\mathrm{d}t}\frac{\partial T}{\partial \dot{\phi}} - \frac{\partial T}{\partial \phi} = m\frac{\mathrm{d}}{\mathrm{d}t}(r^2\dot{\phi}\sin^2\theta).$$

These are the equations of motion for a particle in spherical polar coordinates.

A spherical pendulum consists of a particle free to swing at the end of a string whose other extremity is fixed, as in Figure 12.3.1. Spherical polar coordinates are a natural choice for such a system because r is now constant and the first

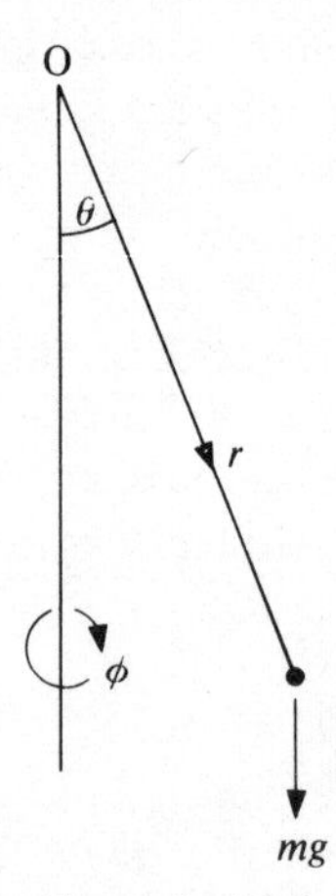

Figure 12.3.1 Illustration for Example 12.3.1: spherical polar coordinates.

equation simplifies to

$$F_r/m = -r\dot{\theta}^2 - r\dot{\phi}^2 \sin^2 \theta.$$

This equation serves to determine F_r, which is the difference between $mg \cos \theta$ and the tension in the string. The other two equations give, with $F_\theta = -mg \sin \theta$ and $F_\phi = 0$,

$$-g \sin \theta = r\ddot{\theta} - r\dot{\phi}^2 \sin \theta \cos \theta,$$

$$0 = \frac{\mathrm{d}}{\mathrm{d}t}(r^2 \dot{\phi} \sin^2 \theta).$$

The last equation verifies that the angular momentum about the vertical is constant, since there is no moment of the forces about the vertical. Between the pair of equations $\dot{\phi}$ can be eliminated to give a second order equation for θ.

Note that in this problem the particle actually has two degrees of freedom, but three equations of motion have been obtained by relaxing the constraint that r is constant until the equations of motion have been formulated. We therefore get more information at the cost of discussing a more complicated situation. It is one of the strengths of Lagrange's theory that this choice is open. Because the system with two degrees of freedom is still holonomic, we can stipulate that r is constant from the beginning, and simplify the kinetic energy to

$$T = \tfrac{1}{2} mr^2 (\dot{\theta}^2 + \dot{\phi}^2 \sin^2 \theta).$$

The effect of this simplification is to yield the equations in θ and ϕ, but not in r. But this is sufficient to determine the motion, and if the calculation of the tension in the string is not of interest, the equation in r is superfluous.

Example 12.3.2

A ring O, of mass m_1, is free to slide along a fixed horizontal smooth wire. An inelastic string, of length l, is fixed to one point A of the wire, is threaded through the ring and has a particle P, of mass m_2, fastened to its other end.

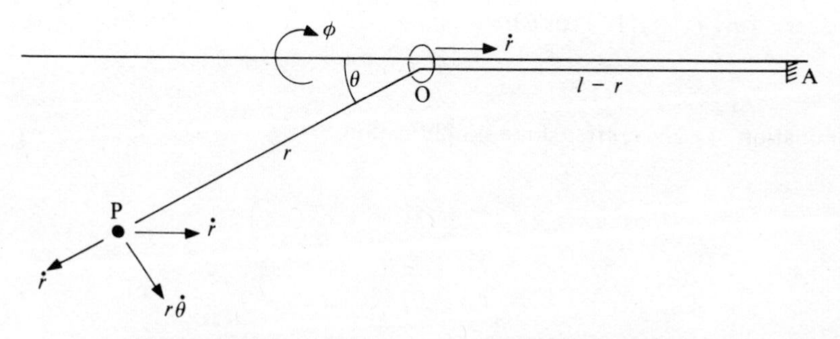

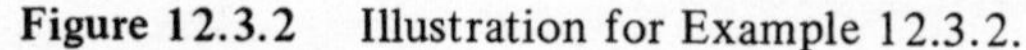

Figure 12.3.2 Illustration for Example 12.3.2.

If the string remains taut, the configuration of the above system is completely specified by the three coordinates r, θ, ϕ, where $r = \text{OP}$, θ is the inclination of OP to the wire and ϕ is the angular displacement from the vertical of the plane through OP and the wire.

The velocity of the ring is $\dot{r}$ along the wire. The velocity of P consists of a horizontal component $\dot{r}$ of the ring together with components relative to the ring of magnitude $\dot{r}$, $r\dot{\theta}$, $r\dot{\phi}\sin\theta$ in the directions of r, θ, ϕ increasing. The kinetic energy is

$$T = \tfrac{1}{2}m_1\dot{r}^2 + \tfrac{1}{2}m_2\{\dot{r}^2(1-\cos\theta)^2 + (r\dot{\theta} + \dot{r}\sin\theta)^2 + r^2\dot{\phi}^2\sin^2\theta\}.$$

If we neglect all frictional forces, only gravity will contribute to the work, and so there is a potential energy given by

$$V = -m_2 gr\sin\theta\cos\phi.$$

Therefore Lagrange's equations in r, θ, ϕ are, respectively,

$$\frac{\mathrm{d}}{\mathrm{d}t}\{m_1\dot{r} + m_2\dot{r}(1-\cos\theta)^2 + m_2\sin\theta(r\dot{\theta} + \dot{r}\sin\theta)\} - m_2\dot{\theta}(r\dot{\theta} + \dot{r}\sin\theta)$$
$$-m_2 r\dot{\phi}^2\sin^2\theta = m_2 g\sin\theta\cos\phi,$$

$$\frac{\mathrm{d}}{\mathrm{d}t}\{m_2 r(r\dot{\theta} + \dot{r}\sin\theta)\} - m_2\{\dot{r}^2\sin\theta(1-\cos\theta) + \dot{r}\cos\theta(r\dot{\theta} + \dot{r}\sin\theta)$$
$$+ r^2\dot{\phi}^2\sin\theta\cos\theta\} = m_2 gr\cos\theta\cos\phi,$$

$$m_2\frac{\mathrm{d}}{\mathrm{d}t}\{r^2\dot{\phi}\sin^2\theta\} = -m_2 gr\sin\theta\sin\phi.$$

A simple analysis of the equations is not feasible in their full generality. We consider the possibility of a solution for which θ is constant. The first two equations then simplify to

$$\{m_1 + 2m_2(1-\cos\theta)\}\ddot{r} - m_2 r\dot{\phi}^2\sin^2\theta = m_2 g\sin\theta\cos\phi,$$
$$\ddot{r} - r\dot{\phi}^2\cos\theta = g\cot\theta\cos\phi.$$

Elimination of $\ddot{r}$ gives, after some simplification,

$$(\cos^2\theta - 2k\cos\theta + 1)\left(r\dot{\phi}^2 + \frac{g\cos\phi}{\sin\theta}\right) = 0,$$

where $$k = 1 + \frac{m_1}{2m_2}.$$

The equations are therefore compatible with the assumed form of solution if

$$\cos\theta = k - (k^2 - 1)^{1/2}.$$

If in addition ϕ is small we have, correct to the first order,

$$\ddot{r} = g\cot\theta$$

so that the acceleration of the ring along the wire is approximately constant.

Example 12.3.3. The double pendulum

Figure 12.3.3 illustrates a double pendulum. Two particles m_1, m_2 are connected to different points of a light string and the system is allowed to oscillate in a vertical plane about a fixed point of the string. The system has two degrees of freedom. Its configuration can be defined by the two angles θ_1 and θ_2 which the two portions of the string make with the vertical. It is assumed that the string remains taut throughout the motion.

For this system we have

$$T = \tfrac{1}{2}m_1 l_1^2\dot{\theta}_1^2 + \tfrac{1}{2}m_2\{l_1^2\dot{\theta}_1^2 + l_2^2\dot{\theta}_2^2 + 2l_1 l_2\dot{\theta}_1\dot{\theta}_2\cos(\theta_2 - \theta_1)\},$$

$$V = -m_1 g l_1\cos\theta_1 - m_2 g(l_1\cos\theta_1 + l_2\cos\theta_2).$$

Therefore Lagrange's equations in θ_1 and θ_2 are

$$\frac{\mathrm{d}}{\mathrm{d}t}\{(m_1 + m_2)l_1^2\dot{\theta}_1 + m_2 l_1 l_2\dot{\theta}_2\cos(\theta_2 - \theta_1)\} - m_2 l_1 l_2\dot{\theta}_1\dot{\theta}_2\sin(\theta_2 - \theta_1)$$

$$= -(m_1 + m_2)g l_1\sin\theta_1,$$

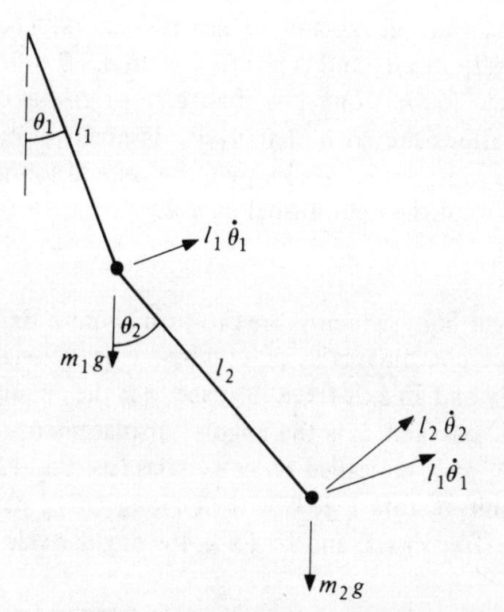

Figure 12.3.3 Illustration for Example 12.3.3: the double pendulum.

$$\frac{\mathrm{d}}{\mathrm{d}t}\{(m_2 l_2^2 \dot\theta_2 + m_2 l_1 l_2 \dot\theta_1 \cos(\theta_2 - \theta_1))\} + m_2 l_1 l_2 \dot\theta_1 \dot\theta_2 \sin(\theta_2 - \theta_1)$$

$$= -m_2 g l_2 \sin\theta_2 .$$

We shall discuss these equations when they are linearised on the basis of small oscillations, so that

$$(m_1 + m_2) l_1^2 \ddot\theta_1 + m_2 l_1 l_2 \ddot\theta_2 = -(m_1 + m_2) g l_1 \theta_1 ,$$

$$m_2 l_2^2 \ddot\theta_2 + m_2 l_1 l_2 \ddot\theta_1 = -m_2 g l_2 \theta_2 .$$

The linear equations describe coupled oscillations of the type discussed in §6.4. We look for the normal modes of oscillation on the assumption that

$$\ddot\theta_1 = -n^2\theta_1, \quad \ddot\theta_2 = -n^2\theta_2 .$$

We then have

$$(m_1 + m_2)(l_1 n^2 - g)\theta_1 + m_2 l_2 n^2 \theta_2 = 0, \quad l_1 n^2 \theta_1 + (l_2 n^2 - g)\theta_2 = 0.$$

For these equations to be consistent we must have

$$(m_1 + m_2)(l_1 n^2 - g)(l_2 n^2 - g) - m_2 l_1 l_2 n^4 = 0.$$

This equation is quadratic in n^2 and so has two roots. The left-hand side is negative when $n^2 = g/l_1$ or g/l_2 and is positive when $n = 0$ or $n \to \infty$. Hence there are two positive roots for n^2, one less than g/l_1 or g/l_2 and one greater. The corresponding oscillations are such that θ_1/θ_2 is positive and negative respectively. When the two roots n_1, n_2 are known, the general solution can be written as a linear combination of the two normal modes.

Example 12.3.4

The position of a rigid body which is free to turn about a fixed point O may be specified by the three coordinates θ, ϕ, ψ, where θ is the angle between an axis OC fixed in the body and an axis fixed in space, ϕ is the angular displacement of OC about the fixed axis and ψ is the angular displacement of the body about OC. The quantities θ, ϕ, ψ, are called Euler's angles (see Fig. 12.3.4).

If we consider unit vectors **i**, **j**, **k** with **k** along OC, **j** perpendicular to the plane of OC and the fixed axis, and $\mathbf{i} = \mathbf{j} \times \mathbf{k}$, the angular velocity of the body is

$$\dot\phi \mathbf{i} \sin\theta - \dot\theta \mathbf{j} + (\dot\psi + \dot\phi\cos\theta)\mathbf{k}.$$

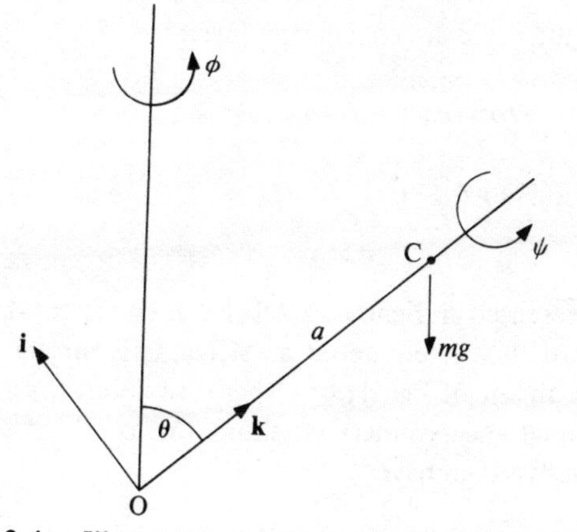

Figure 12.3.4 Illustration for Example 12.3.4: Euler's angles.

Let us consider the motion of an axially symmetric top, whose principal moments of inertia at O are I, I, I_3. The kinetic energy is

$$T = \tfrac{1}{2}I(\dot\theta^2 + \dot\phi^2 \sin^2\theta) + \tfrac{1}{2}I_3(\dot\psi + \dot\phi\cos\theta)^2$$

and the potential energy is

$$V = mga\cos\theta,$$

where a = OC. Lagrange's equations in θ, ϕ, ψ are then, respectively,

$$I\ddot\theta - I\dot\phi^2\sin\theta\cos\theta + I_3(\dot\psi + \dot\phi\cos\theta)\dot\phi\sin\theta = mga\sin\theta,$$

$$\frac{\mathrm{d}}{\mathrm{d}t}\{I\dot\phi\sin^2\theta + I_3(\dot\psi + \dot\phi\cos\theta)\cos\theta\} = 0,$$

$$I_3\frac{\mathrm{d}}{\mathrm{d}t}(\dot\psi + \dot\phi\cos\theta) = 0.$$

The third of these equations gives

$$\dot\psi + \dot\phi\cos\theta = s,$$

where s is the constant spin about the axis of the top. The second equation gives

$$I\dot\phi\sin^2\theta + I_3 s\cos\theta = \text{constant},$$

which is the constant angular momentum of the top about the vertical. The first equation can now be transformed into a second order equation for θ, by elimination of $\dot\phi$ and $\dot\psi$. We note in particular that the condition for steady precession, when θ and $\dot{\phi} = p$ are constant, is given by

$$(Ip^2 \cos\theta - I_3 sp + mga)\sin\theta = 0,$$

which verifies equation (11.6.11).

Example 12.3.5

For the system represented in Figure 12.2.1, let m be the total mass, let I be the moment of inertia of the system about a vertical axis through the mass centre and let I_{w} be the moment of inertia of either disc about the axle. It is assumed that the discs are equal and symmetrical about the axle, and that C is the mass centre of the system. We then have

$$T = \tfrac{1}{2}m(\dot{x}^2 + \dot{y}^2) + \tfrac{1}{2}I\dot{\theta}^2 + \tfrac{1}{2}I_{\mathrm{w}}(\dot{\phi}^2 + \dot{\psi}^2)$$

and Lagrange's equations give

$$m\ddot{x} = Q_x, \quad m\ddot{y} = Q_y, \quad I\ddot{\theta} = Q_\theta, \quad I_{\mathrm{w}}\ddot{\phi} = Q_\phi, \quad I_{\mathrm{w}}\ddot{\psi} = Q_\psi,$$

where the generalised forces are to be calculated from a consideration of the work associated with a small change in the variables. If we assume that the only relevant forces are the horizontal forces at the points of contact, and if these have components F_ϕ, F_ψ in the plane of the discs, and G_ϕ, G_ψ perpendicular to the plane of the discs, the equations of motion are

$$m\ddot{x} = -(F_\phi + F_\psi)\cos\theta + (G_\phi + G_\psi)\sin\theta, \quad m\ddot{y} = (F_\phi + F_\psi)\sin\theta + (G_\phi + G_\psi)\cos\theta,$$

$$I\ddot{\theta} = -l(F_\phi - F_\psi), \quad I_{\mathrm{w}}\ddot{\phi} = aF_\phi, \quad I_{\mathrm{w}}\ddot{\psi} = aF_\psi.$$

These equations could have been written down directly, and the power of Lagrange's method is not fully apparent in this example. It does illustrate, however, that a system with non-integrable constraints can be dealt with if the constraints are ignored until the equations of motion have been derived. The equations are then available for the more general situation of slipping or rolling. For rolling, the above equations are to be solved in conjunction with the constraints given by (12.2.3), (12.2.6) and (12.2.7). Since (12.2.7) is an integrable constraint it could have been used to eliminate θ, say, from the above discussion, but a little care is then required in the calculation of the generalised forces in ϕ and ψ. The neatness of the equations of motion and the ease of calculation of the generalised forces when θ is retained illustrate a situation in which the extra degree of freedom actually simplifies the discussion.

The solution of the equations of motion, subject to the rolling constraints, will not be discussed in detail. It may be verified that they are consistent with the solution $\dot{\phi}$ = constant, $\dot{\psi}$ = constant. Equations (12.2.7) and (12.2.6) then determine the constant values of $\dot{\theta}$ and the speed of the mass centre respectively.

12.4 THE LAGRANGIAN AND HAMILTONIAN

With the introduction of the Lagrangian function

$$L = T - V, \qquad (12.4.1)$$

Lagrange's equations (12.3.16) can be written in the form

$$\frac{\mathrm{d}}{\mathrm{d}t}\frac{\partial L}{\partial \dot{q}_i} - \frac{\partial L}{\partial q_i} = 0. \qquad (12.4.2)$$

This follows because V does not depend on the generalised velocities, so that

$$\partial T/\partial \dot{q}_i = \partial L/\partial \dot{q}_i. \qquad (12.4.3)$$

For the purposes of elementary mechanics, the introduction of L is not essential, but in the more general theory of analytical mechanics, and in modern theoretical physics, the Lagrangian plays a leading role. Indeed, from a more general standpoint, the dynamics of a system is defined in terms of the single Lagrangian function without the intermediate introduction of T and V.

Another function of importance in modern physics is the Hamiltonian, defined by

$$H = \sum_i \frac{\partial L}{\partial \dot{q}_i}\dot{q}_i - L. \qquad (12.4.4)$$

This equation can be written in differential form as

$$\mathrm{d}H = \sum_i \left\{ \frac{\partial L}{\partial \dot{q}_i}\,\mathrm{d}\dot{q}_i + \dot{q}_i \mathrm{d}\left(\frac{\partial L}{\partial \dot{q}_i}\right) - \frac{\partial L}{\partial q_i}\,\mathrm{d}q_i - \frac{\partial L}{\partial \dot{q}_i}\,\mathrm{d}\dot{q}_i \right\} - \frac{\partial L}{\partial t}\,\mathrm{d}t,$$

$$= \sum_i \left(\dot{q}_i \mathrm{d}p_i - \dot{p}_i \mathrm{d}q_i \right) - \frac{\partial L}{\partial t}\,\mathrm{d}t, \qquad (12.4.5)$$

where

$$p_i = \partial L/\partial \dot{q}_i \qquad (12.4.6)$$

and (12.4.2) has been used. The variables p_i are called generalised momenta. At this stage we change our point of view and choose to regard H as a function of the q_i, p_i, t, regarded as independent variables. Note that in order to evaluate H it is necessary to eliminate the $\dot{q}_i$ in (12.4.4) in favour of the p_i defined by (12.4.6). When the differential

$$\mathrm{d}H(q_i, p_i, t) = \sum_i \left(\frac{\partial H}{\partial q_i}\,\mathrm{d}q_i + \frac{\partial H}{\partial p_i}\,\mathrm{d}p_i \right) + \frac{\partial H}{\partial t}\,\mathrm{d}t \qquad (12.4.7)$$

is compared with (12.4.5), it follows that

$$\partial H/\partial t = -\partial L/\partial t, \tag{12.4.8}$$

$$\partial H/\partial p_i = \dot{q}_i, \quad \partial H/\partial q_i = -\dot{p}_i, \tag{12.4.9}$$

since each coefficient of the differentials on the right-hand sides of (12.4.5) and (12.4.7) must separately correspond. For $i = 1, 2, \ldots, n$, relations (12.4.9) form a system of $2n$ first order equations which are equivalent, and can be used as an alternative, to the n second order Lagrange's equations (12.4.2).

One of the useful properties of (12.4.9) is that if a coordinate q_i does not appear in the Hamiltonian, so that $\partial H/\partial q_i = 0$, the corresponding momentum p_i is a constant. The $2(n-1)$ Hamilton's equations for the other coordinates and momenta then form a system of $n-1$ degrees of freedom, and the coordinates not appearing in H can be ignored. For this reason they are called ignorable coordinates. Further, if H (or L) does not depend explicitly on the time, we have that

$$\frac{\mathrm{d}}{\mathrm{d}t} H(q_i, p_i) = \sum_i \frac{\partial H}{\partial q_i} \dot{q}_i + \frac{\partial H}{\partial p_i} \dot{p}_i = 0$$

by virtue of (12.4.9). Hence $H = E$ say, where E is a constant of the motion. This is a generalisation of the concept of energy conservation. Indeed if we are dealing with a system of particles, where the position and velocity of a typical particle can be described in the form

$$\mathbf{r} = \mathbf{r}(q_1, q_2, \ldots, q_n),$$

$$\frac{\mathrm{d}\mathbf{r}}{\mathrm{d}t} = \frac{\partial \mathbf{r}}{\partial q_1} \dot{q}_1 + \frac{\partial \mathbf{r}}{\partial q_2} \dot{q}_2 + \ldots + \frac{\partial \mathbf{r}}{\partial q_n} \dot{q}_n,$$

it follows that the kinetic energy is a homogeneous quadratic function of the generalised velocities and is expressible in the form

$$T = \sum_i \sum_j a_{ij} \dot{q}_i \dot{q}_j.$$

We then have, by direct verification, that

$$\sum_i \frac{\partial L}{\partial \dot{q}_i} \dot{q}_i = \sum_i \frac{\partial T}{\partial \dot{q}_i} \dot{q}_i = 2T,$$

and that

$$H = \Sigma \frac{\partial L}{\partial \dot{q}_i} \dot{q}_i - L = 2T - (T - V) = T + V.$$

Example 12.4.1

The successful exploitation of ignorable coordinates depends on the choice of an appropriate coordinate system. The motion of a single particle in two dimensions can be described, using cartesian coordinates, with the aid of the functions

$$T = \tfrac{1}{2}m(\dot{x}^2 + \dot{y}^2), \quad V = V(x, y),$$

$$L = \tfrac{1}{2}m(\dot{x}^2 + \dot{y}^2) - V(x, y),$$

$$p_x = \partial L/\partial \dot{x} = m\dot{x}, \quad p_y = \partial L/\partial \dot{y} = m\dot{y},$$

$$H = \frac{1}{2m}(p_x^2 + p_y^2) + V(x, y),$$

$$\dot{p}_x = -\frac{\partial H}{\partial x} = -\frac{\partial V}{\partial x}, \quad \dot{p}_y = -\frac{\partial H}{\partial y} = -\frac{\partial V}{\partial y}.$$

There are no ignorable coordinates unless V is independent of x or y.

If V is a function of $r = (x^2 + y^2)^{1/2}$, then it is necessary to use polar coordinates to obtain an ignorable coordinate. We write

$$L = \tfrac{1}{2}m(\dot{r}^2 + r^2\dot{\theta}^2) - V(r),$$

$$p_r = \partial L/\partial \dot{r} = m\dot{r}, \quad p_\theta = \partial L/\partial \dot{\theta} = mr^2\dot{\theta},$$

$$H = T + V = \frac{1}{2m}\left(p_r^2 + \frac{1}{r^2}p_\theta^2\right) + V(r).$$

Here θ is an ignorable coordinate and p_θ is therefore constant and equal to mh (say). Then

$$H = \frac{1}{2m}p_r^2 + \frac{mh^2}{2r^2} + V(r)$$

and the problem is now reduced to one degree of freedom. The remaining equation gives

$$\dot{p}_r = m\ddot{r} = -\frac{\partial H}{\partial r} = \frac{mh^2}{r^3} - V'(r)$$

but in this problem it may be more convenient to use the energy equation

$$H = \frac{1}{2m}p_r^2 + \frac{mh^2}{2r^2} + V(r) = E.$$

Example 12.4.2

In Example 12.3.4 the motion of a spinning top is described in terms of Euler's angles (θ, ϕ, ψ) by the Lagrangian

$$L = T - V = \tfrac{1}{2}I(\dot{\theta}^2 + \dot{\phi}^2 \sin^2 \theta) + \tfrac{1}{2}I_3(\dot{\psi} + \dot{\phi} \cos \theta)^2 - mga \cos \theta.$$

Here

$$p_\theta = \partial L/\partial \dot{\theta} = I\dot{\theta},$$
$$p_\phi = \partial L/\partial \dot{\phi} = I\dot{\phi} \sin^2 \theta + I_3 \cos \theta\, (\dot{\psi} + \dot{\phi} \cos \theta),$$
$$p_\psi = \partial L/\partial \dot{\psi} = I_3(\dot{\psi} + \dot{\phi} \cos \theta),$$

which can be solved to give

$$\dot{\theta} = p_\theta/I,$$
$$\dot{\phi} = \frac{p_\phi - p_\psi \cos \theta}{I \sin^2 \theta},$$
$$\dot{\psi} + \dot{\phi} \cos \theta = p_\psi/I_3.$$

These relations are now used to compute the Hamiltonian

$$H = T + V = \frac{p_\theta^2}{2I} + \frac{(p_\phi - p_\psi \cos \theta)^2}{2I \sin^2 \theta} + \frac{p_\psi^2}{2I_3} + mga \cos \theta.$$

Both ϕ and ψ are ignorable coordinates in this Hamiltonian, implying constant values of p_ϕ and p_ψ which represent the constant angular momenta about the vertical and about the axis of symmetry respectively. The reduced problem, having one degree of freedom, is described by the Hamiltonian

$$H = \frac{p_\theta^2}{2I} + U(\theta)$$

say. Since H is independent of the time, energy conservation gives

$$\frac{p_\theta^2}{2I} + U(\theta) = \tfrac{1}{2}I\dot{\theta}^2 + U(\theta) = E,$$

where E is constant.

EXERCISES

12.1 Two particles P_1, P_2, of masses m_1, m_2 respectively, are connected by an inextensible string which passes over a smooth peg S. Particle P_2 hangs vertically and P_1 slides along a smooth curve which lies in a vertical plane through the peg.

Let $SP_1 = r$ and let the inclination of SP_1 to the downward vertical be θ. Use the principle of virtual work to show that, if the particles are in equilibrium for all positions of P_1 on the curve, the equation of the latter is of the form

$$l/r = 1 - e \cos \theta,$$

where $e = m_1/m_2$ and l is some constant.

12.2 A pentagon ABCDE is formed of rods whose mass is ρ per unit length. The rods are freely jointed together and stand in a vertical plane with the lowest rod AB fixed horizontally. The joints C and E are connected by a light string. The rods BC and EA are each of length a, CD and DE are each of length b, the angles at A and B are each $2\pi/3$ and the angles at E and C are each $\pi/2$.

Show that the tension in the string is

$$\rho g \frac{a + 5b}{2.3^{1/2}}$$

and that the reaction at D is $3^{1/2}\rho g b/2$.

12.3 Four uniform rods of equal length are smoothly jointed at their ends to form a rhombus ABCD, which is suspended from the corner A. The rod AD is of mass $2m$, AB and BC are each of mass m and CD is of negligible mass. The mid points of AB and BC are connected by a string.

Find the tension in the string and the stress in CD.

(Ans.: $3mg$, $mg/2 \cos \frac{1}{2}B\hat{A}D$.)

12.4 A framework consists of six equal uniform rods, each of mass m, which are smoothly jointed to one another at their ends to form a hexagon ABCDEF. The frame is suspended from A and maintained in a rigid configuration by two light rods BF and CE (>BF). The rods AB, BC, CD make angles α, β and γ respectively with the vertical.

Show that $\sin \alpha + \sin \beta = \sin \gamma$.

Show that the stresses R, S in BF and CE are given by

$$R = \tfrac{1}{2}mg(5 \tan \alpha - 3 \tan \beta), \quad S = \tfrac{1}{2}mg(3 \tan \beta + \tan \gamma).$$

12.5 A tripod is composed of three equal uniform rods OA, OB, OC, smoothly hinged at O. The ends A, B, C are joined by three equal strings forming an equilateral triangle, and stand on a smooth horizontal plane. Show that the tension in each string is $(3^{1/2} mg \tan \theta)/6$, where m is the mass of each rod and θ is the inclination of the rods to the vertical.

12.6 A long, uniform, flexible chain, of mass m_c and length l, hangs over a pulley consisting of an axially symmetric disc, of mass m_d and radius a, which can turn freely about a horizontal axis through its centre. A particle of mass m_p is attached to the rim of the pulley. The particle is at the lowest point of the pulley when the lengths of the chain hanging on either side are equal. The angular displacement θ of the disc and the potential energy V of the system are both measured from this position.

(i) Show that

$$V = -\frac{m_c g a^2 \theta^2}{l} + m_p g a(1 - \cos\theta).$$

Show that $\theta = 0$ is a position of stable equilibrium if

$$2m_c a < m_p l.$$

When this inequality is satisfied show that there is a second position of equilibrium in the range $0 < \theta < \pi$ which is unstable.

(ii) Show that if the particle is removed, and the system is displaced slightly from its position of unstable equilibrium, the angular displacement after a time t is given by

$$\theta = e^{\alpha t} - 1, \quad \text{where } \alpha = \left(\frac{2g}{l(1 + m_d k^2/m_c a^2)^{1/2}}\right)^{1/2}$$

and k is the radius of gyration of the pulley about its axis. It is assumed that the chain does not slip on the pulley.

12.7 A smooth circular cylinder, of radius a, is fixed with its axis horizontal and parallel to a smooth vertical wall. The axis is at a distance d from the wall. A smooth uniform rod, of length $2l$, rests on the cylinder with one end on the wall, and in a plane perpendicular to the wall. Show that its inclination θ to the horizontal is given by

$$d + a\sin\theta - l\cos^3\theta = 0.$$

12.8 A uniform solid hemisphere is supported by a string fixed to a point on its rim and to a point on a smooth vertical wall. The curved surface of the hemisphere is in contact with the wall. Show that if θ and ϕ are the inclinations of the string and the plane base of the hemisphere to the horizontal,

$$\tan\phi = \tfrac{3}{8} + \tan\theta.$$

12.9 A uniform sphere, of mass m_s and radius a, lies at the bottom of a fixed, rough, hemispherical bowl, of radius $a(1 + \alpha)$, where $\alpha > 0$. A particle, of mass m_p, is attached to the top of the sphere.

Show that if the sphere rolls without slipping to a position in which the join of the centres of the sphere and the bowl is inclined at an angle θ to the vertical, the increase in potential energy is

$$V = m_s g a\alpha(1 - \cos\theta) + m_p g a(1 - \alpha\cos\theta + \cos\alpha\theta).$$

Hence show that the equilibrium position $\theta = 0$ is stable if $m_s + m_p(1 - \alpha) \geqslant 0$.

12.10 A uniform rod OA, of mass m and length $2l$, can turn freely about the end O, which is fixed. A light elastic string, of modulus $\lambda = \alpha mg$ and natural length $2l$, is attached to the end A and to a fixed point B vertically above O and at a distance $2l$ from A.

Show that, if θ is the angle B$\hat{\text{O}}$C, stable equilibrium is possible when

$$\theta = \pi \text{ for } \alpha \leqslant 1,$$

or
$$\sin \tfrac{1}{2}\theta = \frac{\alpha}{2\alpha - 1} \quad \text{for } \alpha > 1.$$

12.11 A particle of mass m moves in a plane. Its cartesian coordinates are (x, y) and the cartesian components of the force on the particle are (F_x, F_y). Generalised coordinates q_1, q_2 are defined so that $x = x(q_1, q_2)$ and $y = y(q_1, q_2)$.

Derive the generalised forces Q_1, Q_2 and the kinetic energy in terms of $\dot{q}_1, \dot{q}_2$ and the partial derivatives of x and y.

(i) For $x = q_1^2 - q_2^2, y = 2q_1q_2$ show that Lagrange's equations give

$$2m\{(q_1^2 + q_2^2)\ddot{q}_1 + q_1(\dot{q}_1^2 - \dot{q}_2^2) + 2q_2\dot{q}_1\dot{q}_2\} = q_1F_x + q_2F_y,$$
$$2m\{(q_1^2 + q_2^2)\ddot{q}_2 - q_2(\dot{q}_1^2 - \dot{q}_2^2) + 2q_1\dot{q}_1\dot{q}_2\} = -q_2F_x + q_1F_y.$$

(ii) For $x = c \cosh q_1 \cos q_2, y = c \sinh q_1 \sin q_2, (F_x, F_y) = (-\mu x, -\mu y)$ obtain Lagrange's equations and show that q_1 = constant, $\dot{q}_2^2 = \mu/m$ is a solution when $\mu > 0$ and that q_2 = constant, $\dot{q}_1^2 = -\mu/m$ is a solution when $\mu < 0$.

12.12 A conservative dynamical system with two degrees of freedom has kinetic and potential energies given by

$$T = \tfrac{1}{2}a\dot{\theta}^2 + F(\theta)\dot{\phi}^2, \quad V = V(\theta),$$

where a is a positive constant and F is a positive function of θ.

Show that steady motion is possible with $\theta = \alpha, \dot{\phi} = \omega$, where α and ω are constants, provided that

$$\omega^2F'(\alpha) = V'(\alpha).$$

Show that $F(\theta)\dot{\phi}$ is constant for any motion of the system. Show that if the steady motion with $\theta = \alpha, \dot{\phi} = \omega$ is slightly disturbed without altering the value of $F(\theta)\dot{\phi}$,

$$a\ddot{\beta} + \left(\frac{2F'^2}{F} - F'' + \frac{F'V''}{V'}\right)\omega^2\beta = 0$$

to the first order in $\beta = (\theta - \alpha)$.

12.13 Calculate the kinetic and potential energies of a simple pendulum using in turn: (i) the angular displacement; (ii) the horizontal displacement; (iii) the vertical displacement as the generalised coordinate.

Hence obtain the forms of the equations of motion in these coordinates.

12.14 A uniform, smooth rod AB, of mass m_1, hangs from two fixed supports C, D by light inextensible strings AC, BD each of length l. The rod is horizontal and AB = CD. A bead, of mass m_2, can slide freely on the rod and the system moves in a vertical plane.

Let θ be the inclination of the strings to the vertical, and let x be the distance of the bead from a fixed point of the rod. Write down Lagrange's equations and show that

$$l(m_1 + m_2 \sin^2 \theta)\dot{\theta}^2 = 2g(m_1 + m_2)(\cos \theta - \cos \alpha),$$

where $\theta = \alpha$ when $\dot{\theta} = 0$.

12.15 A ring, of mass m_1, slides on a smooth straight wire inclined at an angle α to the horizontal. A particle, of mass m_2, is attached to the ring by a string of length l.

Show that a motion is possible in which the string is inclined at a constant angle α to the vertical and the ring slides down the wire with constant acceleration $g \sin \alpha$.

Show that a small perturbation of this motion is simple harmonic with frequency

$$\left(\frac{(m_1 + m_2)g \cos \alpha}{m_1 l}\right)^{1/2}.$$

12.16 A uniform rod AB, of mass m and length $2a$, is attached at A to a vertical wire. The rod is free to rotate about the end A, and A itself is constrained to move down the wire and is subject to a vertical force mf. Use as generalised coordinates the vertical displacement x of A, the angle θ between the rod and the upward vertical and the angular displacement ϕ of the vertical plane through the rod and the wire. Derive the Lagrangian and the generalised forces and hence obtain Lagrange's equations. Show that $\dot{\phi} \sin^2 \theta = p$ is constant and that

$$\left(1 - \frac{3}{4} \sin^2 \theta\right) \ddot{\theta} - \frac{3}{4} \dot{\theta}^2 \sin \theta \cos \theta - \frac{p^2 \cos \theta}{\sin^3 \theta} + \frac{3f}{4a} \sin \theta = 0.$$

Integrate this relation to get

$$\left(1 - \frac{3}{4} \sin^2 \theta\right) \dot{\theta}^2 + p \operatorname{cosec}^2 \theta - \frac{3f}{4a} \cos \theta = \text{constant}.$$

12.17 Three particles A, B, C, each of mass m, are connected by two inelastic strings AB and BC, each of length l. The particles lie on a smooth horizontal plane initially at the points $(0, l)$, $(0, 0)$, $(0, -l)$ respectively relative to a set of axes Oxy in the plane.

At time $t = 0$ forces $Y\mathbf{j}$, $X\mathbf{i}$, $-Y\mathbf{j}$ begin to act on A, B, C respectively. Use the displacement x of B and the angular displacement θ of either string (assumed to remain taut) as generalised coordinates. Hence show that the generalised forces are

$$Q_x = X, \quad Q_\theta = -2Yl \sin\theta.$$

Calculate the kinetic energy and obtain the equations of motion. Hence show that

$$m(\tfrac{3}{2}\dot{x}^2 + l^2\dot{\theta}^2 - 2l\dot{x}\dot{\theta}\cos\theta) = \int_0^t (X\dot{x} - 2Yl\dot{\theta}\sin\theta)\,dt.$$

Show that, if X and Y are constant,

$$ml\dot{\theta}^2(1 + \sin^2\theta) = 2X\sin\theta + 6Y\cos\theta.$$

12.18 If F is any function of the generalised coordinates and the time, show that the Lagrangians L and $L + \mathrm{d}F/\mathrm{d}t$ yield the same equations of motion.

The Lagrangian for the motion of a system of particles is of the form

$$L = \sum_i \tfrac{1}{2}m_i(\dot{x}_i^2 + \dot{y}_i^2 + \dot{z}_i^2) - V,$$

where V varies only with the relative distances between particles, that is with

$$(x_i - x_j)^2 + (y_i - y_j)^2 + (z_i - z_j)^2$$

for all i, j.

Show that the transformation

$$x_i = x_i' + ut, \quad y_i = y_i' + vt, \quad z_i = z_i' + wt, \qquad i = 1, 2, \ldots, n,$$

where u, v, w are constants, leaves the equations of motion invariant.

12.19 A bead, of mass m, slides round a smooth circular wire, of radius a and moment of inertia I about a diameter. The wire is constrained by a couple M to turn about a vertical diameter.

Let the angular displacement of the wire be ϕ. Let the inclination to the downward vertical of the radius through the particle be θ. Obtain the equations of motion in θ and ϕ.

(i) Show that, if $M = 0$,

$$\dot{\theta}^2 = \frac{2g}{a}\cos\theta - \frac{h^2}{\sin^2\theta + I/ma^2} + c_1,$$

where h and c_1 are constants.

(ii) Show that, if $\dot{\phi}$ = constant = ω,

$$M = ma^2 \omega^2 \sin 2\theta \left(\sin^2 \theta + \frac{2g}{a\omega^2} \cos \theta + c_2 \right)^{1/2},$$

where c_2 is a constant.

12.20 Two discs, each of radius a, are joined by an axle to form an axially symmetric system which rolls without slipping on a plane inclined at an angle α to the horizontal. The mass of the system is m and its radius of gyration about the axle is k. A pendulum is suspended from the mid point of the axle and can swing freely in a plane perpendicular to the axle. The mass of the pendulum is m_p, the distance of its mass centre from the point of support is l and its radius of gyration about an axis through its mass centre parallel to the axle is k_p. The motion is two dimensional with each disc rolling down a line of greatest slope.

Use the displacement x of the axle and the angular displacement θ of the pendulum from the vertical as generalised coordinates. Write down the kinetic energy and potential energy and hence obtain Lagrange's equations.

(i) Show that a motion is possible in which the system rolls down the plane with constant acceleration provided θ is constant and given by

$$\tan \theta = \frac{\tan \alpha}{1 + \kappa \sec^2 \alpha}, \quad \text{where } \kappa = \frac{mk^2}{(m + m_p)a^2}.$$

(ii) If $\alpha = 0$ and θ is small, show that the system oscillates with frequency $2\pi/n$, where

$$n^2 = gl \left(k_p^2 + \frac{ml^2(a^2 + k^2)}{m(a^2 + k^2) + m_p a^2} \right)^{-1}.$$

12.21 A uniform circular drum, of radius a and moment of inertia I about its axis, is free to rotate about its axis, which is horizontal. A light inelastic string is wound round the drum and has one end fixed to the drum. A light spring, of modulus λ and natural length l, is fixed to the free end of the string. A particle, of mass m, is attached to the spring. The system is released from rest in which the string and spring are held in a vertical position with the length of the spring equal to l.

Write down Lagrange's equations in terms of the angular displacement of the drum and the extension of the spring. Hence show that the acceleration of the particle is

$$\frac{g(ma^2 + I \cos nt)}{ma^2 + I},$$

where $$n^2 = \frac{\lambda}{ml}\left(1 + \frac{ma^2}{I}\right)$$

12.22 A sphere, of mass m_1, radius a_1 and moment of inertia I_1 about a diameter, rests on a horizontal plane. A second sphere, of mass m_2, radius a_2 and moment of inertia I_2 about a diameter, rests on the highest point of the first sphere. Both spheres are spherically symmetric.

Suppose the system is slightly displaced from this position of equilibrium. Let ϕ be the angular displacement of the first sphere, and let θ be the inclination of the join of centres to the vertical in the ensuing motion. Show that, provided no slipping takes place at either point of contact,

$$a_1\phi = c \sin\theta \quad \text{and} \quad (d - c\cos^2\theta)\dot{\theta}^2 = 2g(1 - \cos\theta),$$

where

$$c = \frac{m_2 a_1^2(a_1 + a_2)}{I_1 + (m_1 + m_2)a_1^2}, \quad d = (a_1 + a_2)\left(1 + \frac{I_2}{m_2 a_2^2}\right).$$

12.23 A double pendulum consists of two uniform rods OA, AB, each of mass m and length l, smoothly hinged at A. The system is suspended from O and performs small oscillations in a vertical plane. Show that the frequencies of the two normal modes are

$$\left(\frac{3 \pm 6^{1/2}}{2}\frac{g}{l}\right)^{1/2}.$$

12.24 Three particles P_1, P_2, P_3, have masses $2m$, m, m respectively. They are connected by light springs P_2P_3, P_3P_1, P_1P_2. The tensions required to produce unit extensions in the springs are σ, 2σ, 2σ respectively. The system rests on a smooth horizontal plane and when in equilibrium forms an isosceles triangle with P_1, P_2, P_3 at the points $(0, 0)$, (a, a), $(a, -a)$ respectively. Use momentum considerations to show that if the system performs small oscillations in the plane, the displacements of the particles are of the form

$$(q_1, q_2), \quad (-q_1 - q_2, -q_2 - q_3), \quad (q_2 - q_1, q_3 - q_2).$$

Write down the kinetic energy and the potential energy of the system in terms of q_1, q_2, q_3 and their derivatives. Hence obtain Lagrange's equations, and deduce that the frequencies of the three normal modes of oscillation are equal.

12.25 A uniform rectangular plate rests on four equal springs at the corners.

Show that for small vibrations, in which each corner is displaced vertically to first order, the three normal modes have frequencies in the ratios $1 : 3^{1/2} : 3^{1/2}$.

12.26 A uniform rectangular plate has sides of length $2a$, $2b$. It is suspended by two light vertical strings attached at adjacent corners of the plate. The strings are of length l and their distance apart is $2a$.

Find the characters of the four normal modes. For small oscillations about the equilibrium position show that the frequencies are

$$\left(\frac{g}{l}\right)^{1/2}, \left(\frac{g}{3l}\right)^{1/2}, \left(\frac{2g}{l}\right)^{1/2}\left\{\left(1+\frac{3l}{4b}\right) \pm \left(1+\frac{3l}{4b}+\frac{9l^2}{16b^2}\right)^{1/2}\right\}^{1/2}.$$

12.27 A mechanical system is acted upon by impulses. A typical impulse is denoted by $\bar{\mathbf{F}}$. The velocities of a typical particle m immediately before and after the impulse are denoted by $\mathbf{v}_1$ and $\mathbf{v}_2$ respectively.

Deduce the equivalence of the system $\bar{\mathbf{F}}$ and the system $\mathbf{m}(\mathbf{v}_2 - \mathbf{v}_1)$.

From a consideration of the virtual displacement in which the displacement of m is proportional to $(\mathbf{v}_1 + \mathbf{v}_2)$, show that

$$\Sigma\tfrac{1}{2}\bar{\mathbf{F}} \,.\, (\mathbf{v}_1 + \mathbf{v}_2) = \Sigma\tfrac{1}{2}m(v_2^2 - v_1^2),$$

where the summation on the left-hand side is over all the impulses which do work in the virtual displacement and the right-hand side represents the change in kinetic energy of the system.

12.28 The motion of a mechanical system is defined by generalised coordinates q_i. If the motion undergoes an impulsive change, show that

$$[\partial T/\partial q_i] = \bar{Q}_i.$$

The left-hand side represents the change in $\partial T/\partial q_i$ arising from the impulses. The generalised impulse $\bar{Q}_i$ is such that $\bar{Q}_i\,\delta q_i$ is the work of the impulses in a small virtual displacement δq_i.

12.29 Two uniform rods AB, BC, of masses m_1, m_2 and lengths l_1, l_2 respectively, are jointed at B and lie in one straight line on a horizontal table. The system is given a horizontal impulsive blow $\bar{F}$ at A at right angles to ABC.

Formulate the kinetic energy in terms of the initial velocities $\dot{y}_1, \dot{y}_2, \dot{y}_3$ of A, B, C respectively. Deduce the generalised impulses and hence the velocities.

Show that the kinetic energy acquired by the system is

$$\frac{(4m_1 + 3m_2)\bar{F}^2}{2m_1(m_1 + m_2)}.$$

12.30 Two uniform rods AB, BC, each of mass m and length l, are smoothly jointed at B and lie on a horizontal table with AB perpendicular to BC. A horizontal impulse $\bar{F}$ is applied at C in the direction AC.

Show that the kinetic energy produced when A is fixed is 15/16 times the kinetic energy produced when A is free.

12.31 A dynamical system with one degree of freedom has the Hamiltonian $H = q^2(\frac{1}{2}p^2 + 1)$. Write down Hamilton's equations and show that $q = \operatorname{sech} 2^{1/2}\, t$ given that $q = 1, \dot{q} = 0$ at $t = 0$.

12.32 A charged particle of unit mass is in two-dimensional motion under the influence of a dipole. You are given that the Lagrangian of the motion is

$$L = \tfrac{1}{2}(\dot{r}^2 + r^2\dot{\theta}^2) - \mu \cos\theta / r^2,$$

where (r, θ) are the polar coordinates of the particle and μ is the constant strength of the dipole. Derive the generalised momenta p_r, p_θ, write down the Hamiltonian in terms of r, θ, p_r, p_θ and hence obtain Hamilton's equations. Show that $p_\theta^2 = \alpha - 2\mu\cos\theta$, where α is a constant, and hence that $\dot{r}^2 + \alpha/r^2 = 2E$, where E is the constant total energy.

The particle is projected from the point $r = a, \theta = 0$ at time $t = 0$ with a speed v perpendicular to the radius vector. Show that $r^2 = a^2 + (v^2 + 2\mu/a^2)t^2$.

12.33 Find the Hamiltonian for the system whose Lagrangian is given by

$$L = \tfrac{1}{2}(\dot{r}^2 + r^2\dot{\theta}^2 + \dot{z}^2) + \mu r^2\dot{\theta},$$

where (r, θ, z) are cylindrical polar coordinates and μ is a constant. Deduce from Hamilton's equations that p_θ, p_z and $p_r^2 + r^{-2}(p_\theta - \mu r^2)^2$ are constants of the motion.

12.34 A light inextensible string passes over a small light smooth pulley. On one side a mass m_1 is tied to the string. On the other side a mass m_2 slides down the string, its velocity relative to the string being controlled by an inner mechanism to be $u(t)$. Initially m_1 is at rest and each mass is at a depth l below the pulley. Calculate the Lagrangian of the system in terms of the depth x of m_1 below the pulley. Find the Hamiltonian and show that it is constant if $\{(m_1 - m_2)gt - m_1 u - m_2 u_0\}\dot{u} = (m_1 + m_2)gu$, where $u_0 = u(0)$. For the case $m_1 = m_2$, $u_0 = 0$, show that this condition implies that both masses move upwards with acceleration g.

12.35 For a body with kinetic symmetry rotating about a fixed point the Lagrangian is, in the usual notation,

$$L = \tfrac{1}{2}I(\dot{\theta}^2 + \dot{\phi}^2 \sin^2\theta) + \tfrac{1}{2}I_3(\dot{\phi}\cos\theta + \dot{\psi})^2,$$

where the Euler angles θ, ϕ, ψ are the generalised coordinates. Determine the generalised momenta p_θ, p_ϕ, p_ψ. Show that if L_1, L_2, L_3 are defined by

$$L_1 \pm \mathrm{i}L_2 = \mathrm{e}^{\pm \mathrm{i}\psi}(\pm \mathrm{i}p_\theta + p_\phi \operatorname{cosec}\theta - p_\psi \cot\theta), \quad L_3 = p_\psi,$$

then the Hamiltonian H may be expressed in the form

$$H = \frac{1}{2I}(L_1^2 + L_2^2) + \frac{1}{2I_3}L_3^2.$$

We define the Poisson bracket $[F, G]$ of F and G by the relation

$$[F, G] = \sum_{i=1}^{n}\left(\frac{\partial F}{\partial q_i}\frac{\partial G}{\partial p_i} - \frac{\partial F}{\partial p_i}\frac{\partial G}{\partial q_i}\right),$$

where the summation is over all the degrees of freedom. Note that $[F, F] = 0$ and that $[G, F] = -[F, G]$.

Evaluate $[L_1 + iL_2, L_1 - iL_2]$ and $[L_1 \pm iL_2, L_3]$. Hence show that $[L_2, L_3] = L_1$, $[L_3, L_1] = L_2$, $[L_1, L_2] = L_3$.

13 Non-Linear Problems

13.1 CONSERVATIVE FORCES

Many of the problems which occur naturally in mechanics are governed by non-linear differential equations, and yet there is a considerable body of established theory which is developed on the basis of linear differential equations. One reason for this is that most of these problems are too difficult to solve in their full generality, and it is necessary to look for a simplifying procedure. Since linear differential equations are more easily solved than non-linear equations, such a simplification often amounts to a linearisation in some fashion.

A typical procedure is to assume that there is a small perturbation of a known state of a mechanical system, and then to neglect all but the small linear terms in the perturbation because they dominate the non-linear terms. Such a procedure has formed the basis of many successful and established theories. However, apart from the loss of accuracy involved in any approximation, some problems possess qualitative features which arise essentially from the non-linearity and cannot be otherwise explained. Some of these features are described in this chapter.

When the forces involved are conservative, there exists an energy equation. This is, in effect, an integral of the equation of motion and involves only first derivatives (velocities) whereas the equations of motion involve second derivatives (accelerations). If the problem depends on only one space coordinate, the energy equation is sufficient to determine the motion, at least in principle, when combined with appropriate initial conditions. The most familiar situation of this kind is that of one-dimensional motion, where the energy equation is of the form

$$\tfrac{1}{2}\dot{x}^2 + V(x) = E, \tag{13.1.1}$$

where $\frac{1}{2}\dot{x}^2$ is the kinetic energy, $V(x)$ is the potential energy, and E is the total energy, per unit mass. This equation has already been discussed within the context of non-linear oscillations in §6.5. We have also seen in Chapter 7 that even a two-dimensional problem can sometimes be reduced to the form of (13.1.1), as in (7.2.3). This possibility arises when there are two conservation equations, thereby allowing one of the two dependent variables to be eliminated in favour of the other. Equations of the form (13.1.1) are not difficult to integrate although in complicated problems this may have to be done numerically. We summarise here some of the main qualitative features of problems governed by (13.1.1).

Motion is physically possible in any interval for which $E - V(x)$ is positive, since (13.1.1) gives a real value for $\dot{x}$ in such intervals.

Equilibrium is possible for $x = a$ provided $E = V(a)$ and $V'(a) = 0$, since these conditions imply zero velocity and acceleration.

A small perturbation about a position of equilibrium remains small and oscillatory if $V(x)$ has a minimum there. It is approximately a simple harmonic oscillation, with period $2\pi(V''(a))^{-1/2}$, if $V''(a)$ is positive (see §6.5). If $V(x)$ does not have a minimum then the position of equilibrium is unstable.

Example 13.1.1

With a little experience, qualitative information about the motion of a particle subject to a conservative force is quickly deduced from a graph of the potential energy. Figure 13.1.1 shows the variation of

$$V(x) = -n^2x(a^2 - \tfrac{1}{3}x^2)/a, \quad a > 0$$

as a function of x. Bearing in mind the energy equation

$$\tfrac{1}{2}\dot{x}^2 + V(x) = E,$$

we classify the various possibilities in the following way.

(i) $E < V(a) = -2n^2a^2/3$.

Here the motion is restricted to the regime $V(x) < V(a)$, or $x < -2a$. The maximum value of x possible is that for which $E = V(x)$. There the velocity is zero and the acceleration, which is given by

$$-V'(x) = n^2(a^2 - x^2)/a,$$

is negative. Hence, although the particle may initially move in the positive x direction, it will ultimately move in the negative x direction with ever increasing speed.

(ii) $V(a) \leqslant E \leqslant V(-a)$.

Case (i) is still possible, but now there is a second possibility in which the particle oscillates between the two largest roots of the relation $V(x) = E$.

If $E = V(a)$ there is equilibrium since $V'(a) = 0$. The equilibrium is stable since $V''(a) = 2n^2 > 0$. For a value of E slightly greater than $V(a)$, the particle will perform small oscillations with the approximate periodic time $2\pi(V''(a))^{-1/2} = 2^{1/2}\pi/n$.

The point $x = -a$ is also a point of equilibrium but it is unstable, and if the particle starts from rest at a small distance from $x = -a$, it will move further away, at least initially.

(iii) $E > V(-a)$.

Here the whole of the energy curve, as far as the value of x for which $V(x) = E$, is available in the physical problem. This is the only value of x for which the velocity is zero and, since the acceleration there is negative, the

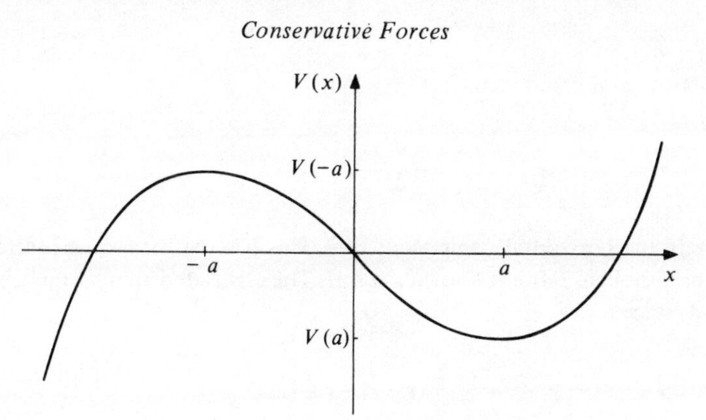

Figure 13.1.1 Illustration for Example 13.1.1: variation of the function $V(x) = -n^2x(a^2 - \frac{1}{3}x^2)/a$.

particle will ultimately be moving in the negative x direction whatever the initial motion might be.

Example 13.1.2. The Spherical Pendulum
In Example 7.4.1 we considered the equation of motion

$$m\frac{\mathrm{d}^2\mathbf{l}}{\mathrm{d}t^2} = m\mathbf{g} - \frac{T\mathbf{l}}{l} \tag{13.1.2}$$

for the spherical pendulum and showed that the pendulum executes elliptic harmonic motion when the amplitude is small. For arbitrary amplitude the problem is non-linear and has certain features in common with the central force problems of Chapter 7.

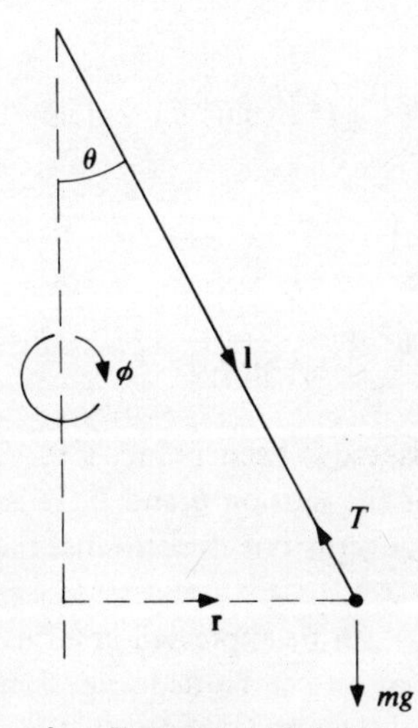

Figure 13.1.2 Illustration for Example 13.1.2: notation for the spherical pendulum.

The horizontal component of (13.1.2) is

$$m\frac{\mathrm{d}^2\mathbf{r}}{\mathrm{d}t^2} = -\frac{T\mathbf{r}}{l}, \tag{13.1.3}$$

where $\mathbf{r}$ is the horizontal component of $\mathbf{l}$. This has the form of a central force problem, although T is not known *a priori*. The associated angular momentum is constant, since

$$\frac{\mathrm{d}}{\mathrm{d}t}(\mathbf{r} \times \dot{\mathbf{r}}) = \mathbf{r} \times \ddot{\mathbf{r}} = \mathbf{0} \tag{13.1.4}$$

by (13.1.3). Hence

$$|\mathbf{r} \times \dot{\mathbf{r}}| = l \sin\theta(l\dot{\phi}\sin\theta) = l^2\dot{\phi}\sin^2\theta = h,$$

where ϕ is the angular displacement about the vertical and h is the constant component of angular momentum per unit mass about the vertical. If the horizontal projection of l sweeps out an area A in time t, it follows that

$$\mathrm{d}A/\mathrm{d}t = \tfrac{1}{2}r^2\dot{\phi} = \tfrac{1}{2}l^2\dot{\phi}\sin^2\theta = \tfrac{1}{2}h, \tag{13.1.5}$$

and the rate of sweep of projected area is constant, as in Example 7.1.1.

There is also an energy equation, since the tension does no work and the gravitational force is conservative. The components of velocity are $l\dot{\theta}$ in the direction of increasing θ and $l\dot{\phi}\sin\theta$ in the direction of increasing ϕ (see Fig. 13.1.2). The potential energy per unit mass is $-gl\cos\theta$ and so the energy equation is

$$\tfrac{1}{2}(l^2\dot{\theta}^2 + l^2\dot{\phi}^2\sin^2\theta) - gl\cos\theta = E$$

or, with the help of (13.1.5),

$$\tfrac{1}{2}l^2\dot{\theta}^2 + \frac{h^2}{2l^2\sin^2\theta} - gl\cos\theta = E. \tag{13.1.6}$$

The governing equation has now been reduced to a first order equation for θ involving two constants of the motion h and E. It is mathematically similar to equation (13.1.1), which governs one-dimensional motion under a conservative force.

The solution of (13.1.6) can be expressed in terms of elliptic functions, from which the details of the motion can be deduced. Some insight can be obtained simply from a calculation of the apsidal angle, that is the angular displacement in ϕ between the minimum and maximum values of θ. This is more readily calculated from the equation

$$\frac{h^2}{2gl^3 \sin^2 \theta}\left(\frac{d\theta}{d\phi}\right)^2 = \left(\frac{E}{gl} + \cos\theta\right)(1 - \cos^2\theta) - \frac{h^2}{2gl^3}$$
$$= (c_1 - \cos\theta)(\cos\theta - c_2)(c_3 + \cos\theta)$$
$$= (c_1 - c)(c - c_2)(c_3 + c), \tag{13.1.7}$$

say, which is obtained from a combination of (13.1.5) and (13.1.6). The right-hand side of (13.1.7) is negative for $\cos\theta = \pm 1$ and must be positive for some real values of θ if a solution exists with a physical interpretation. Hence two of the roots of (13.1.7) must lie in the interval $-1 < \cos\theta < 1$. Let these be $c_1 = \cos\theta_1$, $c_2 = \cos\theta_2$, with $c_1 \geqslant c_2$. Because the expression is negative for $\cos\theta = -1$, it follows that $c_3 > 1$ (see Fig. 13.1.3). The roots also satisfy the relations

$$(1 - c_1)(1 - c_2)(c_3 + 1) = (1 + c_1)(1 + c_2)(c_3 - 1) = h^2/2gl^3, \tag{13.1.8}$$

$$c_2 c_3 + c_3 c_1 - c_1 c_2 = 1. \tag{13.1.9}$$

Relation (13.1.8) follows from (13.1.7) with $\cos\theta = \pm 1$. Relation (13.1.9) follows from a comparison of the coefficients of $\cos\theta$, and implies that

$$1 < c_3 = \frac{1 + c_1 c_2}{c_1 + c_2}, \tag{13.1.10}$$

from which it follows that c_1 and c_2 cannot both be negative. Hence c_1 cannot be negative, since $c_1 \geqslant c_2$. Furthermore c_2 cannot be less than $-c_1$ and we have the following inequalities (see also Fig. 13.1.3):

$$0 \leqslant c_1 < 1, \quad -c_1 \leqslant c_2 \leqslant c_1, \quad 1 < c_3. \tag{13.1.11}$$

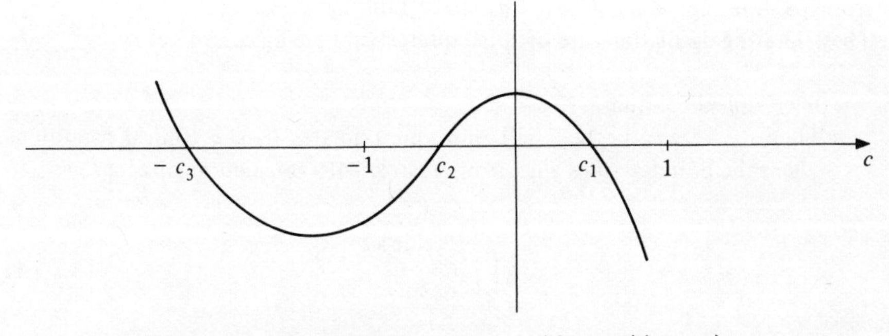

Figure 13.1.3 Variation of $(c_1 - c)(c - c_2)(c_3 + c)$.

For physically possible motion, θ is confined to the interval $\theta_1 \leqslant \theta \leqslant \theta_2$, and the motion is oscillatory between these limits. The apsidal angle ϕ_a is then, from (13.1.7),

$$\phi_a = \left(\frac{h^2}{2gl^3}\right)^{1/2} \int_{\theta_1}^{\theta_2} \frac{d\theta}{\sin\theta\{(c_1 - \cos\theta)(\cos\theta - c_2)(c_3 + \cos\theta)\}^{1/2}}. \quad (13.1.12)$$

This integral can be transformed by the substitution

$$\cos\theta = c_1 \cos^2 \psi + c_2 \sin^2 \psi$$

to give

$$\phi_a = \int_0^{\pi/2} \frac{2\{(1-c_1)(1-c_2)(c_3+1)\}^{1/2} d\psi}{\{1-(c_1\cos^2\psi + c_2\sin^2\psi)^2\}(c_3 + c_1\cos^2\psi + c_2\sin^2\psi)^{1/2}}, \quad (13.1.13)$$

where use has been made of (13.1.8).

Although the motion of the pendulum has been defined in terms of two parameters which arise naturally in the analysis as h and E, more readily observed parameters of the motion are θ_1 and θ_2. It is more useful to regard these, or equivalently c_1 and c_2, as determining the motion. The value of c_3 then follows immediately from (13.1.10). Since the motion of the bob will be a repetition of its motion between adjacent apses, a knowledge of θ_1, θ_2 and the angular separation ϕ_a of adjacent apses is useful information in forming a qualitative impression of the motion.

Figure 13.1.4 shows the result of integrating (13.1.13) numerically. The apsidal angle is shown plotted against the angular amplitude $(\theta_2 - \theta_1)$. Since two parameters are required to define the motion, the different graphs are exhibited for varying values of the minimum angle of inclination θ_1. All these quantities have been normalised so that the range of interest is (0, 1).

The apsidal angle always lies between $\pi/2$ and π. It increases monotonically from its limiting value as $\theta_2 \to \theta_1$ to a limiting value of π as $\theta_2 \to \pi - \theta_1$. These limiting oscillations are of some interest and are discussed below.

(i) The Conical Pendulum $(\theta_2 \approx \theta_1)$

When $\theta_2 - \theta_1 \ll 1$, the oscillation approximates to the conical pendulum where the bob describes a horizontal circle with constant angular speed

$$\omega = \left(\frac{g}{l\cos\theta_1}\right)^{1/2}, \quad (13.1.14)$$

see Example 3.3.3. Note, however, that as $c_2 \to c_1$ the integrand of

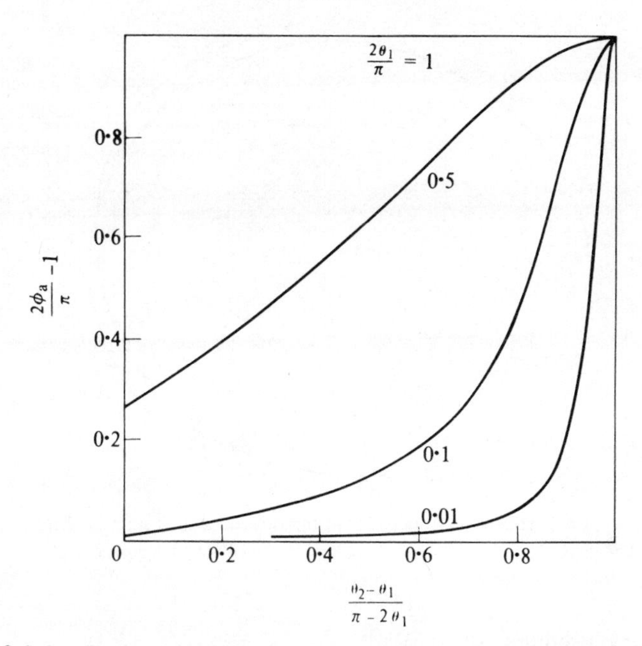

Figure 13.1.4 Variation of the apsidal angle ϕ_a with the minimum and maximum values θ_1, θ_2 of θ for the spherical pendulum.

(13.1.13) tends to become independent of ψ and gives a definite value for ϕ_a, namely

$$\phi_a = \frac{\pi(c_3+1)^{1/2}}{(1+c_1)(c_3+c_1)^{1/2}} = \frac{\pi}{(4-3\sin^2\theta_1)^{1/2}}\,. \tag{13.1.15}$$

This is the apsidal angle in the near circular orbit for which the departure from $\theta = \theta_1$ is always small (see Fig. 13.1.5). Since the angular speed in this orbit is approximately ω, as given in (13.1.14), the period of the small perturbation superimposed on the circular orbit is the time to traverse an angle $2\phi_a$ with angular speed ω, that is

$$\frac{\pi}{\omega(1-\frac{3}{4}\sin^2\theta_1)^{1/2}}\,. \tag{13.1.16}$$

(ii) The Circular Pendulum ($\theta_2 \approx \pi - \theta_1$)

As $\theta_2 \to \pi - \theta_1$, the pendulum tends to rotate rapidly in a circle of radius l, whose centre is the point of suspension and whose plane is inclined at an angle θ_1 to the downward vertical. The apsidal angle in this case is π. Reference to Figure 13.1.4 shows that for a small deviation from the circular motion, the value of π is an overestimate. This means that the positions of the apses are not fixed, but rotate slowly in the sense opposite to that of the motion of the pendulum. The visual impression is that the plane of the pendulum's motion rotates slowly about a vertical axis through the point of support.

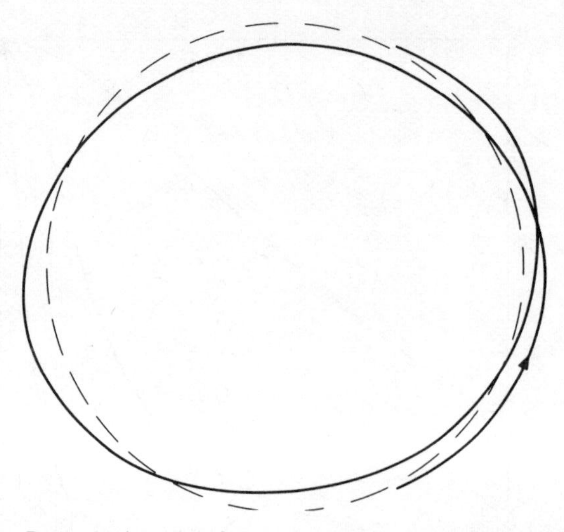

Figure 13.1.5 Perturbed path of a conical pendulum. The full line represents the orbit of the bob as a small departure from the dotted circular orbit.

(iii) The Simple Pendulum ($\theta_1 \approx 0$)

It can be seen from Figure 13.1.4 that there are two principal regimes for small values of θ_1. The first is when $\pi - \theta_2$ is not small, so that the pendulum does not approach the upward vertical position. Then ϕ_a is approximately $\pi/2$ with the positions of maximum amplitude lying in a vertical plane at right angles to the vertical plane through the positions of minimum amplitude. The visual impression is that of a simple pendulum swinging in the vertical plane of maximum amplitude. To the extent that this is in error when θ_1 is small but not zero, Figure 13.1.4 shows that a value of $\pi/2$ for ϕ_a is an underestimate. The effect of the correction is to cause the apparent vertical plane of oscillation to rotate slowly in the direction in which the orbit is described. The phenomenon should be compared with Foucault's pendulum (Example 9.5.2) in which a similar effect arises from the rotation of the earth. One qualitative difference between the two cases is that the effect of the earth's rotation does not disappear when $\theta_1 = 0$, whereas the plane of oscillation ceases to rotate when $\theta_1 = 0$ in the present case. Another is that the sense of rotation of the apse lines for Foucault's pendulum is clockwise in the northern hemisphere and anti-clockwise in the southern hemisphere.

The situation is quite different for the limit $\theta_2 \to \pi - \theta_1$. Then, as we have seen, the pendulum rotates rapidly in a nearly vertical plane with an apsidal angle approximately equal to π and with the plane of rotation itself rotating slowly in a retrograde direction.

Between these two regimes is a transition region, with θ_2 close to $\pi - \theta_1$, where there are substantial changes in ϕ_a for small changes in θ_2. The reason is that, when θ_1 is small, substantial changes in ϕ are confined to those parts of the oscillation for which the pendulum is in the nearly vertical position. There is a rapid change in ϕ of almost π in the vicinity of the downward vertical, and little further change unless the swing of the pendulum carries it near to the upward vertical. When this happens a small increase in θ_2 can produce a substantial increase in ϕ_a until the stage is reached when the pendulum begins to rotate in a nearly vertical circle and the limiting value π of ϕ_a is approached.

13.2 AUTONOMOUS FORCES

Although many of the forces which occur naturally depend on position only, such forces do not exhaust all the possibilities. A more general type of force is one which depends also on the velocity, and perhaps on the time explicitly, so that it is of the form $\mathbf{F}(\mathbf{r}, \dot{\mathbf{r}}, t)$. The discussion in this section is restricted to forces of the form $\mathbf{F}(\mathbf{r}, \dot{\mathbf{r}})$, so that the time does not appear explicitly. For the one-dimensional motion of a particle, the governing equation is then

$$\ddot{x} = f(x, \dot{x}), \tag{13.2.1}$$

say, where f is the force per unit mass. Equations in which the time does not appear explicitly are called autonomous. They have the property that if $x(t)$ is a solution, so is $x(t + t_0)$ and one of the constants of integration will always occur in this fashion as a shift in the origin of time. There are special cases for which (13.2.1) is integrable by simple analysis, but this is not always so. Numerical integration of such an equation is straightforward, but gives only a particular solution to a particular problem. We shall consider here what information of a general character can be obtained about the properties of the solution of (13.2.1).

It is often useful to write the equation in the alternative form

$$v \frac{\mathrm{d}v}{\mathrm{d}x} = f(x, v), \tag{13.2.2}$$

where $v = \mathrm{d}x/\mathrm{d}t$. The plane in which (x, v) are used as cartesian coordinates is called the phase plane. Solution curves in the phase plane are called integral curves or trajectories. Note that whereas (13.2.1) is a second order equation, (13.2.2) is of first order, so that the integral curves form a singly infinite family depending on one parameter of integration. Of course when the solution of (13.2.2) is known, we can identify v with $\mathrm{d}x/\mathrm{d}t$ to get $\mathrm{d}x/\mathrm{d}t = v$ or

$$\int_0^x \frac{\mathrm{d}x}{v(x)} = t + t_0.$$

This relation expresses x implicitly as a function of $t + t_0$ and introduces the parameter t_0 as a shift in the origin of time, as described earlier.

Equation (13.2.2) does not always have a simple integral, but information of a qualitative nature is readily obtainable from it. In particular, since $\mathrm{d}x/\mathrm{d}t = v$, the integral curves are traversed in the direction of x increasing when $v > 0$ and x decreasing for $v < 0$. From equation (13.2.2) they have vertical tangents where $v = 0$ and horizontal tangents where $f(x, v) = 0$, apart from exceptional cases which are discussed below. More generally, (13.2.2) defines the slope of the integral curve at any point of the phase plane as $f(x, v)/v$. This means that if we approximate the curve by its tangent, we can proceed a small distance to a neighbouring point. By repeated application, this process can be made the basis of a scheme of numerical integration, although it may then be desirable to return to (13.2.1) rather than to use (13.2.2). One reason for this is that there is a difficulty in the numerical integration of (13.2.2) when $v = f(x, v) = 0$, for then the slope of the integral curve is not well defined. A point of the phase plane where this is so is called a singularity. In a physical problem it is a point where both the velocity and acceleration (and hence the force) vanish and accordingly it is a point of equilibrium. The neighbourhood of such a point requires special investigation. Numerical integration will then complete the details of the solution at non-singular points.

We shall consider only points of equilibrium, such as $(x_0, 0)$, in the neighbourhood of which $f(x, v)$ can be approximated by an expression of the form

$$f = p(x - x_0) + qv, \tag{13.2.3}$$

where p and q are constants. Many of the examples encountered in applications will be of this form, and they can be classified in the way to be described. Other forms should be investigated on their merits. In particular if either p or q is zero, the higher order terms may have a significant effect on the behaviour of the solution curves, and then a better approximation is required.

It is convenient to transfer the origin of coordinates to the point of equilibrium so that we can, henceforth, take $x_0 = 0$ in (13.2.3). With $f = px + qv$, equation (13.2.1) becomes,*

$$\frac{\mathrm{d}^2 x}{\mathrm{d}t^2} = px + q\frac{\mathrm{d}x}{\mathrm{d}t} \tag{13.2.4}$$

*Phase plane analysis is often discussed with reference to the pair of first order equations

$$\mathrm{d}x/\mathrm{d}t = ax + by, \quad \mathrm{d}y/\mathrm{d}t = cx + dy.$$

With the substitutions

$$v = ax + by, \quad p = bc - ad, \quad q = a + d$$

the above equations become

$$\mathrm{d}x/\mathrm{d}t = v, \quad \mathrm{d}v/\mathrm{d}t = px + qv$$

which are equivalent to (13.2.4).

or, equivalently,

$$\frac{d^2}{dt^2}(xe^{-qt/2}) = (p + q^2/4)xe^{-qt/2}. \tag{13.2.5}$$

Equation (13.2.5) is similar to equation (6.2.6) arising in the theory of damped oscillations. The type of solution will depend on the sign of $(p + q^2/4)$ and we classify solutions in the following way:

(a) $p + q^2/4 = -n^2 < 0, n > 0, q \neq 0,$

$$x = ae^{qt/2} \cos n(t + t_0), \tag{13.2.6}$$

$$v = \dot{x} = ae^{qt/2} \{\tfrac{1}{2}q \cos n(t + t_0) - n \sin n(t + t_0)\}. \tag{13.2.7}$$

To plot the integral curves in the phase plane one can take $\phi = t + t_0$ as the parametric variable and write

$$x = a'e^{q\phi/2} \cos n\phi, \tag{13.2.8}$$

$$v = a'e^{q\phi/2} (\tfrac{1}{2}q \cos n\phi - n \sin n\phi), \tag{13.2.9}$$

say, where $a' = ae^{-qt_0/2}$. Any particular integral curve is obtained by regarding x and v as functions of ϕ, and the different integral curves are obtained by varying the single parameter a'. The resulting curves are spirals as shown in Fig. 13.2.1(a). The equilibrium point in this case is called a focus.

Equations (13.2.6) and (13.2.7) show that when $q > 0$ the amplitudes of x and v increase with t, and the direction of traverse of the spiral is away from the focus. This is described as unstable equilibrium. Conversely the amplitudes of x and v decrease when $q < 0$. The direction of traverse is towards the focus and the situation is stable. This corresponds to the physical interpretation of stable equilibrium in the sense that a particle whose motion is governed by equation (13.2.4) would oscillate with decreasing amplitude if $q < 0$, and so there would be a tendency towards the equilibrium position as is the case for damped harmonic motion (see §6.2).

(b) $p = -n^2 < 0, n > 0, q = 0,$

$$x = a \cos n(t + t_0), \tag{13.2.10}$$

$$v = -an \sin n(t + t_0). \tag{13.2.11}$$

The integral curves are confocal ellipses and the equilibrium point is called a centre. The force is conservative and the integral curves are the curves of constant energy

$$\tfrac{1}{2}v^2 + \tfrac{1}{2}n^2x^2 = \text{constant}.$$

This situation is sometimes described as one of stable equilibrium on the basis that a small displacement from equilibrium will always remain small. In the theory of the phase plane, however, it is more usual to describe as stable only those cases where there is a definite tendency towards the equilibrium point as $t \to \infty$, that is cases for which $q < 0$. The case $q = 0$ is described as neutral and this includes all conservative force fields for which the potential $V(x)$ has a minimum at the singular point, as discussed in §13.1. But if (13.2.3) is only an approximation to a non-linear force field which in its exact form depends on v, it should not be inferred that the non-linear case is also neutral. Such a case requires a more accurate approximation, with the most significant term in v included.

(c) $p + q^2/4 = 0, q \neq 0$,

$$x = (At + B)e^{qt/2}, \tag{13.2.12}$$

$$v = Ae^{qt/2} + \tfrac{1}{2}q(At + B)e^{qt/2}. \tag{13.2.13}$$

The integral curve for $A = 0$ is the straight line $2v = qx$. All other curves pass through the origin and are tangent there to the line $2v = qx$. This is because the origin is approached on all curves as $qt \to -\infty$. Then the first term in equation (13.2.13) for v is small compared with the second. If only the most significant term is retained we have that $2v \sim qx$ as $qt \to -\infty$.

The singular point is called a node and the integral curves are as shown in Figure 13.2.1(c). Equilibrium is stable for $q < 0$ and unstable for $q > 0$.

(d) $p + q^2/4 > 0, p < 0$,

$$x = Ae^{\lambda_1 t} + Be^{\lambda_2 t}, \tag{13.2.14}$$

$$v = A\lambda_1 e^{\lambda_1 t} + B\lambda_2 e^{\lambda_2 t}, \tag{13.2.15}$$

where

$$\lambda_1 = \tfrac{1}{2}\{q + (q^2 + 4p)^{1/2}\}, \tag{13.2.16}$$

$$\lambda_2 = \tfrac{1}{2}\{q - (q^2 + 4p)^{1/2}\}. \tag{13.2.17}$$

There are two integral curves which are straight lines, namely $v = \lambda_2 x$ (A = 0) and $v = \lambda_1 x$ (B = 0). All other curves are tangent to one or other of these straight lines at the origin. If $q < 0$, so that $\lambda_1 < 0, \lambda_2 < 0, |\lambda_2| > |\lambda_1|$, the origin is stable and is approached as $t \to \infty$. Hence

$$x \sim Ae^{\lambda_1 t}, \quad v \sim A\lambda_1 e^{\lambda_1 t}, \quad v \sim \lambda_1 x,$$

as $t \to \infty$, for all the curves except $v = \lambda_2 x$(A = 0) which is called the exceptional direction.

If $q > 0$, then $\lambda_1 > \lambda_2 > 0$. The origin is unstable and the vicinity of the origin represents large negative values of t for which

$$x \sim Be^{\lambda_2 t}, \quad v \sim B\lambda_2 e^{\lambda_2 t}, \quad v \sim \lambda_2 x$$

and the exceptional direction is $v = \lambda_1 x$.

The equilibrium point is again called a node (Fig. 13.2.1(d)).

(e) $p = 0$.
When $p = 0$, equation (13.2.4) is equivalent to

$$\mathrm{d}v/\mathrm{d}x = q$$

since $\mathrm{d}^2x/\mathrm{d}t^2 = (\mathrm{d}v/\mathrm{d}x)(\mathrm{d}x/\mathrm{d}t)$. The integral curves consist of a family of parallel straight lines with slope q. The solution in terms of t is

$$x = Ae^{qt} + B, \quad v = Aqe^{qt} \qquad (13.2.18)$$

if $q \neq 0$, and

$$x = Ae^{qt} + B, \quad v = Aqe^{qt} \qquad (13.2.19)$$

if $q = 0$. Equilibrium is neutral if $q < 0$ and unstable if $q \geqslant 0$.

(f) $p > 0$.
For $p > 0$ the form of the solution is given by (13.2.14)–(13.2.17) but now λ_1 and λ_2 are of opposite sign. Equilibrium is always unstable since one of the exponential factors will tend to infinity as $t \to \infty$. Only two integral curves pass through the origin, namely $v = \lambda_1 x$ $(B = 0)$ and $v = \lambda_2 x$ $(A = 0)$. These straight lines are asymptotes for all the other integral curves. This type of singularity is called a saddle point.

This completes the classification. The various cases are summarised in Figures 13.2.1 and 13.2.2.

Example 13.2.1
As a simple example we consider the integral curves of the equation

$$\mathrm{d}^2x/\mathrm{d}t^2 = v(\mathrm{d}v/\mathrm{d}x) = n^2(a^2 - x^2)/a, \qquad (13.2.20)$$

where $a > 0$.

The force here is conservative and easily integrable, so that an explicit formula is available for the integral curves. In Example 13.1.1 the qualitative aspects of the motion have already been discussed from an analysis of the potential energy. For comparison, the integral curves in the phase plane will now be discussed on the basis of equation (13.2.20).

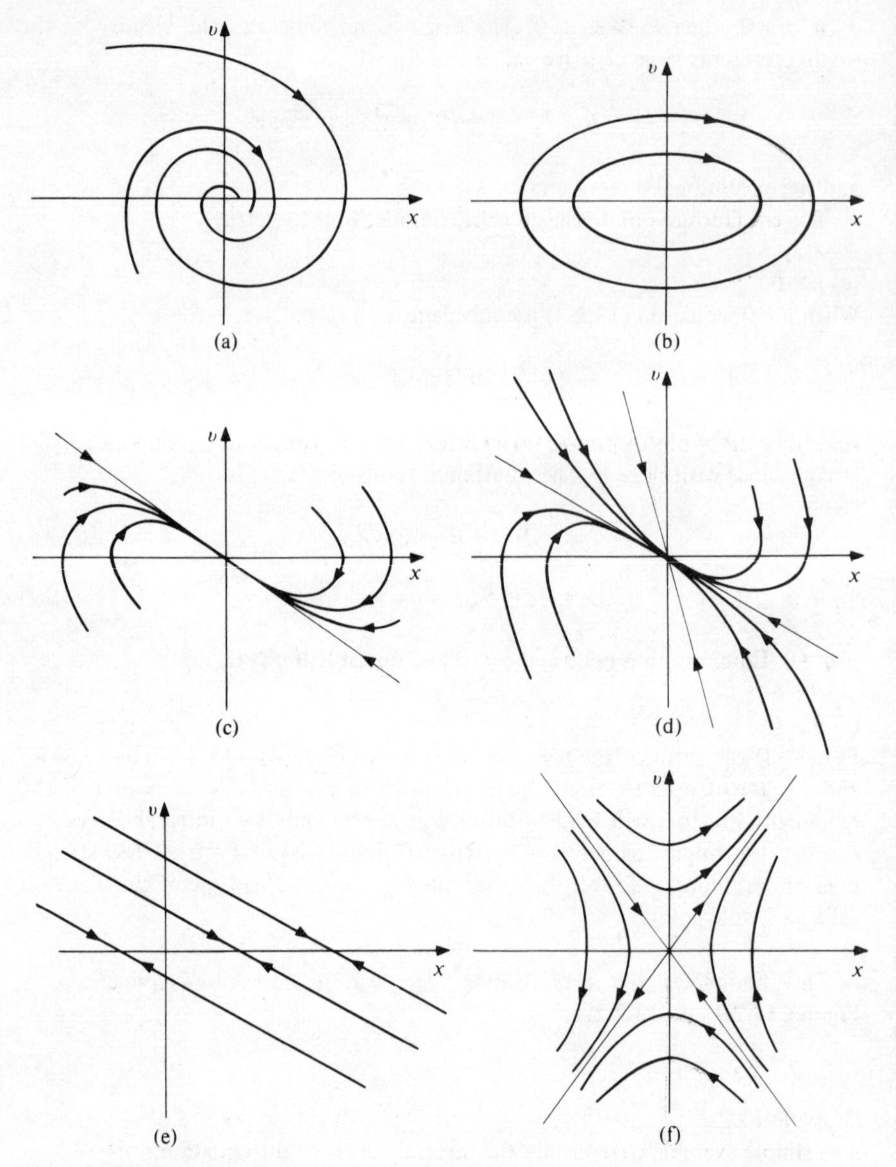

Figure 13.2.1 Classification of the singularities in the phase plane. The arrows show stable equilibrium in (a), (c), (d), neutral equilibrium in (b) and (e) and unstable equilibrium in (f).

(a) Spiral, $p + q^2/4 < 0$, $q \neq 0$.
(b) Centre, $p < 0$, $q = 0$.
(c) Node, $p + q^2/4 = 0$, $q \neq 0$.
(d) Node, $p + q^2/4 > 0$, $p < 0$.
(e) Straight lines, $p = 0$.
(f) Saddle, $p > 0$.

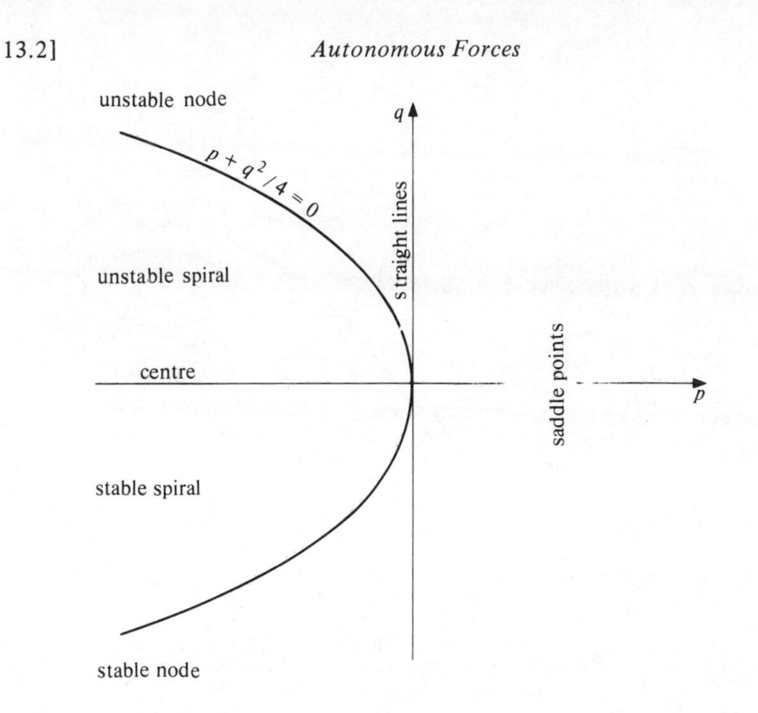

Figure 13.2.2 Dependence of the phase plane singularities on p and q.

The curves have horizontal tangents on $x = \pm a$ and vertical tangents on $v = 0$. These lines separate the regions of the (x, v) plane where $dv/dx > 0$ from those where $dv/dx < 0$.

There are singular points at $x = \pm a$, $v = 0$. Near $x = -a$

$$v(dv/dx) \sim 2n^2(x + a).$$

Since $p = 2n^2 > 0$, $x = -a$ is a saddle point. Near $x = a$

$$v(dv/dx) \sim -2n^2(x - a),$$

so that $p = -2n^2$, $q = 0$ and the singular point is a centre.

The information obtained so far is shown diagrammatically in Figure 13.2.3(a). A little further reflection on the possibilities and some freehand drawing produces Figure 13.2.3(b), which is qualitatively correct. The physical interpretation is that the particle can oscillate between the two extreme values of x on a closed trajectory (see case (ii) of Example 13.1.1). On a trajectory which recedes to infinity the particle will ultimately move in the negative x direction with an ever increasing speed.

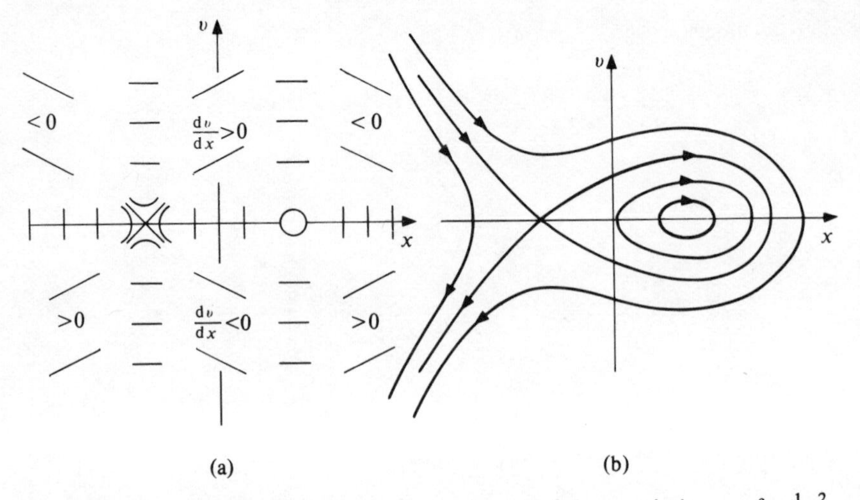

Figure 13.2.3 Illustration for Example 13.2.1: variation of $\frac{1}{2}v^2 - n^2(a^2x - \frac{1}{3}x^3)/a = \text{E}$.

Example 13.2.2

The equation of motion for a pendulum oscillating in a vertical plane with quadratic damping is of the form

$$d^2\theta/dt^2 = \omega(d\omega/d\theta) = -n^2 \sin\theta - k\omega\,|\,\omega\,|, \tag{13.2.21}$$

where n and k are positive constants, θ is the angular displacement and $\omega = \dot{\theta}$. In the (θ, ω) phase plane there are singularities when $\omega = 0$ and $\sin\theta = 0$. At $\theta = \pi$, for example, the linearised approximation is

$$\omega(d\omega/d\theta) = n^2(\theta - \pi)$$

which gives a saddle point singularity. At $\theta = 0$ we have, approximately,

$$\omega(d\omega/d\theta) = -n^2\theta,$$

which gives a centre. This is, however, a situation in which the linearised approximation fails, from lack of sufficient accuracy, to give a true indication of the singular point. Physical intuition suggests that the damping term gives genuine stability with a decay in amplitude, as for a spiral singularity. This is in fact the case. It so happens that equation (13.2.21) is easily integrated, since it is a linear differential equation in ω^2. The result is

$$\frac{\omega^2}{n^2} = \begin{cases} c_1 e^{-2k\theta} + \dfrac{2}{(1+4k^2)^{1/2}} \cos(\theta + \alpha), & \omega > 0 \\[2ex] c_2 e^{2k\theta} + \dfrac{2}{(1+4k^2)^{1/2}} \cos(\theta - \alpha), & \omega < 0 \end{cases} \tag{13.2.22}$$

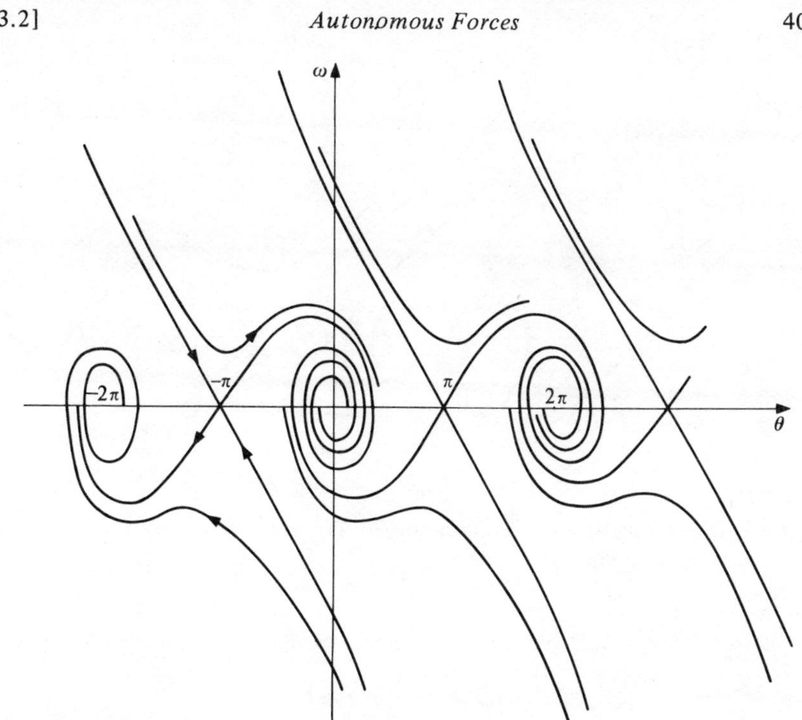

Figure 13.2.4 Illustration for Example 13.2.2: integral curves for pendulum with quadratic damping.

where $\tan \alpha = 2k$ and c_1, c_2 are constants of integration.* The integral curves, as given by equations (13.2.22), are shown in Figure 13.2.4. The neighbourhood of the spiral singularity at the origin represents damped oscillations as expected. The existence of a spiral singularity at each even multiple of π implies that the pendulum can make a number of complete revolutions before finally settling down to damped oscillatory motion. The integral curves which pass through the saddle point singularities at odd multiples of π represent oscillations which come to rest in the unstable equilibrium position.

Example 13.2.3
For a pendulum with linear damping, the equation of motion takes the form

$$\mathrm{d}^2\theta/\mathrm{d}t^2 = \omega(\mathrm{d}\omega/\mathrm{d}\theta) = -n^2 \sin\theta - k\omega, \qquad (13.2.23)$$

which again has singularities on the θ-axis when $\sin\theta = 0$. At $\theta = \pi$ we have

$$\omega(\mathrm{d}\omega/\mathrm{d}\theta) = n^2(\theta - \pi) - k\omega,$$

* On a particular integral curve these constants must be chosen so that the curve is continuous at $\omega = 0$.

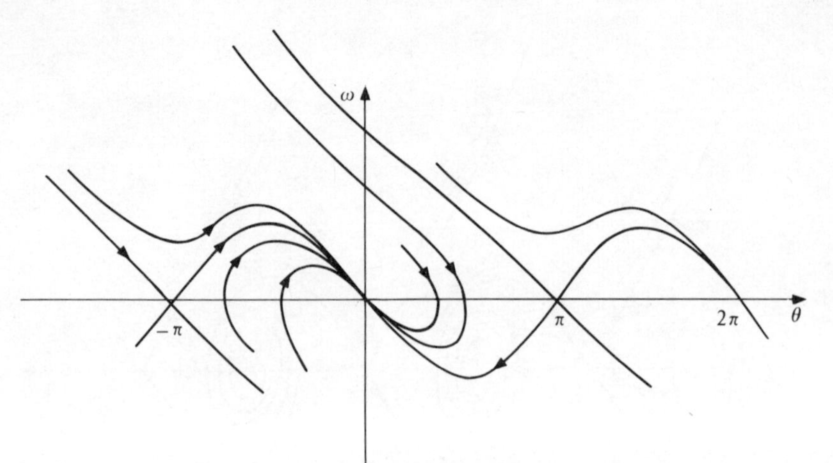

Figure 13.2.5 Illustration for Example 13.2.3: integral curves for pendulum with heavy linear damping.

so that $\theta = \pi$ is a saddle point. But for $\theta = 0$ the linearised approximation now gives

$$\omega(\mathrm{d}\omega/\mathrm{d}\theta) = -n^2\theta - k\omega.$$

If $n^2 - k^2/4 > 0$ there is a spiral singularity and if $n^2 - k^2/4 \leqslant 0$ there is a nodal singularity. These cases correspond respectively to the damped harmonic oscillations and the heavy damping aperiodic motion. The integral curves for the oscillatory case are qualitatively similar to Figure 13.2.4, and Figure 13.2.5 shows the main features when the motion is aperiodic.

13.3 THE METHOD OF AVERAGING

The discussion of damped oscillations in §6.2 shows that, although the solution (6.2.4) of equation (6.2.1) tends to a simple harmonic oscillation in the limit of no damping ($k = 0$), this limiting solution is not a satisfactory approximation when the resisting force is small. The reason is that the amplitude of the oscillation, $a_0 e^{-kt/2}$, will eventually differ substantially from its initial value a_0 for sufficiently large values of the time, no matter how small k might be provided that it is not zero. This is a particular example of a situation in which a small effect is nevertheless cumulative and so can produce substantial changes over sufficiently long periods of time. We say that the approximation is not uniformly valid as $t \to \infty$.

For damped oscillations governed by equation (6.2.1) we have the exact solution (6.2.4), but there are many examples of oscillations which are governed by equations of the type

$$\ddot{x} = -n^2 x + \epsilon f(x, \dot{x}), \tag{13.3.1}$$

where the small positive constant ϵ is included to indicate that the term $\epsilon f(x, \dot{x})$ is small compared with $n^2 x$, but is sufficiently complicated to prevent the integration of the equation in simple terms. An approximate method of solution is therefore to be desired, but it should be one which gives due consideration to the form of the solution for large values of t. One such technique is the method of averaging.

We start from the observation that, since the solution of (13.3.1), when $\epsilon = 0$, is of the form $x = a \cos \phi$ with a and $\dot{\phi}$ $(= n)$ constant, there is something to be said for retaining the same form of solution for the full equation provided allowance is made for possible variations in a and $\dot{\phi}$, which will presumably be slow when ϵ is small. To take again the example of damped harmonic oscillations as illustrated by (6.2.4), the solution does approximate to a simple harmonic oscillation in any range of t for which kt is small. There is a change in amplitude, which can be substantial, but it will take place slowly if k is small and this is the property of the solution which is exploited in the method of averaging.

It is convenient to rewrite equation (13.3.1) as the pair of first order differential equations

$$\dot{x} = v, \qquad \dot{v} = -n^2 x + \epsilon f(x, v), \tag{13.3.2}$$

for the two variables x and v. Next we consider a transformation of variables from x, v to a, ϕ through the relations

$$x = a \cos \phi, \qquad v = -na \sin \phi. \tag{13.3.3}$$

This transformation is suggested by the form of the solution when $\epsilon = 0$ and so $\dot{\phi} = n$, but here a and ϕ are functions of t still to be found. Since

$$\dot{x} = \dot{a} \cos \phi - a\dot{\phi} \sin \phi, \tag{13.3.4}$$

$$\dot{v} = -n\dot{a} \sin \phi - na\dot{\phi} \cos \phi, \tag{13.3.5}$$

we have, by substitution in (13.3.2),

$$\dot{a} \cos \phi - a(\dot{\phi} - n)\sin \phi = 0, \tag{13.3.6}$$

$$-\dot{a} \sin \phi - a(\dot{\phi} - n)\cos \phi = \frac{\epsilon}{n} f(a \cos \phi, -na \sin \phi). \tag{13.3.7}$$

If equation (13.3.6) is multiplied by $\cos \phi$, equation (13.3.7) by $\sin \phi$ and the resulting equations subtracted, $\dot{\phi}$ is eliminated and the following equation for $\dot{a}$ is obtained:

$$\dot{a} = -\frac{\epsilon}{n} f(a \cos \phi, -na \sin \phi)\sin \phi. \tag{13.3.8}$$

Again, by elimination of $\dot{a}$ from (13.3.6) and (13.3.7), we get

$$a(\dot{\phi} - n) = -\frac{\epsilon}{n} f(a \cos \phi, -na \sin \phi) \cos \phi. \tag{13.3.9}$$

At this stage no approximations have been made, and equations (13.3.8) and (13.3.9) are equivalent to the original equations (13.3.2) through the transformation (13.3.3). Equation (13.3.8) verifies the previous conjecture that the rate of change of a is small when ϵ is small. In other words, although it is possible for a to change substantially, the change takes place on a slow time scale compared, say, with the changes in the phase ϕ which take place on a time scale of order n^{-1}. Similarly equation (13.3.9) shows that $\dot{\phi} - n$ is also small, of the order of ϵ.

However, as they stand, equations (13.3.8) and (13.3.9) are not more easily solved than the original equations (13.3.2). They are simplified on the following basis. A substantial change in a is produced only if $\dot{a}$ is on average non-zero. A contribution to $\dot{a}$ which is both small and zero on average will not produce a substantial change in a because the small integrated contributions will continually cancel out. As a simple illustration, consider the equation

$$\dot{a} = -\epsilon a(1 + \cos t) \tag{13.3.10}$$

which is easily integrated to give

$$a = a_0 \mathrm{e}^{-\epsilon(t + \sin t)} \tag{13.3.11}$$

It is clear from the solution (13.3.11) that the substantial changes in a that can arise, do so from the factor $\mathrm{e}^{-\epsilon t}$. This, in turn, comes from that part of $\dot{a}$ which is non-zero on average, namely $-\epsilon a$. But the term $-\epsilon a \cos t$ in $\dot{a}$ contributes the factor $\mathrm{e}^{-\epsilon \sin t}$, which never differs substantially from unity for all t when ϵ is small. We say that

$$a = a_0 \mathrm{e}^{-\epsilon t} \tag{13.3.12}$$

is a uniformly valid first approximation to a because it is valid for all t.

The above argument can be formalised by averaging the right-hand side of (13.3.10) over a period of the oscillatory term, thereby eliminating the term which is zero on average. Explicitly this gives

$$\dot{a} = \frac{1}{2\pi} \int_0^{2\pi} -\epsilon a(1 + \cos t)\, \mathrm{d}t = -\frac{\epsilon a}{2\pi} \int_0^{2\pi} (1 + \cos t)\, \mathrm{d}t = -\epsilon a,$$

if the variation in a over a period is negligible. Hence $\dot{a} = -\epsilon a$ and the integral is given by (13.3.12).

This is the basis of the method of averaging. It brings out clearly the difficulty encountered in approximating to the solution of equations (13.3.2) and, hence, the way to overcome it. To justify the procedure formally is, however, difficult and will not be attempted here. We proceed by analogy and approximate equations (13.3.8) and (13.3.9) by taking the average value of the right-hand sides. Since they are periodic in ϕ with period 2π, it is sufficient to take the average over this interval of ϕ. The result is

$$\dot{a} = -\frac{\epsilon}{2n\pi}\int_0^{2\pi} f(a\cos\phi, -na\sin\phi)\sin\phi\, \mathrm{d}\phi, \qquad (13.3.13)$$

$$\dot{\phi} - n = -\frac{\epsilon}{2n\pi a}\int_0^{2\pi} f(a\cos\phi, -na\sin\phi)\cos\phi\, \mathrm{d}\phi. \qquad (13.3.14)$$

In the evaluation of these average values, the variation of a in the interval $0 \leqslant \phi \leqslant 2\pi$ is regarded as negligible and ignored. The mathematical simplification arises from the disappearance of ϕ from the equations and, in fact, they become uncoupled since equation (13.3.13) states that $\dot{a}$ is some function of a. Accordingly, although it may be non-linear, it is of a simple form and can usually be manipulated without difficulty. When a is known as a function of t, so is the right-hand side of equation (13.3.14) and straightforward integration of this equation then gives ϕ as a function of t.

As a check on the method, consider the special case

$$f = -v = na\sin\phi$$

which gives the equation for linear damping. Since the average value of $\sin^2\phi$ is $\frac{1}{2}$ and the average value of $\sin\phi\cos\phi$ is 0, equations (13.3.13) and (13.3.14) become

$$\dot{a} = -\epsilon a/2 \qquad \text{whence } a = a_0 \mathrm{e}^{-\epsilon t/2},$$

$$\dot{\phi} - n = 0 \qquad \text{whence } \phi = nt + \phi_0.$$

Reference to (6.2.4), with k replaced by ϵ, shows that the damping coefficient is given correctly and that the error in the frequency is of order ϵ^2. The method can be continued to give higher orders of accuracy, but we shall not pursue this.

Note that, although in application it is not essential to introduce a separate distinguishing symbol, the slowly varying quantity in (13.3.9) is

$$\phi - \int n\, \mathrm{d}t.$$

In the present case this is just $\phi - nt$ since n has been assumed constant. The following example discusses the technique when n has a prescribed variation.

Example 13.3.1

The method of averaging can be readily applied to the equation

$$\ddot{x} = -n^2 x, \qquad (13.3.15)$$

or, equivalently,

$$\dot{x} = v, \quad \dot{v} = -n^2 x, \tag{13.3.16}$$

where n is not constant, but is a given function of the time which varies slowly. The equation occurs in the theory of wave propagation and was first studied because of its application in quantum mechanics.

As in the general theory we put

$$x = a \cos \phi, \quad v = -na \sin \phi \tag{13.3.17}$$

and now, in the evaluation of $\dot{v}$, it is necessary to take into account the variation of n with t. We have

$$\dot{x} = \dot{a} \cos \phi - a\dot{\phi} \sin \phi, \tag{13.3.18}$$

$$\dot{v} = -n\dot{a} \sin \phi - na\dot{\phi} \cos \phi - \dot{n}a \sin \phi. \tag{13.3.19}$$

Substitution of (13.3.18) and (13.3.19) into (13.3.16) gives

$$\dot{a} \cos \phi - a(\dot{\phi} - n)\sin \phi = 0, \tag{13.3.20}$$

$$-\dot{a} \sin \phi - a(\dot{\phi} - n)\cos \phi = \frac{\dot{n}a}{n} \sin \phi. \tag{13.3.21}$$

When these equations are solved for $\dot{a}$ and $\dot{\phi}$, the result is

$$\dot{a} = -\frac{\dot{n}a}{n} \sin^2 \phi, \tag{13.3.22}$$

$$\dot{\phi} - n = -\frac{\dot{n}}{n} \sin \phi \cos \phi. \tag{13.3.23}$$

For the method of averaging to be applicable, it is necessary that $\dot{n} \ll n^2$, and this is the condition referred to in the statement that n should be a slowly varying function of the time. Then equations (13.3.22) and (13.3.23) can be simplified by taking the average value of the right-hand sides. The result is

$$\frac{\dot{a}}{a} = -\frac{\dot{n}}{2n} \qquad \text{whence } \frac{a}{a_0} = \left(\frac{n_0}{n}\right)^{1/2}, \tag{13.3.24}$$

$$\dot{\phi} = n \qquad \text{whence } \phi = \int_0^t n \, \mathrm{d}t + \phi_0, \tag{13.3.25}$$

$$x = a \cos \phi = a_0 \, {\frac{n_0}{n}}^{1/2} \cos\left(\int_0^t n \, \mathrm{d}t + \phi_0\right). \tag{13.3.26}$$

In this example, the amplitude of the oscillation is inversely proportional to the square root of the frequency n.

Example 13.3.2

We consider next the plane oscillations of a simple pendulum, of length l and angular displacement θ, when there is a small damping term proportional to the velocity and when the length of the pendulum is allowed to vary in a slow prescribed manner.

The transverse acceleration is $l\ddot{\theta} + 2\dot{l}\dot{\theta}$. In the same direction the components of the resistance and of the gravitational force are, respectively, $-kl\dot{\theta}$ and $-g \sin\theta$ per unit mass. The transverse component of the equation of motion is therefore

$$l\ddot{\theta} + 2\dot{l}\dot{\theta} = -kl\dot{\theta} - g\sin\theta. \tag{13.3.27}$$

Let $n^2 = g/l$ and let the amplitude of the oscillation be small, so that $\sin\theta$ can be replaced by the first two terms of its expansion in powers of θ, namely $\theta - \theta^3/6$. Then equation (13.3.27) can be written as the pair of equations

$$\dot{\theta} = \omega, \qquad \dot{\omega} = -n^2\theta + \frac{n^2\theta^3}{6} + \frac{4\dot{n}\omega}{n} - k\omega. \tag{13.3.28}$$

With the substitutions

$$\theta = a\cos\phi, \qquad \omega = -na\sin\phi, \tag{13.3.29}$$

equations (13.3.28) give, when solved for $\dot{a}$ and $\dot{\phi}$,

$$\dot{a} = -\frac{na^3}{6}\cos^3\phi\sin\phi + 3\frac{\dot{n}}{n}a\sin^2\phi - ka\sin^2\phi, \tag{13.3.30}$$

$$\dot{\phi} - n = -\frac{na^2}{6}\cos^4\phi + 3\frac{\dot{n}}{n}\sin\phi\cos\phi - k\sin\phi\cos\phi. \tag{13.3.31}$$

It is assumed that all the terms on the right-hand sides of equations (13.3.30) and (13.3.31) are small and can be averaged. The only terms which are non zero on average are those proportional to $\sin^2\phi$ and $\cos^4\phi$ whose average values are, respectively, $\frac{1}{2}$ and $\frac{3}{8}$. The averaged equations are therefore

$$\frac{\dot{a}}{a} = \frac{3}{2}\frac{\dot{n}}{n} - \frac{1}{2}k, \quad \text{whence } a = a_0\left(\frac{n}{n_0}\right)^{3/2} \mathrm{e}^{-kt/2}, \tag{13.3.32}$$

and

$$\dot{\phi} = n\left(1 - \frac{a^2}{16}\right), \quad \text{whence } \phi = \int_0^t n\left(1 - \frac{a^2}{16}\right)\mathrm{d}t + \phi_0. \tag{13.3.33}$$

It follows that the uniformly valid approximation is

$$\theta = a_0\left(\frac{n}{n_0}\right)^{3/2} \mathrm{e}^{-kt/2} \cos\left\{\int_0^t n\left(1 - \frac{a^2}{16}\right)\mathrm{d}t + \phi_0\right\}, \tag{13.3.34}$$

where a is given by (13.3.32) as a function of t. In particular we have the interesting result that the amplitude is proportional to $n^{3/2}$ or, equivalently, inversely proportional to the $\frac{3}{4}$ power of the length.

As a check on the approximation we note that the damping term, $\mathrm{e}^{-kt/2}$, is given correctly and that if a is constant ($l = l_0, k = 0$), then

$$\theta = a \cos\left\{n\left(1 - \frac{a^2}{16}\right)t + \phi_0\right\}.$$

The periodic time is therefore

$$T = \frac{2\pi}{n}\left(1 - \frac{a^2}{16}\right)^{-1} = \frac{2\pi}{n}\left(1 + \frac{a^2}{16}\right)$$

approximately since a is assumed small. This result agrees with (6.5.21).

Example 13.3.3

In the previous examples, the method of averaging has been demonstrated when the governing equation is a small modification of the equation $\ddot{x} + n^2 x = 0$ which describes simple harmonic motion. The essence of the technique is to retain the form of solution $x = a \cos \phi$ of the simple harmonic equation, and to allow for a small modification of the equation by appropriate modifications in a and ϕ. It is the calculation of these modifications which is simplified by the method of averaging.

The aim of the present example is to demonstrate that the method of averaging is of more general application, and that the essential property of the reference equation is that it should have a periodic solution. The example is based on the equation

$$\ddot{x} + V'(x) = 0, \tag{13.3.35}$$

which has the energy integral

$$\tfrac{1}{2}\dot{x}^2 + V(x) = a, \tag{13.3.36}$$

say, where a is the constant energy. We know that this equation has a periodic solution when $a - V(x)$ is of the appropriate form (see §6.5). Let us assume that the solution is periodic, with period $2\pi/n$ in t given by

$$\frac{2\pi}{n} = \frac{1}{2^{1/2}} \int \frac{\mathrm{d}x}{\pm(a - V(x))^{1/2}}, \tag{13.3.37}$$

where the integral is to be taken over one complete oscillation. The sign of $(a - V(x))^{1/2}$ is positive over that part of the oscillation where x increases with t and negative where x decreases. Equation (13.3.37) defines n as a function of a.

In equation (13.3.36) a appears as a constant of integration, but in the following analysis a, and therefore n, will be regarded as a variable. Now in the application of the method of averaging it is an essential requirement that the periodic functions which appear should have a fixed period, as we shall see later. We therefore regard the solution $x(a, \phi)$ of (13.3.36) as a function of the parameter a and a variable ϕ which is normalised so that ϕ changes by 2π in one complete oscillation. It follows that $x(a, \phi)$ satisfies the equation

$$\tfrac{1}{2}n^2\left(\frac{\partial x}{\partial \phi}\right)^2 + V(x) = a \tag{13.3.38}$$

where, as already explained, n is chosen to satisfy (13.3.37) so that x is now periodic in ϕ with period 2π.

We shall apply the method of averaging to the equation

$$\ddot{x} + V'(x) = \epsilon f(x, \dot{x}) \tag{13.3.39}$$

or, alternatively, to the pair of equations

$$\dot{x} = v, \quad \dot{v} + V'(x) = \epsilon f(x, v). \tag{13.3.40}$$

We first make a transformation of variables from x, v to a, ϕ such that

$$x = x(a, \phi), \quad v = n(\partial x/\partial \phi), \tag{13.3.41}$$

and $x(a, \phi)$ is the solution of (13.3.38), which may also be written in the form

$$\tfrac{1}{2}v^2 + V(x) = a. \tag{13.3.42}$$

In order that x and v should satisfy (13.3.40), it is necessary to calculate the appropriate variations of a and ϕ as functions of t. The equations which determine these variations can be obtained by the substitution of (13.3.41) in (13.3.40). The resulting equations can then be solved for $\dot{a}$ and $\dot{\phi}$ in a fashion similar to that used in previous examples. A simpler procedure is to note first that, from (13.3.40), (13.3.41) and (13.3.42),

$$\begin{aligned}\dot{a} &= v\dot{v} + V'(x)\dot{x} = v(\epsilon f(x, v) - V'(x)) + V'(x)v \\ &= \epsilon v f(x, v) = \epsilon n f(x, v)\, \partial x/\partial \phi\end{aligned} \tag{13.3.43}$$

and then, from $\dot{x} = v$ in the form

$$\frac{\partial x}{\partial a}\dot{a} + \frac{\partial x}{\partial \phi}\dot{\phi} = v = n\frac{\partial x}{\partial \phi}$$

we have, with the help of (13.3.43),

$$\dot{\phi} - n = -\epsilon n f(x, v)\, \partial x/\partial a. \tag{13.3.44}$$

Equations (13.3.43) and (13.3.44) are now in a form suitable for averaging.* Since the right-hand sides are small and periodic in ϕ with period 2π, the approximate averaged equations are

$$\left.\begin{aligned} \dot{a} &= \frac{\epsilon n}{2\pi} \int_0^{2\pi} f(x, v) \frac{\partial x}{\partial \phi}\, \mathrm{d}\phi, \\ \dot{\phi} - n &= -\frac{\epsilon n}{2\pi} \int_0^{2\pi} f(x, v) \frac{\partial x}{\partial a}\, \mathrm{d}\phi, \end{aligned}\right\} \tag{13.3.45}$$

with x and v given in terms of a and ϕ by (13.3.41). The right-hand sides of (13.3.45) are functions only of a. The first equation becomes uncoupled from the second and can be solved to give a as a function of t. Then the second equation gives $\dot{\phi}$ as a function of a and hence as a function of t, so that ϕ can also be evaluated as a function of t. Note that n varies slowly since it is a function of a given by (13.3.37). In principle the first of equations (13.3.45) can be solved without a knowledge of x and v as functions of a and ϕ. With the help of (13.3.42), it can be written

$$\dot{a} = \frac{\epsilon n}{2\pi} \int f(x, \pm 2^{1/2}(a - V(x))^{1/2})\, \mathrm{d}x. \tag{13.3.46}$$

The integral in (13.3.46) is taken over one complete oscillation and the sign of $2^{1/2}(a - V(x))^{1/2}$ is determined by the sign of v or $\partial x/\partial \phi$, so that it is positive when x is increasing and negative when x is decreasing. Note that a knowledge of a as a function of t is sufficient to give the variation of the amplitude and the approximate period $2\pi/n$ of the oscillation as functions of t.

The analysis can be taken a stage further when the term $\epsilon f(x, v)$ arises from linear damping, so that $f = -v$. If we define

$$A(a) = \int \pm 2^{1/2}(a - V(x))^{1/2}\, \mathrm{d}x,$$

* It is at this stage that the requirement that the period of oscillation be fixed is necessary. In order to apply the method of averaging to (13.3.43) and (13.3.44), the right-hand sides of the equations should be periodic functions of ϕ. If x is periodic with period P, say, where P is a function of a, we have $x(a, \phi) = x(a, \phi + P)$ and $\partial x(a, \phi)/\partial \phi = \partial x(a, \phi + P)/\partial \phi$, so that x and $\partial x/\partial \phi$ are periodic. However

$$\frac{\partial}{\partial a} x(a, \phi) = \frac{\partial}{\partial a} x(a, \phi + P) + \frac{\mathrm{d}P}{\mathrm{d}a} \frac{\partial}{\partial \phi} x(a, \phi + P)$$

so that $\partial x/\partial a$ is not periodic unless $\mathrm{d}P/\mathrm{d}a = 0$ and the period is fixed. To take a simple example, if $x = \sin a\phi$, then $\partial x/\partial a = \phi \cos a\phi$ and the amplitude increases linearly with ϕ.

where the integral is taken over one complete period, relations (13.3.37) and (13.3.46) show that for the case of linear damping

$$\dot{a} = -\frac{\epsilon A}{\mathrm{d}A/\mathrm{d}a} \quad \text{or} \quad \frac{\mathrm{d}}{\mathrm{d}t}\log A = -\epsilon.$$

The integral of this equation gives

$$A(a) = A(a_0)\mathrm{e}^{-\epsilon t},$$

where a_0 is the value of a when $t = 0$. This is an implicit relation for a as a function of t.

13.4 FORCED NON-LINEAR OSCILLATIONS

We saw in §6.3 that the amplitude of forced, linear oscillations increases dramatically near resonance if the resistance is small. In the case of a simple pendulum, for example, the effect could render the linear approximation invalid. We also saw, in Example 6.5.1, that one property of non-linearity is to alter the periodic time and hence the natural frequency of oscillations. This suggests that the increase in amplitude arising near resonance may be limited by the associated change in the natural frequency arising from non-linearity. To investigate this possibility we consider the one-dimensional motion of a particle under the influence of a forcing term $\mathbf{i}f\cos pt$ per unit mass, and a restoring force $-\mathbf{i}(n^2x + \nu x^3)$ per unit mass. The reason for this choice is that in many practical examples the restoring force is an odd function, so that at points at the same distance on either side of the equilibrium position it has the same magnitude but is opposite in sign. The first two terms in the expansion for small x will then be of the form chosen. A familiar example is provided by the small oscillations of a simple pendulum, where the restoring force is

$$-mg\sin\theta = -mg(\theta - \tfrac{1}{6}\theta^3 + \ldots)$$

in the notation of equation (13.3.27).

We take as our equation of motion

$$\ddot{x} = -n^2x - \nu x^3 - k\dot{x} + f\cos pt \tag{13.4.1}$$

or, equivalently,

$$\dot{x} = v, \quad \dot{v} = -n^2x - \nu x^3 - kv + f\cos pt, \tag{13.4.2}$$

with $n > 0$, ν, $k \geqslant 0$, $f \geqslant 0$, $p > 0$ all constant parameters. A term representing linear resistance is also included so that the interplay between resistive and non-linear forces, and their modification of the linear solution near resonance, can be assessed.

The averaging technique, as described in the previous section, suggests that we make the preliminary transformation $x = a \cos \phi$, $v = -na \sin \phi$ and neglect those small terms which are zero on average in the resulting equations for a and ϕ. However, we shall concentrate here on the solutions of (13.4.2) representing forced oscillations. When the linearised equation is used, it is known that the frequency of the forced oscillations is determined by the frequency p of the forcing term, and that the free oscillations ultimately decay because of resistance. It is to be expected that the situation is qualitatively similar when there are small non-linear effects, so that we should look for solutions where $\dot{\phi}$ approximates to p rather than to n. This suggests that the transformation

$$x = a \cos \phi, \quad v = -pa \sin \phi \tag{13.4.3}$$

is a preferable alternative. Of course we are chiefly interested in the resonant regime for which $|n - p| \ll 1$ and then there is little to choose between n and p as a factor in v. However, we shall see that the solution of (13.4.2) for $|n - p| \ll 1$, based on the transformation (13.4.3), happens to agree with the straightforward linearised solution which is valid when $n - p$ is not small. Hence the range of validity of the solution is not restricted to $|n - p| \ll 1$ even though the arguments used to derive it depend on this inequality being satisfied.

Substitution of (13.4.3) in (13.4.2) yields the equations

$$\begin{aligned} &\dot{a} \cos \phi - a\dot{\phi} \sin \phi = -pa \sin \phi, \\ &-p\dot{a} \sin \phi - pa\dot{\phi} \cos \phi = -n^2 a \cos \phi - va^3 \cos^3 \phi + kpa \sin \phi + f \cos pt. \end{aligned} \tag{13.4.4}$$

These equations can be solved for $\dot{a}$ and $\dot{\phi}$ to give

$$p\dot{a} = (n^2 - p^2)a \cos\phi \sin \phi + va^3 \cos^3 \phi \sin \phi - kpa \sin^2 \phi - f \cos pt \sin \phi, \tag{13.4.5}$$

$$p(\dot{\phi} - p) = (n^2 - p^2)\cos^2 \phi + va^2 \cos^4 \phi - kp \sin \phi \cos \phi - fa^{-1} \cos pt \cos \phi. \tag{13.4.6}$$

In order to simplify equations (13.4.5) and (13.4.6), it is assumed that all the terms on the right-hand sides are small. This means that the non-linear terms, the resistive terms and the forcing terms should all be small. In addition the forcing frequency p should approximate to the natural frequency n in order to ensure that the first terms on the right-hand sides are small. This concentrates attention on the oscillations near resonance, where a linear approximation is invalid. With these assumptions, (13.4.5) shows that $\dot{a}$ is small and (13.4.6) verifies the earlier comment that $\dot{\phi} - p$ is small. This is made explicit by the substitution

$$\phi = pt + \chi$$

so that both a and χ are at most slowly varying functions of t.

Finally we approximate the right-hand sides of (13.4.5) and (13.4.6) by averaging over the variations in ϕ. In so doing, changes in the slowly varying quantities a and χ are ignored, as in previous examples. Note that, since $\phi = pt + \chi$, the average values of $\cos pt \sin \phi$ and $\cos pt \cos \phi$ are $\frac{1}{2} \sin \chi$ and $\frac{1}{2} \cos \chi$ respectively.

Equations (13.4.5) and (13.4.6), when averaged, give

$$2p\dot{a} = -kpa - f \sin \chi, \tag{13.4.7}$$

$$2p\dot{\chi} = (n^2 - p^2) + \tfrac{3}{4}\nu a^2 - fa^{-1} \cos \chi. \tag{13.4.8}$$

Unlike the equations in the examples of the previous section, these equations are coupled and cannot be solved independently. They are non-linear and do not have simple solutions. Nevertheless they are much easier to handle than the original equations (13.4.5), (13.4.6) and, most significantly, they are autonomous since t no longer appears explicitly.

We can, for example, examine the equilibrium solutions, for which $\dot{a} = \dot{\chi} = 0$. Physically these solutions represent oscillations for which the amplitude and phase are constant on average. They are of particular interest because the solutions corresponding to stable equilibrium are the non-linear equivalent of the forced oscillations which remain in linear theory after the transient oscillations have decayed.

We consider first the equilibrium solutions when there is no resistance. With $k = 0$, the solutions of (13.4.7), (13.4.8) for $\dot{a} = \dot{\chi} = 0$ are $\sin \chi = 0$ and

$$p^2 - n^2 = \tfrac{3}{4}\nu a^2 - (-1)^r f a^{-1} = A(a) \tag{13.4.9}$$

say, where $r = 0$ or 1 according as $\chi = 0$ or π. We may regard (13.4.9) as a functional relation between the forcing frequency p and the amplitude a of the forced oscillations. The resulting curve showing the variation of a with p is called a response curve. We note that the linear solution ($\nu = 0$) away from resonance is in fact recovered from (13.4.9), which then gives

$$a = f/| n^2 - p^2 |$$

provided we choose $\chi = 0$ if $p < n$ and $\chi = \pi$ if $p > n$ (we do not consider negative values of the amplitude a; such a change of sign is rather taken into account by a change of the phase χ by π). At resonance, however, the non-linear contribution is crucial, for without it the amplitude is singular. Its introduction has several modifying features. Not only is the amplitude finite when $p = n$,

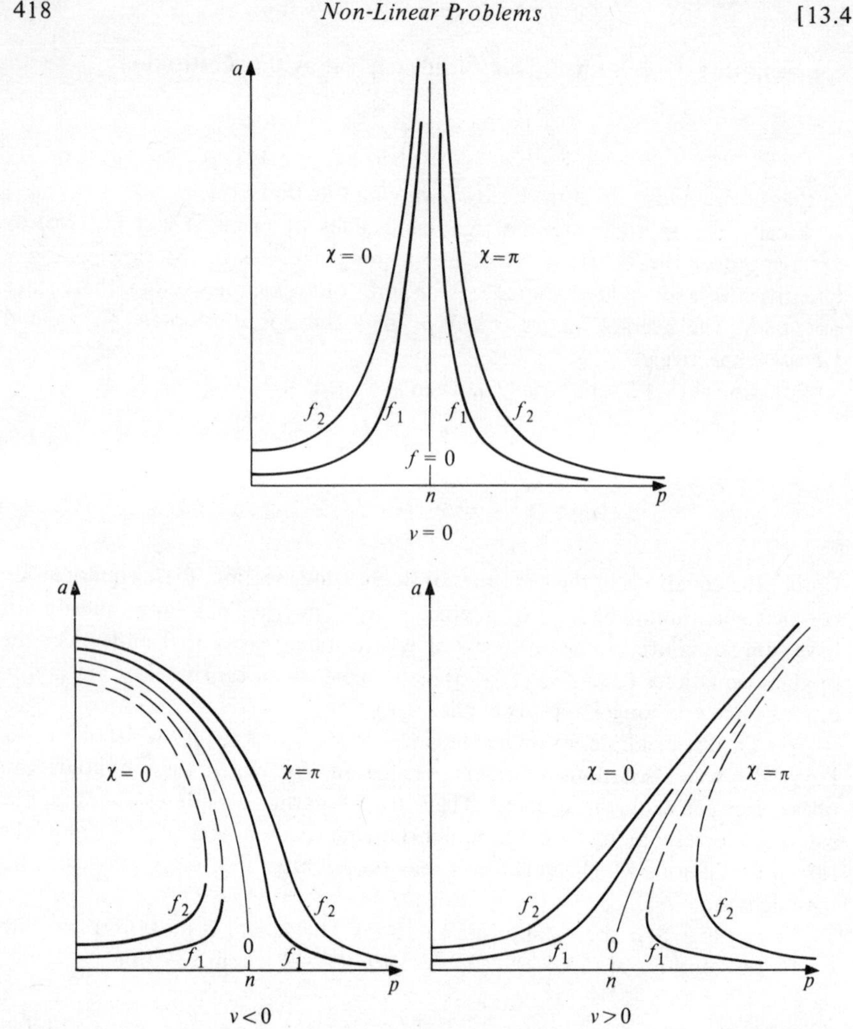

Figure 13.4.1 Response curves for forced, non-linear oscillations with $\nu = 0$, $\nu < 0, \nu > 0$, no damping and three values of f labelled $0, f_1, f_2$.

where $a = (4f/3|\nu|)^{1/3}$, but it is not even a maximum there. The fact that (13.4.9) is a cubic for a means that there are three possible values for a, at least for some values of the parameters, so that equilibrium solutions at a given frequency may not be unique. These features are clearly shown in Figure 13.4.1. It appears that there are three values of a for a given value of p if $p < n$ and $\nu < 0$ or if $p > n$ and $\nu > 0$.

Of equal importance is the question of the stability of the equilibrium solutions, for if they are unstable they will not in practice be observed. To answer this question, we consider a perturbation of (a, χ) from an equilibrium solution (a_0, χ_0) to a neighbouring solution $(a_0 + a_1, \chi_0 + \chi_1)$, where a_0, χ_0 are constant and a_1, χ_1 are relatively small. With $k = 0$, equations (13.4.7), (13.4.8) linearise to

$$2p\dot{a}_1 = -(-1)^r f\chi_1, \tag{13.4.10}$$

$$2p\dot{\chi}_1 = A'(a_0)a_1, \tag{13.4.11}$$

where $A(a)$ is given by (13.4.9) and the fact that χ_0 is either 0 or π has been used. One can now verify from §13.2 that the singular point $a_1 = \chi_1 = 0$ is a centre or a saddle according as $(-1)^r A'(a_0)$ is positive or negative. Alternatively, it follows directly from (13.4.10) and (13.4.11) that

$$4p^2\ddot{a}_1 = -(-1)^r A_1'(a_0)fa_1 \tag{13.4.12}$$

with a similar equation for χ_1. The solution of (13.4.12) is oscillatory, implying neutral stability (centre), if $(-1)^r A'(a_0)$ is positive. If it is negative a will grow exponentially in time leading to instability (saddle).

From (13.4.9), it is seen that $A'(a)$ is just $2p\mathrm{d}p/\mathrm{d}a$ and so has the same sign as the slope $\mathrm{d}a/\mathrm{d}p$ of the response curve. It follows that there is instability if $r = 0(\chi = 0)$ and a decreases with p, or if $r = 1(\chi = \pi)$ and a increases with p. The unstable oscillations are shown dotted in Figure 13.4.1, the net result being that there are either one or two neutrally stable oscillations for any given value of p.

Equation (13.4.9) even has a solution for the free oscillations when $f = 0$, namely

$$p^2 = n^2 + \tfrac{3}{4}\nu a^2. \tag{13.4.13}$$

The response curves for non zero values of f are all asymptotic to (13.4.13). Although the method of solution described here is not strictly appropriate for free oscillations, it is a simple matter to check the validity of (13.4.13) independently, since equation (13.4.1) with $k = f = 0$ is conservative and so can be integrated.

It remains to discuss the further modifications of the small resistive term, so far omitted. When $\dot{a} = \dot{\chi} = 0$, equations (13.4.7), (13.4.8) imply that

$$(n^2 - p^2 + \tfrac{3}{4}\nu a^2)^2 + k^2p^2 = f^2/a^2 \tag{13.4.14}$$

when χ is eliminated. For sufficiently small k it may be expected that the response curves are not markedly different from those given by (13.4.9). There are, however, some significant modifications. Equation (13.4.14), unlike (13.4.9), implies an upper bound on a, since a cannot exceed f/kp. What happens is that the two branches of the response curve obtained when $k = 0$ now join together in the vicinity of this maximum to give one continuous curve, as shown in Figure 13.4.2. For sufficiently small f the modification is more substantial, since it is no longer possible for the response curve to bend back on itself and yield more than one equilibrium solution so that the solution is unique over the whole range of p. Indeed a now tends to zero with f. It is hardly surprising that

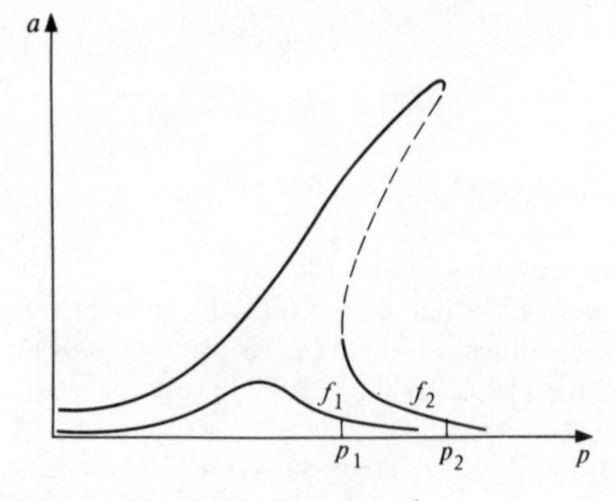

Figure 13.4.2 Response curves for $\nu > 0$ and small damping.

there is no equilibrium solution (other than $a = 0$) without a forcing term, since the resistance will ensure that any initial oscillation will decay to zero.

Finally the resistance changes the neutrally stable oscillations into stable oscillations (the singular point becoming a stable spiral instead of a centre). The effect of this is to ensure that any small perturbation from a stable equilibrium solution will ultimately decay and give a definite tendency to the equilibrium oscillation.

The existence of cut-off frequencies at p_1 and p_2, as shown in Figure 13.4.2, gives rise to jump phenomena, which are easily observed in practice and peculiar to forced non-linear oscillations. Let the particle oscillate with a frequency $p > p_1$ and an amplitude corresponding to a point on the lower branch of the response curve labelled f_2. Suppose now that the frequency is decreased continuously so that the amplitude of the oscillation follows the response curve towards p_1. At p_1 no further continuous increase in amplitude is possible. The oscillation becomes unsteady and finally settles down to an amplitude appropriate to the higher stable branch of the response curve. A similar, and more violent, change takes place if the oscillation initially corresponds to a point on the upper branch of the response curve and the frequency is continuously increased towards p_2.

The preceding discussion has concentrated on the equilibrium oscillations in the vicinity of the principal resonant frequency. For non-linear oscillations there is, however, the possibility of resonant response at other frequencies. With equation (13.4.1), for example, there is an important subharmonic resonance, at a frequency $p/3$, when p is in the neighbourhood of $3n$. Such solutions can be discussed by means of an analysis similar to that described here, but in the case of equation (13.4.1) the subharmonic response only emerges when the approximation is taken to second order.

EXERCISES

13.1 The acceleration of a particle P in a straight line is given by $-n^2x + 3n^2x^2/2a$, where x is the distance of P from a fixed point O of the line and $n > 0$, $a > 0$ are constants.

Sketch the potential energy curve. For what values of the total energy E is an oscillatory motion possible? What is the approximate periodic time time when the amplitude of the oscillations is small?

Find the variation of x with t given that $x = a$, $\dot{x} = 0$ when $t = 0$. Deduce that $x \to \infty$ as $t \to \pi/n$.

(Ans.: $0 < E < 2n^2a^2$, $2\pi/n$, $x = a \sec^2 \frac{1}{2}nt$.)

13.2 The acceleration of a particle P in a straight line is given by $n^2x - 3n^2x^2/2a$, where x is the distance of P from a fixed point O of the line and $n > 0$, $a > 0$ are constants.

Sketch the potential energy curve. For what values of the total energy E is an oscillatory motion possible? What is the approximate periodic time when the amplitude of the oscillations is small?

Find the variation of x with t given that $x = a$, $\dot{x} = 0$ when $t = 0$. Show that the time elapsed when $x = 3a/4$ is $n^{-1} \log 3$.

(Ans.: $-2n^2a^2/27 < E < 0$, $2\pi/n$, $x = a \operatorname{sech}^2 \frac{1}{2}nt$.)

13.3 Let θ be the inclination to the downward vertical of a spherical pendulum of length l. Let θ_1, θ_2 be the minimum and maximum values of θ respectively, and define

$$c_3 = \frac{1 + \cos\theta_1 \cos\theta_2}{\cos\theta_1 + \cos\theta_2}.$$

(i) For $\theta_1 \ll 1$, $\theta_2 \ll 1$ show that, approximately,
the periodic time is $2\pi(l/g)^{1/2}$,
the apsidal angle is $\frac{1}{2}\pi(1 + \frac{3}{8}\sin\theta_1 \sin\theta_2)$,
the angular speed of advance of the apse lines is $\frac{3}{8}\sin\theta_1 \sin\theta_2 (g/l)^{1/2}$.

(ii) For $\theta_2 \approx \pi - \theta_1$ show that, approximately,
the periodic time is $\pi(2l/gc_3)^{1/2}$,
the apsidal angle is $\pi(1 - 3\sin\theta_1/8c_3^2)$,
the angular speed of the apse line is $\frac{3}{4}\sin\theta_1 (g/2lc_3^3)^{1/2}$.

13.4 A particle moves on the inner surface of a paraboloid of revolution whose axis is vertical. Cylindrical polar coordinates r, ϕ, z are defined such that r is the horizontal distance from the axis of the paraboloid, ϕ is the angular displacement about the axis and z is the axial distance measured vertically upwards. The equation of the paraboloid is $r^2 = 4az$.

Show that the periodic time for motion in any horizontal circular path, and for any motion with small amplitude, is $2\pi(2a/g)^{1/2}$.

For any motion in which $z_1 \leqslant z \leqslant z_2$ show that the apsidal angle is given by

$$\phi_a = \left(\frac{z_1 z_2}{a}\right)^{1/2} \int_0^{\pi/2} \frac{(a + z_1\cos^2\psi + z_2\sin^2\psi)^{1/2}\, d\psi}{z_1\cos^2\psi + z_2\sin^2\psi}.$$

Evaluate ϕ_a for a perturbation on motion in a horizontal circle and for motion with small amplitude.
(Ans.: $\frac{1}{2}\pi(1 + z_1/a)^{1/2}, \frac{1}{2}\pi\{1 + (z_1 z_2)^{1/2}/a\}$.)

13.5 Sketch the integral curves in the phase plane for the equation $\ddot{x} = f(x)$ when $f(x)$ is equal to:
(i) x; (ii) $-x$; (iii) x^2; (iv) $-x^2$; (v) x^3; (vi) $-x^3$.
State whether the singularities are stable, neutral or unstable.
(Ans.: (i), (iii), (iv), (v) unstable, (ii), (vi) neutral.)

13.6 Sketch the integral curves in the phase plane for the equation $\ddot{x} = k(V^2 - \dot{x}\,|\,\dot{x}\,|)$, which represents one-dimensional motion with damping and a constant propelling force, where V is the terminal speed. Discuss their physical interpretation.

13.7 Classify the singular points of the equation $\ddot{x} = n^2(a^2 - x^2)/a - k\dot{x}$ for all positive values of k, and discuss their stability.
How would you expect the integral curves for $k = 0$ (see Fig. 13.2.3) to be modified when k is small?

13.8 A particle is connected by a light spring to a point O on a rough horizontal table. Oscillations of the particle are governed by the equation

$$m\frac{d^2x}{dt^2} = -\frac{\lambda x}{l} - \mu mg \operatorname{sgn}\left(\frac{dx}{dt}\right),$$

where

$$\operatorname{sgn}\left(\frac{dx}{dt}\right) = \begin{cases} 1 & \text{if } dx/dt > 0, \\ 0 & \text{if } dx/dt = 0, \\ -1 & \text{if } dx/dt < 0. \end{cases}$$

Let

$$\tau = \left(\frac{\lambda}{ml}\right)^{1/2} t, \quad \xi = \frac{\lambda x}{\mu mgl}, \quad \xi' = \frac{d\xi}{d\tau}.$$

Write down the equation of motion using these non-dimensional variables. Show that the integral curves in the (ξ, ξ') plane are semi-circles. Given that the particle starts from rest at a point where $\xi = \xi_0$, such that

$$4r - 1 < \xi_0 < 4r + 1,$$

where r is an integer, show that the particle makes r complete oscillations.

13.9 The angular displacement θ of a simple pendulum subject to damping satisfies the equation of motion

$$d^2\theta/dt^2 = -n^2 \sin\theta - k\dot{\theta}\,|\,\dot{\theta}\,|.$$

Show that it describes r complete revolutions before oscillating to and fro if the initial angular speed ω satisfies the inequalities

$$1 + \exp\{2(2r - 1)k\pi\} < \frac{1 + 4k^2}{2n^2}\omega^2 < 1 + \exp\{2(2r + 1)k\pi\}.$$

13.10 A particle is in one-dimensional motion along the x-axis and is subject to a resistance equal to $-k\dot{x}\,|\dot{x}|$ per unit mass and a force $f(x)$ per unit mass. Show that, while $\dot{x}$ remains positive,

$$\dot{x}^2 = A(x) + B\mathrm{e}^{-2kx},$$

where B is a constant and

$$A(x) = 2\mathrm{e}^{-2kx} \int_{\infty}^{x} f(\xi)\mathrm{e}^{2k\xi}\,\mathrm{d}\xi.$$

Show that if $f(x)$ is periodic, so is $A(x)$ and with the same period. Let $f(x)$ be the periodic function defined by

$$f(x) = \begin{cases} 0 & x = nX, \\ f_0 & nX < x < (n+\frac{1}{2})X, \\ 0 & x = (n+\frac{1}{2})X, \\ -f_0 & (n+\frac{1}{2})X < x < (n+1)X, \end{cases}$$

where X is constant and n is an integer. Calculate $A(x)$ in the interval $nX \leqslant x \leqslant (n+1)X$ by writing

$$A(x) = \mathrm{e}^{-2kx}\left(C + 2\int_{nX}^{x} f(\xi)\mathrm{e}^{2k\xi}\,\mathrm{d}\xi\right)$$

and using the fact that $A(x)$ is periodic with period X to calculate C. Hence show that

$$A(x) = \begin{cases} \dfrac{f_0}{k}\left(1 - \dfrac{2\exp\{-2k(x-nX)\}}{1+\exp(-kx)}\right), & nX \leqslant x \leqslant (n+\frac{1}{2})X, \\[2ex] -\dfrac{f_0}{k}\left(1 - \dfrac{2\exp\{2k(x-nX-\frac{1}{2}X)\}}{1+\exp(-kx)}\right), & (n+\frac{1}{2})X \leqslant x \leqslant (n+1)X. \end{cases}$$

Deduce that the velocity of the particle will eventually become zero. What are the points of equilibrium, and which of them are stable? (Ans.: $x = nX$, unstable; $x = (n+\frac{1}{2})X$, stable.)

13.11 Verify, by direct substitution, that the solution of the equation

$$\ddot{x} + \frac{t_0^2 x}{(t+t_0)^4} = 0$$

is

$$x = a_0(1 + t/t_0)\cos(1 + t_0/t)^{-1}.$$

Show that the method of averaging, when applied to this equation, gives the exact solution.

13.12 A particle moves under the influence of a restoring force and solid friction. The equation of motion is of the form

$$\ddot{x} = -n^2 x - \epsilon \operatorname{sgn} \dot{x}.$$

Use the method of averaging to show that an approximate solution when $0 < \epsilon \ll a_0 n^2$ is

$$x = \left(a_0 - \frac{4\epsilon t}{n} \right) \cos nt.$$

Deduce that the particle will come permanently to rest in a time of the order of magnitude $a_0 n/4\epsilon$.

13.13 Use the method of averaging to solve the equation for damped harmonic motion in the form

$$\ddot{x} + n^2 x = -k_1 \dot{x} - k_2 \dot{x} \mid \dot{x} \mid,$$

where k_1 and k_2 are small positive constants.

Show that there is an approximate solution of the form $x = a \cos\{n(t + t_0)\}$, where

$$\frac{1}{a} = \frac{1}{a_0} + \frac{4n}{3\pi} k_2 t$$

if $k_1 = 0$ and

$$a = \frac{a_0 \mathrm{e}^{-k_1 t/2}}{1 + \lambda a_0 (1 - \mathrm{e}^{-k_1 t/2})}$$

if $k_1 \neq 0$, where $\lambda = 8nk_2/(3\pi k_1)$.

13.14 Use the method of averaging to show that the equation

$$\ddot{x} + n^2 x = \epsilon \dot{x}(1 - x^2), \quad 0 < \epsilon \ll n$$

has an approximate solution of the form

$$x = a \cos nt,$$

where

$$\dot{a} = \frac{\epsilon a}{2} \left(1 - \frac{a^2}{4} \right).$$

Show that small perturbations about the two solutions $a = 0, a = 2$ are respectively unstable and stable. Verify that this is so by integration of the equation for a. The result is

$$a = a_0 \mathrm{e}^{\epsilon t/2} \{1 + \tfrac{1}{4} a_0^2 (\mathrm{e}^{\epsilon t} - 1)\}^{-1/2}.$$

This differential equation is called the Van der Pol equation. The stable oscillation $x = 2 \cos nt$ is called a self-excited oscillation or a limit cycle. For small values of ϵ all oscillations tend eventually to this as $t \to \infty$.

13.15 For the central force

$$-m\left(\frac{\mu}{r^3}+\frac{\nu}{r^5}\right)\mathbf{r},$$

acting on a particle of mass m, the differential equation for trajectories of the particle can be written in the form

$$\frac{d^2u}{d\theta^2}+u-\frac{\mu}{h^2}=\frac{\nu u^2}{h^2},\qquad u=r^{-1},$$

(see Exercise 7.20).

On the assumption that the term on the right-hand side is small, use the method of averaging to show that an approximate solution is of the form

$$lu = lr^{-1} = 1 + e\cos\{(1-\mu\nu/h^4)\theta\},$$

provided that $l = h^2/\mu$ and $e < 1$. Interpret the trajectory as an ellipse whose axes rotate slowly with the angular speed $\omega\mu\nu/h^4$, where $2\pi/\omega$ is the time to traverse the elliptic orbit.

Deduce that the leading terms agree with the result of Exercise 7.25.

13.16 Let θ denote the angular displacement of a simple pendulum of length l. The energy equation can be written in the form

$$\tfrac{1}{2}\dot{\theta}^2 + 2n_0^2\sin^2\tfrac{1}{2}\theta = 2n_0^2a^2,$$

where $n_0^2 = g/l$ and the energy constant is written as $2n_0^2a^2$. This equation represents oscillatory solutions with period $2\pi/n$, where

$$\frac{2\pi}{n}=\frac{1}{2n_0}\int\frac{d\theta}{\pm(a^2-\sin^2\frac{1}{2}\theta)^{1/2}}.$$

The integral is taken over a complete oscillation and the sign of the denominator is determined by the sign of $\dot{\theta}$.

Investigate the equations

$$\dot{x}=\omega,\quad \dot{\omega}+n_0^2\sin\theta=-\epsilon f(\omega),\qquad 0<\epsilon\ll 1,$$

by means of the transformation from θ, ω to a, ϕ, where

$$\theta=\theta(a,\phi),\quad \omega=n\frac{\partial\theta}{\partial\phi},\quad \frac{1}{2}n^2\left(\frac{\partial\theta}{\partial\phi}\right)^2+2n_0^2\sin^2\tfrac{1}{2}\theta=2n_0^2a^2.$$

Show that the approximate equations obtained by the method of averaging are

$$\dot{\phi}=n,\quad 4n_0^2a\dot{a}=-\frac{\epsilon n}{2\pi}\int f\{\pm 2n_0(a^2-\sin^2\tfrac{1}{2}\theta)^{1/2}\}\,d\theta,$$

where the integral is taken over a complete period of oscillation.

Consider the following cases:

(i) $a < 1$ and $\epsilon f(\omega) = \epsilon_1 \omega$. This case represents the oscillations of a pendulum subject to linear damping. The amplitude θ_m of θ is given by $\sin \frac{1}{2}\theta_m = a$. Show that

$$A(a) = A(a_0)e^{-\epsilon_1 t},$$

where a_0 refers to the value of a at $t = 0$ and

$$A = \int_0^{\theta_m} (a^2 - \sin^2 \tfrac{1}{2}\theta)^{1/2}\, d\theta = 2\{E - (1 - a)K\},$$

$$E = \int_0^{\pi/2} (1 - a^2 \sin^2 \xi)^{1/2}\, d\xi, \quad K = \int_0^{\pi/2} (1 - a^2 \sin^2 \xi)^{-1/2}\, d\xi.$$

The integrals K and E are the complete elliptic integrals of the first and second kinds.

Show that, approximately for $a \ll 1$,

$$\theta_m = \theta_0 e^{-\epsilon_1 t/2} \left(1 - \frac{5}{192}\theta_0^2(1 - e^{-\epsilon_1 t})\right).$$

(ii) $a \gg 1$, $\epsilon f(\omega) = \epsilon_2 \omega \mid \omega \mid$. In this case the pendulum makes rapid rotations in complete circles and is subject to quadratic damping.

Show that, approximately,

$$2a = 3^{1/2} \coth\{3^{1/2} \epsilon_2 n_0 (t + t_0)\},$$

where t_0 is a constant of integration which can be adjusted to give the correct value of a when $t = 0$.

13.17 Consider the equation

$$\ddot{x} + n^2 x = -\nu x^3 - k\dot{x} + f \cos pt$$

where the terms on the right-hand side are assumed to be small compared with the individual terms on the left-hand side. Verify, by direct substitution, that the expression

$$x = a(\cos\theta + \epsilon \cos 3\theta), \quad \theta = pt - \alpha, \epsilon \ll 1$$

is consistent with the equation, provided that all terms of the second order of smallness are neglected. Substitute the assumed form for x into the equation and compare like terms to obtain the following relations:

$$\sin\alpha = kpa/f,$$

$$a(p^2 - n^2) = \tfrac{3}{4}\nu a^3 - f\cos\alpha,$$

$$\epsilon(9p^2 - n^2) = \tfrac{1}{4}\nu a^2.$$

Verify that the leading terms agree with the solution obtained from the method of averaging.

13.18 The point of support of a simple pendulum, of length l, has a small vertical displacement $\epsilon l \sin pt$ $(0 < \epsilon \ll 1)$. The small angular displacement of the pendulum is θ. Show that the approximate equation of motion is

$$\ddot{\theta} + n^2\theta - \tfrac{1}{6}n^2\theta^3 - \epsilon p^2 \theta \sin pt = 0,$$

where $n^2 = g/l$. Use the method of averaging to show that there is a subharmonic resonance when p approximates to $2n$, and that then $\theta = a \cos\{\frac{1}{2}(pt + \chi)\}$ where, approximately,

$$\dot{a} = -\epsilon na \cos \chi, \quad \dot{\chi} = 2n - p - \tfrac{1}{8}na^2 + 2\epsilon n \sin \chi.$$

Show that there are equilibrium solutions when

$$\chi = \pm\pi/2, \quad p = 2n(1 - \tfrac{1}{16}a^2 \pm \epsilon)$$

and that those for $\chi = -\pi/2$ are unstable.

Investigate the solution when the small non-linear term is omitted, that is the term $-\frac{1}{6}n^2\theta^3$ in the original equation which gives rise to the term $-\frac{1}{8}na^2$ in the equation for $\dot{\chi}$. Show that the transformations

$$x = a(\cos \tfrac{1}{2}\chi + \sin \tfrac{1}{2}\chi), \quad y = a(\cos \tfrac{1}{2}\chi - \sin \tfrac{1}{2}\chi)$$

yield the equations

$$\dot{x} = \{(n - \tfrac{1}{2}p) - \epsilon n\}y, \quad \dot{y} = -\{(n - \tfrac{1}{2}p) + \epsilon n\}x.$$

Deduce that the linearisation is not valid when $2(1 - \epsilon)n \leq p \leq 2(1 + \epsilon)n$.

13.19 For the equations

$$\frac{dx}{dt} = v, \quad \frac{dv}{dt} + n^2 x = \epsilon_1 f(x, v) + \epsilon_2 \cos pt$$

make the substitutions

$$x = a \cos \phi, \quad v = -pa \sin \phi, \quad \phi = pt + \chi$$

and use the method of averaging to obtain the following approximate equations near resonance:

$$2p\frac{da}{dt} = -k(a) - \epsilon_2 \sin \chi, \quad 2pa\frac{d\chi}{dt} = -\mu(a) - \epsilon_2 \cos \chi,$$

where

$$k(a) = \frac{\epsilon_1}{\pi}\int_{-\pi}^{\pi} f(a \cos \phi, -pa \sin \phi) \sin \phi \, d\phi,$$

$$\mu(a) = \frac{\epsilon_1}{\pi} \int_{-\pi}^{\pi} f(a \cos \phi, -pa \sin \phi) \cos \phi \, d\phi - (n^2 - p^2)a.$$

Hence show that the amplitude a of the equilibrium oscillations satisfies the relation $\mu^2 + k^2 = \epsilon_2^2$ and that the equilibrium oscillations are stable provided that $\mu^2 + k^2$ and ke^a are increasing functions of a.

13.20 It is given that the equation $\frac{1}{2}n^2(a)(\partial x/\partial\phi)^2 + V(x) = a$ has the periodic solution

$$x(a, \phi) = \sum_{r=1}^{\infty} A_r(a) \cos r\phi.$$

Deduce an expression for $n(a)$ as a function of a.

Investigate the periodic solutions of the equations

$$\dot{x} = v, \quad \dot{v} + V'(x) = f \cos pt, \qquad f > 0,$$

where $f \cos pt$ is a small forcing term. Consider a transformation to a, ϕ, where

$$x = x(a, \phi), \quad v = n \frac{\partial x}{\partial \phi}, \quad \frac{1}{2} n^2 \left(\frac{\partial x}{\partial \phi} \right)^2 + V(x) = a,$$

and show that the equations satisfied by a, ϕ are

$$\dot{a} = nf \cos pt \, (\partial x/\partial\phi), \qquad \dot{\phi} - n = -nf \cos pt \, (\partial x/\partial a).$$

Use the method of averaging to show that there are resonant oscillations when $r\phi = pt + \chi$, where χ is a slowly varying function and r is an integer. Hence show that the approximate equations for a and χ are

$$\dot{a} = -\tfrac{1}{2} rnfA_r \sin \chi, \quad \dot{\chi} = rn - p - \tfrac{1}{2} rnfA_r' \cos \chi.$$

Deduce that there are equilibrium oscillations when $\chi = 0, \pi$ and $p = rn(1 \mp \frac{1}{2} fA_r')$. Show that these equilibrium oscillations are neutrally stable if

$$\chi = 0 \quad \text{and} \quad A_r \frac{d}{da} n(1 - \tfrac{1}{2} fA_r') > 0$$

or

$$\chi = \pi \quad \text{and} \quad A_r \frac{d}{da} n(1 + \tfrac{1}{2} fA_r') < 0.$$

Index